The Endocrine Function of the Human Ovary

The Endocrine Function of the Human Ovary

Proceedings of the Serono Symposia, Volume 7

Edited by

V. H. T. James
Department of Chemical Pathology
St. Mary's Hospital Medical School
London, England

M. Serio
Endocrinology Unit
University of Florence
Florence, Italy

G. Giusti
Endocrinology Unit
University of Florence
Florence, Italy

1976

ACADEMIC PRESS   London New York San Francisco
A Subsidiary of Harcourt Brace Jovanovich, Publishers

ACADEMIC PRESS INC. (LONDON) LTD.
24–28 OVAL ROAD
LONDON NW1 7DX

U. S. Edition published by
ACADEMIC PRESS INC.
111 FIFTH AVENUE
NEW YORK, NEW YORK 10003

Copyright © 1976 by Academic Press Inc. (London) Ltd.

All rights reserved

NO PART OF THIS BOOK MAY BE REPRODUCED IN ANY FORM BY PHOTOSTAT,
MICROFILM, OR ANY OTHER MEANS, WITHOUT WRITTEN PERMISSION FROM
THE PUBLISHERS

Library of Congress Catalogue Card Number: 76-16975
ISBN: 0-12-380150-8

'PRINTED PHOTOLITHO IN GREAT BRITAIN BY J. W. ARROWSMITH LTD.
BRISTOL

PREFACE

In 1972 and 1973, the Postgraduate School of Endocrinology in Florence held two courses on "The Endocrine Function of the Human Testis", sponsored by I.F. Serono, and the proceedings were subsequently published in two volumes under that title, by Academic Press. The success of the meeting and the response to the publication encouraged the organization of a similar course of lectures on the human ovary. Again, the meeting was most generously supported by Serono, which made it possible for the Organizing Committee to invite leading investigators in this field to Florence to participate. The papers which were read are presented in this volume, and it is the hope of the editors that the reader will profit from the opportunity to peruse the proceedings.

The members of the Organizing Committee are very grateful to many people who in various ways assisted and supported the meeting. Professor U. Teodori and Professor L. Martini have offered continuous help with the planning and organization of the post-graduate endocrinology courses and, as on the previous occasions, Dr. S. Rossetti and Dr. B. Casanova of I.F. Serono, Rome, gave generously of their time and assistance. Not least, we thank Mrs. Sonia Ciraolo of the Organizzazione Internazionale Congressi, Florence, for her invaluable assistance with the scientific and secretarial work associated with the meeting.

V.H.T. JAMES
M. SERIO
G. GIUSTI

CONTENTS

Preface	v
Histology of the Ovary	
by G.C. Balboni	1
The Interrelationship of Follicular Cyclic AMP, Steroids and Prostaglandins during the Ovulatory Cycle	
by J.M. Marsh and W.J. LeMaire	25
Steroidogenic Capacities of the Different Compartments of the Human Ovary	
by J.M. Marsh, K. Savard and W.J. LeMaire	37
The Role of Prostaglandins in Gonadotropin Secretion	
by C. Patrono, G.B. Serra, G. Ciabattoni, D. Grossi-Belloni and R. Canete Soler	47
Metabolism of Mammalian Oocyte During Growth, Maturation and Activation	
by F. Mangia, C.J. Epstein, R.P. Erickson, F. Palombi and G. Siracusa	63
The Role of Blood Flow in Regulating Ovarian Function	
by G.D. Niswender, A.M. Akbar and T.M. Nett	71
Female Sex Steroids and Gonadotrophin Secretion	
by L. Martini	81
Selected Aspects of Control of LH and FSH Secretion in Women	
by W.D. Odell, R.S. Swerdloff and F. Wollesen	89
Some Aspects of Hypothalamus-Hypophyseal-Ovarian Feedback Mechanisms in Women	
by P. Franchimont, U. Gaspard, J.R. Van Cauwenberge and E. Depas	109
Ovarian Steroid Secretion and Metabolism in Women	
by D.T. Baird	125
The Role of Extraglandular Estrogen in Women in Health and Disease	
by C.D. Edman and P.C. MacDonald	135

The Role of Sex Hormone Binding Globulin in Health and Disease
    by D.C. Anderson 141
Diversity in Androgenic Effect as Studied in Rat and Mouse
    by P. De Moor, G. Verhoeven, W. Heyns and H. Van Baelen 159
Hormonal Control of Progesterone Receptor in the Guinea Pig Uterus
    by M.T. Vu Hai Lun Thi and E. Milgrom 175
Specific Binding of Human Luteinizing Hormone and Human Chorionic Gonadotropin to Ovarian Receptors
    by C.Y. Lee and R.J. Ryan 181
Some Aspects of the Activity of the Hypothalamo-Pituitary Gonadal System in Children
    by J. Girard and P.W. Nars 195
The Effect of Hypothalamic and Pituitary Tumours on the Development of Puberty
    by J.S. Jenkins 223
Postmenopausal Ovarian Function
    by A. Vermeulen 237
Female Hypogonadism – Therapy Orientated Diagnosis of Secondary Amenorrhoea
    by M.G.R. Hull, M.A.F. Murray, S. Franks, B.A. Lieberman and H.S. Jacobs 245
Amenorrhoea – Its Investigation and the Assessment of the Role of Prolactin
    by G.M. Besser 261
Classification and Procedure for the Evaluation of Amenorrhea; The Clinician's View
    by G. Giusti, M.P. Piolanti, M. Pazzagli, F. Bassi and C. Vigiani 267
Modification of Gonadotropin Release After Multiple or Prolonged Stimulation with Synthetic LH-RH
    by G. Lafuenti, G.C. Giraldi, S. Fazioli, P. Muscatello, M. Signa, R. Caniglia and G.B. Serra 275
Disorders of the Hypothalamic-Pituitary-Ovarian Axis in Anorexia Nervosa
    by A. Borghi, G. Forti, S. Fusi, C. Vigiani, E. Calabresi and G. Cattani 291
Prolactin and Ovarian Function
    by R. Rolland, J.M. Hammond, L.A. Schellekens, R.M. Lequin and F.H. de Jong 305
Prolactin and Ovulation
    by P.G. Crosignani, A. D'Alberton, G.C. Lombroso, A. Attanasio, E. Reschini and G. Giustina 323
Recent Advances in the Neuroendocrine Regulation of Gonadotropin Secretion
    by S.S.C. Yen, B.L. Lasley, C.F. Wang and A. Lein 333
Pituitary-Ovarian Relationships in Disorders of Menstruation
    by D.T. Baird 349

Hormonal Regulation in Normal and Abnormal Menstrual Cycles
    by S.G. Korenman and B.M. Sherman 359
Functional Aberrations of the Hypothalamic-Pituitary System in Polycystic Ovary Syndrome: A Consideration of the Pathogenesis
    by S.S.C. Yen, C. Chaney and H.L. Judd 373
Corticoid Treatment of the Short Luteal Phase: A Case Report
    by E. Friedrich, E. Keller and A.E. Schindler 387
Using Antibodies to Study the Role of LH-RH in Ovulation
    by S.L. Jeffcoate 393
The Biochemistry of Ovarian Cancer
    by K. Griffiths and W.J. Henderson 397
Hormonal Studies in a Virilizing "Luteoma" of Pregnancy ("Nodular theca-luteinic hyperplasia of pregnancy")
    by G. Forti, R. Buratti, M. Colafranceschi, G. Fiorelli, M. Pazzagli, G. Lombardi, M. Marchionni and M. Serio 417
Release of C_{21}, C_{19} and C_{18} Steroids by Human Ovaries Stimulated *In Vivo* by HCG
    by S. Dell'Acqua, G.B. Serra, A. Lucisano, A. Montemurro, B. Cinque, E. Arno and A. Bompiani 427
Androgenic Hormones in Hirsutism
    by G. Giusti, E. Roncoli, G. Forti, S. Cattaneo, G. Fiorelli, M. Pazzagli and M. Serio 437
Ovarian and Adrenal Vein Catheterization Studies in Women with Idiopathic Hirsutism
    by M.A. Kirschner, I.R. Zucker and D.L. Jespersen 443
Plasma Androgens in Patients with Hirsutism
    by V.H.T. James, A.E. Rippon and H.S. Jacobs 457
Androgen Secretion by the Human Ovary: Measurement of Androgens in Ovarian Venous Blood
    by M. Serio, S. Dell'Acqua, E. Calabresi, G. Fiorelli, G. Forti, S. Cattaneo, A. Lucisano, G. Lombardi, M. Pazzagli and D. Borrelli 471
Testosterone 5α-Reduction in Human Skin as an Index of Androgenicity
    by P. Mauvais-Jarvis, F. Kuttenn and F. Gauthier-Wright 481
Androgen Metabolism in the Skin of Hirsute Women
    by R.J. Oake and J.P. Thomas 495
Subject Index 509

HISTOLOGY OF THE OVARY

G. C. Balboni

Institute of Human Anatomy, University of Florence, Italy

INTRODUCTION

It is not possible to sum up in a relatively short chapter all the knowledge concerning the histology of the human ovary. The aim of the present work is only to introduce the recent developments in this field and to point out some morphological problems related to ovarian steroidogenic activity.

As is well known, the ovary has two main functions in the body: 1) the gametogenic function, i.e. the production of the female germinal cells; 2) the endocrine function, consisting in the elaboration of female hormones (estrogens, progesterone and relaxine) and of some androgens.

From a morphological point of view, the following facts must be stressed:

a) The above-mentioned functions are both carried on, to a large extent, by the same ovarian organelles: the maturation of germinal cells occurs in the evoluting follicles; the elaboration of estrogens is accomplished by the theca interna layer of evoluting and involuting (atretic) follicles and by the theca lutein cells of corpora lutea; progesterone is produced by the granulosa lutein cells of the corpora lutea, and the same source produces relaxine. As regards androgens, they are thought to be produced by the hilus cells and by stroma cells of the ovary.

b) The ovarian organelles, and to some extent the whole ovary undergo, during the sexual life, continuous structural and functional changes, due to the cyclic action of hypophyseal hormones.

c) The stromal components of the ovary (cells, intercellular substance and capillaries) are able to undergo morphological and functional adaptation in the various phases of the menstrual cycle and in the different sites of the organ. The stroma cells are able to transform themselves into endocrine cells.

d) All the endocrine cells in the ovary have the structural and ultrastructural characteristics of steroid-producing cells.

DEVELOPMENT OF THE OVARY

The first anlage of the sex gland, either testis or ovary, appears in the fourth week as a ridge, called the *genital ridge*, which forms a projection on each side between the dorsal mesentery and the Wolffian body. The genital ridge derives from a thickening of the coelomic epithelium and a condensation of the underlying mesenchyme. These two components participate in some way in the formation of the gonad. A third component is represented by the *primordial germ cells*. These cells of a special type originate from the endoderm of the yolk sac and migrate through the hindgut wall and its mesentery to penetrate the undifferentiated sex gland, where they are well demonstrable at the sixth week (Witschi, 1948; Pinkerton et al., 1961). The germ cells may be identified on the basis of their morphological features and of their content of alkaline phosphatase and intracytoplasmic glycogen (McKay, 1953; Pinkerton et al., 1961; Falin, 1969; Fuyuta et al., 1974). A yet unknown chemotactic factor is supposed to favour the migration of these cells into genital ridge (Dubois, 1964, 1965, 1966).

No general agreement exists about the role that the two first components (coelomic epithelium and subjacent mesenchyme) play in the formation and development of the ovary. The problem has an importance from a morphogenetic point of view, but may be of some interest also in determining the relation between the histogenetic significance of the cells involved in the gonad formation and their functional properties in adult life.

According to the traditional view, around the time when the primordial germ cells are arriving and proliferate in the genital ridge, some cords of cells, which have an epithelial appearance similar to the covering epithelial cells, appear among mesenchymal cells. Germinal cells penetrate the cords, which are called *primary sexual cords* or *medullary cords* and are supposed to develop from the covering epithelium. Up to the seventh week the genital gland does not present differentiating features. In the male, all germinal cells are incorporated in the cords, these latter being destined to become enlarged and hollowed out to form the seminiferous tubules. The medullary cords in the male link up with the mesonephric structures, forming the *urogenital connections* or Mihalcovicz's organ, to complete the framework of the testicle (rete testis).

In the female, sexual differentiation starts at the eighth week and the sexual cords very quickly undergo fragmentation, degenerate and disappear. Some urogenital connections may develop, but they also degenerate simultaneously with the primary cords. However, some remnants may persist in the region of the hilus, forming the "rete ovary". It is believed that some cells of these remnants may be male-directed cells, capable of giving rise in later life to masculinizing tumors, such as arrhenoblastomas (Novak). Connective tissue and blood vessels penetrate in the region of medullary cords giving rise to the ovarian medulla.

The disappearance of the primary cords is immediately followed in the female by a further proliferation from the covering epithelium: a second group of cellular columns, the so-called *secondary or cortical cords of Valentin-Pfluger*, penetrate in the region of the cortex and contain the gonocytes, now transformed into ovogonia, actively proliferating. These secondary cords also undergo fragmentation, so that small cellular groups arise, in which the ovogonia are contained. Around the oogonia a layer of encapsulating cells develops, representing the progenitor of the follicular epithelium.

The traditional view, therefore, emphasizes the importance of the covering epithelium (*germinal epithelium*, Waldayer, 1870) and of the secondary sex cords in the formation of the ovary (De Winiwarter, 1910; Brambell, 1927); the follicular cells would have an epithelial origin, while other ovarian cellular components would derive from the mesenchyme.

Many researches, however, have raised some doubts about the origin of the sex cords from the germinal epithelium (Fischel, 1930; Politzer, 1933; Gruenwald, 1942; Bejdl, 1952; Novak, 1953, etc.), drawing attention to the fact that the underlying mesenchyme is firmly involved in the formation and differentiation of the gonad.

On the basis of observations in the Batracians, a double origin of the undifferentiated gonad has been postulated by Witschi (1936): the cortical portion derives from the proliferation of the covering epithelium, while the core or medullary portion of the gonad is formed by a condensation of the underlying mesenchyme. The primary sex cords are supposed to arise not from the coelomic epithelium, but from the mesenchyme in the medullary zone, here forming the so-called *paramesonephrotic blastema*. When sex differentiation begins, a proliferation of cortex occurs, giving rise to the cortical cords, that now have the significance of secondary cords, from which the follicular cells arise in the female. According to this view, a sort of cortico-medullary antagonism exists in sex differentiation, so that a prevalence of the cortex allows the development of the ovary, while if the medulla predominates a testicle develops.

However, further observations of Witschi (1965) seem to demonstrate that the proliferation of the covering epithelium is not so important in forming the cortex, which, in fact, is constituted by the gonocytes accumulating beneath the epithelium.

On the other hand, Gruenwald (1942) had already demonstrated that during the initial phase of gonadal development, a basal membrane does not exist under the covering epithelium and the gonadal blastema is formed by contributions both from mesothelium and mesenchyme.

More recently, Gropp and Ohno (1966), utilizing the histochemical reaction for alkaline phosphatase (AP), demonstrated that not only are the germinal cells positive to the reaction, but also that in early development an alkaline-phosphatase positive blastema appears in the gonadal anlage. In the earlier stages, the cells of the covering epithelium also are AP-positive and probably participate with the mesenchymal cells in the formation of this so-called common blastema, but afterwards they are not involved in the formation of the sex cords, which are assumed to arise from the common blastema. When sex differentiation occurs, AP-positive cells from the common blastema proliferate in the direction of the cortex, where germinal cells are penetrated. By a process of continuous incorporation of germinal cells, the blastema cords transform into ovigerous cords, where germinal cells begin to enlarge and to lose their AP-positivity, becoming oogonia and later oocytes, while the blastema cells transform into follicular cells. An intimate contact between follicular cells and oocytes is thought to be important in order to guarantee the normal maturation of oocytes (Ohno and Smith, 1964). According to this view, production of secondary sex cords (of Valentin-Pfluger) is not demonstrable and there is no reason to assume a cortico-medullary antagonism in the gonadal differentiation; the follicular cells do not derive from the cortex and may be considered homologous with the interstitial cells of the testicle, both deriving from the same common blastema. The only difference in female or male development would be in the direction into which the common blastema differentiate: in a cortical direction in the case of ovarian differentiation, remaining in the medullary zone if a testis is going to develop (Czyba and Coll, 1973). It is important to point out that, according to the opinion of Witschi, Gropp and Ohno believe that in early stages the common blastema derives from the perimesonephrotic mesenchyme, belonging to the perimesonephrotic streak, which in its cranio-anterior region comprises also the adrenal cortical anlage. In such a way "both the gonadal and the adrenal cortical blastema have in common an alkaline phosphatase activity and both give rise to cellular systems which later produce steroid hormones" (Gropp and Ohno).

In any case, according to the modern view of gonadal development, the components involved would be only two: the germ cell and the common blastema. Obviously, it is difficult to attribute a different significance to the cellular compartments in the ovary in relation to their embryogenetic origin. In Novak's opinion the fact that both granulosa cells (which have an epithelial appearance) and thecal cells (that are connective tissue cells) may derive from mesenchymal cells, would explain the close functional kinship of these cells and certain characteristics of granulosa and thecal cell tumors in later life. A similar belief is expressed by Gropp and Ohno in relation to the possibility that cellular buds, derived from the former cellular cords unstuffed by gonocytes and persisting in the deeper cortical zone of the adult ovary and probably corresponding to the so-called "Granulosa-cell ballen" (R. Mayer, 1951), may give rise to gonadal stromal tumors, as described by Hertig and Gore (1961).

On the other hand, some authors (cf. Ham, 1973) do not believe that the histogenetic significance of the various cellular components of the ovary presents a major problem, on account of their common mesoblastic origin. However, in our opinion, it is not completely correct to consider as homologous the mesoblastic layers and the mesenchyme: in fact, this latter derives in the most part from the mesoblast, but not only from it.

An original interpretation of the origin of the gonads is furnished by Hadjioloff (1969), who supposed the existence of a sexual tissue to be considered as the sixth fundamental tissue in the body. This fundamental sexual tissue forms the undifferentiated gonad and later differentiates into the male or female sexual tissue, giving rise to the characteristic structures of the sex differentiated gonads. Within this concept, the follicular cells surrounding the degenerating oocytes can transform into interstitial endocrine cells.

SEX DIFFERENTIATION

Sex differentiation is a very complex phenomenon, essentially due to genetic factors. The genetic sex is established at the moment of fertilisation: 46, XY chromosomes in the male, 46, XX in the female. Hormonal factors may be involved in the development and differentiation of the gonads and the sexual organs, but they are always secondary to genetic influences. The main problems in sexual differentiation are the following:

1) genetic mechanism
2) cells on which the differential genetic influence acts
3) differential genetic activity and hormone production
4) hormonal influence on the development of the genital tract.

First, the evidence available supports the view that only an X chromosome

Histology of the Ovary

is genetically active and carries the structural genes for testicular as well as ovarian determination, the former only functioning in the presence of a Y chromosome or suitable autosomal modifier (Hamerton, 1968). In the female, the second X chromosome remains inactive or heterochromatic, as does the Y chromosome in the male. However, it is believed that these heterochromatic chromosomes contain some "controlling elements" (regulatory genes), which affect both sex determination and some somatic characters (Hamerton). The effect of the controlling elements in the Y chromosome are stronger than those on the X heterochromatic chromosome: the presence of a Y chromosome results always in testicular development, probably by a process of inactivation of the female genetic sequence in the euchromatic X chromosome. Mittoch (1969) observed that testicular differentiation is associated with a rate of gonadal growth considerably greater than that of the ovary and supposed that this increased growth may be a prerequisite to testicular development, probably due to an action of the Y chromosome in increasing the rate of mitosis in the cells of the male gonad. On the other hand, Barlow (1973) believes that any difference in growth rate between embryonic ovaries and testicles could be explained also by the heterochromatic X chromosome decreasing the mitotic rate in the female cells. Recent research of Buehr and Petzoldt (1974) has pointed out that the increased growth rate in the testicle is a result and not a cause of differentiation: therefore it cannot be considered a primary effect of chromosomes in sex differentiation.

The X heterochromatic chromosome does not affect ovarian differentiation, but may have some action in regulating the rate of atresia and the oestrogen production by the ovary: in fact, it has been observed that in the 45, X state atresia is complete and no germ cells survive to puberty, while production of oestrogens is reduced and hence sexual infantilism occurs (Hamerton).

As regards the cells on which the genetic influence acts in determining the development either of an ovary or a testis, it seems to be ascertained by experimental observations (Hemsworth, 1963) and studies of genetic abnormalities (Hamerton), that they are the *somatic cells* of the gonadal blastema. Therefore, the germ cells would not play any role in the sex differentiation of the gonad (Cziba, Giroud and Laurent, 1973). According to Hamilton and Mossman (1972), however, the germ cells must have some inductive effect on the gonadal blastema, so that reciprocally "these cells cannot persist unless they come to be included in the appropriately differentiated, or prepared, mesenchyme of the gonadal ridge".

Thirdly, many studies have given information about the influence of steroid hormones on gonadal differentiation in amphibians, birds, as well as in mammals. The literature on this subject is very extensive (cf. Zuckerman, 1962; Czyba *et al.*, 1973). In the present chapter, it must be pointed out only that, in physiological conditions, the genetic activity in sex differentiation (i.e. genetic sequences in the sexual chromosomes) determines in the cells of the gonadal anlage the elaboration of particular protein molecules, enzymatic in nature, that are involved in the biosynthesis of specific sexual steroid hormones. The phenomenon occurs very early in the developing male gonad. Baillie *et al.* (1966) were able to demonstrate the presence of 3 β- and 17 β-hydroxysteroiddehydrogenase in human male embryos of six weeks, before the histological sex differentiation of the gonad Goldman *et al.* (1966) have found 3 β-hydroxysteroiddehydrogenase in the testis of human embryos during the second month. It seems therefore, that the undifferentiated mesenchyme of the future testicle may be able to produce steroid hormones. *In vitro* experiments have shown that the embryonic testicle, incubated in the presence of radioactive precursors, synthesizes testosterone and androstenedione (Acevedo *et al.*, 1961-1963; Rice *et al.*, 1966; Bloch, 1962, 1964-1967, etc.).

On the other hand, in the female gonad the presence of enzymatic activities involved in steroidogenesis, such as 3 β-hydroxysteroiddehydrogenase, appear much later, in advanced stages of morphological differentiation (Goldman, 1966; Schlegel, 1967; Motta, 1970). Ovaries of human embryos, cultured *in vitro* in the presence of labelled precursors have not shown any ability to synthesize steroid hormones (Bloch, 1964). It seems, therefore, that the foetal ovary does not have an early endocrine activity as does the testicle, probably because of the fact that ovarian hormones are not required in the differentiation of the female genital tract.

As regards this problem, i.e. the hormonal influence on the differentiation of the genital tract, it is enough to say here that a lot of research has demonstrated that in the male the Wolffian ducts transformation into male excretory ducts is due to the testosterone produced by the interstitial cells of the testis, while the regression of the Mullerian ducts is brought about by the testicle secreting a second hormonal substance, still unknown in nature (Stoll and Maraud, 1974).

In the female, the differentiation of the Mullerian ducts and the regression of Wolffian ducts occur in a spontaneous and automatic way, without hormonal influences. Only later, after the third month, does the oestrogenic activity of the foetal ovary probably act in perfecting the development of the Mullerian tract.

BASIC MICROSCOPIC FEATURES OF THE OVARY

The surface of the ovary is covered by a layer of cells that is cuboid in young females and tends to become flattened later in life. This epithelium, currently called *"germinal epithelium"* (see also: Development of the Ovary), is of coelomic origin and does not present during sexual life any true germinal function. It is continuous with the flattened mesothelium of the peritoneum at the Farre-Waldayer line. Recent research has demonstrated that the germinal epithelium covering preovulatory follicles contains numerous lysosome-like bodies, that become very abundant some hours before follicle rupture (Biersing and Cajander, 1974c). This fact and the observation that pronounced degenerative changes occur in the underlying tunica albuginea and gradually proceed inward affecting the theca externa of the preovulatory follice, suggest that lytic substances may emanate from the covering epithelium and play an important role in effecting weakening of the follicular wall and rupture of the follicle (Bjersing and Cajander, 1974).

Two regions may be distinguished in the ovary:

1) The cortex that is peripheral and contains the ovarian organelles (i.e. evoluting and involuting follicles, functioning and involuting corpora lutea). The connective tissue of the *cortical stroma* is essentially cellular, but its components (cells, intercellular substance and capillaries) can modify in relation to the phases of the menstrual cycle and the stages of the ovarian organelles that they surrounnd (Balboni, 1968, 1970-1973-1974). Immediately beneath the germinal epithelium, the cortical stroma is condensed and forms the so-called "false albuginea", or "tunica albuginea". This consists of a few cell layers embedded in an abundant

intercellular substance with compact interwoven collagen bundles. The tunica albuginea increases in density with advancing age and can undergo a sharp thickening in some pathological conditions (Stein-Leventhal syndrome); when it covers preovulatory follicles, it presents, after injection of HCG, some degeneration of the cells and dissociation and fragmentation of the collagen, that favour follicular rupture (Bjersing and Cajander).

2) The *medulla* represents the core of the organ and is formed by a loose connective tissue with many elastic fibers and strands of muscle cells, derived from the mesovarium. From the hilus the blood vessels penetrate into or leave the medulla, where their branches run a spiral course (spiral vessels) and form the so-called "bulb" of the ovary, which is thought to have some importance in determining ovulation. Well-developed perivascular nervous networks are present in the medulla and they are cholinergic and adrenergic in nature (Sassu and Colombino, 1969; Balboni, 1971). Some isolated nerve fibers without relation to vessels run in the medulla and can be followed in the cortical stroma. The medullary zone is rich also in lymphatic vessels and spaces, these latter being often wide. Some embryonic remnants (tubules of rete ovarii) and hyaline wavy membranes (vascular corpora albicantia) may occur, these latter derived from degenerate vessels.

Near the hilus of the human ovary, some small groups of cells may be found (*hilus cells*), which are very similar to interstitial cells of the testicle (Berger, 1922; Kohn, 1928; Stemberg, 1949; Dhom, 1955; etc.). It seems that these cells are especially evident during pregnancy and at the menopause (Stemberg, 1949). The hilus cells are very often intimately associated with non-myelinated nerve fibers and, at first, have been interpreted as sympathicotropic or chromaffin cells (Berger, 1924; Brannan, 1927; etc.). In more recent years they have been considered identical to the interstitial cells of the testis on the basis of their morphological appearance, histochemical reactions and the possible presence in their cytoplasm of crystalloids of Reinke. They are supposed to produce androgens and, in fact, these cells present all the ultrastructural characteristics of steroid-producing cells (Unsicker, 1970). The hilus cells are sensitive to gonadotrophic stimulation (Stemberg, Segaloff and Gaskill, 1953) and tumors arising from them produce virilizing effects.

As regards their relation to the hilus nerves, Unsicker observed that some axons may penetrate the basement membrane and come into close contact with the hilus cells, but typical synapses do not occur.

THE OVARIAN STROMA

The ovarian stroma presents a great evolutive potentiality (Dubreuil, 1946), as is shown by the fact that it undergoes important changes in the different phases of the menstrual cycle (Duke, 1947; Catchpole, Gersh and Pan, 1950; Balboni, 1959, 1962; etc.) and is able to give rise to endocrine structures such as the theca interna and the so-called "interstitial tissue" (Dubreuil, 1945-1948; Mossman, 1964; Motta, 1966-1968; Balboni, 1974; etc.).

From this point of view, the stroma of the ovary is peculiar, because it has not only a general metabolic function, but its cells may be directly involved in the endocrine activity of the organ. This fact matches well with the above-mentioned recent view on the histogenesis of the female gonad; in fact, it is not easy to apply to the ovary the classic distinction between the parenchymal and stromal components!

Particular attention was paid by the author to the behaviour of the stroma in the different zones of the ovary (peri and interfollicular zone, pericapillary spaces, corpora lutea, etc.) and to the evolutive possibilities of the stroma cells in the ovary (Balboni, 1968, 1970, 1973).

On the basis of the findings, obtained both with the light and the electron microscope, it was possible to postulate that the connective tissue and the capillaries pursue a particular fate depending on the sites of the organ where they exist and act. To some extent, the behaviour of the stromal components is influenced by the hypophyseal hormones, but at a certain moment other factors, probably local metabolic conditions and vegetative innervation, may play a role in conditioning it.

As concerns the stroma cells, their evolutive possibilities, as studied in physiological and experimental conditions, may be summarized as follows.

Obviously, the most common type of stromal cell is represented by fibrocytes and fibroblasts. However, these cells present very variable morphological characteristics, that appear to be related to local conditions more than the general hormonal situation. In the peripheral zone of the cortex, beneath the tunica albuginea, fibrocytes are generally spindle-shaped cells, regularly dispersed with collagen fibrils to form bundles, that have a characteristic swirly appearance. Elsewhere, fibrocytes are very irregular in shape, with many thin cytoplasmic prolongations, that run in a loose intercellular substance. Generally, they are poor in organelles; in some instances, near the capillaries and in the estrogenic phase, they present very numerous pinocytotic vesicles, that give evidence of their role in nutritive exchange.

Fibroblasts are characterized by a well-developed rough endoplasmatic reticulum and Golgi apparatus. They are often found near follicles undergoing atresia and in the regressing corpora lutea, probably providing production of collagen for the fibrotic involution of these structures.

Histiocytes and macrophages are more frequently observed near the blood vessels and their number increases in the later estrogenic phase. Macrophages are also numerous inside recent corpora lutea, where they are actively phagocytosing cellular debris and hematic pigment.

Mastocytes are not so frequent in the cortical stroma and only occur where it is loose and proliferating; in the medulla they are numerous around the blood vessels. In the ovulatory period the mastocytes increase in number and many of them appear degranulating (Balboni, 1960).

The existence of smooth cells in the theca externa of some mammals has been reported in several studies (O'Shea, 1970-1973; Osvaldo-Decima, 1970; Fumagalli *et al.*, 1971; Burden, 1972; Okamura *et al.*, 1972; Bjersing and Cajander, 1974), but it is not yet well ascertained in the human ovary. In our own research, some cells have been observed in the perifollicular stroma of the human ovary, that are characterized by the presence in the cytoplasm of small strands of microfilaments, glycogen accumulations and by areas of close membrane contact with adjacent cells. They have been interpreted as intermediate cells, deriving from stroma cells and evolving into smooth muscle cells.

It is well known that the stroma cells of the human ovary may evolve into steroidogenic cells in forming the theca interna of maturing follicles. This transformation is preceded by a "blastic phase" (Balboni, 1970, 1973), well shown by the light microscope by a cytoplasmic basophilia due to RNA and by a developed nucleolar apparatus. With the E.M. this blasic phase is characterized by the appearance in the cytoplasm of groups of flattened or dilated vesicles of the rough endoplasmatic reticulum. It is generally accepted that this blastic phase is necessary for the elaboration of the cellular organelles (smooth E.R.) and the enzymes involved in steroidogenesis. When fully developed, these cells present all the cytological, cytochemical and ultrastructural characteristics of steroid-producing cells. (See also, Ovarian Organelles.) According to Mossman et al. (1964), these cells begin to differentiate around secondary and growing vesicular follicles from "embryonic" rapidly multiplying fibroblast-like cells of the surrounding stroma, but they become true steroid secretors only during the final preovulatory growth period. Then they form a functional *thecal gland*, that is thought to be an important source of estrogens (Corner, 1938; Dubreuil, 1953; Falk, 1959; Harrison, 1962). When the follicle ovulates the thecal gland cells undergo degeneration and disappear. In Mossman's opinion, the so-called *paraluteal cells*, that correspond to the cells that other authors (Dubreuil, Novak, Balboni, etc.) consider thecal lutein cells, or *luteal cells of thecal origin*, do not derive from the thecal cells of the ruptured follicle, but differentiate from the stroma adjacent to the developing corpus luteum and they are probably intermediate stages in the differentiation of luteal cells from stroma cells. If the follicles undergo atresia before ovulation and the formation of a true thecal gland, the still embryonic theca cells may differentiate into the *interstitial gland cells*. The interstitial gland formed from the theca interna cells of the atretic follicles is present in the ovary from infancy to old age and may be considered the most important gland in the ovary (Mossman et al., 1964; Motta et al., 1970).

The volume of this interstitial gland tissue in the human ovary is less than 1% of the total ovarian volume in non-pregnancy and early pregnancy, but in the last trimester of gestation its relative amount increases rapidly to 4%-6% (Mossman). On the other hand, great variability exists in the development of the interstitial gland in the ovary of the different mammals. In some mammals, i.e. rabbit, the interstitial gland accumulates and forms the major portion of the ovarian stroma. In the human ovary, the interstitial cells are of transient nature and do not accumulate: as stated by Guraya (1974), "there is some sort of balance between the formation of new cells and their reversion to original stromal tissue". In fact, a true interstitial gland does not exist in the human ovary, its equivalent being represented by the hypertrophied theca internal of the atretic follicles (Dubreuil). No general agreement exists about the above-mentioned distinction between thecal gland cells, interstitial gland cells and paraluteal cells, as stated by Mossman et al.: in the opinion of several authors, the so-called paraluteal cells or theca lutein cells derive directly from the theca layer of the ruptured follicle (Gillman and Stein, 1941; Dubreuil, 1953, 1962; Novak, 1952; Balboni, 1956; Gillim et al., 1969; Leavitt et al., 1973; etc.). This fact is important from a functional point of view, because these cells are believed to produce estrogens as well as the theca cells of the follicles. In addition to the above-described endocrine compartments of the ovarian stroma (theca interna and its derivates), that form definite structures, some isolated cells directly deriving from mesenchymal stroma cells may develop into steroidogenic cells and form small clusters scattered in the cortical stroma (Balboni, 1970, 1973).

The involutive pattern of these steroidogenic cells is dramatic: they appear as large epithelioid-like cells, whose cytoplasm is filled with many autophagic structures (lysosomes, residual bodies, etc.), fragmented membranes of the endoplasmatic reticulum, few swollen mitochondria, etc.; the nucleus presents an irregular condensation of chromatin or may be sharply pycnotic.

Much evidence suggests that gonadotrophins control the morphological development and function of the stromal endocrine compartments (Claesson, 1954; Balboni, 1960; Savard *et al.*, 1965; Armstrong, 1968; Merker and Diaz Encinas, 1969; Guraya, 1971-1973; etc.), but a neural influence cannot be excluded (Dahl, 1970; Unsicker, 1970; Fink and Schofield, 1971; Balboni, 1971-1973).

If the steroidogenic nature of the stroma-derived endocrine cells of the ovary is now well ascertained, the problem exists about the significance of the individual cell types in relation to the production of the various ovarian steroids.

The main cytological, cytochemical and ultrastructural features of these cells are typical of all the steroidogenic cells in the body: their cytoplasm contains diffuse sudanophilic lipids, interpreted as lipoproteins corresponding to the abundant smooth endoplasmic reticulum, deeply sudanophilic lipid droplets consisting of either phospholipids, or triglycerides and phospholipids, or triglycerides, cholesterol or cholesterol esters and phospholipids, depending upon the physiological situation (Guraya, 1974), and some enzymes, as 3 β-hydroxysteroid-dehydrogenase, directly involved in steroidogenesis. With the E.M. the prominent features are the well-developed Golgi apparatus and smooth endoplasmic reticulum, the mitochondria with tubular or vesicular christae, the presence of lipid droplets and lysosomes. These features, however, do not permit any conclusion to be drawn about the type of the steroid produced.

On the basis of biochemical and experimental observations it is generally accepted that the theca interna cells and the interstitial cells produce estrogens. The theca interna cells of developing follicles probably synthesize most of the estrogens secreted by the preovulatory ovary (Ryan and Smith, 1965), while the interstitial gland (in the female the hypertrophied theca interna of atretic follicles) synthesizes most of the estrogens of the postovulatory ovary.

No agreement exists about the activity of the theca lutein cells or paraluteal cells: if they are intermediate cells between the stroma cells and the luteal cells (Mossman), they may be supposed to participate in the production of progesterone; if they are, as many authors believe, true theca-derived cells, they might go on to produce estrogens and contribute to estrogen plasma level in the luteal phase.
In Short's opinion (1967), a degree of collaboration may occur within the corpus luteum between granulosa and theca cells; "acting in union they might be able to produce a product that neither could synthesize in isolation". On the other hand, the possibility that recombination or new formation of compartments in the ovary may play a role in hormonal production has been emphasized recently also by Smith and Zuckerman (1973).

The ovary produces some androgens. In the female, the secretion of androgens by the ovary presents cyclical variations, probably due to gonadotrophic influence (Czyba *et al.*, 1969): it is higher in the middle of the cycle, during gestation and the postmenopausal period, while it decreases in the estrogenic and estrogenic-progestogenic phase. The hilus cells are probably involved in the production of androgens from the ovary (Unsicker, 1970), but *in vitro* biochemical studies (Savard *et al.*, 1965; Rice and Savard, 1966) and the results of studies of tissue culture of separated cell types (Channing, 1969) suggest that the ovarian stroma produce androgens in humans. It must be pointed out, however, that the presence of androgens in a cell type is not evidence of secretion, because androgens may be only precursors of estrogens. In addition, the stroma tissue used for these studies is a pool of cell types such as theca cells, interstitial cells, undifferentiated stroma cells, etc.. Recently, Jong, Bair and van der Molen (1974) confirmed the ability of ovarian stroma to produce androgens. No difference in concentration of testosterone and dehydroepiandrosterone in venous plasma from the "active" and "inactive" ovary was observed: this fact suggests that the follicular contribution to the secretion of these steroids is insignificant *in vivo* and that they may be produced only by the ovarian stroma. The problem, however, remains because this androgenic activity cannot be attributed to any cell type.

In the author's opinion, it is necessary to admit either that the stromal endocrine components (i.e. theca interna cells of maturing and regressing follicles) may secrete androgens and/or estrogens in relation to the phase of the cycle, pregnancy, etc. (under gonadotrophic as well as neural influences), or that such androgen secretion is accomplished only by specialized cells such as the hilus cells and probably the isolated steroidogenic cells that develop directly from the undifferentiated stroma cells, as described previously.

OVARIAN ORGANELLES

A) Follicles

These are scattered throughout the cortical zone and in ovaries of children and young women they occur in large numbers, mostly at the stage of primordial follicles; during the reproductive life they progressively decrease in number and may be seen in the various stages of their maturation or regression.

Primordial follicle. This consists of a central germ cell (primary oocyte blocked in the first prophase of meiosis), encircled by a flattenered layer of follicular epithelium. The ultrastructure of the human oocyte and its relationship with follicular cells has been the object of several researches (Tardini *et al.*, 1960-1961; Zamboni, 1966-1967; Hertig and Adams, 1967; Hertig, 1968; Motta *et al.*, 1971-1974; etc.). With the E.M., the oocyte presents a spherical nucleus with granular chromatin and, in suitable sections, an evident nucleolus. The cytoplasmic organelles are grouped near the nucleus; they consist of closely spaced mitochondria, Golgi vesicles, annulate lamellae, etc. This complex of organelles is located in the centre of the cell and forms the so-called "Balbiani's vitelline body". The annulate lamellae are smooth or poorly granular membrane systems, that are believed to derive directly from the nuclear membrane and to be typical of young proliferating cells. The plasma membranes of oocytes and follicular cells are closely applied to each other; only a few small cytoplasmic projections of follicular cells may penetrate the ovum. The follicular cells are poor in organelles and their membranes are joined to each other by junctions of the type of desmosomes. An evident basal membrane separates the follicular cells from the surrounding perifollicular stroma.

Primary follicle. It consists of the primary oocyte and a layer of cuboidal or low columnar follicular cells, which tend to proliferate. The oocyte begins to grow and its volume increases. With the E.M., the mitochondria appear to increase in number and to be located in the outer cytoplasm, and the Golgi apparatus is represented by several groups of smooth vesicles and tubules. Annulate lamellae decrease in number and some vesicles appear, which contain electron-dense material. The plasma membrane of the oocyte gives rise to microvilli, that interdigitate with the cytoplasmic projections of follicular cells; between the epithelium and the egg some PAS-positive material begins to be secreted and forms a particular layer, the "zona pellucida". The follicular cells contain now more organelles than are present in primordial follicles. The intercellular junctions are mainly of the adherens type. The basal membrane (in terms of light microscopy), that encircles the follicle, tends to become thinner.

Secondary follicle. The oocyte continues to grow, while the epithelium proliferates and forms a thick stratified columnar layer, called "granulosa". The "zona pellucida" is now more prominent. With the L.M. it appears as a translucent, structureless membrane, intensely PAS-positive; with the E.M. it may present a granular or filamentous appearance. It is formed by mucopolysaccarides secreted by the follicular cells and the oocyte. This latter is believed to play some role in the process of polymerization of these mucopolysaccarides (Hope, 1965; Motta, 1971; etc.). According to Motta et al. (1971) an identity exists between zona pellucida, PAS-positive membrane of the Call-Exner bodies and the basal membrane of the follicle, all being the product of the secretory activity of the follicular cells. Cytoplasmic processes from the innermost layer of the granulosa extend through the zona pellucida and may either come into close relationship to the plasma membrane of the oocyte (adherens type junction may be present), or invaginate it. On the other side, microvilli from the oocyte penetrate the zona pellucida. It is generally believed that these microvilli have some importance in increasing the absorptive capacity at the surface of the oocyte, while the cytoplasmic projections of the follicular cells may facilitate the transport of nutritive material to the egg (Chiquoine, 1960; Adams and Hertig, 1966; Hope, 1965; Baca and Zamboni, 1967; Tedde Piras, 1971; Motta, 1971; etc.). The follicular cells are now rich in organelles: the Golgi apparatus is well developed and mitochondria with linear cristae are abundant. A smooth and rough endoplasmic reticulum is present as well as numerous, scattered, free ribosomes. Between the follicular cells, small spaces are developing, into which some fluid accumulates. The plasmic membranes of adjacent follicular cells are associated with each other by both desmosomes and gap-junctions.

Graafian follicle. The fluid-filled spaces amongst the granulosa cells tend to coalesce to form a single large follicular cavity or "antrum", in which the follicular fluid is contained. The increasing amount of the follicular fluid expands the follicle, which becomes vesicular. In relation to their stage of development, the Graafian follicles may be distinguished in small, medium and large size follicles. The granulosa cells are displaced peripherally, where they form a 4-6 layer thick wall, and accumulate around the oocyte to form the "culumus oophorus". The cells that encircle the egg directly with its zona pellucida (that now is 20-30 micron thick) form the so-called "corona radiata". The granulosa cells present the characteristics of active secretory cells (Golgi apparatus, RER, free ribosomes, etc.). The intercellular junctions are represented now mainly by gap-junctions. The presence of gap-junctions or nexuses was studied by several authors in the granulosa cells of developing follicles (Anderson, 1971; Merck et al., 1973; Fletcher, 1973; Albertini and Anderson, 1974; etc.), because of the physiological significance of this junction type in transporting ions and low molecular weight substances. Unpublished research, performed in our Institute by Kujawa and Tedde, has shown that in the vesicular follicles of the rat ovary, a high percentage of the surface of the cell membrane of the granulosa cells is occupied by nexus-type junctions (6.7%), while a very low percentage corresponds to adherens-type junctions (0.54%). The basal membrane in the Graafian follicles becomes very thin and in the perifollicular stroma the differentiation of the theca layers begins: the

Histology of the Ovary

theca interna and the *theca externa*. The transformation of the stroma cells into theca interna cells is preceded by a "blastic phase" (Balboni), in which the cells present with the L.M. significant amounts of RNA, and with the E.M. a well-developed RER. When fully developed, the theca interna cells appear as epithelioid-like cells, rich in sudanophilic material and 3 β-hydroxysteroiddehydrogenase-positive cells. A network of sinusoidal capillaries permeates this layer. With the E.M., the theca interna cells show the characteristics of steroidogenic cells (highly developed smooth endoplasmic reticulum, vesicular mitochondria, lipid droplets, etc.). The theca externa is fibrous in character; in the early stages of development of the thecal sheaths, its cells are in a blastic phase and represent a transitional stage between stroma cells and theca interna cells.

Mature follicle. Generally, in humans, only one follicle in the two ovaries is destined to rupture and ovulate. A fully-developed follicle is about 12-15 mm in diameter and it causes a bulge, the so-called "stigma" on the surface of the ovary; the tissue layers between the outermost part of the follicle and the ovarian surface become very thin. The oocyte, which was in the late prophase of the first meiotic division, completes this division, giving rise to the first polar body and to the secondary oocyte. Then the second meiotic division begins, but it may be accomplished only after ovulation and fertilization has occurred. According to Moricard (1954), the hormones present in the follicular fluid, particularly the gonadotrophins, would act on the oocyte determining the completion of the first meiotic division (meiogenic function of the follicular fluid). Direct contact of the oocyte with the follicular fluid occurs in the maturation stage, because the cells of the corona radiata and of the cumulus oophorus tend to separate and create intercellular spaces, into which the follicular fluid can penetrate. Many changes occur in the preovulatory follicle and concern both the follicular components (granulosa cells, theca interna layer and capillaries) and the perifollicular tissues, particularly those overlying the follicle at the stigma.

Bjersing and Cajander (1974) have studied these changes in the vesicular follicles of the rabbit ovary prior to ovulation induced by the injection of an ovulatory dose of HCG. In such conditions, the granulosa cells undergo a dissociation around the whole follicle; the linear gap-junctions (abutment nexuses) between these cells decrease in number, probably causing a decrease in cohesive force between the cells facilitating follicular expansion. On the other side, the annular nexuses increase significantly: this fact may be related to the subsequent transformation of granulosa cells into lutein cells. It is known that nexuses are permeable to small biological active molecules such as cyclic $3',5'$-AMP and that, according to Merk *et al.* (1972) an intercellular passage of these molecules might "facilitate a more uniform response to gonadotrophins by granulosa tissue". In fact, granulosa cells may present, a short time before ovulation, morphological signs of luteinization, such as an increase of SER, forming typical whorls, the appearance of numerous lipid droplets and some mitochondria with vesicular cristae.

In the meantime, the theca interna cells become hypertrophied and they also show an increase in the SER, lipid droplets, mitochondria with vesicular cristae and annular nexuses. Important changes occur in the thecal capillary network that becomes well developed; close to the time of ovulation, large gaps appear in the thin cytoplasmic portions of the endothelium, while the capillary basal membrane undergoes fragmentation. Leakage of fluid occurs, producing an interstitial oedema, that is prominent in the apical region of the follicle. Prostaglandins, the level of which increases strongly in the Graafian follicles prior to ovulation (Yang *et al.*, 1973), may be involved in promoting vascular permeability. Because of the formation from the granulosa cells of cytoplasmic protrusions that pass through the follicular basal membrane, this latter presents some large openings, that permit easy passage of fluid to the follicular antrum (secondary follicular fluid). All these facts, and the previously mentioned changes in the theca externa and albuginea covering the preovulatory follicle play a role in the mechanism of ovulation (Bjersing and Cajander).

Morphological-functional Considerations on the Endocrine Activity of the Developing Follicles

It is common knowledge that the developing follicles produce estrogens. The cells involved in this endocrine activity are generally believed to be theca interna cells (cf. Zuckerman). These cells present all the histochemical and histological characteristics of steroid-producing cells under both the light and the electron microscope. The thecal tissue demonstrates a capability to synthesize estrogens *in vitro* from endogenous precursors or from pregnenolone, so that the thecal elements of the human follicles or corpus luteum may be considered as the main source of estrogens (Channing, 1969).

As regards the follicular (granulosa) cells, the questions arise: are they in some way involved in estrogen production during the period of follicular maturation, and if and when they acquire the ability to produce progesterone, before morphological transformation into luteal cells? The follicular cells of growing follicles appear with the L.M. to be rich in cytoplasmic RNA and with the E.M. present an abundant rough E.R., many free ribosomes, a well-developed Golgi apparatus, numerous small mitochondria with linear cristae. These morphological features indicate an active protein-synthetic function that is probably related to both cellular multiplication and the construction of new cytoplasm that occur in the granulosa layer. On the other hand, follicular cells are involved also in the production of the follicular fluid, which contains protein and mucopolysaccharides, as well as in the formation of the mucopolysaccharides of the zona pellucida. From a morphological point of view, it seems, therefore, that until the preovulatory phase the follicular cells of growing follicles are not equipped for any important steroidogenic function and do not participate directly in estrogen production. It cannot be excluded, however, that they might collaborate in some way with the theca interna cells. It must be remembered that the granulosa is an avascular layer and that all the exchanges in these cells occur by diffusion from the theca interna blood vessels. Ryan and Petro (1966) have incubated *in vitro* with radioactive progesterone and pregnenolone isolated granulosa and theca cells, harvested from ovarian follicles of Pergonal-treated women, and have observed that the conversion of pregnenolone to progesterone is much more active in the granulosa cells, but they are not able to produce estradiol. Ryan *et al.* (1968) have studied steroid formation by isolated and recombined granulosa and theca cells, incubated *in vitro* with C^{14} acetate. Both the cell types were able to produce ex-novo steroids, but isolated granulosa cells failed to produce estradiol. Labelled estradiol was obtained only from the theca cells or from experiments with recombined thecal and granulosa cells. On the other hand, the importance of the combination of the various ovarian compartments in the formation of estrogens, particularly estradiol, is emphasized also by Smith and Zuckerman (1973).

As regards the ability of the granulosa cells to produce progesterone, it must be pointed out that, in physiological conditions, prior to ovulation the ovary secretes only a small amount of this hormone; probably, as stated by Short (1967), the enzymatic systems for progesterone production are to some extent already present in the granulosa cells (the 3β-hydroxysteroiddehydrogenase is detectable in the granulosa cells of growing follicles - Motta, 1970), but only after the follicle is ruptured can blood vessels penetrate the granulosa layer, whose cells undergo full luteinization. As observed by Deane (1952) in the rat and by Balboni (1956) in the human, the granulosa cells retain a high content of cytoplasmic RNA in the first days after ovulation, when they begin to transform into granulosa lutein cells. This fact is probably due to the need for building up the enzymes and the structures (i.e. smooth E.R.) for steroid production. The sudden elevation of the progesterone level at the ovulatory phase, before the corpus luteum is formed, may be explained by the fact that the transformation of the granulosa cells into lutein cells does not occur simultaneously for all the cells, but some of them may be affected by LH very early, even before ovulation. From a morphological point of view, the findings of Bjersing and Cajander confirm the increase of the annular nexuses in the granulosa cells of preovulatory follicles under the influence of HCG and the appearance, in these conditions, of electron microscopic signs of luteinization in some of these cells. Recent research of Canning (1974) has furnished interesting data on the temporal effects of factors such as LH, HCG, FSH and dibutyryl c-AMP upon luteinization of Rhesus monkey granulosa cells in culture. Abel (1975) has made an ultrastructural study of the granulosa-luteal cell transformation in the ovary of the dog, in order to establish temporal and functional correlations between the appearance of the typical subcellular organelles and the hormonal changes that precede or succeed the steps of this cellular differentiation; the role that each subcellular compartment in the luteal cells plays in progesterone synthesis and release has also been investigated.

Involutive (atretic) follicles. In the period between birth and puberty, a large number of primordial follicles degenerates and disappears. During sexual life, only a very small proportion of follicles runs the entire course of maturation and only one each month completely matures and ruptures. The majority degenerate; this degenerative process is called *atresia* and may affect the follicles in all stages of their development.

Follicular atresia consists essentially of: 1) degeneration of the germ cell; 2) degeneration of the follicular epithelium and its replacement by connective tissue; 3) thickening of the basal membrane that encircles the follicle (*atresia membrane*); 4) hypertrophy of the thecal interna sheath of the Graafian follicles and formation of the so-called interstitial gland.

The atretic process, however, may present different morphological patterns, in relation to the stage of follicular evolution in which it begins.

According to Dubreuil (1953) one may distinguish: 1) *degenerate follicles*, that are primordial or primary follicles degenerating without leaving traces; 2) *hemorrhagic follicles*: during the atretic process affecting Graafian follicles of medium and large size, an intrafollicular hemorrhage occurs; the remnants of these follicles are represented by a *corpus nigrum*; 3) *thecogenic follicles*, that are characterized by the hypertrophy and persistence of the theca interna. This group includes the atretic follicles, the cystic follicles, the folded follicles (follicules plissés).

Two main types of follicular atresia have been distinguished by Schroeder (1930): 1) *obliterans atresia*, that affects Graafian follicles of small and medium size and is characterized by connective tissue proliferation that fills up the cavity and by the formation of a *corpus fibrosus*; 2) *cystic atresia*, that affects Graafian follicles of large size and gives rise to cystic cavities with a fibrous wall.

The most important histo-physiological phenomenon related to the atretic process is, of course, the hypertrophy of the theca interna and the formation of the so-called interstitial tissue. This problem has been previously discussed in relation to the endocrine cells derived from ovarian stroma.

B) Corpus luteum

After the rupture, the collapsed follicle begins to transform into the corpus luteum. If fertilization occurs, a *corpus luteum of pregnancy* develops, that lasts and functions for about three months and afterwards begins to regress very slowly; if fertilization does not occur a menstrual corpus luteum is formed, that lasts only about 14 days. In both instances, luteal cells produce progesterone, which is responsible for the growth of the endometrium and, if pregnancy takes place, for its early maintenance. Several stages may be distinguished in the life of the corpus luteum.

1) Stage of proliferation and hyperemia. Immediately after rupture, the histological picture is practically the same as of the mature follicle. The granulosa cells preserve yet their appearance, but undergo some proliferation. The theca interna layer is prominent and hyperemic.

2) Stage of vascularization. Very early, thin-walled blood channels and some connective tissue, arising from the theca interna, penetrate through the granulosa into the lumen, into which some vessels may open, giving rise to hemorrhage.

3) Stage of transformation. Simultaneously with vascular penetration or immediately after, the granulosa cells undergo rapid hypertrophy and complete their transformation into *granulosa lutein cells*. This transformation is preceded by a "blastic phase" (Balboni), characterized by the presence in the cells of a well-developed rough endoplasmic reticulum. The theca cells also participate in the formation of the corpus luteum; they arrange themselves in strands of *theca-lutein cells* (paraluteal cells) at the periphery of the corpus luteum and in the septa between the granulosa lutein cells.

4) Stage of maturity. By the fourth day after ovulation, the granulosa lutein cells have reached their full development and they all present the cytological and physiological properties of progesterone-producing cells. A wide network of sinusoidal capillaries penetrate them. Theca lutein cells are still present and are believed to produce estrogen. According to Dubreuil (1953), the corpus luteum may, therefore, be considered a gland with a double endocrine function, i.e. a progesterone and estrogen producing gland.

5) Stage of regression. This consists of degeneration of the lutein cells and proliferation of connective tissue, which gives rise to a *corpus fibrosus*. If hyaline transformation occurs in the connective tissue, a *corpus albicans* takes place.

Menstrual corpus luteum. In the female, the menstrual corpus luteum reaches maturity in the period between the 7th and 10th day from rupture and this period corresponds to the peak of its secretory activity (Balboni, 1956; Corner, 1956). In this phase of maturity the *granulosa lutein cells* are 15-30 microns in diameter and polygonal in shape; their nuclei are in most cases vesicular with an obvious nucleolus. The L.M. shows that the cytoplasm presents some basophilia, due to the presence of RNA. As already stated, the RNA content is more prominent in the transformation stage in relation to the synthesis of the cytoplasmic enzymes and structures involved in steroidogenesis and RNA content when lipid droplets appear and accumulate. Mature luteal cells contain important enzymatic components: 3β-hydroxysteroiddehydrogenase represents the most significant finding. They are rich also in ascorbic acid and contain lipid inclusions such as cholesterol, phospholipids, etc.. With current histological methods it is possible to distinguish light and dark cells, these latter with a PAS-positive cytoplasm and hyperchromatic nuclei. Probably they represent different functional stages. The ultrastructure of the human menstrual corpus luteum as well as of the corpus luteum of pregnancy has been studied by several authors (Green and Maque, 1965; Green et al., 1967, 1968; Lennep and Madden, 1965; Tokida, 1965; Pedersen and Larsen, 1968; Adams and Hertig, 1969a; Motta, 1969; Gillim et al., 1969; Crisp et al., 1970; etc.).

In the E.M., the granulosa lutein cells of the menstrual corpus luteum in the mature stage present a very irregular cellular outline; the free surface bordering on the pericapillary spaces bears many microvilli or cytoplasmic expansions. Where the cellular surface of adjacent cells is apposed, protoplasmic projections may intertwine in a complicated fashion; tight junctions exist between the apposed cell membranes. Annular gap-junctions may be present. In some sites the cell membranes of adjacent cells separate, giving rise to channels (intercellular canaliculi), that are continuous with the pericapillary spaces and are bounded by microvilli (Green et al., 1968). These canaliculi are thought to serve as channels for the transfer of steroids to the vascular bed. A similar significance might be attributed to the intracellular canaliculi, that are also encountered in these cells, but are not so extensive as those found in the granulosa lutein cells of pregnancy (Crisp et al., 1970). These canaliculi are thought to increase the surface area for secretion of steroids as well as for absorption of hormone precursors. The presence of mitochondria in proximity to the canaliculi might supply the energy to facilitate transport across membranes, regardless of the direction of transport (Crisp et al.). Microfilaments, approximately 50 Å in thickness, are present in the peripheral cytoplasm of lutein cells and extend into the matrix of microvilli. These microfilaments might compartmentalize the luteal cells in a central and a peripheral region (Adams and Hertig) and might also aid and regulate the absorption of extracellular material (Crisp et al.) probably by producing in vivo contraction movements (Motta and Di Dio, 1974). The cytoplasm of lutein cells contain both granular and agranular E.R., this latter being prominent and forming folded membrane complexes. A different degree of development of the E.R. might explain the distinction in light and dark cells. Mitochondria are numerous, rod-shaped, and have tubular or villiform cristae. Osmiophilic inclusions of various shape and size may be found within several mitochondria and are interpreted as accumulations of steroid substrates in analogy with the findings in the interstitial cells of the testis (Christensen, 1965; De Kretser, 1967). The Golgi complex is represented by scattered regions of flattened membranes and small vesicles, some of which are coated. Lipid droplets are numerous and may be closely associated with elements of the smooth E.R.. Microtubules, multivesicular bodies, and lysosomes are scattered in the cytoplasm.

The *theca lutein cells* are prominent in the first stages of formation of the corpus luteum and they persist in the maturity stage. However, they tend to decrease in number and to disappear before the full involution of the corpus luteum is accomplished. Probably they transform again into fibrocytes of the surrounding connective tissue.

With the L.M. their size is about one half of that of granulosa lutein cells. The nucleus contains obvious nucleoli and the cytoplasm is light and vacuolated.

3β-Hydroxysteroiddehydrogenase is demonstrable.

With the E.M. these cells show the typical features of steroid-producing cells. The smooth tubular endoplasmic reticulum is generally abundant; some scattered cisternae of the rough E.R. are also present. Probably in relation to different amounts of endoplasmic reticulum, dark, light and intermediate cells are distinguishable (Gillim et al., 1969). The Golgi elements are not dispersed through the cytoplasm, but are concentrated at one pole of the nucleus. The lipid droplets are usually extracted in tissue preparations. In comparison to the granulosa lutein cells, the thecal lutein cells have a more regular outline and do not present the well-organized patches of microvilli and the cytoplasmic microfilaments as well as the complicated intertwined plasma membrane arrangement of adjacent cells. Lipofuscin pigment granules, probably lysosomes in nature, are present in theca lutein cells. They contain dense pigment, a granular matrix and lipid inclusion, all enveloped by a unit membrane (Crisp et al.).

Corpus luteum of pregnancy. This grows more than the menstrual corpus luteum and attains a diameter of about 4-5 cm in the third month of pregnancy. It is generally believed that it functions only in the first trimester of gestation and then it begins to involute slowly. Gillman and Stein (1941) stated that the critical period of the secretory activity of the corpus luteum of pregnancy, i.e. the maximum of its endocrine function, corresponds to days 50-60. According to Neelson and Greene (1958) the corpus luteum of pregnancy functions only during the first 6 weeks and it deteriorates rapidly during the 8th-16th weeks. In fact, the first morphological evidence of regression is apparent only by the third month and it cannot be excluded that some activity might continue for some time.

The *granulosa lutein cells* present the general features described for the menstrual corpus luteum. However some particular features must be pointed out. With the L.M. some granules and colloidal droplets have been described in the luteal cells of pregnancy. Some authors (Portes et al., 1938; Simmonet and Robey, 1939) suggest that these granules and droplets are typical of the corpus luteum of pregnancy and may have diagnostic value. According to Gillman and Stein the granules might be an expression of secretory activity, while colloidal droplets might represent stagnant regression material. In my observations (Balboni, 1956) the colloidal droplets were more frequent and larger in the corpus luteum of tubal pregnancy with a dead foetus.

With the E.M. the granulosa lutein cells in early pregnancy present a well-developed S.E.R., that is organized to form many folded membrane complexes and systems of concentric membranous whorls (Crips et al., 1970). These whorls are not observed in the luteal cells during the active phase of menstrual corpus luteum and they are supposed to be related to increased synthesis of free cholesterol or to high titers of human chorionic gonadotrophin which occur in the maturity phase of the corpus luteum of pregnancy (Crisp et al.). Cisternae of the rough E.R. and free ribosomes are also present. As described for the menstrual corpus luteum, the luteal cells in pregnancy contain an important system of intercellular channels as well as of intracellular canaliculi. The presence of gap-junctions was described in the luteal cells of the rat corpus luteum by Albertini and Anderson (1975). They observed that the gap-junctions are frequently associated with tubular elements of smooth E.R. and believed that this relationship might "provide for expedient interchange of intercellular messengers such as cAMP to enzymes in the S.E.R. in order to coordinate its synthetic activity". Recently Brassart and Maillet (1975) have observed that also in the human and rabbit corpora lutea gap-junctions exist in the interdigitations between luteal cells, while septate-like zones are present between the microvilli of one cell or of adjacent cells. Abel et al. (1975) suggest that the extensive gap-junctions found in the luteal cells of the dog are probably responsible for spreading the impulse induced by gonadotrophins first along the surface of the target cell and subsequently on to adjacent cells. They think also that gap-junctions might be responsible for diffusion or transport of the products elaborated in one compartment of a luteal cell to a complementary compartment in the adjacent cell. The mitochondria of luteal cells of the human corpus luteum of pregnancy are generally larger and pleomorphic in comparison to the mitochondria of the cells of menstrual corpus luteum. They contain osmiophilic inclusions and sometimes small myelin figures. Granules have been described in the luteal cells: some of them appear as membrane bound granules of variable size and configuration, frequently associated with the Golgi complex and contain an electron-dense, heterogenous material. Other membrane bound granules are smaller (15-20 millimicrons in diameter), with a moderate electron-dense matrix and are closely associated with or trapped between cisternae of the rough E.R.. It was suggested that these granules represent a proteinaceous secretory product, probably *relaxin*. As a matter of fact, the observations of Zarrow and O'Connor (1966) in the rabbit, using the indirect fluorescent antibody technique, indicate the site of production of relaxin in the luteal cells. In Crisp's opinion, the release of the polypeptide hormone relaxin from storage granules might occur during cell regression at term by the action of hydrolytic enzymes within the lysosomes.

The *theca lutein cells* are numerous in the first two months of pregnancy, but afterwards they tend to decrease in number and after the 5th month they have completely disappeared (Gillman and Stein, 1941; Balboni, 1956). During early pregnancy they present the characteristics of steroid-producing cells already described in relation to the menstrual corpus luteum.

In 1951, White et al. described in the corpus luteum of human pregnancy some cells, irregular in shape, with a homogenous acidophilic cytoplasm, rich in phospholipids and in ketosteroids. They called these cells "K" cells, because of their supposed function in elaborating ketosteroids. These cells seem to be similar to those that Momigliano (1927) supposed to be responsible for the formation of the human corpus luteum and they derive from the wall of the theca interna vessels. Probably they correspond to the macrophages described in the pericapillary spaces of the corpus luteum by Gillim et al. (1969).

It is well known that the capillaries of the corpus luteum of several mammals present many pores in the endothelial wall.

In the human, it was observed (Balboni and Tedde, 1973) that the endothelial wall of the capillaries in the corpus luteum is very irregular and may present zones of important thinning, but only seldom some true pores. The cellular body of the endothelial cells, where the nucleus is contained, is rich in organelles; the cytoplasmic expansions present pinocytotic vesicles and larger vacuoles, which give to the cytoplasm labyrinthic appearance. Some irregular cytoplasmic processes project in the lumen, into which some free cytoplasmic fragments may be observed. The basal lamina is thin, but continuous also under the pores.

Some problems exist in regard to the secretory mechanism of steroids by the luteal cells. In fact, luteal cells lack secretory structures such as secretory granules, derived from the Golgi apparatus. The dense bodies that some authors (Yamada and Ishikawa, 1960) have interpreted as secretory granules, are now thought to be lysosomes (Motta, 1969; Crisp et al., 1970; etc.), that are mainly involved in the autophagic processes that occur in the involutive phase of the corpus luteum as stated by Lennep and Madden (1965). According to Crisp et al., lysosomes may also participate in the releasing mechanism of relaxin, but it seems that they do not act in steroid secretion. No evidence was found by more recent authors for secretion by reverse pinocytosis, as suggested by Tokia (1965), or by an apocrine mechanism as supposed by Yamada and Ishikawa (1960) and by Green and Maqueo (1965). Much importance is ascribed now to the presence of microvilli on the side of luteal cells facing subendothelial spaces, because they may provide additional surface area for the passage of steroids through the cell membrane to the extracellular spaces probably by active transport or by diffusion (Motta, 1969; Gillim et al., 1969).

An extensive ultrastructural study of the compartmentalization of luteal cells in the dog ovary has been carried out by Abel et al. (1975) in order to establish the subcellular components responsible for the individual steps in progesterone synthesis and release. The results of this study appear to be of interest in interpreting the significance of the subcellular structures in the human luteal cells, where compartmentalization is not as evident as in the dog cells.

Involution of the corpus luteum. After a period of activity both the menstrual corpus luteum and the corpus luteum of pregnancy begin to involute. The general process of luteolysis of the two types of corpus luteum is similar and consists mainly of degeneration of luteal cells with nuclear picnosis, accumulation in the cytoplasm of lipids, especially neutral fats and cholesterol, appearance of enzymatic activities such as acid phosphatase and aminopeptidase activity. The connective stroma, which during the maturation phase consists of a thin network of collagen fibrils, when involution starts, tends to become more and more fibrous and to compress progressively the degenerating lutel cells, whose size decreases until complete cellular lysis occurs. The corpus luteum is transformed into a *corpus fibrosus*. If the collagen forming the corpora fibrosa undergoes hyalinization, *corpora albicantia* are found, which survive a long time and may be observed in great numbers in sections of ovaries of adult women. If some hematic pigment persists, remaining from the hemorrhage which occurred at rupture, the corpus fibrosus is called a *corpus nigrum*.

With the electron microscope the involution of luteal cells is characterized as follows:

1) Vacuolization of the cytoplasm, consequent to the extraction of the accumulated lipid material.

2) Swelling and alterations of mitochondria, with the appearance within them of evident myelin figures.

3) Enlargement and fragmentation of the S.E.R..

4) Presence of many autophagic structures such as lysosomes, phagosomes, residual bodies, etc..

5) Increase of the collagenous intercellular component.

6) Capillaries in the involuting corpus luteum also present evidence of regression (Balboni and Tedde, 1973). Their wall is formed by both vacuolated light cells and by irregularly-shaped dark cells. Sometimes the dark cells seem to project and to be expelled into the lumen. The basal lamina that surrounds the capillary wall is dense and thick.

CONCLUDING REMARKS ON STEROIDOGENESIS IN THE OVARY

If any conclusion can be drawn from all the organogenetic and histological information concerning the steroidogenic cells in the ovary, this conclusion may only be a general one.

The ovary is a peculiar organ, in which all the histological compartments (i.e. stroma and specific organelles) participate in an endocrinological function, which is, in the main, regulated by the hypophyseal hormones. However, the possibility cannot be excluded that some influence may be exerted by nervous stimulation. If one considers the ovulatory mechanism, for example, the presence of smooth muscle in the ovary of many species and its apparent motor adrenergic innervation suggests that neurogenic contractions of these muscle cells may play a role in the ejection of the oocyte (cf. Bell, 1972). According to Fink and Schofield (1971), the adrenergic nerve fibers which are in intimate relationship with follicular cells, that they have observed in the cat, have a permissive or synergistic action with respect to gonadotrophins in processes which require the presence of cyclic AMP. According to the same authors, ovarian nerves in the cat also play a role in participating in the selection of those follicles which are to undergo maturation. In research on the adrenergic innervation of human and rabbit ovaries (Balboni, 1971) isolated fibers were observed in the cortical stroma, presenting along their course some varicosities, interpreted as nerve terminals. These terminals were thought to have some physiological significance in relation to the extensive and changing capacity of the stroma connective tissue in the ovary. On the other hand, Unsiker (1970), on the basis of his studies of the adrenergic innervation of the interstitial gland of the mouse, has suggested that adrenergic axons might affect the transformation of fibrocytes in interstitial gland cells. In addition, he was able to observe that many oxons come in close contact with interstitial cells, forming nerve terminals, which may be deeply embedded in the cytoplasm of the innervated cells. Dahl (1970) demonstrated that in the fowl ovary numerous axon terminals are in membranous contact with the steroidogenic cells of the theca gland.

The data mentioned above support the idea that some degree of neural regulation may exist also for the endocrine cells in the ovary. Clearly, this neural influence is quite different in the various species; at present, it is not clear to what extent it may act in the human ovary.

Another point is that a sharp distinction of different compartments in relation to the production of the various steroid hormones (estrogens, progesterone, androgens) cannot always be made in the different physiological and experimental conditions, the cells involved in steroidogenesis may be able to utilize different metabolic pathways, that lead to the elaboration of different steroids or their precursors. From a morphological point of view this fact matches well with the common origin of ovarian somatic cells from a common blastema and with the fact that the cellular equipment for steroidogenesis is similar to the various cell types involved in the production of ovarian hormones.

REFERENCES

Abel, J.H. Jr., McClellan, M.C., Verhage, H.G. and Niswender, G.N. (1975a). *Cell Tiss. Res.* **158**, 461-480.

Abel, J.H. Jr., McClellan, M.C., Verhage, H.G. and Niswender, G.N. (1975b). *Cell Tiss. Res.* **160**, 155-176.

Acevedo, H.F., Axelrod, L.R., Ishikawa, E. and Takaki, F. (1961). *J. clin. Endocr. Metab.* 21, 1611-1613.
Acevedo, H.F., Axelrod, L.R., Ishikawa, E. and Takaki, F. (1963). *J. clin. Endocr. Metab.* 23, 885-890.
Adams,,E.C. and Hertig, A.T. (1969a). *J. Cell. Biol.* 41, 696-715.
Adams, E.C. and Hertig, A.T. (1969b). *J. Cell. Biol.* 41, 716-735.
Albertini, D.F. and Anderson, E. (1975). *Anat. Rec.* 181, 171-194.
Anderson, E. (1971). *Anat. Rec.* 169, 473a.
Armstrong, D.T. (1968). *Rec. Prog. Horm. Res.* 24, 255-319.
Baca, M. and Zamboni, L. (1966). *J. Ultrastr. Res.* 19, 354-381.
Baillie, A.M., Ferguson, M.M. and Hart, D. McK. (1966). *J. clin. Endocr. Metab.* 26, 738-741.
Balboni, G.C. (1956). *Arch. ital. Anat. Embriol.* 61, 373-400.
Balboni, G.C. (1959). *Arch. ital. Anat. Embriol.* 64, 40-73.
Balboni, G.C. (1960). *Arch. ital. Anat. Embriol.* 65, 115-136.
Balboni, G.C. (1962). "Scritti in onore del Prof. E. Maurizio", *SAGA, Genova* 1, 201-219.
Balboni, G.C. (1968). *Bull. Ass. Anat.* 140, 468-477.
Balboni, G.C. (1970). *Bull. Ass. Anat.* 171, 106-114.
Balboni, G.C. (1971). *Boll. Soc. it. Biol. sper.* 48, 84-86.
Balboni, G.C. (1973). *Arch. ital. Anat. Embriol.* 78, 37-58.
Balboni, G.C. and Tedde, G. (1973). *Bull. Ass. Anat.* 156, 41-50.
Barlow, P.W. (1973). *Humangenetik* 17, 105-136.
Bejdl, W. (1952). *Wien. Klin. Wochinschrift.* 7, 484.
Bell, C. (1972). *Farmacol. Rev.* 24, 657-736.
Berger, L. (1922). *C.R. Acad. Sci. (Paris)* 175, 907-909.
Berger, L. (1924). *C.R. Soc. Biol. (Paris)* 90, 267-268.
Bjersing, L. and Cajander, S. (1974c). *Cell. Tiss. Res.* 149, 313-327.
Bjersing, L. and Cajander, S. (1974d). *Cell. Tiss. Res.* 153, 1-14.
Bjersing, L. and Cajander, S. (1974e). *Cell. Tiss. Res.* 153, 15-30.
Bjersing, L. and Cajander, S. (1974f). *Cell. Tiss. Res.* 153, 31-44.
Bloch, E. (1964). *Endocrinology* 74, 833-845.
Bloch, E. (1967). *Steroids* 9, 415-430.
Bloch, E., Tissenbaum, B. and Bernirschke, K. (1962). *Biochem. biophys. Acta* 60, 182-184.
Brambell, F.W.R. (1927b). *Proc. Roy. Spc. Med.* 101, 391-409.
Brannan, D. (1927). *Am. J. Path.* 3, 343-358.
Brassart, B. (1975). *59th Congrès Ass. Anat.*, Bordeaux (comun. verbale).
Buehr, M. and Petroldt, U. (1974). *Verh. Anat. Ges.* 68S, 163-167.
Burden, H.W. (1972). *Am. J. Anat.* 133, 125-142.
Catchpole, H.R., Gersh, I. and Pan, C. (1950). *J. Endocrinol.* 6, 277-281.
Channing, C.P. (1969). *J. Endocrinol.* 45, 297-308.
Channing, C.P. (1974). *Endocrinology* 94, 1215-1223.
Chiquoine, A.D. (1960). *Am. J. Anat.* 106, 149-196.
Claesson, L. (1954). *Acta Physiol. Scand.* 31, **Suppl.** 113, 23-52.
Corner, G.W. (1938). *Physiol. Rev.* 18, 154-172.
Corner, G.W. Jr. (1956). *Am. J. Anat.* 98, 377-401.
Christensen, A.K. (1965). *J. Cell. Biol.* 26, 911-935.
Crisp, T.M., Dessouky, A. and Denys, F.R. (1970). *Am. J. Anat.* 127, 37-70.
Czyba, J.C., Cosnier, J., Giroud, C. and Laurent, J.L. (1973). *SIME Ed.*, Villeurbanne.
Dahl, E. (1970). *Zeit. Zellforsch.* 109, 212-226.
Deane, H.W. (1952). *Am. J. Anat.* 91, 363-413.
Dohn, G. (1955a). *Z. Geburtsch. Gynäk.* 142, 182-228.
Dohn, G. (1955b). *Z. Geburtsch. Gynäk.* 142, 289-313.
Dubois, R. (1964). *C.R. Acad. Sci.* 258, 3904-3907.
Dubois, R. (1965). *C.R. Acad. Sci.* 260, 5885-5887.
Dubois, R. (1966). *C.R. Acad. Sci.* 262, 2623-2626.
Dubreuil, G. (1945). *C.R. Soc. Biol.* 139, 850.
Dubreuil, G. (1946). *Bull. Histol. appl.* 23, 17-34.
Dubreuil, G. (1948). *Ann. Endocr.* 9, 434.
Dubreuil, G. (1953). *C.R. Ass. Anat. 40a. Reun.*, Bordeaux, 1-27.
Dubreuil, G. (1962). *Ann. Endocr.* 23, 1-14.
Duke, K.L. (1947). *Anar. Rec.* 98, 507-526.

Falin, L.I. (1969). *Acta Anat.* 72, 195-232.
Falk, B. (1959). *Acta Physiol. Scand.* 47, **Suppl.** 163, 1-101.
Fink, G. and Schofield, G.C. (1971). *J. Anat.* 109, 115-126.
Fischel, A. (1930). *Zeit. f. Anat. Entw.* 92, 34-72.
Fletcher, W.H. (1973). *J. Cell. Biol.* 59, 101a.
Fumagalli, Z., Motta, P. and Calvieri, S. (1971). *Experientia (Basel)* 27, 682-683.
Fuyata, M., Miyayama, Y. and Fujimoto, T. (1974). *Okajimas Folia Anat. Jap.* 51, 251-262.
Gillim, S.W., Christensen, A.K. and McLennan, C.E. (1969). *Am. J. Anat.* 126, 409-428.
Gillman, J. and Stein, H.B. (1941). *Surg. Gynec. Obstet.* 72, 129-140.
Goldman, A.S., Yakovac, W.V. and Bongiovanni, A.M. (1966). *J. clin. Endocr. Metab.* 26, 14-22.
Green, J.A. and Maqueo, M. (1965). *Am. J. Obstet. Gynec.* 92, 946-957.
Green, J.A., Garcilazo, J.A. and Maqueo, M. (1967). *Am. J. Obstet. Gynec.* 99, 855-863.
Green, J.A., Garcilazo, J.A. and Maqueo, M. (1968). *Am. J. Obstet. Gynec.* 102, 57-64.
Groop, A. and Ohno, S. (1966). *Zeit. f. Zellforsch.* 74, 505-528.
Gruenwald, P. (1942). *Am. J. Anat.* 70, 359-397.
Guraya, S.S. (1971). *Physiol. Rev.* 51, 785-807.
Guraya, S.S. (1974). *Acta Anat.* 90, 250-284.
Hadjioloff, A.I. (1969). *54th Congres Ass. Anat.*, Sofia.
Ham, A.W. (1971). "Histology", Lippincott Comp., Philadelphia and Montreal.
Hamerton, J.L. (1968). *Nature* 219, 910-914.
Hamilton, W.J. and Massman, H.W. (1972). "Human Embryology", W. Heffer and Son Ltd., Cambridge.
Harrison, R.J. (1962). *In* "The Ovary" (Zuckerman, ed), Vol. I, Academic Press, New York and London.
Hemsworth, B.N. and Jackson, H. (1963). *J. Reprod. Fertil.* 6, 229-233.
Hertig, A.T. (1968). *Am. J. Anat.* 122, 107-137.
Hertig, A.T. and Adams, E.C. (1967). *J. Cell. Biol.* 34, 647-675.
Hertig, A.T. and Gore, H. (1961). "Atlas of Tumor Pathology", Washington (cited by Groop and Ohno).
Hope, J. (1965). *J. Ultrastr. Res.* 12, 592-610.
Jong, F.H., Bard, D.T. and Van der Molen, H.S. (1974). *Acta Endocr.* 77, 575-587.
Kohn, A. (1928). *Endokrinologie* 1, 3-10.
Kretser, D.M. (1967). *Z. Zellforsch.* 83, 344-358.
Leavitt, W.W., Basom, C.R., Bagwell, J.N. and Blaha, G.C. (1973). *Am. J. Anat.* 136, 235-250.
van Lennep, E.W. and Madden, L.M. (1965). *Z. Zellforsch.* 66, 365-380.
McKay, D.G., Hertig, A.T., Adams, E.C. and Danziger, S. (1953). *Anat. Rec.* 117, 201-220.
Meyer, R. (1915). *Z. Geburtsh. Gynäk.* 77, 505-525.
Merk, F.B., Botticelli, C.R. and Albright, J.T. (1972). *Endocrinology* 90, 992-1007.
Merk, F.B., Albright, J.T. and Botticelli, C.R. (1973). *Anat. Rec.* 175, 107-125.
Merker, H.J. and Diaz-Encima, J. (1969). *Z. Zellforsch.* 94, 605-623.
Mittwoch, U. (1969a). *Nature* 221, 446-448.
Mittwoch, U. (1969b). *Nature* 224, 1323-1325.
Momigliano, E. (1927). *Ric. Morfol.* 6, 1-78.
Moricard, M. (1954). *C.R. Ass. Anat.* 40, 762-770.
Mossman, H.W., Koering, M.J. and Ferry, D. (1964). *Am. J. Anat.* 115, 235-256.
Motta, P. (1966). *Biol. Lat.* 18, 107-137.
Motta, P. (1969). *Z. Zellforsch.* 98, 233-245.
Motta, P., Takeva, Z. and Bourneva, V. (1970). *Experentia* 26, 1128-9.
Motta, P., Takeva, Z. and Nesci, E. (1971). *Acta Anat. (Basel)* 80, 537-562.
Motta, P. and Di Dio, L.J.A. (1974). *J. Submicrosc. Cytol.* 6, 15-27.
Motta, P. and van Blerkom, J. (1974). *J. Submicrosc. Cytol.* 6, 297-310.
Nelson, W.W. and Greene, R.R. (1958). *Am. J. Obstet. Gynec.* 76, 66-90.
Novak, E. (1952). "Gynecologic and Obstetric Pathology", W.B. Saunders Comp., Philadelphia.
Ohno, S. and Smith, J.B. (1964). *Cytogenetics* 3, 324-333.
Okamura, H., Virntamasen, P., Wright, K.H. and Wallach, E.E. (1972). *Am. J. Obstet. Gynec.* 112, 183-191.
O'Shea, J.D. (1970). *Anat. Rec.* 167, 127-140.
O'Shea, J.D. (1971). *J. Reprod. Fertil.* 24, 283-285.

O'Shea, J.D. (1972). *J. Reprod. Fertil.* 28, 138-139.
Osvaldo-Decima, L. (1970). *J. Ultrastr. Res.* 29, 218-237.
Pedersen, P.H. and Larsen, J.P. (1968). *Acta Endocr.* 58, 481-496.
Pinkerton, J.H.M., McKay, D.G., Adams, C. and Hertig, A.T. (1961). *Obstet. Gynecol.* 18, 165-181.
Politzer, G. (1933). *Zeit. f. Anat. Entwicklung* 100, 331-361.
Portes, L., Ascheim, S. and Robey, M. (1938). *Gynec. Obstet.* 37, 100-121.
Rice, B.F. and Savard, K. (1966). *J. clin. Endocr. Metab.* 26, 593-609.
Ryan, K.J. and Smith, O.W. (1965). *Rec. Prog. Horm. Res.* 21, 367-407.
Ryan, K.J. and Petro, Z. (1966). *J. clin. Endocr. Metab.* 26, 46-58.
Ryan, K.J., Petro, Z. and Kaiser, J. (1968). *J. clin. Endocr. Metab.* 28, 355-358.
Sassu, G. and Colombino, R. (1969). *Studi Sassaresi* 47, 115-126.
Savard, K., Marsh, J.M. and Rice, B.F. (1965). *Rec. Prog. Horm. Res.* 21, 285-365.
Schlegel, R.J., Farias, F., Russo, N.C., Moorn, J.R. and Garduer, L. (1967). *Endocrinology* 81, 565-572.
Schroeder, R. (1930). "Weibliche Genitalorgane" in von Möllendorff: "Handbuch der Mikroskopischen Anatomie des Menschen" T.7 - I P., Springer-Verlag, Berlin.
Short, R.U. (1967). *Arch. Anat. Microsc.* 56, Suppl. 3-4, 258-272.
Simmonet, H. and Kobey, M. (1939). "Le Corps Jaune", Masson Ed., Paris.
Smith, O.W. and Zuckerman, N.G. (1973). *Steroids* 22, 379.
Stemberg, W.H. (1949). *Am. J. Path.* 25, 493-521.
Stemberg, W.H., Segoloff, A. and Gaskill, C.J. (1953). *J. clin. Endocr. Metab.* 13, 139-153.
Stoll, R. and Maraud, R. (1974). *Bull. Ass. Anat.* 58, 699-764.
Tardini, A., Vitali Mazza, L. and Mansani, F.E. (1960). *Arch. De Vecchi* 33, 281-305.
Tardini, A., Vitali Mazza, L. and Mansani, F.E. (1961). *Arch. De Vechhi* 35, 25-71.
Tedde Piras, A. (1971). *Arch. ital. Anat. Embr.* 76, 67-86.
Tokida, A. (1965). *Mic. Med. J.* 15, 27-76.
Unsicker, K. (1970a). *Zeit. Zellforsch.* 109, 46-54.
Unsicker, K. (1970b). *Zeit. Zellforsch.* 109, 495-516.
Waldayer, W. (1870). "Eustock un Ei", Engelman Ed., Leipzig.
White, R.F., Hertig, A.T., Rock, J. and Adams, E.C. (1951). *Contr. Embryol. Carneg. Inst.* 34, 55-74.
Winiwarter, H. de (1910). *Arch. Biol.* 25, 683.
Winiwarter, H. de and Sainmont, G. (1909a). *Arch. Biol.* 24, 1-96.
Winiwarter, H. de and Sainmont, G. (1909b). *Arch. Biol.* 24, 165-276.
Witschi, E. (1948). *Contr. Embryol. Carneg. Inst.* 32, 67-80.
Witschi, E. (1951). *Rec. Prog. Horm. Res.*, 1-27.
Witschi, E. (1963). *In* "The Ovary" (H.G. Grady and D.E. Smith, eds) Williams and Wilkins Ed., Baltimore.
Witschi, E., Bruner, J.A. and Segal, S.J. (1953). *Anat. Rec.* 115, 381 **Abs**.
Yamada, E. and Ishikawa, T.M. (1960). *J. Med. Sci.* 11, 235-259.
Yang, N.S.T., Marsh, J.M. and Le Maire, W.J. (1937). *Prostaglandins* 4, 395-404.
Zamboni, L. and Mastroianni, L. Jr. (1966a). *J. Ultrastr. Res.* 14, 95-117.
Zamboni, L. and Mastroianni, L. Jr. (1966b). *J. Ultrastr. Res.* 14, 118-132.
Zarrow, M.X. and O'Connor, W.B. (1966). *Proc. Soc. exp. Biol. Med.* 121, 612-614.
Zuckerman, S. (1962). "The Ovary", Academic Press, New York and London.

THE INTERRELATIONSHIP OF FOLLICULAR CYCLIC AMP, STEROIDS AND PROSTAGLANDINS DURING THE OVULATORY PROCESS*

J.M. Marsh and W.J. LeMaire

*The Endocrine Laboratory, University of Miami Medical School
P.O. Box 875, Biscayne Annexe, Miami 33152, Florida, USA*

I. INTRODUCTION

Luteinizing hormone (LH) causes a number of changes in the synthesis of cyclic AMP, prostaglandins and steroids in the mature Graafian follice. Furthermore, the response to LH in terms of each of these parameters changes between the time of the LH peak and the rupture of the follicular wall. This chapter will describe the changes which occur in these three biochemical parameters of follicular function around the time of ovulation, and attempt to show how they may be interrelated and involved in the process of ovulation itself. The work from our laboratory has been carried out over a period of time by several investigators who are listed alphabetically as follows: Peter J.A. Davies M.D.; William J. LeMaire M.D.; John M. Marsh Ph.D.; Thomas M. Mills Ph.D.; Kenneth Savard D.Sc.; and Norman S.T. Yang Ph.D..

II. THE EXPERIMENTAL MODEL

In these studies we have used the rabbit Graafian follicle as our experimental model because the rabbit is a reflex ovulator and the time of ovulation can be reasonably accurately predicted after mating or the injection of a hormone preparation with LH activity. Walton and Hammond (1929) showed that ovulation occurred about 10 h after mating and Harper (1961, 1963) shows that it occurred between 10 and 12 h after the injection of human chorionic gonado-

---

* Dedicated to Dr. Karl H. Slotta (Emeritus Professor of Biochemistry and Medicine of the University of Miami) on the occasion of his eightieth birthday.

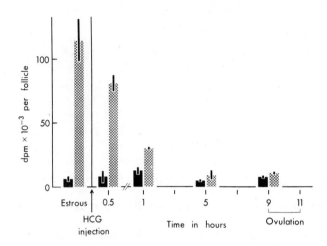

Fig.1 Preovulatory changes in the accumulation of cyclic [8-^{3}H] AMP in isolated rabbit Graafian follicles. The solid bars represent the results of the control incubations and the cross-hatched bars represent results of incubations with 5 μg/ml of luteinizing hormone. The height of the bars of the estrous rabbits represent the mean of 10 experiments and the bracket the standard error of the mean. The height of the other bars represents the average of 3 experiments and the brackets the range. HCG, human chorionic gonadotropin (from Marsh et al., 1973).

tropin (hCG). Adult nonpregnant New Zealand white rabbits between 6 and 8 months of age were used in all experiments. The animals were housed in individual cages in air conditioned quarters and fed a standard rabbit chow and water *ad libitum*. The lighting conditions consisted of 15 h of light (7.00-22.00) and 9 h of darkness (22.00-7.00). Ovulation was induced by mating or by the injection of LH (50 μg) of hCG (100 i.u.) into the marginal ear vein. The animals were sacrificed at appropriate times by a blow to the head and the follicles were dissected from the adjacent ovarian tissue according to the method described by Mills, Davies and Savard (1971).

III. PREOVULATORY CHANGES IN STEROIDOGENESIS

The first assessment in our laboratory of biochemical changes in follicles during the process of ovulation involved the measurement of steroidogenesis and was carried out by Mills *et al.* (1971), Mills and Savard (1972) and Mills and Savard (1973). Rabbits were sacrificed before or at various intervals after mating. The large mature Graafian follicles were dissected from the ovary, and their ability to synthesize steroids was assessed by incubating them *in vitro* and measuring the incorporation of acetate-1-^{14}C into several steroids. The labelled steroids were purified to radiochemical purity by standard techniques. The *in vitro* response of these isolated follicles to LH was also assessed in these incubations.

It was found that follicles obtained from rabbits before mating were capable of synthesizing a wide variety of steroids from acetate-1-^{14}C including estrone, estradiol, progesterone, 17 hydroxy-progesterone, testosterone and androstenedione. The synthesis observed under control conditions was relatively low but the addition of LH to the incubation medium resulted in a marked increase of ^{14}C acetate incorporation into all of the steroids (Mills et al., 1971; Mills and Savard, 1972). Follicles isolated from rabbits 2 h after mating exhibited high levels of steroidogenic activity both in the control and LH treated follicles. In fact, there was no significant effect of LH *in vitro* in these follicles. This is understandable, since these follicles had already been exposed to high concentrations of endogenous LH *in vitro*. At 12 h after mating (the time of the expected ovulation) the isolated follicles showed very little steroidogenic capacity and no apparent response to exogenous LH (Mills and Savard, 1973).

Thus it seemed that the mature ovarian follicle undergoes marked changes in steroidogenesis between the time of coitus and ovulation. First there is an increase in steroidogenesis followed by a decrease to low levels at the time of ovulation. There is also a refractoriness to *in vitro* LH stimulation which generally coincides with this decline in steroidogenesis. These changes in *in vitro* steroidogenesis and LH responsiveness correlate well with the *in vivo* measurements of ovarian steroid secretion (Hilliard and Eaton, 1971; Hilliard, Scaramuzzi, Pang, Penardi and Sawyer, 1974) and the measurements of steroids in follicular fluid (Younglai, 1972).

IV. PREOVULATORY CHANGES IN CYCLIC AMP

Drawing from our studies on the corpus luteum we expected cyclic AMP was probably a mediator of this section of LH on steroidogenesis in the Graafian follicle, and we undertook an assessment of cyclic AMP synthesis in this tissue (Marsh, Mills and LeMaire, 1972; Marsh, Mills and LeMaire, 1973). At this time we were unable to measure the mass of cyclic AMP in these small amounts of tissue, but we could measure the incorporation of radioactive adenine into cyclic AMP just as we did previously in the corpus luteum studies (Goldstein and Marsh, 1973). Graafian follicles were isolated from ovaries of untreated rabbits, and from rabbits at various intervals after the injection of hCG. The follicles were preincubated with [^{3}H] adenine, washed and incubated under control conditions or with LH or other test substances. After the test incubation the follicles were homogenized and [^{3}H] cyclic AMP isolated and purified to radiochemical purity, by Dowex chromatography, BaSO$_4$ precipitation, and thin layer chromatography.

As shown at the left-hand side of Fig.1, control follicles from untreated estrous rabbits accumulated a low level of [^{3}H] cyclic AMP, but there was a marked increase in this accumulation when LH was included in the *in vitro* incubations. This effect appeared to be specific for LH in that bovine serum albumin, prolactin or follicle stimulating hormone did not cause an increase in cyclic AMP synthesis (Marsh et al., 1972). The minimal effective dose of LH was between 5 and 50 ng/ml, which is similar to the minimal effective dose of LH found earlier to be necessary to stimulate follicular steroidogenesis (Mills et al., 1971).

When the follicles were obtained from rabbits ½, 1, 5 and 9 h after injection of hCG, as shown in Fig.1, there was a progressive decline in the *in vitro* responsiveness of these follicles to LH until at 9 h the *in vitro* effect of LH was no longer detectable. Tsafriri, Lindner, Zor and Lamprecht (1972) and Lamprecht, Zor, Tsafriri and Lindner (1973) independently obtained similar results with isolated rat follicles. They examined the effect of LH on cyclic AMP accumulation by the

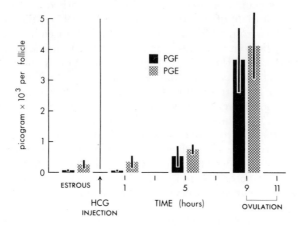

Fig.2 Preovulatory changes in the concentration of prostaglandins in rabbit Graafian follicles. The follicles were isolated at estrus or at 5 and 9 h after the injection of 100 i.u. of hCG. PGF (solid bars) and PGE (crosshatched bars) were isolated and measured by radioimmunoassay. The height of the bars for PGF and PGE of estrous rabbits and PGF at 9 h after HCG represent the mean of five experiments and the brackets the standard deviation (SD). The bars for PGF and PGE at 1 and 5 h represent the mean of four experiments with the SD. The bar for PGE at 9 h represents the mean of three experiments with the SD. (From LeMaire *et al.*, 1973.)

incorporation of [^{3}H] adenine and by the measurement of mass and they found that LH caused a marked increase in cyclic AMP during a 20 min incubation. Furthermore Lamprecht *et al.* (1973) reported a refractoriness of isolated rat follicles to stimulation by LH if the follicles had been cultured previously with LH for 18 h. Accordingly, there is a good correlation in the changes in follicular cyclic AMP with the changes in follicular steroidogenesis; indicating that cyclic AMP is a mediator of this action of LH and that the decrease in steroidogenesis in follicles just prior to ovulation might be due to their inability to synthesize cyclic AMP. Recently Mills (1975) confirmed this proposed mediatory role for cyclic AMP in follicular steroidogenesis when he reported that exogenous cyclic AMP would stimulate steroidogenesis when this cyclic nucleotide was added to incubating Graafian follicles obtained from estrous rabbits. A maximal amount of exogenous cyclic AMP was just about as effective as a maximal amount of LH in stimulating steroidogenesis. Furthermore, LH did not produce an additive effect upon that already caused by a maximal amount of cyclic AMP. Thus it appears that LH stimulates steroidogenesis in the Graafian follicle solely via a mechanism involving this cyclic nucleotide.

V. PREOVULATORY CHANGES IN PROSTAGLANDIN

We turned then to the measurement of prostaglandins in these follicles. Other investigators had implicated prostaglandins in the process of ovulation by showing that inhibitors of prostaglandin synthesis, such as indomethacin, blocked ovulation. We wished to carry out a direct measurement of prostaglandins in

Graafian follicles to see if there were changes in the concentration of these substances during the ovulatory process. Again rabbits were sacrificed before and after the injection of hCG or LH or after mating. The mature follicles were isolated and the amount of prostaglandin E (PGE) and prostaglandin F (PGF) present in them were measured. There were no *in vitro* incubations in this aspect of the work. The prostaglandins were separated into E and F types by silicic acid column chromatography and measured by a radioimmunoassay using antibodies generously given to us by Dr. Harold Behrman of the Merck Institute for Therapeutic Research.

We found, as shown in Fig.2, that follicles isolated from estrus rabbits had low, but measurable amounts of E and F prostaglandins (LeMaire, Yang, Behrman and Marsh, 1973). We have not yet determined if they are E_1 or E_2 or F_{1a} or F_{2a}, since our radioimmunoassay method does not distinguish between prostaglandins which only differ by the number of double bonds.

As the time of ovulation approached there was a marked increase in both the E and F prostaglandins. At 9 h, PGF had increased 60 fold and PGE 15 fold. This rise in prostaglandin could also be brought about by an LH injection or by mating. The concentration of prostaglandins is represented in this figure as picograms per follicle, but the same type of changes can be demonstrated if the concentrations are represented as the picograms per mg wet weight. This rules out the possibility that the increases in prostaglandin concentration were simply due to increases in follicular size.

The pretreatment of the rabbits with indomethacin completely abolished this rise in prostaglandins at 5 or 9 h after hCG injection (Yang, Marsh and LeMaire, 1973). This helps to confirm the validity of the radioimmunoassay method and also supports the interpretation that follicular prostaglandins are directly involved in ovulation.

We have also measured PGE and PGF in rabbit Graafian follicles after ovulation (Yang, Marsh and LeMaire, 1974). Figure 3 illustrates the changes in PGF and PGE in two types of follicles: the follicles which ovulate (as shown by the solid dots and the two upper lines); and the large follicles which do not ovulate (as shown by the open dots and the two lower lines). Turning first to the changes in follicles that do ovulate, the left portion of the upper dashed and solid lines depict the ovulatory rise in PGE and PGF respectively that was previously illustrated in Fig.2. Ovulation takes place as a rule between 10 and 12 h after hCG. At 10 h or more you can see that the concentration of PGE in ovulated follicles continues to increase to a maximum at 15 h. Then it slowly declines to baseline values at about 48 h after the injection of hCG. The concentration of PGF in ovulated follicles on the other hand begins to decline almost immediately after ovulation. This difference in the time course of the changes in PGE and PGF suggest the possibility of a different source and different role for these two types of prostaglandins in follicular function, but there is insufficient data as yet to come to a firm conclusion.

At the bottom of Fig.3 we have illustrated the changes in the concentration of PGE and PGF in a number of large follicles that failed to ovulate. It is evident that the prostaglandin concentrations in these follicles are far below the concentrations in the ovulated follicles shown above. This indicates that the marked increase in PGE and PGF during the preovulatory period is only limited to those large follicles which will eventually go on to ovulate.

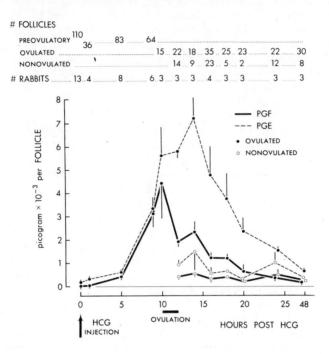

Fig.3 Pre- and postovulatory changes in prostaglandin levels in rabbit Graafian follicles. The ovulatory process was initiated by the intravenous injection of 100 i.u. of hCG. At 5 and 9 h the large preovulatory follicles were dissected and used for the measurements. At 10 h and thereafter the follicles that had clearly ovulated by morphological evidence were used for measurement in the group of ovulated follicles. From 12 h post hCG on, there was in each animal a number of follicles that did not show evidence of ovulation and they were dissected also and grouped as nonovulated follicles. The type and number of follicles analysed and the number of animals at each time are indicated at the top of the figure. The points represent the mean and the brackets the standard error of the mean. When no brackets are shown, the point represents the average of only two determinations. (From Yang *et al.*, 1974.)

We have also measured the changes in PGE and PGF in Graafian follicles obtained from a mature rat around the time of ovulation (LeMaire, Leidner and Marsh, 1975). This was done to see if the changes in endogenous prostaglandins which we had observed in a reflex ovulator such as the rabbit also occurred in a spontaneous ovulator such as the rat. In this study the estrous cycle of the rat was carefully followed by daily vaginal smears and the animals were sacrificed at appropriate times. The maximal value of plasma LH in these animals was at 4 pm on the day of proestrus and the time of ovulation was expected to occur between 1 and 4 am on the morning of estrus. It is apparent from these measurements (shown in Fig.4) that there is a marked rise in the concentration of PGE and PGF in the rat Graafian follicles between the time of the LH peak and the time of ovulation and a rapid fall thereafter. These changes are similar to the changes which were found to occur in rabbit Graafian follicles and indicate that prostaglandins are involved in the process of ovulation in spontaneous ovulators as well as reflex ovulators.

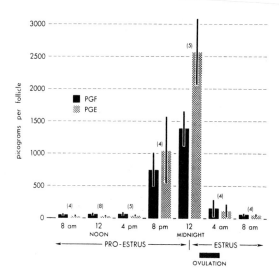

Fig.4 Pre- and postovulatory changes in the concentration of PGF (full bars) and PGE (crosshatched bars). Pre- and postovulatory Graafian follicles were isolated from normal adult rats during part of a 4 day estrous cycle and the prostaglandins measured by radioimmunoassay. The height of the bars represents the mean and the vertical lines the standard error of the mean. The numbers in parenthesis above the bars indicate the numbers of animals at each time interval. (From LeMaire et al., 1975.)

We are interested in determining how hCG or LH brings about these rises in prostaglandin and have attempted to study the mechanism of this action of these gonadotropins. In order to carry out this study, however, we thought it would be advisable to set up an *in vitro* system in which LH or other test substances could be incubated with Graafian follicles and the effects of these substances assessed on prostaglandin synthesis. The *in vitro* methodology we used involved the dissection of the large Graafian follicles of estrous rabbits and the incubation of single follicles for 1 h in Krebs-Ringer bicarbonate buffer. After this preincubation the medium was removed and new medium containing various test substances added back. The test incubation was usually run for 5 h, after which the contents of the vessels were homogenized and assayed for prostaglandins as described by Yang *et al.* (1973).

It was found as shown in Fig.5 that the incubation of these follicles for 5 h with LH resulted in a highly significant ($p < 0.0001$) stimulation of both PGF and PGE (Marsh, Yang and LeMaire, 1974). This stimulation was also specific in that other proteins such as bovine serum album, prolactin or follicle stimulating hormone had no effect. The only protein, other than LH, which caused a significant stimulation was hCG, indicating that this stimulation of prostaglandin accumulation was due to a specific LH effect and not a general protein effect. The lowest dose used, 5 ng of LH/ml, caused a significant stimulation, which is in the same order of magnitude as the lowest effective dose of LH required to increase steroidogenesis and cyclic AMP accumulation. Some stimulation of prostaglandin

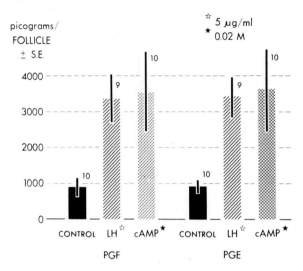

Fig.5 Effect of cyclic AMP on prostaglandin accumulation in isolated rabbit Graafian follicles. The follicles were isolated from estrous rabbits according to the procedure described by Mills et al. (1971). They were preincubated for 1 h in Krebs-Ringer bicarbonate buffer, (KRBB) pH 7.4, 37° under an atmosphere of 95% O_2 and 5% CO_2. After rinsing the follicles were incubated for 5 h under the same conditions. Control incubations were carried out in KRBB. The incubations with LH contained 5 µg/ml of NIH-LH-S 15 and the incubations with cyclic AMP (cAMP) contained 0.02 M concentration of this cyclic nucleotide. Each experiment was carried out with single follicles. The numbers above the bars indicate the number of experiments in each group, the height of the bars the mean prostaglandin synthesis and the vertical lines the SE of the mean. (From Marsh et al., 1974.)

synthesis could be observed at 3 h of incubation but the highly significant stimulation was only apparent after 5 h. This time course is similar to that observed when LH or hCG is injected *in vivo*, and indicates that we are probably dealing with the same kind of phenomenon.

In a series of experiments, cyclic AMP was tested at a concentration of 0.02 M to see if exogenous cyclic AMP could mimic the effect of LH on prostaglandin synthesis. As you can see in Fig.5 this concentration of the cyclic nucleotide had a stimulatory effect on both PGF and PGE accumulation which was about equal to that of a saturating level of LH, indicating that cyclic AMP could be a mediator of this effect of LH.

Figure 6 is a review of some of the points of the three studies I have discussed. At the bottom right we have illustrated the preovulatory changes in steroidogenesis, at the bottom left the preovulatory changes in cyclic AMP and at the top the changes in prostaglandins E and F. As you can see, there is an apparent reciprocal relationship between the rise in concentration of prostaglandin F and E and the fall off of responsiveness to LH in terms of cyclic AMP accumulation and steroidogenesis. These time course relationships have suggested to us that the prostaglandins might be the cause of the observed refractoriness to LH. In this regard it has been reported by others that prostaglandins play a negative feedback role, inhibiting the effects of tropic hormones on adenylate cycles in adipose tissue and the kidney.

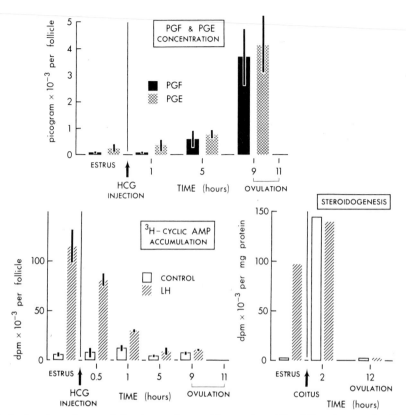

Fig.6 A review of some of the changes in steroidogenesis, cyclic AMP and prostaglandin accumulation during the preovulatory period in the rabbit Graafian follicle. In the two lower panels the open bars represent the amount of steroidogenesis or cyclic AMP accumulation which occurs in follicles incubated under control conditions. The shaded bars represent the measurements carried out on follicles incubated with LH. In the upper panel the solid black bar represents the follicular concentration of PGF and the crosshatched bar the follicular concentration of PGE. (From Mills and Savard, 1973; Marsh *et al.*, 1973; LeMaire *et al.*, 1973.)

VI. SUMMARY

Figure 7 is a summary slide in which we have tried to put most of our data together as a working model. It illustrates how LH might produce its effects on a hypothetical cell of the Graafian follicle. First of all we believe that LH causes its increase in steroidogenesis by stimulating adenylate cyclase and increasing cyclic AMP. Cyclic AMP in turn then activates a limiting step in steroidogenesis, probably between cholesterol and pregnenolone. In view of our *in vitro* studies with exogenous cyclic AMP (Marsh *et al.*, 1974) we also believe that cyclic AMP is a mediator of the action of LH which brings about the rise in prostaglandins. The mechanism of this action of cyclic AMP might be mediated by the increase in estrogens or progestins, since it has been reported that estrogens increase the concentration of prostaglandins in the uterus (Caldwell, Tillson, Brock and

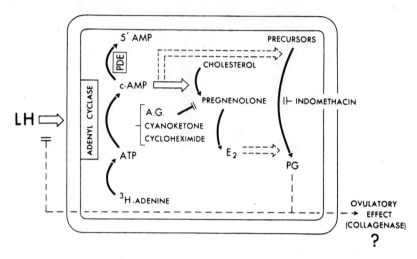

Fig. 7 A hypothetical model of some aspects of ovulation. The abbreviations used are: LH = luteinizing hormone; cAMP = cyclic AMP; E_2 = estradiol; PG = prostaglandins; AG = aminoglutethimide. The solid arrows indicate biochemical conversions. The open and dashed arrows indicate stimulations. The symbol $\bar{T}$ indicates inhibition.

Speroff, 1972). Another possibility shown at the top of the slide is that cyclic AMP may activate a protein kinase which in turn could stimulate a lipase. The increased lipase activity could convert an esterified form of a precursor unsaturated fatty acid such as arachidonic into its free form which would be quickly converted to prostaglandin. It has been proposed that the limiting step in prostaglandin synthesis might be such a lipase type reaction and cyclic AMP has been shown to activate the hormone sensitive triglyceride lipase of adipose tissue (Steinberg and Huttunen, 1972).

Finally, we believe that the increase in prostaglandins may be involved in the rupture of the follicle wall possibly by activating a collagenase-like ovulatory enzyme, by causing ovarian contraction or by causing changes in the vasculature of the follicle. In addition, we are considering the possibility that the massive increase in prostaglandins which occurs prior to ovulation may be responsible for the decrease in the *in vitro* response to LH. The symbol $\bar{T}$ indicates inhibition in the model. Some of the aspects of this hypothetical model may eventually prove to be incorrect in greater or lesser degree, but we feel that it does serve a purpose in that it pin-points several sites for further investigation.

VII. REFERENCES

Caldwell, B.V., Tillson, S.A., Brock, W.A. and Speroff, L. (1972). *Prostaglandins* 1, 217-228.
Goldstein, S. and Marsh, J.M. (1973). *In* "Protein Phosphorylation in Control Mechanisms" (F. Huijing and E.Y.C. Lee, eds) Vol. V, pp.123-144. Academic Press, New York and London.
Harper, M.J.K. (1961). *J. Endocr.* 22, 147-152.
Harper, M.J.K. (1963). *J. Endocr.* 26, 307-316.
Hilliard, J. and Eaton, W., Jr. (1971). *Endocrinology* 89, 522-527.

Hilliard, J., Scaramuzzi, R.J., Pang, C.N., Penardi, R. and Sawyer, C.H. (1974). *Endocrinology* **94**, 267-271.
Lamprecht, S.A., Zor, U., Tsafriri, A. and Lindner, H.R. (1973). *J. Endocrinol.* **57**, 217-233.
LeMaire, W.J., Yang, N.S.T., Behrman, H.R. and Marsh, J.M. (1973). *Prostaglandins* **3**, 367-376.
LeMaire, W.J., Leidner, R. and Marsh, J.M. (1975). *Prostaglandins* **9**, 221-229.
Marsh, J.M., Mills, T.M.,and LeMaire, W.J. (1972). *Biochim. biophys. Acta* **273**, 389-394.
Marsh, J.M., Mills, T.M. and LeMaire, W.J. (1973). *Biochim. biophys. Acta* **304**, 197-202.
Marsh, J.M., Yang, N.S.T. and LeMaire, W.J. (1974). *Prostaglandins* **7**, 269-283.
Mills, T.M. (1975). *Endocrinology* **96**, 440-445.
Mills, T.M. and Savard, K. (1972). *Steroids* **20**, 247-262.
Mills, T.M. and Savard, K. (1973). *Endocrinology* **92**, 788-791.
Mills, T.M., Davies, P.J.A. and Savard, K. (1971). *Endocrinology* **88**, 857-862.
Steinberg, D. and Huttunen, J.K. (1972). *Advances in Cyclic Nucleotide Research* **1**, 47-62.
Tsafriri, A., Lindner, H.R., Zor, U. and Lamprecht, S.A. (1972). *J. Reprod. Fertil.* **31**, 39-50.
Walton, A. and Hammond, J. (1929). *Brit. J. Exp. Biol.* **6**, 190-205.
Yang, N.S.T., Marsh, J.M. and LeMaire, W.J. (1973). *Prostaglandins* **4**, 395-404.
Yang, N.S.T., Marsh, J.M. and LeMaire, W.J. (1974). *Prostaglandins* **6**, 37-44.
Younglai, E.V. (1972). *J. Reprod. Fert.* **30**, 157-159.

STEROIDOGENIC CAPACITIES OF THE DIFFERENT COMPARTMENTS OF THE HUMAN OVARY

J.M. Marsh, K. Savard* and W.J. Lemaire

The Endocrine Laboratory, University of Miami Medical School
P.O. Box 875, Biscayne Annexe, Miami 33152, Florida, USA

I. INTRODUCTION

Ovarian steroidogenesis is rather complex compared to this process in the adrenal cortex or the testis, because the normal adult ovary undergoes cyclic changes in the morphology of its steroidogenic tissues. These changes include the growth and maturation of the Graafian follicles, ovulation, and the formation and regression of the corpus luteum. The steroid output from the ovary likewise changes in terms of the types and quantities of steroids secreted, depending upon the ovarian structures which are present. One approach to the study of the steroidogenic capacities of the different compartments of the human ovary has been to dissect these different structures from the intact ovary, and to determine their ability to synthesize steroids under *in vitro* incubation conditions. This approach has led to the identification of three steroidogenic compartments in the ovary: the follicle; the corpus luteum; and the stroma. In this review an attempt will be made to give an overview of the steroidogenic capacities, and the biosynthetic pathways in each of these compartments. In addition the mechanism of the gonadotropic response of the corpus luteum of the menstrual cycle and of pregnancy will be evaluated. In order to remain close to the theme of the symposia, this review will be confined most to *in vitro* studies carried out on normal human ovarian tissues.

---

* Dedicated by K. Savard, D Sc. to Dr. Karl H. Slotta (Emeritus Professor of Biochemistry and Medicine of the University of Miami) on the occasion of his eightieth birthday.

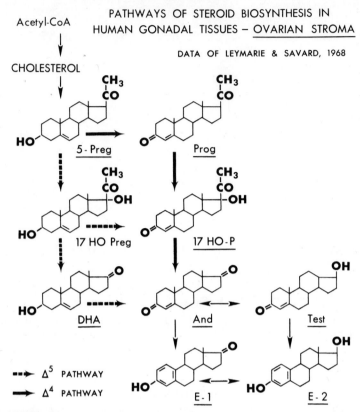

Fig.1 Pathways of steroid biosynthesis in human ovarian tissues. The Δ^5-3β-hydroxysteroid pathway is shown by the dashed arrows and the Δ^4-3-ketosteroid pathway by the solid arrows. The abbreviations are as follows: 5-preg = pregnenolone; 17HO-preg = 17-hydroxy-pregnenolone; Prog = progesterone; 17-HO-P = 17-hydroxy-progesterone; DHA = dehydroepiandrosterone; And = androstenedione; Test = testosterone; E-1 = estrone; E-2 = estradiol. The underlined abbreviations represent the steroid products formed by ovarian stroma from acetate-1-^{14}C in the experiments by Rice and Savard (1966).

II. STROMA

The ovarian compartment which is probably the least understood in terms of steroidogenesis, is the stroma. Its name suggests that it is a biochemically inert tissue which acts as a supportive structure for the other elements of the ovary, but the work of several investigators has indicated that this tissue does have steroidogenic activity (Forleo and Collins, 1964; Rice and Savard, 1966; Somma et al., 1969).

In the Endocrine Laboratory at the University of Miami, Rice and Savard (1966) obtained stromal tissue from normal ovarian specimens and incubated slices of this tissue with the radioactively labelled steroidogenic precursor, acetate-1-^{14}C. After the incubation the steroid products were isolated, purified and rigorously identified by quantitative reverse isotope dilution methods. In these incubations acetate-1-^{14}C was incorporated into as many as eight steroid products:

androstenedione; testosterone; dehydroepiandrosterone; 17-hydroxy-progesterone; pregnenolone; estrone and estradiol. The greatest incorporation of radioactivity was found in androstenedione, which was unique compared to the profiles of radioactive steroids formed by the follicle and the corpus luteum (Ryan and Smith, 1965; Savard, Marsh and Rice, 1965). Human chorionic gonadotropin (hCG) increased the incorporation of ^{14}C-acetate into each of the steroid products indicating that this compartment of the ovary was under gonadotropic control.

The appearance of androstenedione as the principal steroidal product suggested that the stroma might be a prominent site of androgen synthesis in the ovary. Unfortunately, no one has yet measured the mass amounts of androgens produced by the stroma in this type of incubation. The closest experiment to this, has been the measurement of the secretion of steroids by human stromal tissue in culture. Channing (1969) obtained human stromal tissue from normal ovaries, incubated it in tissue culture for 3 days, and found that 17-hydroxy-progesterone, progesterone and androstenedione were the major secretory products. At the present time, it is still uncertain, however, what the exact contribution of the stromal compartment might be to the steroidal output of the ovary *in vivo*.

The pathway for the formation of the eight steroids from acetate-1-^{14}C can be described by the classical steroid biosynthetic sequence shown in Fig.1. Cholesterol is formed from acetyl-CoA and is then converted into pregnenolone by the cholesterol side chain cleavage system in the mitochondrion. Cholesterol from the plasma can also serve as a precursor of the steroids and it is uncertain at this time which source, the *de novo* synthesized cholesterol or the plasma cholesterol is the more important precursor. The pregnenolone which is formed is believed to leave the mitochondrion and to be acted upon by the enzymes of either of two pathways: the Δ^5-3β-hydroxysteroid pathway (shown in Fig.1 by the dashed arrows) or the Δ^4-3-ketosteroid pathway (shown as large solid arrows). Both pathways are thought to involve enzymes present in the microsomes and both eventually lead to the production of androgens and estrogens. The incorporation of ^{14}C-acetate into the eight steroids alone, indicated that both pathways were utilized in the stromal tissue, but a very careful study was also carried out by Lemarie and Savard (1968) who incubated homogenates of stromal tissue with ^{3}H labelled pregnenolone and ^{14}C labelled progesterone. From the ratios of ^{3}H to ^{14}C, found in the steroid products, they concluded that both pathways were operable in the stroma and contributed to the synthesis of androstenedione.

III. THE FOLLICLE

The first direct evidence indicating that isolated ovarian follicles were capable of synthesizing steroids *in vitro* was presented in a series of publications by Ryan and Smith (1961a, 1961b, 1961c, 1961d) which described the ability of minced follicular tissue to synthesize estrogens and other steroids from radioactively labelled precursors such as acetate, cholesterol and progesterone. In the experiments with acetate-1-^{14}C they found that the major steroid hormones produced were estradiol and estrone. Androgens such as androstenedione and dehydroepiandrosterone were synthesized to the next highest extent and C^{19} steroids such as pregnenolone 17, hydroxy progesterone and progesterone to the least extent. Both pathways, the Δ^5 and Δ^4 were probably used in the synthesis of these follicular steroids, but the work of Ryan and Smith (1965), Lemarie and Savard

(1968) and Somma et al. (1969), indicated that the Δ^5 pathway was the preferred route for estrogen synthesis.

There is considerable uncertainty regarding the cellular site of steroidogenesis in the mammalian Graafian follicle. Histological and histochemical evidence indicates that the theca interna cells are responsible for steroidogenesis and that the granulosa cells are probably not involved (Guraya, 1974). However, *in vitro* biochemical studies using separated follicular cells have not led to such clear cut answers. Ryan, Petro and Kaiser (1968), separated human ovarian follicles into theca and granulosa cells, and incubated each cell type separately with acetate-1-^{14}C. Both the theca and the granulosa cells were capable of synthesizing steroids including estrogens, but there were quantitative differences. Granulosa cells accumulated mostly progesterone and 17 hydroxyprogesterone, while the theca cells converted most of the intermediates into estrone and estradiol. When the two populations of cells were recombined, they appeared to act synergistically producing more estrogen than either alone. This result is compatible with the two-cell theory for estrogen synthesis. This theory, which was first introduced by Falck (1959), suggests that there is a functional relationship between the theca and the granulosa cells and that both cell types are necessary for normal estrogen secretion. At this time, however, there is not enough information available to come to a firm conclusion and the problem of the precise cellular site of follicular steroidogenesis remains unresolved.

Ryan and Petro (1966) also incubated these separated granulosa and theca cells with ^{3}H labelled pregnenolone and ^{14}C labelled progesterone to determine the extent to which the Δ^5 and the Δ^4 pathways were used in each cell type. They found that the theca cells preferentially used the Δ^5 pathway while the granulosa cells used the Δ^4 pathway. It appears then, that the theca cells behave like the whole follicle in terms of their pathway of steroidogenesis and this gives some support to the argument that theca cells alone might be responsible for steroidogenesis in the follicle.

The previous studies measured steroidogenesis in terms of the incorporation of radioactive precursors, but one investigator has measured the steroidogenic capacity of theca and granulosa cells in terms of the mass of steroid secreted in tissue culture. Channing (1969) obtained follicles from patients with normal ovarian function and prepared separated theca and granulosa cells. These cells were then cultured separately for several days and the amounts of the different steroids secreted into the medium were determined. Both cell types were capable of secreting mass amounts of steroids. The granulosa cells predominantly secreted progesterone, while the theca cells secreted much larger amounts of estrogens and androgens. On the whole, this profile of steroids secreted by theca and granulosa cells in culture was very similar to the pattern of acetate-1-^{14}C incorporation into steroids found by Ryan et al. (1968).

IV. THE CORPUS LUTEUM

A. *Steroidogenic Capacity*

Our laboratory began an investigation of the steroidogenic capacity of the human corpus luteum and its *in vitro* response to gonadotropins several years ago. The general experimental approach involved the dissection of the corpus luteum from the adjacent ovarian tissue and the preparation of thin tissue slices. These

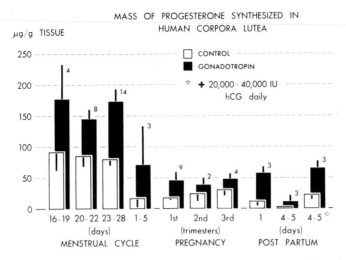

Fig.2 The synthesis of mass amounts of progesterone by human corpora lutea *in vitro*. The solid bars represent control incubations and the black bars represent incubations with hCG. The height of the bars represent the means and the brackets the standard deviations. Prepared from the data of LeMaire *et al.* (1968).

slices were then incubated with radioactive precursors such as acetate-1-^{14}C, and the steroid products isolated and the amount of acetate-1-^{14}C incorporation determined. In the case of progesterone, the mass of steroid synthesized was usually also measured. When corpora lutea of the menstrual cycle and of pregnancy were studied in this manner, it was found that seven steroids were formed from acetate-1-^{14}C (Hammerstein, Rice and Savard, 1964; Savard *et al.*, 1965). Progesterone was the most prominent product with the other steroids such as 17-hydroxyprogesterone, 20 α-hydroxy-Δ^4-pregnen-3-one, pregnenolone, androstenedione, estradiol and estrone being formed to smaller extents.

Gonadotropins with luteinizing hormone activity uniformly increased the synthesis of all the steroid products and there was no apparent difference between the effect of hCG and human luteinizing hormone (LH) (Rice, Hammerstein and Savard *et al.*, 1965). The uniform increase of all the steroid products by gonadotropins indicates that the limiting step in steroidogenesis probably involves the conversion of cholesterol into pregnenolone. When gonadotropins accelerate this limiting step, they cause an increase in the synthesis of all the steroids after cholesterol in the biosynthesis pathway. The pattern of acetate incorporation into steroids in these *in vitro* incubations of human corpora lutea correlates well with the microgram amounts of steroids isolated from pooled unincubated human corpora lutea (Zander, 1958; Zander, Brendle, von Munstermann, Diczfalusy, Martinsen and Tillinger, 1959) and the quantities of these steroids which have been found in the ovarian veins of ovaries containing a corpus luteum (Mikhail, Zander and Allen, 1963; Mikhail, 1967). Parenthetically it is not generally recognized that the corpus luteum of pregnancy continues to secrete progesterone throughout pregnancy. In fact, Dr. LeMaire, of our laboratory and his associates have shown that the corpus luteum at term secretes measurable amounts of progesterone into the ovarian vein (LeMaire, Conly, Moffett and Cleveland, 1970;

LeMaire, Conly, Moffett, Spellacy, Cleveland and Savard, 1971b).

Judging from the pattern of ^{14}C-acetate incorporation in these experiments, it seems that the Δ^4-3 ketosteroid pathway would be sufficient to account for the production of all of the radioactive products. This interpretation is supported by the study of Ryan and Smith (1965) in which slices of a human corpus luteum of pregnancy was incubated with ^{3}H labelled pregnenolone and ^{14}C labelled progesterone. The ratio of ^{3}H to ^{14}C in the products which were formed, indicated that the Δ^4 pathway was probably the exclusive route for steroidogenesis in the corpus luteum.

The mass of progesterone synthesized in *in vitro* incubations of slices of human corpora lutea has also been measured. In our laboratory LeMaire, Rice and Savard (1968) surveyed the capacity of 50 human corpora lutea obtained throughout the reproductive cycle to synthesize mass amounts of progesterone and to respond to gonadotropins. One of the more important findings to come from this survey (shown in Fig.2) was that corpora lutea of the menstrual cycle synthesized much larger mass amounts of progesterone than did corpora lutea of pregnancy. There is a relatively high level of synthesis and a very significant response to LH throughout the luteal phase of the menstrual cycle. When the corpora lutea were obtained after the onset of menses (days 1-5) there was a decline in the synthesis and the response to LH was too inconsistent to be statistically significant. Corpora lutea obtained during all 3 trimesters of pregnancy also produced a lower level of mass amounts of progesterone during *in vitro* incubations and exhibited smaller responses to hCG *in vitro*. Corpora lutea obtained one day post partem displayed a similar low level of steroidogenesis and gonadotropin responsiveness. When corpora lutea were obtained 4-5 days post partum, the level of steroidogenesis was found to have declined even further. This further decline seemed to be due to the disappearance of circulating chorionic gonadotropin, for, as shown in Fig.2 by the bars at the extreme right, when the concentration of hCG in the circulation was kept elevated by daily injections of exogenous hCG the biosynthetic capacity of the post partum corpus luteum could be maintained at the level of the corpus luteum of pregnancy for as long as 5 days post partum (LeMaire *et al.*, 1971b).

B. The Mechanism of Action of hCG and LH in Human Corpora Lutea

This difference in steroidogenic capacity between corpora lutea of the menstrual cycle and corpora lutea of pregnancy, could have been due to differences in the concentration of precursor cholesterol. The level of stored cholesterol might have been depleted in this corpus luteum of pregnancy by the constant action of hCG leading to a smaller production of mass amounts of progesterone in the *in vitro* incubations. It was possible, however, that there might be other differences between these two types of corpora lutea related to steroidogenesis and their responsiveness to gonadotropin, so we undertook a study of the mechanism of action of LH and hCG in the human corpus luteum and compared the responses of these two types of corpora lutea (Marsh and LeMaire, 1974).

The mechanism of action of LH in accelerating steroidogenesis in corpora lutea has been extensively studied in bovine tissue (Marsh, 1968), and it has been shown that cyclic AMP is a mediator of this action. LH increases the endogenous cyclic AMP prior to its effect on steriodogenesis, and exogenous cyclic AMP mimics the effect of the hormone. LH produces its accumulation of cyclic AMP by stimulation of the adenyl cyclase enzyme system rather than by causing an inhibition of the cyclic nucleotide phosphodiesterase enzyme system (Marsh, 1970). Furthermore, it has been found that prostaglandins will stimulate both adenyl cyclase and steroidogenesis in this tissue (Marsh, 1971). In single experiments in human corpora lutea it was also shown that hCG increased the endogenous concentration of cyclic AMP and that exogenous cyclic AMP increased steroid synthesis in this tissue (Marsh, Butcher, Savard and Sutherland, 1966; LeMaire, Askari and Savard, 1971a).

One of the first experiments to be undertaken in the further investigation of the mechanism of action of hCG and LH in human corpora lutea was to examine the effect of exogenous prostaglandins on steroidogenesis in incubating slices of this tissue. Both hCG and prostaglandin E_2 produced a marked increase in the synthesis of progesterone in the corpora lutea obtained during the menstrual cycle. The corpora lutea of pregnancy, however, exhibited a smaller control synthesis, as mentioned before, and showed a very minimal response to both hCG and PGE_2. Next we undertook to see if hCG would increase the accumulation of cyclic AMP in incubated slices of corpora lutea. The method we used involved the measurement of the incorporation of tritium labelled adenine into cyclic AMP. hCG produced a large increase in the accumulation of tritium cyclic AMP when it was added to incubating slices of human corpora lutea of the menstrual cycle. Prostaglandins also caused an increase in tritium cyclic AMP in these tissue slices. Again when corpora lutea of pregnancy were used there was a marked reduction in response. Prostaglandin E_2 produced a small increase, but hCG had hardly any effect at all.

As we had done previously in the bovine corpus luteum study, we attempted to determine, if these effects of hCG and PGE_2, in increasing cyclic AMP accumulation were due to a stimulation of the adenylate cyclase system or to an inhibition of the phosphodiesterase. hCG and PGE_2, as well as NaF, increased the adenylate cyclase activity of homogenates of human corpora lutea of the menstrual cycle. NaF was also tested in these homogenates since it has been shown to be a potent stimulator of the adenylate cyclase of many other tissues, including the bovine corpus luteum (Marsh, 1970). There was very little effect of hCG on the adenylate cyclase of homogenates of corpora lutea of pregnancy, but PGE_2 did produce a stimulation. NaF produced a marked stimulation of adenylate cyclase in the homogenates of corpora lutea of pregnancy; about equal to that observed with the corpora lutea of the cycle. This indicated that there was not an absolute deficiency of adenylate cyclase activity in the corpus luteum of pregnancy, but rather a deficiency in some part of the mechanism by which gonadotropins stimulated this enzyme system.

The former data indicated that hCG and PGE_2 increased the accumulation of cyclic AMP by stimulating the adenylate cyclase enzyme system, but to confirm this conclusion we also assessed the possible effect of these two agents on the phosphodiesterase enzyme in homogenates of corpora lutea of the menstrual cycle. When the enzyme was assayed under *maximal substrate* conditions, there was no apparent effect of hCG or PGE_2 on this enzyme. We concluded, therefore, that hCG and PGE_2 produced their increase in cyclic AMP solely by activating the adenylate cyclase enzyme.

Thus when we examined several of the components of our concept of the mechanism of action of LH *in human* corpora lutea, we found (as shown in Fig.3) that in the corpus luteum of the menstrual cycle, the situation appeared to be the same as that of the bovine tissue (Marsh and LeMaire, 1974). LH activates the

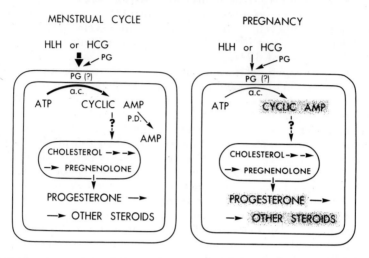

Fig. 3 Schematic representation of the mechanism of action of hCG and prostaglandins in stimulating steroidogenesis in human corpora lutea of the menstrual cycle and of pregnancy.

adenylate cyclase enzyme which increases the concentration of the endogenous cyclic AMP and this cyclic nucleotide, in turn, increases steroidogenesis in an as yet unknown way. Prostaglandin E$_2$, as was the case of bovine corpora lutea, also stimulated the adenylate cyclase of corpora lutea of the cycle, and cyclic AMP probably also mediated the steroidogenic effect of this prostaglandin.

The situation with human corpora lutea of pregnancy was, however, quantitatively quite different. The response to hCG and PGE$_2$ in terms of steroidogenesis was very small compared to that of corpora lutea of the cycle, and we believe that a part of the cause of this diminished response is due to an inability of the adenylate cyclase system of this tissue to respond to this hormone (Marsh and LeMaire, 1974). This diminished responsiveness could be due to a decrease in the number or quality of hormone receptor sites and in this regard it has been reported by Cole, Weed, Schneider, Holland, Geary and Rice (1973) and by Rao, Griffin, Carman, Yussman and Queenan (1975), that corpora lutea of pregnancy could not specifically bind I^{125} labelled hCG or I^{125} labelled human LH while corpora lutea of the menstrual cycle showed a marked specific binding capacity.

V. SUMMARY

In summary, we have attempted to outline the pathways of steroid biosynthesis in three compartments of the ovary, the stroma, the Graafian follicle and the corpus luteum. The stroma appears to utilize both the Δ^5-3β-hydroxysteroid pathway and the Δ^4-3-ketosteroid pathway while the follicle preferentially uses the Δ^5 pathway and the corpus luteum the Δ^4 pathway.

We have also attempted to relate to you our current understanding of the mechanism of action of luteinizing hormone in the human corpus luteum. Finally, we have presented some of our data which indicates that the corpora lutea of pregnancy have a greatly diminished response to gonadotropins with luteinizing hormone activity. The mechanism by which these corpora lutea lose their responsiveness is unknown and will probably be the subject of many future studies.

VI. REFERENCES

Channing, C.P. (1969). *J. Endocrinol.* 45, 297-308.
Cole, F.E., Weed, J.C., Schneider, G.T., Holland, J.B., Geary, W.L. and Rice, B.F. (1973). *Am. J. Obstet. Gynec.* 117, 87-95.
Falck, B. (1959). *Acta Physiol. Scand.* 47, Suppl. 163, 1-101.
Forleo, R. and Collins, W. (1964). *Acta Endocrinologica* 46, 265-278.
Guraya, S.S. (1974). *In* "Gonadotropins and Gonadal Function" (N.R. Moudgal, ed) pp.220-236. Academic Press, New York and London.
Hammerstein, J., Rice, B.F. and Savard, K. (1964). *J. clin. Endocr.* 24, 597-605.
LeMaire, W.J., Rice, B.F. and Savard, K. (1968). *J. clin. Endocr.* 28, 1249-1256.
LeMaire, W.J., Conly, P.W., Moffett, A. and Cleveland, W.W. (1970). *Am. J. Obstet. Gynec.* 108, 132-134.
LeMaire, W.J., Askari, H. and Savard, K. (1971a). *Steroids* 17, 65-84.
LeMaire, W.J., Conly, P.W., Moffett, A., Spellacy, W.N., Cleveland, W.W. and Savard, K. (1971b). *Am. J. Obstet. Gynec.* 110, 612-618.
Lemarie, P. and Savard, K. (1968). *J. clin. Endocr.* 28, 1547-1554.
Marsh, J.M. (1968). *Adv. Expt. Med. and Biol.* 2, 213-222.
Marsh, J.M. (1970). *J. biol. Chem.* 245, 1596-1603.
Marsh, J.M. (1971). *Ann. N.Y. Acad. Sci*, 180, 416-425.
Marsh, J.M. and LeMaire, W.J. (1974). *J. clin. Endocr.* 38, 99-106.
Marsh, J.M., Butcher, R.W., Savard, K. and Sutherland, E.W. (1966). *J. biol. Chem.* 241, 5436-5440.
Mikhail, G. (1967). *Clin. Obstet. Gynec.* 10, 29-39.
Mikhail, G., Zander, J. and Allen, W.L. (1963). *J. clin. Endocr.* 23, 1267-1270.
Rao, Ch.V., Griffin, L.P., Carman, F.R., Yussman, M.A. and Queenan, J.T. (1975). *Fert. Steril.* 26, 198.
Rice, B.F. and Savard, K. (1966). *J. clin. Endocr.* 26, 593-609.
Rice, B.F., Hammerstein, J. and Savard, K. (1964). *J. clin. Endocr.* 24, 606-615.
Ryan, K.J. and Smith, O.W. (1961a). *J. biol. Chem.* 236, 705-709.
Ryan, K.J. and Smith, O.W. (1961b). *J. biol. Chem.* 236, 710-714.
Ryan, K.J. and Smith, O.W. (1961c). *J. biol. Chem.* 236, 2204-2206.
Ryan, K.J. and Smith, O.W. (1961d). *J. biol. Chem.* 236, 2207-2212.
Ryan, K.J. and Smith, O.W. (1965). *Rec. Prog. Horm. Res.* 21, 367-409.
Ryan, K.J. and Petro, Z. (1966). *J. clin. Endocr.* 26, 46-52.
Ryan, K.J., Petro, Z. and Kaiser, J. (1968). *J. clin. Endocr.* 28, 355-358.
Savard, K., Marsh, J.M. and Rice, B.F. (1965). *Rec. Prog. Horm. Res.* 21, 285-365.
Somma, M., Sandor, T. and Lanthier, A. (1969). *J. clin. Endocr.* 29, 457-466.
Zander, J. (1958). *J. biol. Chem.* 232, 117.
Zander, J., Brendle, E., von Munstermann, A.M., Diczfalusy, E., Martinsen, B. and Tillinger, K.G. (1959). *Acta Obstet. Gynec. Scand.* 38, 724.

THE ROLE OF PROSTAGLANDINS IN GONADOTROPIN SECRETION

C. Patrono, G.B. Serra, G. Ciabattoni, D. Grossi-Belloni
and R. Canete Soler*

*Departments of Pharmacology and Obstetrics and Gynecology
Università Cattolica del S. Cuore, Rome, Italy*

Prostaglandins (PGs) are being implicated at key sites in a wide variety of physiologic processes, including several areas of the reproductive cycle (see Labhsetwar, 1974, for a review). In particular, several recent findings suggest the involvement of PGs in gonadotropin secretion, at the pituitary and/or hypothalamic level. Part of the evidence is based on the stimulatory effects of exogenous PGs on gonadotropin secretion, and part on the effects of inhibitors of PG synthesis, such as the non-steroid antiphlogistic drugs. This evidence will be reviewed in the present chapter, in an attempt to define its physiological significance, and newer experimental approaches developed in our laboratory will be described. The clinical implications of PG participation in gonadotropin secretion will also be discussed.

I. CYCLIC CHANGES OF CIRCULATING PGs IN RELATION TO GONADOTROPINS

Although PGs are unlikely to function as true circulating hormones, in view of their rapid and extensive enzymatic degradation, some reports have appeared indicating the existence of significant cyclic changes in their peripheral plasma concentrations. Thus, Caldwell *et al.* (1972) first reported an increase of peripheral PGF levels on day 14 of the sheep estrous cycle, but no significant change at the time of ovulation. Patrono *et al.* (1974a) recently reported a significant rise of peripheral plasma concentrations of $PGF_{2\alpha}$, but not of $PGF_{1\alpha}$, during the luteal phase of the human menstrual cycle. However, as shown in Fig.1, both

---

* Fellow of the Fundacion Endocrino-Metabolica "Fernandez Cruz", Madrid, Spain.

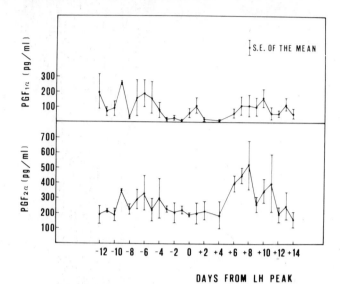

Fig. 1 Mean (± SEM) peripheral plasma concentrations of PGF_{2a} and PGF_{1a} in six normal menstruating females during the menstrual cycle. The day of the maximum concentration of plasma LH was arbitrarily designated as day 0.

PGF_{1a} and PGF_{2a} - as measured by specific radioimmunoassays (Patrono, 1973; Patrono et al., 1974b) - are at their nadir at the time of the LH peak. No cyclic changes of peripheral PGE concentrations have been reported. Thus, on the basis of these findings, it seems unlikely that circulating PGs are directly responsible for gonadotropin release.

II. EFFECTS OF EXOGENOUS PGs ON GONADOTROPIN SECRETION

In view of the different PGs having quite distinct biological significance and, possibly, physiologic roles, their effects will be considered separately.

A. Prostaglandin E_1

1. In vitro

PGE_1 was first shown by Zor et al. (1970) to increase rat anterior pituitary adenyl cyclase activity and cyclic AMP levels, but no LH release. However, in a perfusion system, PGE_1 increased LH release in a dose-dependent manner (Dowd et al., 1973). More recently, Ratner et al. (1974) have reported that the addition of PGE_1 (10^{-5} M) to the incubation medium of rat anterior pituitaries produced a significant increase in pituitary cyclic AMP and LH release. LH release induced by addition of PGE_1 to the incubation medium of rat anterior pituitaries has been confirmed by Sato et al. (1975).

2. In vivo

Spies and Norman (1973) found that 5 to 20 μg of PGE_1 infused into the third ventricle of rats on the afternoon of vaginal proestrus reversed the pentobarbital blockade of ovulation and increased plasma LH levels; however, it proved ineffective if infused into the pituitary or injected subcutaneously. Batta et al.

(1974) reported that the intracarotid injection of PGE_1 (10 µg) into spayed rats primed with estradiol benzoate caused a prompt and sustained rise in plasma LH levels, and a transient rise in plasma FSH; in adult male rats, PGE_1 caused a rise in plasma LH levels, but did not affect FSH levels. Ratner et al. (1974) have reported that the intravenous (i.v.) administration of 5 µg of PGE_1 to adult male rats caused a significant increase in pituitary cyclic AMP 10 minutes after injection, but no change in serum LH; when 20 µg of PGE_1 was administered, a significant increase in both cyclic AMP and serum LH levels was observed. Harms et al. (1974) reported a lack of effect on plasma LH and FSH levels of i.v. injections of PGE_1 (up to 50 µg/100 g) to ovariectomized rats. The same authors (Harms et al., 1973) had previously shown that PGE_1 (5 µg) injected into the 3rd ventricle of ovariectomized rats significantly increased plasma prolactin but not LH or FSH levels; the intrapituitary injection of the same amount of PGE_1 failed to alter LH, FSH or prolactin. When these experiments were repeated in ovariectomized estrogen-primed rats, it was found that, in contrast to its lack of effect in ovariectomized rats, 3rd ventricular injections of PGE_1 significantly increased plasma LH levels; intrapituitary injections of PGE_1 had again no effect on plasma LH or FSH (Harms et al., 1974). PGE_1 (5 µg) injected into the 3rd ventricle of intact conscious male rats was found to increase slightly but significantly plasma LH and FSH levels; no effect was noted in castrated animals (Ojeda et al., 1974a). Castracane and Saksena (1974) reported failure of a single subcutaneous injection of a rather massive dose of PGE_1 (1 mg) to modify plasma LH concentrations in fertile male rats. Sato et al. (1975) have reported a significant increase of plasma LH 10 minutes after the i.v. injection of PGE_1 (500 µg/animal) in ovariectomized mature rats with hypothelamic lesions; following microinjections of 50 µg of PGE_1 into the pituitary of castrated rats plasma LH levels were significantly increased at 45 minutes. Ojeda et al. (1974b) found no significant changes in plasma prolactin after the i.v. injection of PGE_1 (5-50 µg/100 g) in ovariectomized rats; on the contrary, a dose-related increase of plasma prolectin could be demonstrated 15 minutes after intraventricular injections (0.5-5 µg); PGE_1 injected bilaterally into the anterior pituitary gland (2.5 µg in each lobe) induced a small but significant rise in plasma prolactin levels 15 minutes later; estrogen pretreatment inhibited the effect of intrapituitary but not of intraventricular injections of PGE_1. Sato et al. (1974a) reported that a single i.v. injection of PGE_1 (2-200 µg) in spayed rats primed with estrogen and progesterone significantly increased the serum prolactin level.

B. Prostaglandin E_2
   1. In vitro
      In a perfusion system, PGE_2 increased LH release in a dose-dependent fashion (Dowd et al., 1973). Sato et al. (1975) have reported that addition of PGE_2 (0.002-0.02 µg) to the incubation medium of rat anterior pituitaries produced a significant increase in LH release.
   2. In vivo
      Tsafriri et al. (1973) reported that subcutaneous injections of PGE_2 (2 × 750 µg/rat) in the early afternoon of proestrous elicited a rise in serum LH levels in Nembutal-blocked rats. Harms et al. (1973) found that PGE_2 (5 µg) injected into the 3rd ventricle of ovariectomized rats increased plasma LH dramatically and FSH slightly, without affecting plasma prolactin; intrapituitary injections proved to be completely ineffective. The same authors (Harms et al.,

1974) later reported that i.v. PGE_2 (5-50 µg/100 g) significantly increased plasma LH but not FSH in ovariectomized rats; in ovariectomized estrogen-primed rats, 3rd ventricular injections of PGE_2 (0.5-5 µg) produced a dose-related increase in plasma LH and a small but significant rise in plasma FSH; intrapituitary injections (5 µg) produced a small increase in LH in the estrogen-primed rats. Ojeda et al. (1974a) have recently reported that PGE_2 (5 µg) injected into the 3rd ventricle of intact or castrated conscious male rats markedly increased plasma LH levels; they have also described a small but significant increase in plasma FSH in intact but not in castrated animals. Castracane and Saksena (1970) reported that a single subcutaneous injection of 250 µg PGE_2 caused a slight but significant increase in plasma LH of fertile male rats. Failure of i.v., intraventricular and intrapituitary injections of PGE_2 to modify plasma prolactin levels in both ovariectomized and estrogen-treated ovariectomized rats has been reported by Ojeda et al. (1974b). On the contrary, Sato et al. (1974a) found that a single i.v. injection of PGE_2 (2-20 µg) into spayed rats primed with estrogen and progesterone significantly increased the serum prolactin level; they also reported that a single i.v. injection of PGE_2 (500 µg/rat) into male rats significantly increased serum FSH concentrations. The same authors (Sato et al., 1975) have more recently reported a significant increase of plasma LH 10 minutes after the i.v. injection of PGE_2 (500 µg/animal) in ovariectomized mature rats with hypothalamic lesions; moreover, following micro-injections of 50 and 100 µg of PGE_2 into the pituitary of castrated rats, plasma LH levels rose significantly at 10 and 45 minutes.

C. Prostaglandin F_{2a}
  1. In vitro

Sato et al. (1975) have reported significantly increased LH release from rat anterior pituitaries following the addition of PGF_{2a} (200 µg) to the incubation medium.

  2. In vivo

Labhsetwar (1970) first found that luteolytic doses of PGF_{2a} increased pituitary LH stores significantly in pregnant rats. Plasma LH increased in rats (unpublished data of Labhsetwar) and in hamsters (Labhsetwar, 1972) treated with PGF_{2a}. Moreover, a single intraventricular injection of 10 µg PGF_{2a} in spayed rats pretreated with estradiol benzoate caused a sharp rise in plasma LH within a few minutes (Labhsetwar, 1974). Batta et al. (1974) have reported that a single intracarotid injection of PGF_{2a} into spayed rats primed with estradiol benzoate caused a rise in plasma LH between 6-8 minutes, and thereafter a significant drop to undetectable levels at 30 minutes; moreover, PGF_{2a} had a prompt and long lasting effect on plasma FSH in the same animals, a finding which could not be confirmed by Labhsetwar (1973); PGF_{2a} proved ineffective in male rats. McCann's group found no significant change of plasma LH and FSH after i.v. injection of PGF_{2a} (up to 50 µg/100 g) in ovariectomized rats, nor after 3rd ventricular injection in ovariectomized, estrogen-primed ovariectomized, intact or castrated conscious male rats (Harms et al., 1973; Harms et al., 1974; Ojeda et al., 1974a). Sato et al. (1975) have recently reported a significant increase of plasma LH 10 minutes after the i.v. injection of 500 µg PGF_{2a} in ovariectomized mature rats with hypothalamic lesion; however, plasma LH was unaffected by the intrapituitary injection of 100 µg PGF_{2a} in castrated rats. Carlson et al. (1973) reported an increase in LH concentrations in jugular vein plasma of cyclic ewes following the subcutaneous (25 mg) or continuous intracarotid infusion of PGF_{2a}

(6 μg/hour or more) given at mid-cycle; by contrast, intracarotid infusions of PGF_{2a} (5-100 μg/hour) failed to cause any significant increase in plasma LH levels during anoestrous. Chamly and Christie (1973) found no change in either the pattern of tonic LH secretion or the LH surge evoked by estradiol, following the intracarotid infusion of PGF_{2a} at a lower rate (5 μg/hour) in ovariectomized ewes. Vermouth and Deis (1972) reported that the mean serum prolactin concentration in rats at 8, 10 and 12 hours after the intraperitoneal injection of PGF_{2a}, on day 18 of pregnancy, was significantly greater than the control values with a peak at 12 hours. Sato et al. (1974a) observed an increase in serum prolactin after a single i.v. injection of PGF_{2a} (20-200 μg) in spayed rats primed with estrogen and progesterone. Ojeda et al. (1974b) found no significant change of plasma prolactin levels after intraventricular injections of PGF_{2a} (5 μg) in ovariectomized and estrogen-treated ovariectomized rats. Yue et al. (1974) reported that administration of PGF_{2a} via the transcervical route for termination of 2nd trimester human pregnancies was associated with a significant increase in serum prolactin, an effect which could not be demonstrated with hypertonic saline or PGF_{2a} administered intra-amniotically.

From a rapid analysis of these data, it is evident that conflicting results are at least as common as consistent ones, although the latter are seldom obtained under the same experimental conditions. The many variables that we have attempted to underline, such as animal species, sex, castration, treatment with ovarian steroids, dose of PG injected and route of injection, make any general conclusions very hard to be drawn. Notwithstanding these difficulties, what these findings clearly indicate is that different exogenous PGs - *at pharmacologic doses* in most cases - can interact with the physiologic mechanisms responsible for gonadotropin secretion, under a wide variety of experimental circumstances. On the basis of results obtained by the same authors with different routes of administration, it appears as if a major site of action of PGs is represented by the central nervous system (most likely the hypothalamus), although several experiments indicate another site of action in the anterior pituitary (Harmes et al., 1974; Sato et al., 1975). PGE_2 has been most consistently reported to induce LH release, while PGE_1 seems to affect LH and, more convincingly, prolactin secretion. The rather inconsistent effects of both PGs on FSH release are compatible with the involvement in such a response of a different releasing hormone. The effects of PGF_{2a} are rather contradictory, although its stimulation of LH and prolactin release has been observed in several species. If specific receptors for the different PGs indeed exist at the hypothalamic or pituitary level, it is likely that at high concentrations of a given PG both specific and non specific interactions occur, possibly accounting for some degree of overlapping of the stimulatory effects.

III. EFFECTS OF INHIBITORS OF PG SYNTHESIS ON GONADOTROPIN SECRETION

1. In vivo

The discovery of the inhibitory activity of non-steroid antiphlogistic drugs (aspirin-like drugs) on PG synthesis and release (Vane, 1971; Ferreira et al., 1971; Smith and Willis, 1971) has provided investigators with a new and exciting tool for studying the interactions of PGs with gonadotropin secretion. Orczyk

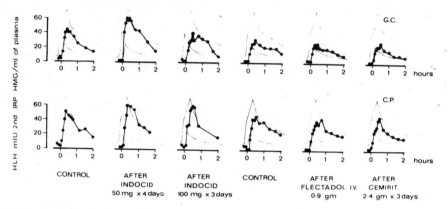

Fig. 2 Plasma LH levels in two normal men receiving an i.v. injection of 50 μg synthetic Gn-RH at time 0, before and after two different chronic regimens of oral indomethacin; before and after an acute and chronic treatment with acetylsalicylic acid. The shaded area represents the normal range of responses in our laboratory.

and Behrman (1972) first showed that indomethacin reduces plasma, pituitary and hypothalamic concentrations of PGF, and blocks PMS-induced ovulation in immature rats; injection of either LH or a mixture of $PGF_{2\alpha}$ and PGE_2 at the time of the expected ovulatory surge of LH was effective in reversing ovulation blockade, that occurred after a single does of indomethacin given 3 hours before the expected LH surge; however, LH was not effective in reversing the blockade of ovulation induced by chronic administration of indomethacin. These data seemed to indicate an interaction of indomethacin at the hypothalamic-pituitary axis, in constrast with concomitant results reported by Armstrong and Grinwich (1972) suggesting that there might be an additional block at the ovarian level, since indomethacin proved effective in blocking the ovulatory response to exogenous LH. Subsequently, Behrman et al. (1972) have provided additional evidence indicating that indomethacin may block ovulation at the ovarian level, by interfering with the rupture of follicle and release of the ovum. Aspirin was also shown by the same authors to reduce plasma, pituitary and hypothalamic concentrations of PGF and to block PMS-induced ovulation; however, this effect could be reversed by either LH or Gn-RH, suggesting that aspirin blocked ovulation at the hypothalamic level. In the rabbit, O'Grady et al. (1972) and Grinwich et al. (1972) showed that indomethacin is able to suppress ovulation induced either by the mating stimulus or by exogenous gonadotropin, again suggesting a block at the ovarian level. Tsafriri et al. (1973) reported that indomethacin (10 mg/rat) does not interfere with the proestrous surge of LH secretion, when given at a time and dose level effective in preventing ovulation. Sato et al. (1974b) have confirmed that indomethacin (2 mg/rat) does not interfere with the proestrous surge of LH and FSH secretion, although blocking ovulation in the majority of animals. Labhsetwar and Zolovick (1973) found that micro-injections of aspirin (120 μg) into the anterior hypothalamic area (AHA) interfered with progesterone-induced ovulation in immature rats primed with PMSG; restoration of ovulation could be obtained with simultaneous injections of either $PGF_{2\alpha}$ (1.0 μg) or dopamine (180 μg) into the AHA; intrapituitary injec-

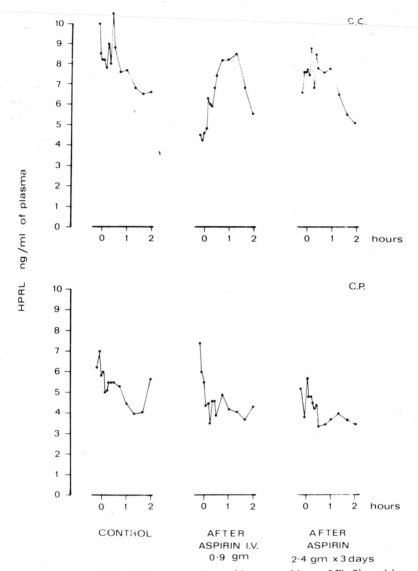

Fig.3 Plasma prolactin levels in two normal men (the same subjects of Fig.2) receiving an i.v. injection of 50 μg synthetic Gn-RH at time 0, before and after an acute and chronic treatment with acetylsalicylic acid.

tion of aspirin (80 μg) proved ineffective in blocking ovulation. These results were taken to suggest that catecholamines and $PGF_{2\alpha}$ interact to potentiate the adrenergic transmission in the hypothalamus, which mediates progesterone-induced ovulation. Patrono and Serra (1974) reported failure of chronic indomethacin treatment to modify the LH response to synthetic Gn-RH in normal men. Carlson et al. (1974) reported that indomethacin (600 mg in two days) blocked the estradiol-induced LH release in anestrous sheep, but not the LH release induced by synthetic Gn-RH. Sato et al. (1975) have recently reported that chronic indomethacin treatment (6 mg in three days) significantly reduced the LH response to synthetic Gn-RH in ovariectomized rats primed with estradiol and progesterone.

We have extended our own original observations in humans (Patrono and Serra, 1974), to include the effects of acetylsalicylic acid on LH and prolactin plasma levels following the i.v. administration of synthetic Gn-RH. As shown in Fig.2, similarly to indomethacin, the drug either given i.v. - as lysine acetylsalicylate - 5 minutes before Gn-RH or administered orally - as microincapsulated acetylsalicylic acid - for three days at therapeutically active dosage failed to modify the LH response appreciably. Both regimes did not affect plasma prolactin levels of the same subjects to any significant extent, as shown in Fig.3.

2. *In vitro*

Very recently, Noar et al. (1975) have shown that incubation of rat hemipituitaries with aspirin or indomethacin (100 µg/ml) decreased both PG-synthetase activity and PGE_2 content, but did not affect the stimulatory action of synthetic Gn-RH (25 ng/ml) on either cyclic AMP accumulation or LH release; basal LH release was not altered by aspirin or indomethacin, but was stimulated by flufenamic acid, another non-steroid antiphlogistic drug; strangely enough, the same drug abolished the effect of Gn-RH on cyclic AMP accumulation.

As we have outlined elsewhere (Patrono and Serra, 1974; Patrono, 1974), while positive results obtained with inhibitors of PG synthesis may indeed provide clues to possible physiologic roles of PGs, negative results are more difficult to interpret. Thus, when using indomethacin or aspirin, failure to observe any change in a given physiologic process - which we suspect to be somehow mediated via PGs - may well be due to insufficient deprivation of the particular PG involved, at the key site responsible for that process. This may in turn reflect a) poor access of the drug to this site, b) low drug sensitivity of the PG-synthetase involved, c) lower efficacy of the inhibitor against synthesis of PGE_1 and PGF_{1a} from eicosatrienoic acid than against synthesis of PGE_2 and PGF_{2a} from arachidonic acid, or *vice versa*. As for the first point, the above cited results of Orczyk and Behrman (1972) and of Behrman et al. (1972) seem to indicate that both indomethacin and aspirin gain access to the hypothalamus and pituitary, in concentrations adequate to reduce tissue PGF content. Although no direct evidence of inhibited PG synthesis in these tissues has been reported by other investigators, in connection with inhibited LH release, the additional data also imply that these drugs gained access to either the hypothalamus or the pituitary. As for the second point, it has been clearly demonstrated that PG-synthetases of different tissues display wide variations in drug sensitivity, a fact which has led Vane to propose that this enzyme system exists in multiple molecular forms, each having its own pharmacological profile (see Ferreira and Vane, 1974, for a review). Evidence for the existence of a differential sensitivity of human tissues to the PG-inhibitory action of aspirin has recently been presented (Patrono et al., 1975a).

Therefore, we have investigated the drug sensitivity of the human pituitary PG-synthetase in a preliminary fashion. Fragments of human pituitary tissue were obtained at surgery and studied *in vitro* with the technique of continuous superfusion (Serra and Midgley, 1970). As shown in Fig.4, the addition of fenoprofen - a potent inhibitor of PG synthesis and release (Patrono et al., 1974c) - to the superfusion medium resulted in approximately 20% inhibition of the Gn-RH induced PGF_{2a} release. The same concentration of drug (quite comparable with the

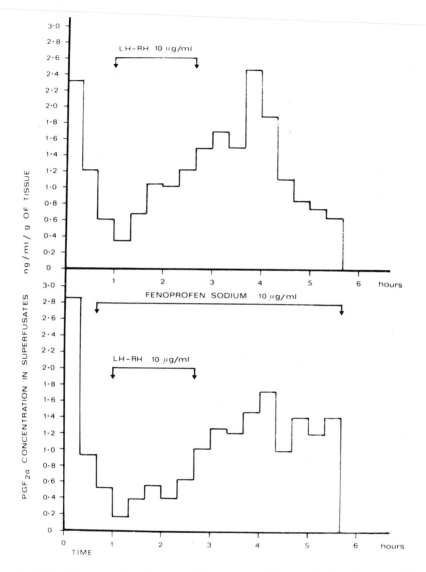

Fig.4 Reduced $PGF_{2\alpha}$ release from surgical specimens of human pituitary tissue superfused with fenoprofen (10 µg/ml). The upper portion of the figure represents spontaneous and LH-RH stimulated release of $PGF_{2\alpha}$ from the control chambers; the lower portion depicts $PGF_{2\alpha}$ concentrations in the superfusates collected from the chambers exposed to fenoprofen.

plasma level attained in man after therapeutic dosage) is capable of completely suppressing $PGF_{2\alpha}$ release from superfused human or rat platelets (Patrono *et al.*, 1975b) and from superfused rat uterus (Patrono *et al.*, 1974c). These studies need to be confirmed and extended to include other drugs; however, they raise the possibility that the PG-synthetase of the human pituitary has a lower sensitivity - as compared to other tissues - to the inhibitory action of non-steroid antiphlogistic agents.

As for the third point, we have very recently demonstrated that the conversion of eicosatrienoic acid into $PGF_{1\alpha}$ by rat platelets is inhibited by indomethacin and fenoprofen to the same extent as the conversion of arachidonic acid into $PGF_{2\alpha}$ (Patrono et al., 1975b). It remains, however, to be established whether this is also true in the case of hypothalamic and pituitary PG-synthetases.

Finally, we would like to draw attention to two additional factors which have to be taken into account in any interpretation of the effects of non-steroid antiphlogistic drugs:

a) at high concentrations, these drugs can also inhibit other enzymes besides PG-synthetase. This is particularly relevant in the case of indomethacin, which is a potent inhibitor of phosphodiesterase (Flores and Sharp, 1972), a finding which casts some doubts on the interpretation of results obtained with high doses of this particular drug.

b) aspirin-like drugs simultaneously inhibit the formation of PGE and PGF, which have been shown to exert integrated inhibitory and facilitatory effects on important physiologic processes, such as the nerve terminal release of norepinephrine (Brody, 1973). Therefore, negative results could also reflect simultaneous suppression of opposing actions of PGE and PGF.

With all these limitations in mind, the general conclusion which can be drawn from the effects of inhibitors of PG-synthetase on gonadotropin secretion is that synthesis of endogenous PGs seems to take part - although in an as yet indefinite way - in the process of gonadotropin secretion at the pituitary and, more convincingly, at the hypothalamic or higher level.

It is perhaps not unappropriate at this stage to suggest that future studies employing aspirin-like drugs as inhibitors of PG synthesis should support their conclusions with the pertinent answers to the following questions: a) which PGs were being inhibited under the experimental conditions employed? b) to what extent? c) was any other enzyme system - such as phosphodiesterase - being inhibited to any significant extent?

IV. THE RELEASE OF PGs FROM THE PITUITARY AND HYPOTHALAMUS

In view of the difficulties and limitations inherent to the use of exogenous PGs or aspirin-like drugs for studying the role of PGs in gonadotropin secretion, a more direct approach would be highly desirable. Evidence of *in vivo* PG release from the pituitary or the hypothalamus is very difficult to obtain, due to their ubiquitous nature and to the extreme facility with which they are synthetized and released upon the slightest tissue manipulation. McCracken and Roberts (1975) have recently reported that $PGF_{2\alpha}$ is released from the brain of the anestrous sheep in a pulsatile manner, and that the pattern of these pulses is significantly altered prior to the release of LH induced by 17β-estradiol; however, its precise source could not be established.

Using an *in vitro* superfusion technique, first described by Serra and Midgley (1970) for the study of LH release from single rat anterior pituitaries, we have found that both rat and human pituitaries release $PGF_{2\alpha}$ with a pattern similar to that demonstrated for rat and human LH (Serra and Midgley, 1970; Serra, 1973). As shown in Fig.4, a marked and sustained release of $PGF_{2\alpha}$ from super-

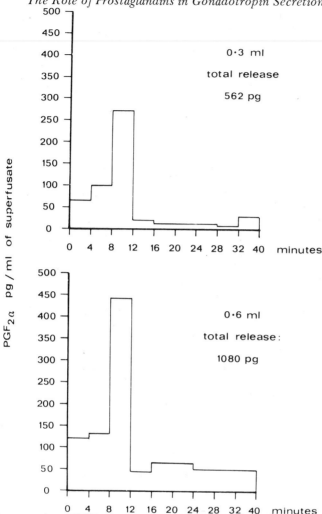

Fig.5 Concentration of $PGF_{2\alpha}$ in superfusates of rat hypothalamic synaptosomes, using two different volumes of synaptosome suspension. Each horizontal bar represents the mean concentration measured in the superfusates from four chambers, collected simultaneously at 4 min intervals.

fused fragments of human pituitary tissue followed the addition of a pharmacologic dose of synthetic Gn-RH to the superfusion medium. The possibility that a PG receptor system is involved in the secretion of anterior pituitary hormones had been originally proposed by Guillemin's group (Amoss *et al.*, 1971; Vale *et al.*, 1971), following the demonstration that a PG antagonist (7-oxa-13-prostynoic acid) blocks LRF stimulation of LH secretion and TSH response to TRH. On the other hand, Naor *et al.* (1975) failed to observe any stimulatory effect of synthetic LH-RH, added to the incubation medium of rat hemipituitaries, on PGE_2 level or PG-synthetase activity.

As indicated in previous sections of this chapter, more abundant and less controversial evidence points to the hypothalamus as a major site for PG partici-

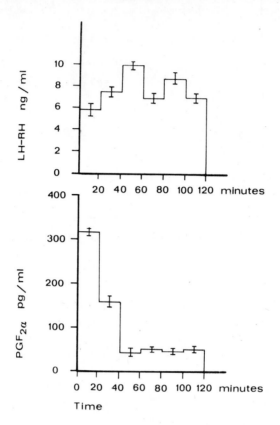

Fig.6 Concentrations of $PGF_{2\alpha}$ and LH-RH in superfusates of rat hypothalamic synaptosomes. Each horizontal bar represents the mean concentration (± SEM) measured in the superfusates from six chambers, collected simultaneously at 20 min intervals.

pation in gonadotropin secretion. Previous reports had appeared demonstrating the *in vitro* release of LRF and CRF in subcellular fractions of bovine median eminence which contain structures resembling nerve vesicles and terminals (Fink *et al.*, 1972), and of CRF and PIF from synaptosomes isolated from the sheep hypothalamus (Edwardson *et al.*, 1972). More recently, Pelletier *et al.* (1974), using an immunohistochemical technique at the electron microscopic level, have demonstrated that LH-RH is present in the nerve endings of the rat median eminence. In addition, secretion of LH-RH and TRH with identical immunochemical, chromatographic and enzymic properties as the synthetic peptides, from synaptosomes prepared from rat and sheep hypothalmi has been reported by Jeffcoate (1974).

Synaptosomes show a variety of metabolic properties characteristic of whole cells, and have been shown to synthetize proteins and phospholipids. Therefore, they appear to be an ideal model for studying the interactions of PGs with hypothalamic hormones. The recent development of a superfusion technique for studying the release of neurotransmitters from isolated synaptosomes (Raiteri *et al.*, 1974) provides a most useful tool for a dynamic approach to this problem. As shown in Fig.5, isolated synaptosomes from rat hypothalami do release PGF_{2a} during continuous superfusion, and this release is proportional to the volume of the suspension employed. As shown in Fig.6, the simultaneous release of PGF_{2a} and LH-RH (as determined by radioimmunoassay) can be evaluated in the same superfusates of rat hypothalamic synaptosomes. The interactions between these two processes and the pharmacologic influences upon them are presently being investigated in our laboratory.

V. CLINICAL IMPLICATIONS OF A PG PARTICIPATION IN GONADOTROPIN SECRETION

The issue of a physiologic role of PGs in the process of gonadotropin secretion is not merely of academic interest, since it raises at least two important practical questions: 1) are the currently used aspirin-like drugs responsible for unrecorded side-effects in the reproductive area? 2) is it possible to develop more specific inhibitors of pituitary and/or hypothalamic PG-synthetases, and thereby obtain a new pharmacologic tool to block gonadotropin secretion?

As for the first point, we are puzzled over the idea that a vast population is acutely or chronically receiving anti-inflammatory therapy without apparently showing any dramatic symptomatology - which might be referred to inhibited PG formation - other than the well-known gastric complaints. Undoubtedly, our present understanding of the possible facets of interaction of these drugs with PG synthesis and metabolism is far from being satisfactory, and it might well turn out to be an oversimplification of a complex phenomenon. It is also quite possible that minor modifications of the endocrine balance do indeed occur in people taking non-steroid antiphlogistic drugs, and that these have simply not been looked for; nor had we any rationale for seeking them until Vane's report in 1971. In any event, in extrapolating from animal studies to clinical situations, one should consider that at doses currently employed in the treatment of chronic inflammatory diseases - doses which are usually lower than those proven effective in blocking ovulation under experimental conditions - aspirin-like drugs only partially inhibit PG synthesis and release, even in "target" tissues such as the synovial membrane of patients with rheumatoid arthritis (Patrono et al., 1975c); therefore, the residual PG-synthetase activity might still be adequate to maintain physiologic processes at a "normal" level.

As for the second question, we would tentatively answer "yes", in view of the demonstrated existence of a differential sensitivity of PG-synthetases from different human tissues. Hopefully, this might lead to more specific aspirin-like drugs in which the desired effects have been enhanced and the unwanted ones have been considerably reduced. In addition, more sophisticated pharmacological tools might eventually become available to interfere with single steps of the "prostaglandin system" within a given tissue, such as the pituitary or the hypothalamus, as we obtain a deeper insight into the intimate mechanism of PG synthesis.

ACKNOWLEDGEMENTS

The authors acknowledge with gratitude the help of Dr. M. Signa in performing the clinical studies, and of Dr. A. Bertollini in preparing the rat hypothalamic synaptosomes.

The synthetic Gn-RH was generously supplied by the Istituto Farmacologico Serono, Rome. Samples of authentic prostaglandins were kindly provided by the Upjohn Company, Kalamazoo, through the courtesy of Dr. John E. Pike. The reagents for the radioimmunoassay of human LH were generously provided by the National Pituitary Agency (University of Maryland, School of Medicine), Endocrine Study Section and National Institute of Arthritis and Metabolic Diseases. The antiserum to LH-RH was obtained through the generosity of Dr. S.L. Jeffcoate.

Dr. D. Grossi-Belloni is the recipient of a Fellowship from Eli Lilly and Co., Indianapolis.

REFERENCES

Amoss, M., Blackwell, R., Vale, W., Burgus, R. and Guillemin, R. (1971). *Intern. Cong. Phys. Sci.* 9, 17.
Armstrong, D.T. and Grinwich, D.L. (1972). *Prostaglandins* 1, 21-28.
Batta, S.K., Zanisi, H. and Martini, L. (1974). *Neuroendocrinology* 14, 224-232.
Behrman, H.R., Orczyk, G.P. and Greep, R.O. (1972). *Prostaglandins* 1, 245-258.
Brody, M.J. (1973). *Population Report* G, 25-28.
Caldwell, B.V., Tillson, S.A., Brock, W.A. and Speroff, L. (1972). *Prostaglandins* 1, 217-228.
Carlson, J.C., Barcikowski, B. and McCracken, J.A. (1973). *J. Reprod. Fert.* 34, 357-361.
Carlson, J.C,, Barcikowski, B., Cargill, V. and McCracken, J.A. (1974). *J. clin. Endocr. Metab.* 39, 399-402.
Castracane, V.D. and Saksena, S.K. (1974). *Prostaglandins* 7, 53-55.
Dowd, A.J., Hoffman, D.C. and Speroff, L. (1973). *Endocr. Soc. 55th Ann. Meeting*, Chicago, 173.
Edwardson, J.A., Bennett, G.W. and Bradford, H.F. (1972). *Nature* 240, 554-556.
Ferreira, S.H., Moncada, S. and Vane, J.R. (1971). *Nature New Biol.* 231, 237-239.
Ferreira, S.H. and Vane, J.R. (1974). *A. Rev. Pharmacol.* 14, 57-73.
Fink, G., Smith, G.C., Tibbals, J. and Lee, V.W.K. (1972). *Nature New Biol.* 239, 57-59.
Flores, A.G.A. and Sharp, G.W.G. (1972). *Am. J. Physiol.* 233, 1392-1397.
Grinwich, D.L., Kennedy, T.G. and Armstrong, D.T. (1972). *Prostaglandins* 1, 89-96.
Harms, P.G., Ojeda, S.R. and McCann, S.M. (1972). *Science* 181, 760-761.
Harms, P.G., Ojeda, S.R. and McCann, S.M. (1974). *Endocrinology* 94, 1459-1464.
Jeffcoate, S.L. (1974). *In* "International Symposium on Hypothalamic Hormones: Chemistry, Physiology, Pharmacology and Clinical Uses" Milan, Italy, October 14-16.
Labhsetwar, A.P. (1970). *J. Reprod. Fert.* 23, 155-159.
Labhsetwar, A.P. (1972). *J. Endocrinol.* 53, 201-213.
Labhsetwar, A.P. (1973). *Prostaglandins* 4, 627-628.
Labhsetwar, A.P. (1974). *Fed. Proc.* 33, 61-77.
Labhsetwar, A.P, and Zolovick, A. (1973). *Nature New Biol.* 246, 55-56.
McCracken, J.A. and Roberts, J.S. (1975). *In* "Proceedings of the Second International Conference on Prostaglandins" (B. Samuelsson and R. Paoletti, eds). Raven Press, New York (*in press*).
Naor, Z., Koch, Y., Bauminger, S. and Zor, U. (1975). *Prostaglandins* 9, 211-219.
O'Grady, J.P., Caldwell, B.V., Auletta, F.J. and Speroff, L. (1972). *Prostaglandins* 1, 97-106.
Ojeda, S.R., Harms, P.G. and McCann, S.M. (1974a). *Prostaglandins* 8, 545-552.
Ojeda, S.R., Harms, P.G. and McCann, S.M. (1974b). *Endocrinology* 95, 613-618.
Orczyk, G.P. and Behrman, H.R. (1972). *Prostaglandins* 1, 3-20.
Patrono, C. (1973). *J. Nucl. Biol. Med.* 17, 25-29.
Patrono, C. (1974). *Prostaglandins* 8, 83-84.
Patrono, C. and Serra, G.B. (1974). *Prostaglandins* 6, 345-348.
Patrono, C., Grossi-Belloni, D., Ciabattoni, G., Serra, G.B. and Dell'Acqua, S. (1974a). *J. Steroid Biochem.* 5, 375.
Patrono, C., Grossi-Belloni, D. and Ciabattoni, G. (1974b). *Clin. Res.* 22, 346A.
Patrono, C., Ciabattoni, G. and Grossi-Belloni, D. (1974c). *Pharmacol. Res. Commun.* 6, 509-518.

Patrono, C., Ciabattoni, G., Greco, F. and Grossi-Belloni, D. (1975a). *In* "Proceedings of the Second International Conference on Prostaglandins" (B. Samuelsson and R. Paoletti, eds). Raven Press, New York *(in press)*.

Patrono, C., Ciabattoni, G. and Grossi-Belloni, D. (1975b). *Prostaglandins* **9**, 557-568.

Patrono, C., Bombardieri, S., Di Munno, O., Pasero, G.P., Greco, F., Grossi-Belloni, D. and Ciabattoni, G. (1975c). *In* "Prostaglandins in Inflammation" (G.P. Lewis, ed) *(in press)*.

Pelletier, G., Labrie, F., Puviani, R., Arimura, A. and Schally, A.V. (1974). *Endocrinology* **95**, 314-317.

Raiteri, M., Angelini, F. and Levi, G. (1974). *European J. Pharmacol.* **25**, 411-414.

Ratner, A., Wilson, M.C., Srivastava, L. and Peake, G.T. (1974). *Prostaglandins* **5**, 165-171.

Sato, T., Jyujo, T., Iesaka, T., Ishikawa, J. and Igarashi, M. (1974a). *Prostaglandins* **5**, 483-490.

Sato, T., Taya, K., Jyujo, T. and Igarashi, M. (1974b). *J. Reprod. Fert.* **39**, 33-40.

Sato, T., Hirono, M., Jyujo, T., Iesaka, T., Taya, K. and Igarashi, M. (1975). *Endocrinology* **96**, 45-49.

Serra, G.B. (1973). *In* "The Endocrine Function of the Human Testis" (V.H.T. James, M. Serio and L. Martini, eds) Vol. I, pp.395-411. Academic Press, New York and London.

Serra, G.B. and Midgley, A.R. Jr. (1970). *Proc. Soc. exp. Biol. Med.* **133**, 1370-1374.

Smith, J.B. and Willis, A.L. (1971). *Nature New Biol.* **231**, 235-237.

Spies, H.G. and Norman, R.L. (1973). *Prostaglandins* **4**, 131-141.

Tsafriri, A., Koch, Y. and Lindner, H.R. (1973). *Prostaglandins* **3**, 461-467.

Vale, W., Rivier, C. and Guillemin, R. (1971). *Fed. Proc.* **330**, 363.

Vane, J.R. (1971). *Nature New Biol.* **231**, 232-235.

Zor, U., Kaneko, T., Schneider, H.P.G., McCann, S.M. and Field, J. (1970). *J. biol. Chem.* **245**, 2882-2888.

METABOLISM OF MAMMALIAN OOCYTE DURING GROWTH, MATURATION AND ACTIVATION

F. Mangia*, C.J. Epstein[†], R.P. Erickson[†], F. Palombi* and
G. Siracusa*

*Institute of Histology and General Embryology, Faculty of Medicine
University of Rome, Rome, Italy

[†]Department of Pediatrics and of Biochemistry and Biophysics
University of California, San Francisco, California, USA

I. INTRODUCTION

Experimentally-derived information on the metabolism of mammalian female germ cells is still very fragmentary as compared with the wealth of data on oogenesis in lower forms. In fact, the life history of mammalian germ cells was uncertain until 1962, when it was definitely shown that in most mammals, including humans, the number of female germ cells is fully established at birth, with no stem cells remaining (Lima-de-Faria and Borum, 1962; Peters et al., 1962; Rudkin and Griech, 1962). This paucity of information about female mammalian gametogenesis is not surprising in view of the technical difficulties in obtaining viable oocytes free from contaminating somatic cells and in reasonable amounts for biochemical assays, at early stages of development. Recently, these difficulties have been overcome, and in the last two years information on growing mammalian oocytes obtained by direct biochemical assays on isolated germ cells has become available.

We shall report here on the metabolism of the mammalian oocyte from the "resting" stage to fertilization. It was felt that particular attention is due to this period of oocyte development which may bear great relevance to later embryo quality. The present review, which is based in part on the authors' experimental data on rodent oogenesis, will mainly deal with macromolecular synthetic activities and their control in oocytes and eggs, and will consider the phases of 1) oocyte growth during folliculogenesis, from the resting stage to germinal vesicle

(G.V.) breakdown; 2) oocyte maturation, i.e. the resumption of meiosis up to the metaphase II block; and 3) egg activation, i.e. the metabolic events leading to the overcoming of this block and the completion of meiosis.

II. OOCYTE DEVELOPMENT

A. RNA

Oocyte growth has long been known to represent a phase of high genetic activity in amphibians and echinoderms, forms in which the accessibility of the material allows for direct analysis. During this period the chromosome configuration is of the lampbrush type, and ribonucleoproteins (RNP) are made which are partly stored and will be inherited by the embryo (see Davidson, 1968, for a review).

In 1967 Baker and Franchi demonstrated, in different mammalian species including rodents and man, that the chromosomes of the growing oocyte bear lateral loops and branches, configurations which suggest high transcriptional activity.

RNA synthetic activity in mammalian oocytes was tentatively evaluated after *in vivo* labelling with H^3-uridine and autoradiography at both the light (Oakberg, 1967, 1968; Moore *et al.*, 1974) and the electron microscopic level (Baker *et al.*, 1969). According to the observed patterns of incorporation, increasing amounts of RNA are synthesized in the oocyte nucleus throughout oocyte growth. However, H^3-uridine incorporation rapidly decreases with the beginning of antrum formation, and no transcription is detectable in oocytes that are contained in fully-grown antral follicles (Oakberg, 1967, 1968; Moore *et al.*, 1974). The labelled product was first observed to appear in the cytoplasm 12 h after nuclear incorporation and to persist in the cytoplasm for up to three weeks (Oakberg, 1967).

A biochemical study of oocyte RNA's preserved through ovulation was recently performed by Bachvarova (1974) and a transcription pattern consistent with earlier autoradiographic observations was found. In her study H^3-adenosine was injected *in vivo* into cycling mouse ovaries. Ovulated eggs were subsequently collected at definite times thereafter, and incorporation per egg was measured. While preantral oocytes actively incorporated the precursor, no RNA synthesis could be detected in fully-grown antral oocytes. Ribosomal RNA (rRNA) was found to be the main component at all the stages examined, although small amounts of heterogeneously labelled RNA were also seemingly present.

Unfortunately, the data obtained by *in vivo* labelling procedures can only be taken as indicative, since the possibility cannot be ruled out that the low incorporation of the precursor into antral oocytes might result from a reduced uptake at these stages of oocytes development. To overcome this problem, Moore *et al.* (1974) have evaluated by histochemical methods the activity of endogenous DNA-dependent RNA polymerase at various stages of oocyte development and have found that enzymatic activity can be detected in isolated growing oocytes but is absent from the nuclei of the fully-grown cells.

As an alternative direct approach, the ultrastructural cytochemistry of the nuclear RNP particles present at various steps of oocyte development has been studied by F. Palombi and A. Viron (manuscript in preparation) with the EDTA preferential method of Bernhard (1969) as modified for ultrathin frozen sections (Puvion and Bernhard, 1975). It was observed that numerous perichromatin granules - extranucleolar RNP-containing particles - are contained in the growing oocyte nucleus, whereas they seem to be absent from the nucleus of preovulatory oocytes.

A detailed ultrastructural analysis of the nucleolar machinery by Chouinard (1971) on the prepubertal mouse shows that the nucleolar morphology specifically changes during oocyte growth: the nucleolus undergoes a progressive degranulation that is accompanied by accumulation of fibrous material of unknown nature which eventually results in a compact, homogeneously fibrillar "nucleolus". The findings are interpreted as indicating a storage role for the fully-grown oocyte nucleolus. The ultrastructure of these nucleoli, known to be inactive in RNA production (Oakberg, 1967; Moore et al., 1974), is similar to that of the "nucleoli" of zygote pronuclei (Szollosi, 1955) in which, again, RNA synthesis does not appear to occur. The nature of the fully-grown preovulatory nucleolus has also been investigated by acridine-orange staining of frozen semi-thin section (Palombi and Viron, unpublished data), and it was found that the stored material does not contain appreciable amounts of RNA. A similar result has also been reported for the nucleoli of mouse pronuclei (Austin and Braden, 1953).

B. Protein

1. Quantitative evaluation of protein synthetic activity

Biochemical studies of RNA synthesis in growing mammalian oocytes require amounts of isolated germ cells that at present are very difficult to obtain. However, as an alternative experimental strategy, information of oocyte gene expression can be more easily obtained by characterization and quantitation of the oocyte protein synthetic activity.

The volume of growing mouse oocytes increases about twentyfold during folliculogenesis (Brambell, 1928) and the protein content also increases (Alfert, 1950; Flax, 1953). Growing mouse oocytes actively incorporate exogenous amino acids in vivo (Baker et al., 1969), and this strongly suggests that their growth is due to the accumulation of proteins made by the oocyte itself, although a possible role of follicle cells cannot be excluded. However, protein synthesis has recently been directly demonstrated to occur in isolated growing mouse oocytes, freed from follicle cells and incubated in vitro in the presence of labelled amino acids (Cross and Brinster, 1974; Mangia and Canipari, unpublished data).

The protein synthetic machinery in growing oocytes has been recently studied by electron microscopy (Garcia et al., 1975), and it was shown that the total number of ribosomes per oocyte increases during the growth period, while ribosome density apparently decreases. Moreover, the absolute number of ribosomes forming polysomes per oocyte increases about eightfold throughout oocyte development.

Unfortunately, morphological observations do not give any information about the possible existence of inactive polysomes among translating ones, as already described in eggs of lower forms (Spirin and Nemer, 1965; Kaulenas and Fairbain, 1966). Therefore, the actual rate of protein synthesis in growing mammalian oocytes can only be measured by direct biochemical studies on the isolated germ cells. This evaluation requires characterization of several parameters, including the kinetics of precursor transport across cell membrane and measurement of the actual size of the intracellular pool. These problems are presently under investigation in the laboratory of some of us (University of Rome) and preliminary results suggest that the rate of protein synthesis increases throughout oocyte development (Mangia and Canipari, unpublished data).

2. *Analysis of total translation products*

Electron microscopic observations (Garcia *et al.*, 1974) indicate that the number of ribosomes per polysome remains constant in the mouse of dictyate oocyte, during both resting and growth stages. By contrast, a reduction from 15-50 to 4-12 ribosomes per polysome has been observed in going from predictyate diplotene to the dictyate stage. Although purely morphological, these data suggest that proteins made by pre-dictyate diplotene and by dictyate oocytes may be different and that oocyte gene products may not undergo major variations from resting and/or growing stages up to maturation.

Similar conclusions have been obtained by direct analysis of the proteins made *in vitro* by isolated mouse oocytes at several stages of development using SDS-polyacrylamide gel electrophoresis (Mangia *et al.*, 1976b). In these experiments, peaks of protein radioactivity from small, medium and large-size oocyte are consistently superimposable on each other. However, the relative proportions of high molecular weight components decrease during growth, while those of the small molecular weight components increase. It has been tentatively concluded that, within the limits of resolution of the technique used, protein species made by developing mouse oocytes are qualitatively but not quantitatively homogeneous during the whole growth period.

3. *Accumulation and synthesis of specific proteins*

So far, only two enzymatic activities have been directly studied by biochemical assays of isolated female germ cells and their developmental patterns followed during growth period: these enzymes are glucose-6-phosphate dehydrogenase (G6PD), an X-linked enzyme, and lactate dehydrogenase (LDH), an autosomally-controlled enzyme (Mangia and Epstein, 1975).

Both activities have been shown to increase continuously during oocyte growth. G6PD activity increases about sevenfold with growth of oocyte cytoplasm diameter from 44 to 95 μ, with a linear relationship between enzyme activity and oocyte volume. LDH activity increases even more dramatically (more than tenfold) in the same developmental interval and displays a power function relationship to volume. As a result the specific activity of the enzyme undergoes a fourfold increase during oocyte growth rather than remaining constant as it does for G6PD.

High levels of LDH activity are present in mature mouse ova and early embryos: the LDH is in the form of LDH isozyme-1 (a tetramer of B subunits, B_4) (see Erickson *et al.*, 1975, for a review). In addition, in the rat, treatment of ovum lysate with triton X-100 has been claimed to solubilize small amounts of membrane-bound LDH-5 (tetramer of A subunits, A_4), probably derived from follicle cells (Poznakhirkina *et al.*, 1975).

We have studied the actual rate of synthesis of LDH-1 by immunoprecipitation assays on isolated growing mouse oocytes pulse-labelled *in vitro* with H^3-tyrosine (Mangia *et al.*, 1976a). The rate of LDH-1 synthesis, expressed as a percentage of incorporation of ^{3}H-tyrosine into total oocyte protein, has been found to increase significantly from 3.9% in 45-55 μ diameter oocytes to 4.9% in 60-70 μ diameter oocytes. The percentage of LDH-1 synthesis did not significantly differ between medium sized and large preovulatory oocytes, and a mean value of 5.1% was observed. LDH-1 synthesis was no longer detectable in either unfertilized or fertilized eggs or in early embryos. These findings indicate that the availability for translation of LDH-B-subunit mRNA (as compared to total mRNA population) relatively increases in the cytoplasm of the newly growing oocyte and reaches a plateau in later stages of development. Moreover, translation of LDH-B-subunit mRNA apparently ceases at maturation. Unfortunately, since information on the actual rate of total protein synthesis in growing mouse oocytes is still lacking, quantitative data on the absolute synthetic rate of LDH-B-subunits in the growing female germ cells cannot yet be obtained.

Other enzyme activities have been studied in growing mouse oocytes by histochemical methods, including succinic dehydrogenase (SDH) activity as a specific marker of mitochondrial activity, and DNA-dependent RNA polymerase. SDH activity has been found to increase throughout oocyte growth and fully grown oocytes apparently show a higher enzyme concentration than earlier stages (Vivarelli et al., 1976). The findings on RNA polymerase activity obtained by Moore et al. (1974) have been already discussed in the RNA section of the present chapter.

C. Other Biochemical Parameters

Aside from the synthesis of macromolecular products and their characterization, very few other biochemical parameters have as yet been studied in growing mammalian oocytes. Among these are ATP content and DNA repair in responponse to U.V. irradiation.

The ATP content of isolated mouse oocytes at several stages of growth, has been assayed by a luciferin-luciferase method, and a linear increase of about fivefold has been observed throughout oocyte growth (Mangia, Kwok and Epstein, unpublished data): while the 40-50 μ oocyte contains about 0.4 p mole of ATP, this value increases to about 2 p moles in fully grown antral oocytes (85-90 μ in diameter). ATP content does not vary any farther during maturation.

Masui and Pedersen (1976) have recently studied, by autoradiography, DNA repair in isolated mouse ovarian oocytes following U.V. irradiation. They found that DNA repair capability seems to be a common property of dictyate mouse oocytes at all stages of growth, and of maturation as well.

III. THE MATURING OOCYTE

The surge of luteinizing hormone (LH) prior to ovulation has the effect among others of promoting the resumption of meiosis in the graafian follicle oocyte. Besides, large oocytes freed from granulosa cells spontaneously undergo maturation *in vitro* in a variety of mammals, including humans (see Donahue, 1972; and Kennedy, 1972, for reviews).

The nuclear progression during maturation *in vitro* has been accurately timed by Donahue (1968a) in mouse oocytes and has been studied more recently by time lapse cinematography (Sorensen, 1973). Metaphase I first appears after 4-5 hours of *in vitro* culture, and metaphase II is reached after 11-17 hours.

Pyruvate is the only energy source which can support mouse oocyte maturation *in vitro* (Biggers et al., 1967), although lactate can substitute for pyruvate if NAD is added to the culture medium (Zeilmaker et al., 1972; Sorensen, 1972). Moreover, in the rat the completion of the first meiotic division is dependent on aerobic metabolism (Zeilmaker et al., 1972; Zeilmaker and Verhamme, 1974). However, the completion of the second meiotic division does not appear to require phosphorylative oxidation, since spontaneous formation of the second polar body may occur in the presence of cyanide or in the absence of oxygen

(Zeilmaker and Verhamme, 1974).

It was hypothesized by classical authors that dictyate oocytes acquire the capacity to mature shortly before antrum formation (Brambell, 1928; Pincus and Enizmann, 1937). This problem has been more recently investigated (Szibeck, 1972; Erickson and Sorensen, 1972) and it has been definitely shown that only oocytes from antral follicles can undergo meiosis. Moreover, no strict correlation has been found between oocyte size and capacity to resume meiosis (Erickson and Sorensen, 1974).

Synthesis of Macromolecules and Control of Maturation

As early as 1968 it was shown that mouse oocytes cannot mature *in vitro* in the presence of puromycin, and it was suggested that meiotic progression is dependent on oocyte protein synthesis (Donahue, 1968b; see Cross and Brinster, 1974, for a review of the effect of drugs on maturation). However, only in 1972 was direct evidence first reported that *in vitro* maturing oocytes actually incorporate labelled amino acids into their proteins. In these experiments, isolated maturing mouse oocytes were labelled with C^{14}-valine and the patterns of protein synthetic activity and uptake of precursor were found to be constant throughout the whole maturation period (Stern *et al.*, 1972). Leucine uptake and incorporation by mouse oocytes during maturation has been studied (Cross and Brinster, 1974), and uptake was found to be affected by the presence of cumulus cells at dictyate and metaphase I, but not at metaphase II.

More recently the proteins made by mouse oocytes maturing *in vitro* in the presence of labelled lysine have been analysed on SDS-polyacrylamide gel electrophoresis and a large peak of incorporation co-migrating with very lysine-rich histones (f.1) from calf thymus, has been observed. This lysine-labelled peak is specifically synthesized during maturation, since it is virtually absent in the protein made by non-maturing 40-70 μ diameter oocytes (Wassermann and Sorensen, 1974).

The mechanisms which regulate resumption of meiosis in dictyate oocytes have been studied by Cho *et al.* (1974). According to these authors, addition to the culture medium of dibutyryl cyclic AMP (dbcAMP) and/or theophylline reversibly inhibits germinal vesicle breakdown (GVB), although these drugs have no effect on further meiotic progression if added after GVB (Cho *et al.*, 1974). The electrophoretic patterns of the proteins made by maturing or dbcAMP-blocked oocytes, cultured overnight *in vitro*, do not differ appreciably (Stern and Wassermann, 1974). These results, similar to those obtained on amphibian oocytes, suggest that the pattern of the protein synthesized during maturation is regulated by the oocyte cytoplasm (Stern and Wassermann, 1974).

IV. ACTIVATION AND PARTHENOGENESIS

In most organisms meiosis in the female gamete does not proceed to completion and the egg remains in a state of suspended development, which is only resumed if fertilization occurs. Sperm penetration induces in the egg structural and metabolic changes which are known as "activation". In most mammals, including man, the ovulated egg is arrested at the metaphase of the second meiotic division, and the most evident aspect of activation is the resumption of meiosis. The biochemical mechanisms by which the male gamete induces activation are still unknown. The activating stimulus would, however, seem to be rather nonspecific, since release from the metaphase II block and parthenogenetic activation can be obtained with a variety of different stimuli, both physical (thermal, electric, osmotic shocks) and chemical (enzyme or anaesthetic treatments) (for a recent review, see Graham, 1974).

The development of mammalian parthenotes is very limited. Early claims (Pincus, 1939) that rabbit eggs could parthenogenetically develop to birth have not been confirmed by authorities in this field (Thibault, 1949; Chang, 1954). The only mammalian parthenotes which have been observed to implant are those derived from mouse eggs (Tarkowski et al., 1970; Graham, 1970), and the most advanced stage of development reached was the 8-somite stage (Witkowska, 1973).

A. Control of Activation

In our laboratory (University of Rome) the parthenogenetic activation of mouse eggs has been studied with the aim of clarifying the molecular mechanisms that regulate the metaphase II block and the activation process (Siracusa et al., 1976). In general terms, these mechanisms might consist either in a negative control (the egg prevents meiotic progression by synthetizing an inhibiting factor), or in a positive one (meiosis is resumed when the egg synthesizes an activating factor). Our results are compatible with the first hypothesis. We have found that in vitro incubation of metaphase II eggs with inhibitors of protein synthesis (cycloheximide and puromycin) at concentrations which cause at least 80% inhibition of protein synthesis induces parthenogenetic activation (i.e. formation of the female pronucleus). If, together, with these inhibitors, dibutyryl cyclic AMP or theophylline (an inhibitor of phosphodiesterase and, therefore, of the endogenous degradation of cAMP) are also added to the culture medium, a higher yield of activation is obtained. Treatment with dbcAMP or theophylline alone has however no effect (although these drugs penetrate equally well in the egg in the presence or absence of cycloheximide).

These data suggest that the maintenance of the metaphase block requires the continuous synthesis of a protein factor, which probably interferes with a cAMP-dependent activating mechanism. When protein synthesis is inhibited, endogenous cAMP might switch on the activating mechanism, and exogenous cAMP (or theophylline) might further increase the activation.

ACKNOWLEDGEMENT

This work was supported by grant no. 730-0208 from the Ford Foundation, by grant HD-03132 from the National Institute of Child Health and Human Development and by grant no. 73.00673.04 from the Consiglio Nazionale della Ricerche.

REFERENCES

Alfert, M. (1950). *J. Cell. Comp. Physiol.* **36**, 381-409.
Austin, C.R. and Braden, A.W.H. (1953). *Austr. J. Exp. Biol. Med. Sci.* **6**, 324-333.
Bachvarova, R. (1974). *Develop. Biol.* **40**, 52-58.
Baker, T.G. and Franchi, L.L. (1967). *Chromosoma* **22**, 358-377.
Baker, T.G., Beaumont, H.M. and Franchi, L.L. (1969). *J. Cell Sci.* **4**, 655-675.
Bernhard, W. (1969). *J. Ultrastr. Res.* **27**, 250-265.
Biggers, J.D., Whittingham, D.G. and Donahue, R.P. (1967). *Proc. natn. Acad. Sci. U.S.A.* **58**, 560-567.

Brambell, F.W.R. (1928). *Proc. Roy. Soc.* **B103**, 258-272.
Chang, M.C. (1954). *J. Exp. Zool.* **121**, 351-381.
Cho, W.K., Stern, S. and Biggers, J.D. (1974). *J. Exp. Zool.* **187**, 383-386.
Chouinard, L.A. (1971). *J. Cell Sci.* **9**, 637-663.
Cross, P.C. and Brinster, R.L. (1974). *Exp. Cell Res.* **86**, 43-46.
Davidson, E.H. (1968). "Gene Activity in Early Development". Academic Press, New York and London.
Donahue, R.P. (1968a). *J. Exp. Zool.* **169**, 237-250.
Donahue, R.P. (1968b). "Maturation of the Mouse Oocyte *In Vitro*". Ph.D. Dissertation, Johns Hopkins University, Baltimore.
Donahue, R.P. (1972). *In* "Oogenesis" (J.D. Biggers and A.W. Schuetz, eds) pp.413-438. University Park Press, Baltimore and Butterworths, London.
Erickson, G.F. and Sorensen, R.A. (1974). *J. Exp. Zool.* **190**, 123-127.
Erickson, R.P., Spielmann, H., Mangia, F., Tennenbaum, D. and Epstein, C.J. (1975). *In* "The Third International Conference on Isozymes" (C.L. Markert, ed) pp.313-324. Academic Press, New York, San Francisco and London.
Flax, M.H. (1953). "Ribose Nucleic Acid and Protein during Oogenesis and Early Embryonic Development in Mouse". Ph.D. Dissertation, Columbia University, New York.
Garcia, R.B., Benech, C. and Sotelo, J.R. (1974). *In* "Physiology and Genetics of Reproduction", Part A (E.M. Coutinho and F. Fuchs, eds) pp.307-322. Plenum Press, New York.
Graham, C.F. (1970). *Nature* **226**, 165-167.
Graham, C.F. (1974). *Biol. Rev.* **49**, 399-422.
Kaulenas, M.S. and Fairbairn, D. (1966). *Develop. Biol.* **14**, 481-490.
Kennedy, J.F. (1972). *In* "Oogenesis" (J.D. Biggers and A.W. Schuetz, eds) pp.439-457. University Park Press, Baltimore and Butterworths, London.
Lima-de-Faria, A. and Borum, K. (1962). *J. Cell Biol.* **14**, 381-388.
Mangia, F. and Epstein, C.J. (1975). *Develop. Biol.* **45**, 211-220.
Mangia, F., Erickson, R.P. and Epstein, C.J. (1976a). *Submitted for publication.*
Mangia, F., Smith, S.A. and Epstein, C.J. (1976b). *Submitted for publication.*
Masui, Y. and Pedersen, R.A. (1976). *Science (in press).*
Moore, G.P.M., Lintern-Moore, S., Peters, H. and Faber, M. (1974). *J. Cell Biol.* **60**, 416-422.
Oakberg, E.F. (1967). *Arch. Anat. Microsc. Morphol. Exp.* **56** (**Suppl.** 3-4), 171-184.
Oakberg, E.F. (1968). *Mutat. Res.* **6**, 155-165.
Peters, H., Levy, E. and Crone, M. (1962). *Nature* **195**, 915-916.
Pincus, G. (1939). *J. Exp. Zool.* **82**, 85-130.
Pincus, G, and Enzmann, E.V. (1937). *J. Morph.* **61**, 351-384.
Poznakhirkina, N.A., Serov, O.L. and Korochkin, L.I. (1975). *Biochem. Genet.* **13**, 65-72.
Puvion, E. and Bernhard, W. (1975). *J. Ultrastruct. Res. (in press).*
Rudkin, G.T. and Griech, H.A. (1962). *J. Cell Biol.* **12**, 169-175.
Siracusa, G., Molinaro, M. and Coletta, M. (1976). *Develop. Biol. (in press).*
Sorensen, R.A. (1972). *Biol. Reprod.* **7**, 139.
Sorensen, R.A. (1973). *Amer. J. Anat.* **136**, 265-276.
Spirin, A.S. and Nemer, M. (1965). *Science* **150**, 214-217.
Stern, S., Rayyis, A. and Kennedy, J.F. (1972). *Biol. Reprod.* **7**, 341-346.
Stern, S. and Wassarman, P.M. (1974). *J. Exp. Zool.* **189**, 275-281.
Szollosi, D. (1965). *J. Cell Biol.* **25**, 545-562.
Szybek, K. (1972). *J. Endocr.* **54**, 527-528.
Tarkowski, A.K., Witkowska, A. and Nowicka, J. (1970). *Nature* **226**, 162-165.
Thibault, C. (1949). *Ann. Sci. nat. Zool.* **11**, 136-219.
Vivarelli, E., Siracusa, G. and Mangia, F. (1976). *J. Reprod. Fertil. (in press).*
Wassarman, P.M. and Sorensen, R.A. (1974). *Fed. Proc. Am. Soc. Biol. Chem.* **33**, 1412.
Witkowska, A. (1973). *J. Embryol. exp. Morph.* **30**, 547-560.
Zeilmaker, G.H., Hülsmann, W.C., Wensinck, F. and Werhamme, C.M. (1972). *J. Reprod. Fertil.* **29**, 151-152.
Zeilmaker, G.H. and Werhamme, C.M. (1974). *Biol. Reprod.* **11**, 145-152.

THE ROLE OF BLOOD FLOW IN REGULATING OVARIAN FUNCTION

G.D. Niswender, A.M. Akbar* and T.M. Nett

*Department of Physiology and Biophysics, Colorado State University
Fort Collins, Colorado, USA*

I. INTRODUCTION

There has been considerable controversy regarding the roles of the gonadotropic hormones in regulation of ovarian function. Luteinizing hormone (LH) has been shown to stimulate synthesis of progesterone *in vitro* by luteal tissue obtained from women (Rice *et al.*, 1964), rhesus monkeys (Channing, 1970), cows (Marsh and Savard, 1966) and ewes (Kaltenbach *et al.*, 1967). In addition, passive immunization with anti-LH or anti-HCG sera results in decreased luteal function in monkeys (McDonald, 1971), cows (Hoffman *et al.*, 1974) and ewes (Reimers and Niswender, 1975). These observations indicate that LH is the primary luteotropic stimulus in these species. Yet, in each of these species LH levels are at their nadir during the phase of the cycle when progesterone secretion is maximal (Midgley and Jaffe, 1968; Niswender and Spies, 1973; Hendricks *et al.*, 1970; Niswender *et al.*, 1968). Similarly, in these species concentrations of follicle-stimulating hormone (FSH) remain constant, or decrease significantly, during that phase of the reproductive cycle when the follicles are developing most rapidly (Midgley and Jaffe, 1968; Niswender and Spies, 1973; Akbar *et al.*, 1974; L'Hermite *et al.*, 1972).

These observations led to the hypothesis that the ovary may "utilize" large quantities of these gonadotropins for its normal function. However, when this hypothesis was tested using the ewe as the experimental model there was no "utilization" of LH or FSH as judged by arterial-venous differences across the ovary (Scaramuzzi *et al.*, 1968; Cicmanec and Niswender, 1973) or by the rate of clearance of these hormones from the body (Akbar *et al.*, 1974).

---

* Present address: Department of Endocrinology, Cook County General Hospital, Chicago, Illinois.

Since the quantity of tropic hormone reaching a target organ is the product of blood concentrations of hormone and the quantity of blood reaching the target organ, a logical extension of previous studies was to quantify flow of blood to the ovaries throughout the estrous cycle. The ewe was used as a model for these studies since the required surgery could be performed easily and because of the considerable information available regarding luteal function in this species.

II. MATERIALS AND METHODS

Western range ewes of mixed breeding weighing between 40 and 70 kg and between four and eight years of age which had exhibited at least two consecutive, normal estrous cycles (14-18 days) were used for these studies. The ewes were checked for estrus twice daily using vasectomized rams. The first day an ewe stood to be mounted by a ram was designated as day 0 of the cycle.

Doppler ultrasonic blood flow transducers were used to measure the velocity of the blood flowing in the ovarian artery. This procedure has been shown to quantify reliably blood flow in the ovarian artery and to agree well with estimates obtained using radioactive microspheres (Niswender et al., 1975). In the first experiment blood flow transducers were implanted around each ovarian artery of four ewes on day 3 or 4 of the cycle. In these ewes one to three corpora lutea were present on one ovary and no corpora lutea were present on the opposite ovary. Two arteries served the left ovary of the fourth ewe so it was removed and the blood flow transducer was implanted around the artery leading to the right ovary which contained a corpus luteum. On the first day of the experiment (day 7 to 9 of the cycle) each ewe was fitted with an indwelling jugular cannula. Blood flow velocity recordings and jugular venous blood samples were obtained at 6-h intervals for 3 days, at 4-h intervals for 2 days, at 2-h intervals for 6 days and again at 6-h intervals for 3 days. This sampling regimen was designed so that samples were obtained most frequently around estrus. Levels of LH (Niswender et al., 1969), FSH (L'Hermite et al., 1972), prolactin (Davis et al., 1971) and progesterone (Niswender, 1973) were quantified by radioimmunoassay. All samples were assayed in duplicate, with triplicate standard curves included with each assay. At each sampling period, audio output from the Doppler velocity meter was recorded on magnetic tape for 2.5 min from each probe. The frequency for each 2.5 min sampling period was averaged using a Beckman Universal Eput Meter and Timer.

At the termination of the experiment the ewes were anesthetized with sodium pentobarbital, mid-ventral laparotomy was performed, and No.382 Medical Grade Silastic Elastomer (Dow Corning) was injected into the ovarian artery near its origin, allowing blood pressure to carry the elastomer through the vessel to the ovary. The sheep were killed, and the ovaries and their associated blood vessels were removed. The liquid silastic was allowed to harden for at least 7 days at 4°C. Most of the tissue was dissected free and the casts were cleaned with concentrated hydrochloric acid. The diameter of the silastic cast, for that portion of the ovarian artery contained within the transducer, was determined in three places using a dissecting microscope and the mean diameter was calculated. This method has been shown to be reliable for estimating the diameter of small blood vessels (Neumaster and Krovetz, 1964; Wolf, 1966). Blood flow is calculated as the product of velocity times cross-sectional area of the artery.

In a second experiment a radioactive microsphere method (Wagner et al., 1969; Rosenfeld et al., 1973) was used to measure the relative flow of blood to the different compartments within the ovary. Ewes were anesthetized with 20-25 ml (65 mg/ml) sodium pentobarbital and surgical anesthesia was maintained with a closed circuit inhalation apparatus using nitrous oxide, oxygen and halothane. Mid-ventral laparotomy was performed and a polyvinyl cannula was inserted into the dorsal aorta via the femoral artery and advanced until the tip was approximately 3 cm cranial to the origin of the ovarian arteries. At 1700 h on day 12, 50 microcuries of ^{169}Yb microspheres (50 ± 10 micron, S.A. 11.4 mCi/g) were injected into the dorsal aorta of all ewes, at 1700 h on day 14 all ewes received 50 microcuries of ^{51}Cr microspheres (50 ± 10 micron, S.A. 11.4 mCi/g) and on day 16, 50 microcuries of ^{141}Ce microspheres (50 ± 10 micron, S.A. 11.9 mCi/g) were injected. The microspheres were suspended in 4 ml of 20% dextran and injected at a rate of 2 ml/min. The ewes were killed at 2000 h on day 17, the ovaries were removed, and the radioactivity in each ovary was determined by differential counting in a Nuclear Chicago Model 1185 Automatic gamma spectrometer. The data were corrected for channel overlap and expressed as counts per minute per gram wet weight of tissue.

In order to evaluate arterial-venous shunting within the ovary eight ewes were treated exactly as described in the previous experiment except that 50 microcuries of 15 ± 4 micron microspheres containing ^{85}Sr (S.A. 11.1 mCi/g) were also injected into 4 of the ewes on day 12 and the remaining 4 ewes were injected on day 16 of the cycle. The ovaries were collected and counted as described previously.

In a final experiment the effect of exogenous prostaglandin (PG) $F_2 a$ on blood flow to the ovary was studied. Five ewes, each with a single corpus luteum, had blood flow transducers placed around each ovarian artery on day 5 of the estrous cycle. On day 9 each ewe received two intrauterine injections of 5 mg of $PGF_2 a$ at 4-h intervals. Beginning 2 h prior to the first $PGF_2 a$ injection blood flow recordings and jugular blood samples were collected hourly for 20 h. Serum was obtained and progesterone was quantified. The diameter of each ovarian artery was determined and blood flow was calculated as in the first experiment.

The data regarding blood flow to the ovaries and levels of reproductive hormones in serum were subjected to multiple correlation analysis. All data from each ewe were normalized to the preovulatory LH peak and plotted as the mean ± one standard error of the mean for each 4-h interval. The computer program, RSUM, used for this purpose has been described (Duddleson et al., 1972). The values obtained using radioactive microspheres were evaluated using a one-way analysis of variance and Duncan's Multiple Range test (Steel and Torrie, 1960). A paired "t" test was used to compare the uptake of 50 micron v. 15 micron microspheres.

III. RESULTS

Serum levels of the reproductive hormones and the flow of blood to each ovary throughout the estrous cycle are depicted in Fig.1. Peak levels of all three gonadotropic hormones occurred at estrus, and serum levels of LH and FSH ($P < 0.05$), LH and prolactin ($P < 0.05$) and FSH and prolactin ($P < 0.01$) were correlated (Table I). Levels of progesterone in serum and blood flow to the ovaries

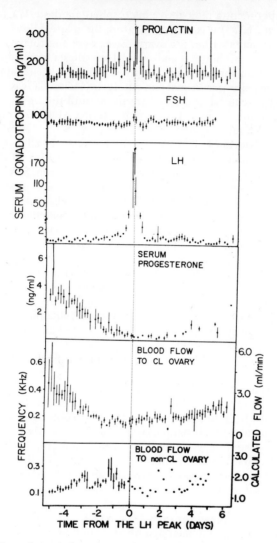

Fig.1 Levels of gonadotropic hormones and progesterone in serum and blood flow to the ovaries throughout the estrous cycle in ewes. All data are normalized to the preovulatory peak of LH. Each point represents the mean ± one standard error (n = 4).

with corpora lutea were also correlated ($P < 0.05$). In fact, most of the outlying values for progesterone were associated with outlying values for blood flow to ovaries which contained corpora lutea. There were no apparent cyclic changes in flow of blood to ovaries without corpora lutea. The average diameter of the ovarian artery was $0.520 ± 0.124$ mm ($\bar{x} ± S.E.$) for the six vessels for which silastic casts were obtained successfully.

The results obtained based on the uptake of radioactive microspheres are depicted in Fig.2. On day 12 of the cycle 98% of the microspheres in the luteal ovary were present within the corpus luteum. The number of microspheres taken

Table I. Multiple correlations between serum levels of reproductive hormones and blood flow to the ovary bearing the corpus luteum.

Parameter	FSH	Prolactin	Progesterone	Ovarian Blood Flow†
LH	0.2772*	0.2366*	-0.1124	-0.1363
FSH		0.4316**	0.0827	0.1053
Prolactin			-0.0545	-0.1337
Progesterone				0.2543*

* $P < .05$: ** $P < .01$: † The data included is blood flow to the ovaries bearing corpora lutea prior to the preovulatory LH peak.

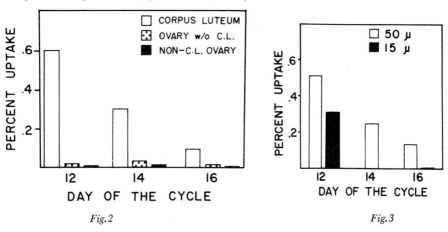

Fig. 2

Fig. 3

Fig. 2 Uptake of radioactive microspheres by the ovaries of cyclic sheep. The corpus luteum, the extra-luteal component of the ovary which had contained the corpus luteum and the opposite ovary are depicted separately. The uptake of microspheres by the corpus luteum was not different ($P < .1 > .05$) between days 12 and 14 or between days 14 and 16 but was different ($P < .01$) between days 12 and 16. There were no significant differences in the uptake by the extra-luteal portion of the lutel ovary or the opposite ovary.

Fig. 3 Uptake of radioactive microsphere by the corpus luteum in cycling ewes. The uptake of 50 micron microspheres was less ($P < .05$) at days 14 and 16 than on day 12 but days 14 and 16 did not differ from each other. There was no difference between the uptake of 50 micron v. 15 micron microspheres on day 12 ($P < .1 > .05$) but the difference on day 16 was significant ($P < .001$).

up by the corpus luteum decreased by day 14 and was even less by day 16 of the cycle. There were no differences in the uptake of microspheres by the extra-luteal portion of the luteal ovary or by the non-luteal ovary on any of the days studied.

Data obtained from injections of mixtures of 50 micron and 15 micron microspheres are summarized in Fig. 3. The uptake of 50 micron microspheres by the corpus luteum was similar on all days to that noted in the previous experiment. There were 70% ($P > 0.05$) as many 15 micron as 50 micron microspheres taken up on day 12, but only 3% ($P < 0.01$) as many 15 micron as 50 micron microspheres were taken up on day 16.

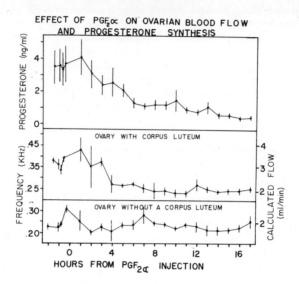

Fig.4 Systemic progesterone levels and blood flow to both ovaries following intrauterine injections of 5 mg of $PGF_2\alpha$ at 0 and 4 h.

As can be seen in Fig.4, serum levels of progesterone declined to baseline within 7 h of the initial intrauterine injection of $PGF_2\alpha$. Blood flow to the ovary which contained the corpus luteum declined to baseline within 4 h following the initial injection of $PGF_2\alpha$ while blood flow to the opposite ovary was not affected.

IV. DISCUSSION

The objective of these studies was to determine if blood flow to the ovary increased during the luteal phase of the cycle. Data obtained using the Doppler ultrasound procedure suggested that blood flow to the ovary bearing the corpus luteum increased 4- to 6-fold during the luteal phase of the cycle and further indicated that blood flow to this ovary was correlated with systemic progesterone levels. These findings suggest that the total amount of LH and prolactin reaching the luteal ovary increased during the luteal phase of the cycle. However, the quantity of oxygen, acetate, glucose, cholesterol and all other blood borne substances reaching the ovary are also increased during the luteal phase of the cycle.

The findings obtained using the radioactive microsphere procedures suggested that the corpus luteum received nearly all of the blood flowing to the luteal ovary. Therefore, the high correlation between blood flow and systemic progesterone levels seems even more physiologically significant. In addition, when combinations of 50 micron and 15 micron microspheres were injected on days 12 and 16, there were fewer ($P < 0.01$) 15 micron microspheres trapped within the corpus luteum on day 16. This observation suggests that arterial-venous shunts, between 50 micron and 15 micron, occur within the corpus luteum as this gland regresses. Thus, total blood flow to the luteal ovary declines dramatically during the late luteal phase of the cycle, blood flow to the corpus luteum decreases, and in addition, arterial-venous shunts develop so that essentially no blood (as judged by uptake of 15 micron microspheres) gets into the capillary bed. These findings indicate that changes in luteal blood flow may play a very important role in regulation of the life span and function of this structure.

The changes in blood flow to the ovary bearing the corpus luteum cannot be explained merely by changes in the weight of the corpus luteum. If the blood flow is adjusted for the weight of the corpus luteum at different times of the cycle (Karsch, 1970), the changes are slightly less dramatic, but the patterns are still similar. This is not surprising, since the corpus luteum weighs 450-800 mg during the mid-luteal phase of the cycle, while the rest of the ovary weighs 1-2 g.

Although changes in blood flow to the ovary during the cycle are quite dramatic, the physiological factors responsible for these changes are unknown. There were no apparent hormonal changes correlated with the mid-cycle elevation in blood flow to the ovary bearing the corpus luteum. However, McCracken et al. (1971) have suggested that infusions of LH increased blood flow to the autotransplanted ovary in ewes. Our own studies have demonstrated that both blood flow to the luteal ovary and systemic levels of progesterone decrease significantly within hours following injection of antisera to LH on day 10 of the cycle (Reimers and Niswender, 1975). These data strongly suggest that LH is involved in the regulation of ovarian blood flow.

The dramatic decline in blood flow to the corpus luteum during the late luteal phase of the cycle appears to be a direct result of a luteolytic agent. Pharriss (1970) suggested that this factor was $PGF_2\alpha$. We have been able to mimic the hemodynamic events which occur during the late luteal phase of the cycle in sheep by injections of $PGF_2\alpha$ (Nett et al., 1974). Within hours of administration of $PGF_2\alpha$ there is a dramatic decline in ovarian blood flow coincident with decreased systemic progesterone levels. Others have reported similar findings (Novy and Cook, 1973; Brown et al., 1974; Bruce and Hillier, 1974).

The estimates of blood flow to the ovary in these studies (0.3-6 ml/min) are somewhat lower than the 10-20 ml/min reported by McCracken et al. (1969) but agree with the 6-8 ml/min reported by Cook et al. (1969) and with the estimates of 1.5-12 ml/min reported by Domanski et al. (1967). Since McCracken et al. (1969) were using the ovarian transplant model for their studies and collected blood from the constricted jugular vein, it is possible that a significant amount of blood was contributed by tissues other than the ovaries.

These results indicate that either the ultrasonic method or injection of radioactive microspheres can be used to estimate the relative flow of blood to the ovary. By precisely determining the diameter of the ovarian artery, the Doppler ultrasonic procedure could be made quantitative, while the microsphere procedure could only reflect relative differences in ovarian blood flow from day to day. The limitations of the radioactive microsphere method for measuring ovarian blood flow resulted from the unique blood vascular system which supplies this organ. In sheep each ovarian artery originates directly from the dorsal aorta at a 90° angle. Therefore, if laminar flow of the microspheres occurs they will not reach the ovary. This problem was not apparent for other tissues such as the uterus. Although there was considerable variation between animals, the uptake of microspheres within an animal was quite consistent when injections were performed during the mid-luteal phase of the cycle (days 6, 10 and 12). Therefore, it appears that this procedure is reliable for estimating relative blood flow. The major advantage of the Doppler method is that measurements can be made over long periods of time and can be repeated at frequent intervals. The major disadvantages of the Doppler procedure are that it is difficult to measure precisely the diameter of the ovarian artery, and it is limited to the quantification of blood flow in an individual blood vessel.

The radioactive microsphere technique potentially allows estimation of blood flow to individual compartments within the ovary; however, the number of determinations which can be made is limited to the number of isotopes which can be counted differentially. This technique has also been criticized because it depends upon blockage of blood vessels by the microspheres. In practice this is usually not a problem since only a small number of microspheres of high specific activity is injected, resulting in blockage of an insignificant number of capillaries.

IV. SUMMARY

Blood flow to both ovaries of sheep was measured using Doppler ultrasonic transducers. Flow to the luteal ovary was elevated 2-6 fold during the mid-luteal phase of the cycle and diminished rapidly as the corpus luteum regressed. When radioactive microspheres were used to assess the relative flow of blood to different compartments within the luteal ovary it was observed that 98% of the microspheres were localized within the corpus luteum on day 12 of the cycle while only 40% of the microspheres were found in the corpus luteum by day 16. In addition, the uptake of 50 micron and 15 micron radioactive microspheres was not different on day 12 of the cycle but only 3% as many 15 micron microspheres were taken up on day 16 ($P < 0.01$). These observations suggest: 1) that blood flow to the corpus luteum is highly correlated with progesterone synthesis; 2) that blood flow to the corpus luteum declined dramatically as this structure regresses; and 3) that arterial-venous shunts develop within the corpus luteum which shunts blood away from the capillary network as this structure regresses. Thus, blood flow to the corpus luteum appears to play an important role in the regulation of luteal function.

In a final experiment $PGF_2\alpha$ was administered to sheep during the luteal phase of the estrous cycle. Systemic progesterone levels fell to baseline within 7 h of the injection and blood flow to the luteal ovary was diminished to basal levels within 4 h. There was no effect on blood flow to the non-luteal ovary. These findings suggest that it is possible to mimic the normal hemodynamic changes which occur within the corpus luteum at the end of the luteal phase of the cycle with an intrauterine injection of large doses of $PGF_2\alpha$.

ACKNOWLEDGEMENTS

This work was supported by NIH Contract 69-2134 and a grant from G.D. Searle and Company.

The authors wish to express their gratitude to the Hormone Distribution Office, National Institutes of Health for supplying the standard used for the radioimmunoassay of gonadotropic hormones and to Dr. Leo E. Reichert, Jr. for supplying the purified LH (LER-1056-C2), FSH (LER-1563) and prolactin (LER-860-2) for radioiodination.

REFERENCES

Akbar, A.M., Nett, T.M. and Niswender, G.D. (1974). *Endocrinology* **94**, 1318-1324.
Akbar, A.M., Reichert, L.E., Jr., Dunn, T.G., Kaltenbach, C.C. and Niswender, G.D. (1974). *J. Anim. Sci.* **39**, 360-365.
Brown, B.W., Hales, J.R.S.,and Mattner, P.E. (1974). *Experientia* **30**, 914-915.
Bruce, N.W. and Hillier, K. (1974). *Nature* **249**, 176-177.
Channing, C.P. (1970). *Endocrinology* **87**, 49-60.
Cicmanec, J.L. and Niswender, G.D. (1973). *Proc. Soc. exp. Biol. Med.* **144**, 99-105.
Cook, B., Kaltenbach, C.C., Niswender, G.D., Norton, H.W. and Nalbandov, A.V. (1969). *J. Anim. Sci.* **29**, 711-718.
Davis, S.L., Reichert, L.E., Jr. and Niswender, G.D. (1971). *Biol. Reprod.* **4**, 145-153.
Domanski, E., Skrzeczkowski, L., Stupnicki, E., Fitko, R. and Dobrowolski, W. (1967). *J. Reprod. Fertil.* **14**, 365-372.
Duddleson, W.G., Midgley, A.R., Jr. and Niswender, G.D. (1972). *Computers and Biomedical Res.* **5**, 205-217.
Hendricks, D.M., Niswender, G.D. and Dickey, J.F. (1970). *Biol. Reprod.* **2**, 346-351.
Hoffman, B., Schams, D., Bopp, R., Ender, M.L., Gimenez, T. and Karg, H. (1974). *J. Reprod. Fertil.* **40**, 77-85.
Kaltenbach, C.C., Cook, B., Niswender, G.D. and Nalbandov, A.V. (1967). *Endocrinology* **81**, 1407-1409.
Karsch, F.J. (1970). Ph.D. Thesis, University of Illinois.
L'Hermite, M., Niswender, G.D., Reichert, L.E., Jr. and Midgley, A.R., Jr. (1972). *Biol. Reprod.* **6**, 325-332.
Marsh, J. and Savard, K. (1966). *J. Reprod. Fertil.* (Suppl. 1), 113.
McCracken, J.A., Uno, A., Goding, J.R., Ichikawa, Y. and Baird, D.T. (1969). *J. Endocrinology* **45**, 425-440.
McCracken, J.A., Baird, D.T. and Goding, J.R. (1971). In "Recent Progress in Hormone Research" (E.B. Astwood, ed). Academic Press, New York.
McDonald, G.J. (1970). *Proc. IV Ann. Meet. Soc. Study Reprod.* Abst. 16.
Midgley, A.R., Jr. and Jaffe, R.B. (1968). *J. clin. Endocr. Metab.* **28**, 1699-1703.
Nett, T.M., Stiagmiller, R.B., Diekman, M.A., Akbar, A.M., Ellinwood, W.E. and Niswender, G.D. (1975). *J. Anim. Sci. (submitted)*.
Neumaster, T. and Krovetz, L.J. (1964). *J. Applied Physio.* **19**, 1184-1186.
Niswender, G.D., Roche, J.F., Foster, D.L. and Midgley, A.R., Jr. (1968). *Proc. Soc. exp. Biol. Med.* **129**, 901-904.
Niswender, G.D., Reichert, L.E., Jr., Midgley, A.R., Jr. and Nalbandov, A.V. (1969). *Endocrinology* **84**, 1166-1173.
Niswender, G.D. (1973). *Steroids* **22**, 413-424.
Niswender, G.D. and Spies, H.G. (1973). *J. clin. Endocr. Metab.* **37**, 326-328.
Niswender, G.D., Moore, R.T., Akbar, A.M., Nett, T.M. and Diekman, M.A. (1975). *Biol. Reprod. (submitted)*.
Novy, M.J. and Cook, M.J. (1973). *Am. J. Obstet. Gynec.* **117**, 381-385.
Pharriss, B.B. (1970). *Perspectives in Biol. Med.* **14**, 434-444.
Reimers, T.J. and Niswender, G.D. (1975). In "Immunization with Hormones in Reproductive Research" (J. Geelen, ed). ASP Biological and Medical Press, B.V. *(in press)*.
Rice, B.F., Hammerstein, J. and Savard, K. (1964). *J. clin. Endocr. Metab.* **24**, 606-615.
Rosenfeld, C.R., Killam, A.P., Battaglia, F.C., Makowski, E.L.,and Meschia, G. (1973). *Am. J. Obstet. Gynec.* **115**, 1045-1052.
Scaramuzzi, R.J., Caldwell, B.V. and Moor, R.M. (1970). *Biol. Reprod.* **3**, 110-119.
Steel, R.C.D. and Torrie, J.H. (1960). "Principles and Procedures of Statistics". McGraw-Hill, New York.
Wagner, H.N., Rhodes, B.A., Sasaki, Y. and Ryan, J.P. (1969). *Investigative Radiology* **4**, 374-386.
Wolf, K.B. (1969). *Am. J. Med. Tech.* **32**, 211-217.

FEMALE SEX STEROIDS AND GONADOTROPHIN SECRETION

L. Martini

Department of Endocrinology, University of Milan, Italy

I. INTRODUCTION

The present review will summarize some recent data obtained in the author's laboratory regarding the mode of operation of the feedback mechanisms which control gonadotrophin secretion in the female rat. The following topics will be underlined:

a) Effects of oestrogens on LH and FSH secretion.

b) Effects of progesterone and of its 5a-reduced metabolites on LH and FSH secretion.

II. EFFECTS OF OESTROGENS ON LH AND FSH SECRETION

The experiments reported here were planned in order to discover if exogenous oestrogens exert a quantitatively similar inhibitory effect on the secretion of LH and FSH, when given at different intervals following ovariectomy.

The study has been performed on female rats of the Sprague-Dawley strain (initial body weight 150 ± 20 g). Animals were castrated while etherized and killed at different intervals after the operation (from 7 to 28 days). At each time, a group of control and a group of experimental animals were killed. The experimental animals were given a daily subcutaneous injection of 50 μg of oestradiol benzoate in the two days preceding sacrifice. This experimental design permitted each experimental animal to receive the same amount of oestrogens at the same time prior to being killed, but at different intervals following castration. Serum levels of LH (Niswender *et al.*, 1968) and of FSH (Daane and Parlow, 1971) were measured with specific radioimmunoassay procedures.

It is apparent from Table I that oestradiol exerts a strong inhibitory effect on the secretion of LH. It is interesting to note that the dose of oestradiol selected

Table I. Effect of castration and of treatment with oestradiol benzoate (OEB 50 µg/rat/day for 2 days) on serum levels of LH and FSH of adult female rats killed at different intervals after gonadectomy.

Groups[a]	Time after operation (days)	Treatment	ng LH/ml (NIH-LH-S16)	ng FSH/ml (NIAMD-RAT-FSH-RP-1)
1. (7)	7	Oil	10.00 ± 1.24[b]	980.00 ± 37.65
2. (6)		OEB	2.50	390.00 ± 10.00
3. (7)	14	Oil	20.18 ± 3.71	975.00 ± 25.00
4. (6)		OEB	2.50 ± 0.11	518.33 ± 12.50[c]
5. (7)	21	Oil	62.56 ± 4.78	1106.00 ± 142.40
6. (6)		OEB	4.37 ± 0.89	687.50 ± 2.31
7. (7)	28	Oil	191.50 ± 22.12	1168.50 ± 94.30
8. (6)		OEB	8.94 ± 1.32	941.67 ± 46.39[d]

[a] Number of animals in brackets: [b] Values are means ± SE. [c] 4 Vs 3 P < 0.0005: [d] 8 Vs 7 P < 0.025.

for these experiments suppressed LH secretion almost completely both in animals which have rather low levels of serum LH (animals castrated for 7 or 14 days), and in animals which have extremely high levels of serum LH (animals castrated for 21 or 28 days). Oestrogens exhibit a completely different pattern of activity on FSH. First of all oestradiol, even if given in the large amounts used in the present experiments, is not able to suppress totally serum FSH at any post-castration time considered. Moreover, the inhibitory effect exerted by oestradiol on FSH release shows a progressive decline as post-castration time increases. It is apparent from Table I that the same dose of oestradiol becomes less effective in reducing serum FSH levels at the time following ovariectomy progresses (Zanisi and Martini, 1975a).

The present experiments confirm that oestrogens are good suppressors of LH release in castrated female rats (Blake et al., 1972; Ajika et al., 1972; Swerdloff and Walsh, 1973). Moreover, they indicate that the time elapsed between ovariectomy and administration of oestrogens does not play a significant role in this phenomenon. On the contrary FSH secretion is less critically influenced by exogenous oestrogens, since serum levels of FSH have never been brought close to zero by the treatment at any post-castration time considered. Moreover, after ovariectomy, there is a progressive decline in the sensitivity of the oestrogen-sensitive feedback system which controls FSH secretion. This indicates that duration of castration prior to treatment is an important factor in oestrogen-induced FSH inhibition.

The observation that FSH is not totally suppressed by oestrogens is surprising, since serum levels of FSH (in contrast to what happens to serum levels of LH) rise immediately following castration (Zanisi and Martini, 1975b). This fact might indicate that FSH release is controlled by a strict ovary-dependent feedback mechanism. Moreover, after ovariectomy, serum levels of FSH rise only 3 or 4 times over control levels. Consequently, one would assume that it should be easier to suppress them, than to suppress LH levels, which, after castration, rise up to 70 times normal values (Zanisi and Martini, 1975b). The loss of responsiveness to oestrogens of the feedback elements which control FSH secretion as post-castration time progresses is also surprising. The reasons for this phenomenon are not understood at the moment. It is not clear either why such a progressive decline of responsiveness to oestrogen should exist only in the case of FSH and not in the case of LH.

It has not been fully determined so far whether oestrogens exert their feedback effects on the hypothalamus, on the anterior pituitary or on both sites. Receptor proteins binding oestrogens have been found to be present in both structures (McGuire and Lisk, 1969; Friend and Leavitt, 1972; Ginsburg et al., 1972; Kulin and Reiter, 1972; Luttge and Whalen, 1972; Kato, 1973). It is obvious that if one accepts the view that the feedback effect of oestrogens is mainly exerted at hypothalamic level, the data presented here are difficult to reconcile with the theory which postulates the existence of one single hypothalamic releasing hormone for the control of both pituitary gonadotrophins (Schally et al., 1971; Matsuo et al., 1971a,b). On the other hand, if one postulates that the elements sensitive to oestrogens are exclusively located in the anterior pituitary, it will become difficult to accept the theory that one single type of pituitary cell manufactures both gonadotrophins (Phifer et al., 1971, 1973).

III. EFFECTS OF PROGESTERONE AND OF ITS 5α-REDUCED METABOLITES ON LH AND FSH SECRETION

It has recently been reported that in several of its target structures (uterus, anterior pituitary, hypothalamus, etc.) progesterone is metabolised into the following 5α-reduced metabolites: 5α-pregnan-3,20-dione (dihydroprogesterone, DHP) and 5α-pregnan-3α-ol-20-one (3α-ol) (Armstrong and King, 1971; Karavolas and Herf, 1971; Massa et al., 1972a; Robinson and Karavolas, 1973). The hypothesis has been put forward that the formation of these compounds represents an essential step for progesterone to exert its biological effects. Such a hypothesis is similar to the one proposed for explaining the mode of action of testosterone (Baulieu et al., 1968; Gloyna and Wilson, 1969; Robel, 1971; Massa et al., 1972b). Progesterone (P) is known to influence the secretion of anterior pituitary gonadotrophins. Depending on the circumstances, P may exert either inhibitory (negative feedback effect) or stimulatory (positive feedback or facilitatory effect) actions (Motta et al., 1970; Neill and Smith, 1974). The experiments to be described here have been planned in order to validate the hypothesis that DHP and 3α-ol might represent the mediators of P actions in the structures where the steroid exerts its feedback effects on LH and FSH secretion. The activities of DHP and 3α-ol have been evaluated in the rat, using two different experimental approaches, designed to test respectively the inhibitory and the facilitatory activities of these steroids on gonadotrophin secretion. The 3β epimer 3α-ol (5α-pregnan-3β-ol-20-one, 3β-ol) was also included in this study. Normally P-sensitive structures convert P into this metabolite to a very limited extent (Karavolas and Herf, 1971; Mickan, 1972; Robinson and Karavolas, 1973; Tabei and Heinricks, 1974). However, this steroid may play some physiological role since it is produced in the ovaries and secreted into the ovarian vein (Holzbauer, 1971).

In order to analyse whether DPH, 3α-ol and 3β-ol are able to stimulate gonadotrophin release, these steroids have been administered to castrated oestrogen-pretreated female rats. It is a well documented phenomenon that P and several physiological or synthetic progestogens (20α-dihydro-progesterone,

Table II. Effect of 100 µg/rat of progesterone (P), 5a-pregnane-3, 20-dione (dihydroprogesterone) DHP, 5a-pregnan-3a-ol-20-one (3a-ol), 5a-pregnan-3β-ol-20-one (3β-ol) on serum levels of LH and FSH of castrated female rats pretreated for 5 days with 0.4 µg ethinyl oestradiol (E-OE) and sacrificed 8 h after progestogen administration.

Treatment			ng LH/ml (NIH-LH-S17)	ng FSH/ml (NIAMD-RAT FSH-RP-1)
1. E-OE	(0.4 µg)	(18)^a	3.43 ± 0.39$^{b\ c}$	283.87 ± 7.95^j
2. E-OE + P	(100 µg)	(20)	8.49 ± 0.82^d	1209.46 ± 31.75^k
3. E-OE + DHP	(100 µg)	(18)	5.87 ± 1.06$^{e\ h}$	566.74 ± 35.01^l
4. E-OE + 3a-ol	(100 µg)	(16)	8.45 ± 1.16^f	603.11 ± 19.84^m
5. E-OE + 3β-ol	(100 µg)	(11)	16.33 ± 1.71$^{g\ i}$	515.93 ± 12.04^n
6. OIL		(24)	27.36 ± 1.73	1027.24 ± 40.00

a Number of animals in brackets: b Values are means ± SE: c 1 Vs 6: P < 0.0005: d 2 Vs 1: P < 0.0005: e 3 Vs 1: P < 0.0125: f 4 Vs 1: P < 0.0005: g 5 Vs 1: P < 0.0005: h 3 Vs 2: P < 0.025: i 5 Vs 2: P < 0.0005: j 1 Vs 6: P < 0.005: k 2 Vs 1: P < 0.0005: l 3 Vs 1: P < 0.0005: m 4 Vs 1: P < 0.0005: n 5 Vs 1: P < 0.0005.

17a-hydroxy-progesterone, etc.) exert a positive feedback effect on LH and/or FSH secretion in animals so prepared (Taleisnik et al., 1969; Taleisnik et al., 1970; Swerdloff et al., 1972; Kalra et al., 1972; Brown-Grant, 1974). In order to test their inhibitory activity on gonadotrophin secretion, DHP, 3a-ol and 3β-ol were administered in rather large amounts to castrated female rats without any oestrogen pretreatment.

All experiments have been performed in adult female rats of the Sprague-Dawley strain. They were submitted to castration when weighing 150 ± 10 grams. For studying the positive feedback effect, an approach similar to that adopted by Swerdloff et al. (1972) was used. Adult female rats were ovariectomised (regardless of the phase of the oestrous cycle) fifteen days prior to the beginning of the study and divided into six groups. Beginning on the fifteenth day after castration, five groups were submitted to treatment with Ethinyl Oestradiol (EOE). EOE was dissolved in sesame oil, and injected subcutaneously for four consecutive days at 8.30 a.m. in a daily dose of 0.4 µg/rat. The sixth group was similarly injected for 4 days with sesame oil. Treatment with these doses of EOE has been previously shown to bring back to precastration levels the elevated gonadotrophin titres of castrated animals (Swerdloff et al., 1972). On the fifth day, the six groups of rats received the following subcutaneous treatments: Group 1 : 0.4 µg EOE; Group 2 : 0.4 µg EOE + 100 µg P; Group 3 : 0.4 µg EOE + 100 µg DHP; Group 4 : 0.4 µg EOE + 100 µg 3a-ol; Group 5 : 0.4 µg EOE + 100 µg 3β-ol; Group 6 : sesame oil. The animals were sacrificed by decapitation the same day at 5.30 p.m. Blood was collected in order to evaluate serum levels of LH and FSH according to the procedures of Niswender et al. (1968) and of Daane and Parlow (1971). In order to study the negative feedback effect of the various steroids, adult female rats were ovariectomised (regardless of the phase of oestrous cycle) 21 days prior to the beginning of the study, and divided into 5 groups. On day 21 after castration, the animals were injected subcutaneously at 8.30 a.m. in the following fashion: Group 1 : 2 mg P; Group 2 : 2 mg DHP; Group 3 : 2 mg 3a-ol; Group 4 : 2 mg 3β-ol; Group 5 : sesame oil. The animals were killed 24 hours later by decapitation. Blood was collected in order to evaluate plasma levels of LH and FSH.

The results in Table II summarise the data obtained in the experiments aimed at evaluating the possibility that DHP, 3a-ol and 3β-ol might facilitate the release

Table III. Effect of 2 mg/rat of progesterone (P), 5α-pregnane-3, 20-dione (dihydroprogesterone, DHP), 5α-pregnan-3α-ol-20-one (3α-ol), 5α-pregnan-3β-ol-20-one (3β-ol) on serum levels of LH and FSH of female rats castrated since 3 weeks and sacrificed 24 h after progestogen administration.

Treatment			ng LH/ml (NIH-LH-S17)	ng FSH/ml (NIAMD-RAT FSH-RP-1)
1. P	(2 mg)	(26)[a]	21.06 ± 0.97[b]	939.99 ± 26.82
2. DHP	(2 mg)	(21)	15.21 ± 0.79[c]	647.03 ± 17.78[g]
3. 3α-ol	(2 mg)	(25)	11.48 ± 0.97[d][f]	545.28 ± 12.75[h]
4. 3β-ol	(2 mg)	(20)	16.94 ± 0.91[e]	1017.47 ± 27.27
5. OIL		(23)	21.95 ± 0.83	916.89 ± 27.15

[a] Number of animals in brackets: [b] Values are means ± SE: [c] 2 Vs 5: $P < 0.005$: [d] 3 Vs 5: $P < 0.0005$: [e] 4 Vs 5: $P < 0.0005$: [f] 3 Vs 2: $P < 0.0005$: [g] 2 Vs 5: $P < 0.0005$: [h] 3 Vs 5: $P < 0.005$.

of pituitary gonadotrophins. It appears that castrated oil-treated female rats have rather elevated levels of serum LH. In agreement with previous data these are significantly depressed by the injection of EOE (Swerdloff et al., 1972). The injection of P on the fifth day of oestrogen priming is followed by a significant increase of serum levels of LH. In animals given P serum levels of LH are more than twice as high as those found in animals not treated with P. DHP, 3α-ol and 3β-ol are all able to significantly elevate serum levels of LH over those observed in animals given EOE alone. In this test, 3α-ol appears to be at least as effective as P. DHP, on the contrary, is less effective than P. The activity of 3β-ol is higher than that of any other progestational agent tested.

Table II summarises also the data obtained in the same group of animals by measuring plasma FSH. First of all it is apparent that the elevated levels of FSH found after ovariectomy can be depressed by the administration of five consecutive daily injections of small doses of EOE. This is confirmatory of previous data (Swerdloff et al., 1972). It is also clear that under the experimental conditions selected, P is able to exert a strong positive feedback effect on the release of FSH. The FSH values in EOE + P-treated animals are even higher than those found in control animals submitted to castration and treated with oil. DHP, 3α-ol and 3β-ol all exert a stimulating effect on FSH release; such an effect is less pronounced than that of P, and is quantitatively similar for the three steroids.

Table III summarises the data obtained in the experiments in which the negative feedback effect of the different progestational agents on gonadotrophin secretion has been evaluated. It is evident that P, under the conditions of the present experiment, does not inhibit LH release. On the contrary, DHP, 3α-ol and 3β-ol are all able to depress significantly serum levels of LH. 3α-ol seems to be more active than either DHP or 3β-ol. P and 3β-ol do not diminish serum levels of FSH in castrated female rats. On the contrary, DHP and 3α-ol exert a rather strong inhibiting effect on FSH release, 3α-ol being more active than DHP in this respect.

The results have indicated that DHP and 3α-ol, the 5α-reduced metabolites formed from P in the central structures (anterior pituitary, hypothalamus, etc.), exert a positive feedback effect on LH and FSH release in castrated female rats primed with small doses of oestrogen. These observations are compatible with the hypothesis that P might exert its facilitatory effect on gonadotrophin secretion after the local (intrahypothalamic, intra-pituitary, etc.) conversion into DHP and/or 3a-ol. The process of 5a-reduction seems to be more important for the appearance of the positive feedback effect of P on LH than on FSH secretion. This conclusion is based on the observation that 3a-ol is as effective as P on LH release, while neither DHP nor 3a-ol equals the activity of P on FSH release.

The observation that 3β-ol (the 3β-epimer of 3a-ol) shows a considerable LH-releasing activity and exerts some effect also on FSH secretion is interesting, and may have physiological relevance even if, as previously mentioned, this steroid is not normally formed from P by the central target structures sensitive to the hormone (Karavolas and Herf, 1971; Mickan, 1972; Robinson and Karavolas, 1973; Tabei and Heinricks, 1974). It must be recalled, however, that 3β-ol is synthesized in rather large amounts by the rat ovary, and that its concentration in the gland and its secretion rates into the ovarian vein, vary under different physiological conditions. In particular the secretion of this steroid seems to exhibit a peak in the afternoon of prooestrus at the time of the gonadotrophin surge (Holzbauer, 1971; Zmigrod et al., 1972). It is consequently likely that this compound might play an important role in the processes leading to gonadotrophin release reaching the hypothalamic-pituitary complex via the systemic circulation.

In the studies devoted to analysing the negative feedback activity of the different steroids it has been found that P is totally ineffective in suppressing LH and FSH secretion in castrated rats when given in the absence of oestrogens. This finding agrees with previous evidence (McCann, 1962; Nallar et al., 1966; Caligaris et al., 1971; Swerdloff et al., 1972; Tapper et al., 1972; McPherson et al., 1974). On the contrary DHP and 3a-ol exert some inhibitory effect on gonadotrophin release. There are several possible interpretations of these results. One possibility is obviously that P operates in the negative feedback system controlling gonadotrophin secretion only after conversion into DHP and 3a-ol. If one accepts such an interpretation, the inactivity of P here reported might be explained by the fact that the rather short interval (24 h) used in the present experiments did not allow sufficient conversion of the hormone into its "active" 5a-reduced metabolites. However, such a possibility appears to be a rather remote one, since it has been demonstrated that the 5a-reductase activity of the anterior pituitary and of the hypothalamus increases several-fold three weeks after castration (Massa et al., 1972b; Denef et al., 1973, 1974). Consequently, in the present experimental conditions a rather large conversion of P into DHP and 3a-ol could be expected. Another possibility is that P exerts its negative effects (in the experimental conditions in which they appear) (Davidson, 1969) acting as such and without needing the reduction into DHP and 3a-ol. For the reasons just mentioned, the authors would favour such a view. In case such an interpretation should be proved valid by the additional experiments which are presently in progress in the Milan laboratory, the data presented here might provide a preliminary answer to the question of the opposite effects P exerts on gonadotrophin secretion. Apparently, the positive effect seems to be due to the conversion into the 5a-reduced metabolites, while the negative feedback effect appears to be a property of P as such.

IV. SUMMARY AND CONCLUSIONS

A. Oestrogens exert a strong suppressory effect on serum LH in castrated female rats. Such an effect does not seem to be influenced by the interval of time which has elapsed between castration and administration of the hormone. On the contrary, oestrogens are not able to suppress totally serum FSH in castrated females. Moreover, the inhibitory effect exerted by oestrogens on FSH release shows a progressive decline as post-castration time increases.

B. In castrated oestrogen-primed female rats DHP, 3α-ol and 3β-ol exert a positive feedback effect on LH and FSH secretion qualitatively similar to that of P. This observation is compatible with the hypothesis that P exerts its facilitatory effects on gonadotrophin secretion after the local (intra-hypothalamic and intra-pituitary) conversion into DHP and 3α-ol. The LH- and FSH-releasing activity of 3β-ol is particularly interesting, since this steroid is not a regular metabolite of P in the P-sensitive structures. However, 3β-ol is formed in rather large amounts in the ovary. Consequently it is possible to postulate that this compound might play a physiological role in the processes leading to gonadotrophin release reaching the hypothalamic-pituitary complex via the systemic circulation.

C. P is ineffective in suppressing LH and FSH secretion in castrated rats when given without oestrogens. On the contrary, DHP, 3α-ol and 3β-ol exert some inhibitory effect on gonadotrophin secretion under the same experimental conditions. A possible interpretation of these results is presented and discussed in the text.

D. The experiments presented underline once more the fact that the mechanisms which control LH and FSH secretion are basically different.

ACKNOWLEDGEMENTS

The experimental work described in this paper has been supported by grants from the Ford Foundation, New York, N.Y., the Population Council, New York, N.Y., and the Consiglio Nazionale delle Ricerche, Rome, Italy.

The assistance of Drs. A.R. Midgley, Jr., L.E. Reicher, Jr., G.D. Niswender, A.F. Parlow and of the National Institute of Arthritis Metabolism and Digestive Diseases, Rat Pituitary Hormone Program, in providing LH and FSH radioimmunoassay materials is gratefully acknowledged.

Thanks are also due to Mr. M. Vigliani and Mr. L. Guadagni for their skilful technical assistance.

REFERENCES

Ajika, K., Krulich, L., Fawcett, C.P. and McCann, S.M. (1972). *Neuroendocrinology* 9, 304-315.
Armstrong, D.T. and King, E.R. (1971). *Endocrinology* 89, 191-197.
Baulieu, E.E., Lasnitzki, I. and Robel, P. (1968). *Nature* 219, 1155-1156.
Blake, C.A., Norman, R.L. and Sawyer, C.H. (1972). *Proc. Soc. exp. Biol. Med.* 141, 1100-1103.
Brown-Grant, K. (1974). *J. Endocrinol.* 62, 319-332.
Caligaris, L., Astrada, J.J. and Taleisnik, S. (1971). *Endocrinology* 89, 331-337.
Davidson, J.M. (1969). *In* "Frontiers in Neuroendocrinology" (W.F. Ganong and L. Martini, eds) pp.343-388. Oxford University Press, New York.
Daane, T.A. and Parlow, A.F. (1971). *Endocrinology* 88, 653-663.
Denef, C., Magnus, C. and McEwen, B.S. (1973). *J. Endocrinol.* 59, 605-621.
Denef, C., Magnus, C. and McEwen, B.S. (1974). *Endocrinology* 94, 1265-1274.
Friend, J.P. and Leavitt, W.W. (1972). *Acta Endocr.* 69, 230-240.

Ginsburg, M., Morris, I.D., MacLusky, N.J. and Thomas, P.J. (1972). *J. Endocrinol.* **55**, XX.
Gloyna, R.E. and Wilson, J.D. (1969). *J. clin. Endocr. Metab.* **29**, 970-977.
Holzbauer, M. (1971). *Br. J. Pharmacol.* **43**, 560-569.
Kalra, P.S., Kalra, S.P., Krulich, L., Fawcett, C.P. and McCann, S.M. (1972). *Endocrinology* **90**, 1168-1174.
Karavolas, H.J. and Herf, S.M. (1971). *Endocrinology* **89**, 940-942.
Kato, J. (1973). *Acta Endocr.* **72**, 663-670.
Kulin, H.E. and Reiter, E.O. (1972). *Endocrinology* **90**, 1371-1374.
Luttge, W.G. and Whalen, R.E. (1972). *J. Endocrinol.* **52**, 379-395.
Massa, R., Stupnicka, E. and Martini, L. (1972a). *Excerpta Medica Intern. Congr. Ser.* **256**, 118.
Massa, R., Stupnicka, E., Kniewald, Z. and Martini, L. (1972b). *J. Steroid Biochem.* **3**, 385-399.
Matsuo, H., Arimura, A., Nair, R.M.G. and Schally, A.V. (1971a). *Biochem. Biophys. Res. Commun.* **45**, 822-827.
Matsuo, H., Baba, Y., Nair, R.M.G., Arimura, A. and Schally, A.V. (1971b). *Biochem. Biophys. Res. Commun.* **43**, 1334-1339.
McCann, S.M. (1962). *Am. J. Physiol.* **202**, 601-604.
McGuire, J.L. and Lisk, R.D. (1969). *Neuroendocrinology* **4**, 289-295.
McPherson, J., Costoff, A., Eldridge, J.C. and Mahesh, V.B. (1974). *Fertil. Steril.* **25**, 1063-1070.
Mickan, H. (1972). *Steroids* **19**, 659-668.
Motta, M., Piva, F. and Martini, L. (1970). *Bull. Swiss Acad. Med. Sci.* **25**, 408-418.
Nallar, R., Antunes-Rodriguez, J. and McCann, S.M. (1966). *Endocrinology* **79**, 907-911.
Neill, J.D. and Smith, M.S. (1974). *In* "Current Topics in Experimental Endocrinology" (V.H.T. James and L. Martini, eds) Vol.II, pp.73-106. Academic Press, New York and London.
Niswender, G.D., Midgley, A.R., Jr., Monroe, S.E. and Reichert, L.E., Jr. (1968). *Proc. Soc. exp. Biol. Med.* **128**, 807-811.
Phifer, R.F., Midgley, A.R., Jr. and Spicer, S.S. (1971). *In* "Gonadotropins" (B.B. Saxena, C.G. Belling and H.M. Gandy, eds) pp.9-25. Wiley Interscience, New York.
Phifer, R.F., Midgley, A.R., Jr. and Spicer, S.S. (1973). *J. clin. Endocr. Metab.* **36**, 125-141.
Robel, P. (1971). *Acta Endocr.*, Suppl. **153**, 279-294.
Robinson, J.A. and Karavolas, H.J. (1973). *Endocrinology* **93**, 430-435.
Schally, A.V., Arimura, A., Baba, Y., Nair, R.M.G., Matsuo, H., Redding, T.W., Debeljuk, L. and White, W.F. (1971). *Biochem. Biophys. Res. Commun.* **43**, 393-399.
Swerdloff, R.S. and Walsh, P.C. (1973). *Acta Endocr.* **73**, 11-21.
Swerdloff, R.S., Jacobs, H.S. and Odell, W.D. (1972). *Endocrinology* **90**, 1529-1536.
Tabei, T. and Heinricks, W.L. (1974). *Neuroendocrinology* **15**, 281-289.
Taleisnik, S., Caligaris, L. and Astrada, J.J. (1969). *J. Endocrinol.* **44**, 313-321.
Taleisnik, S., Caligaris, L. and Astrada, J.J. (1970). *J. Reprod. Fertil.* **22**, 89-97.
Tapper, C.M., Naftolin, F. and Brown-Grant, K. (1972). *J. Endocrinol.* **53**, 47-57.
Zanisi, M. and Martini, L. (1975a). *Acta Endocr.* **78**, 689-694.
Zanisi, M. and Martini, L. (1975b). *Acta Endocr.* **78**, 683-688.
Zmigrod, A., Lindner, H.R. and Lamprecht, S.A. (1972). *Acta Endocr.* **69**, 141-152.

SELECTED ASPECTS OF CONTROL OF LH AND FSH SECRETION IN WOMEN

W.D. Odell, R.S. Swerdloff and F. Wollesen

*Harbour General Hospital, UCLA School of Medicine
1000 Carlson Street, Torrence, California 90509, USA*

I. THE SYSTEM

Figure 1 depicts a schematic presentation of the central nervous system-pituitary-ovarian relationships. This simplified schematogram can be used to understand present concepts of control of ovarian function. Some of the enclosed facets or boxes of the schematogram may be further subdivided into subanatomical or subendocrine units, based upon present understanding. This is a closed-loop feedback system, with each part of the system relating to other parts in a controlled fashion.

II. HYPOTHALAMIC CONTROL OF PITUITARY FUNCTION

It is currently believed that the ten amino acid, Gonadotrophin Releasing Hormone (GnRH), modulates control over pituitary synthesis and secretion of both Luteinizing Hormone (LH) and Follicle Stimulating Hormone (FSH). For example (Amoss *et al.*, 1971; Matsuo *et al.*, 1971; Milhaud *et al.*, 1971; Schally *et al.*, 1971; Besser *et al.*, 1972; Job *et al.*, 1972; Job *et al.*, 1972a; Kastin *et al.*, 1972; McNeilly *et al.*, 1972; Von zur Muhlen and Kobberling, 1972; Yen *et al.*, 1972; and Nakano *et al.*, 1973), all of these publications indicate that GnRH releases LH in much greater quantities than FSH (using the International Reference Preparation, HMG No.2, LH:FSH ratio = >1.0). Separate and specific Luteinizing Hormone Releasing Hormone and Follice Stimulating Releasing Hormone have not been described to date. Yet, certain data in normal and pathophysiological observations, and under experimental conditions, indicate that the ratio of FSH to LH may vary, so that LH:FSH = <1.0. Such variations do not appear

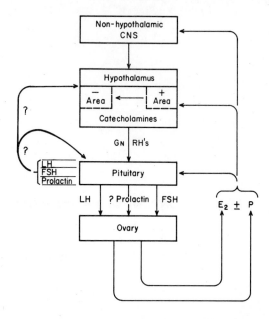

Fig.1 Simplified schematogram of the endocrine reproductive system in women. E_2 indicates estradiol, P indicates progesterone, $E_2 \pm P$ indicates estradiol acts at times synergistically with progesterone. Within the hypothalamus,- area indicates negative feedback area, i.e., $E_2 \pm P$ suppresses LH and FSH secretion; + area indicates positive feedback area, i.e. $E_2 \pm P$ stimulates LH and FSH secretion. The short-loop feedback of LH and FSH appears to exist, but it is not clear whether LH acts only via the hypothalamus or also directly on the pituitary. Prolactin *per se* appears to modify LH and FSH secretion, but again its locus of action is uncertain. Lastly, the role (if any) of prolactin in controlling ovarian function in women is uncertain.

to be observed when infusion or injection of the decapeptide is performed.

Figures 2AB depict the response for three doses of GnRH administered to women during the follicular phase, mid-cycle phase, and luteal phase of the menstrual cycle on the secretion of LH and FSH, respectively (Wollesen *et al.*, 1975). Note that under these conditions in women (as well as men) GnRH releases predominately LH, and to a much less extent, FSH. The ratio of LH and FSH remains > 1 throughout these studies performed in different phases of the menstrual cycle. In separate and extensive studies in men, we have shown that the dose-response curve for GnRH exists between 1 and 3,000 μg of GnRH (Wollesen *et al.*, 1975a). Figures 3AB depict these studies in men.

Figures 4AB depict dose-response curves for LH and FSH at these three phases of the menstrual cycle. It is to be noted from these figures that the slopes of the lines describing the dose-response relations of GnRH versus LH response, are different for the follicular phase, luteal phase, and mid-cycle phase of the cycle. For the FSH response, slopes are different for follicular phase compared to luteal phase. Furthermore, the LH:FSH ratios remain relatively constant after GnRH administration during these three phases. These data indicate that although present information suggests that the ten amino acid, GnRH, is the only hypothalamic peptide controlling pituitary gonadotrophin secretion, either modulation

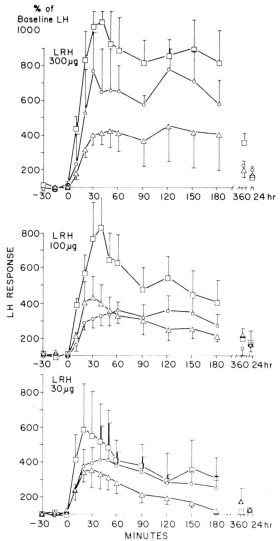

Fig. 2A The LRH-LH time response curves for normal, fertile women at three different doses. For each subject, the responses have been corrected for covariance of body surface area and expressed as a percent of the mean of three preinjection values. N for each mean and standard error is 4. (△ = follicular phase; ○ = preovulatory phase; □ = luteal phase.)

of pituitary function or other hypothalamic hormones controlling gonadotrophin secretion exist. However, these studies were done in normal women at different phases of the menstrual cycle and those conditions which should modulate pituitary function (e.g., changing blood steroid concentrations) must exist at these three phases of the cycle, when the exogenous GnRH was injected.

The observed ratios of LH and FSH after administration of GnRH are very similar to those observed after the ovulatory surge at mid-cycle in normal women

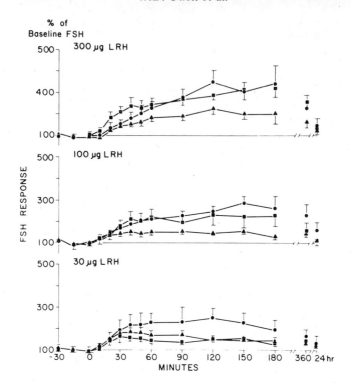

Fig. 2B The LRH-FSH time-response curves for normal, fertile women at three different phases of the menstrual cycle and at three different doses. For each subject the responses have been corrected for covariance of body surface area and expressed as percent of the mean of 3 preinjection values. N for each mean and standard error is 4. (▲ = follicular phase; ● = preovulatory phase; ■ = luteal phase.)

or animals. Thus, it is attractive to believe that the mid-cycle ovulatory surge is caused by a brief sustained secretion of GnRH. However, how the early follicular phase pattern of LH and FSH is produced remains unknown. These data show that GnRH secretion is unlikely to be the cause. Lastly, since sensitivity to GnRH is greatest during the luteal phase, and yet LH and FSH concentrations in blood are lowest, endogenous GnRH secretion must be minimal during the luteal phase.

One might ask whether response to a sustained infusion of GnRH might modify the response in terms of LH:FSH. In separate studies we and others have shown that infusions do not appear to modify the relative amounts of LH and FSH; i.e., an infusion does not result in LH:FSH ratios of = 1 or of < 1.

In summary of this portion, it is evident that hypothalamic control of pituitary function is exerted by means of the secretion of hypothalamic neural humors into the portal circulation which cause stimulation or inhibition of secretion of pituitary gonadotrophins. The singly identified GnRH, the ten amino acid peptide, may not be the only neural hormone controlling pituitary secretion of LH and FSH; other hormones having *inhibitory function*, thus modifying the response to GnRH, or alternatively, separate hypothalamic stimulatory hormones

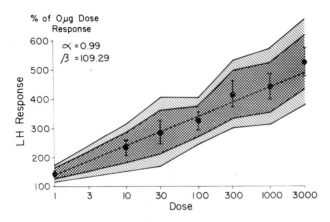

Fig.3A The LRH-LH dose-response curve for normal human adult males. For each subject the response has been defined as the area under the time-response curve from 0 to 180 minutes and has been expressed as the area under the time-response curve at the 0 μg dose. N for each mean and standard error is 8.

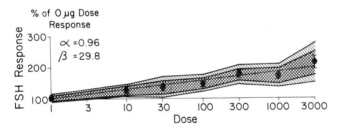

Fig.3B The LRH-FSH dose-response curve for normal human adult males. For each subject the response has been defined as the area under the time-response curve from 0 to 180 minutes and has been expressed as percent of the area under the time-response curve at 0 μg dose. N for each mean and standard error is 8.

for LH and FSH may well exist, based upon present data. It does not appear that ovarian modulation of pituitary response to the single known LHRH may explain known phenomena during the menstrual cycle.

III. LH AND FSH SECRETION

LH and FSH (and TSH) are glycopeptide hormones with molecular weight of approximately 28,000, and consisting of two peptide chains, designated, alpha and beta (Papkoff and Samy, 1967; Pierce *et al.*, 1971; Parlow and Shome, 1974).

These studies show that the alpha chain of LH, FSH, and TSH has an identical amino acid composition, while the beta chain of each is biochemically unique. LH and possibly FSH are secreted in a pulsatile fashion (Midgley and Jaffe, 1971; Yen *et al.*, 1972a) (see Fig.5). Apparently, the ovary responds to

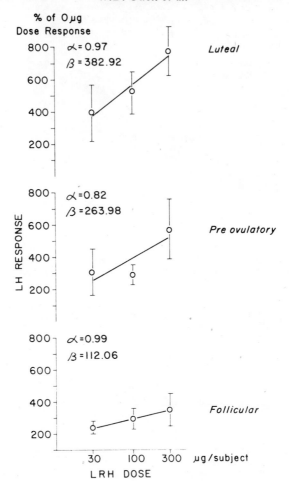

Fig. 4A The LRH-LH dose-response curves for normal fertile women at three different phases of the menstrual cycle. For each subject the responses have been corrected for covariance of body surface area, calculated as the area under the time-response curve from 0 to 180 minutes, and expressed as percent of the area under the baseline. N for each mean and standard error is 4. a = regression coefficient. β = slope.

the average or mean blood concentrations of both LH and FSH. The metabolic clearance rate (MCR) of LH is more rapid than that of FSH (Kohler *et al.*, 1968; Coble *et al.*, 1969). Disappearance curves for each hormone are not linear, and thus, strictly speaking, half-times of disappearance (T-1/2) may not be calculated. However, with this reservation, and based upon metabolism of radioiodinated human LH, LH has a very approximate half-time of disappearance of about one hour, while that for FSH is three to four hours. Thus, it is possible that pulsatile secretion of both LH and FSH occurs, but the smaller MCR of FSH results in dampening of such secretion making the pulsations less easily observed than those of LH. Furthermore, it is not known whether pulsatile GnRH secretion causes the pulsatile secretion of LH. It does not appear likely that the pulsatile secretion

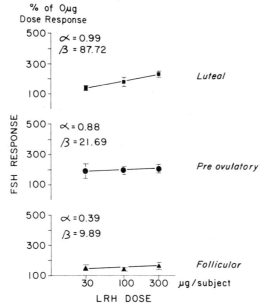

Fig. 4B The LRH-FSH dose-response curves for normal fertile women at three different phases of the menstrual cycle. For each subject the responses have been corrected for covariance of body surface area, calculated as the area under the time-response curve from 0 to 180 minutes, and expressed as percent of the area under the baseline. N for each mean and standard error is 4. a = regression coefficient. β = slope.

is a pituitary modulation of constant GnRH secretion, because infusions of GnRH do not result in pulsatile LH secretion (Edmonds et al., 1974).

Laburthe et al. (1973) and Rabinowitz et al. (1973) have demonstrated that free alpha chain of glycopeptide hormones is present in blood of postmenopausal women, and after GnRH administration to normal subjects, such alpha chain might result from direct secretion of alpha chain or from formation of alpha chain by metabolism of the intact hormone in the periphery. Figures 6AB show studies performed in our laboratory demonstrating that direct secretion of free alpha chain occurs (Edmonds et al., 1974 1974a; Edmonds et al., 1975, 1975a). In brief, when intact purified *human LH* was infused into eugonadal men, no changes in blood alpha chain concentrations were observed. On the other hand, when GnRH was administered to the same subjects at a different time to elevate blood LH, free alpha chain appeared in blood in relatively large concentrations. These studies demonstrate that free alpha chain is secreted in response to GnRH and that such alpha chains are not formed by degradation of the intact LH and/or FSH in peripheral tissues.

IV. FEEDBACK CONTROL OF THE HYPOTHALAMIC PITUITARY GONADAL AXIS

For many years it was assumed that an inherent central nervous system rhythmicity existed which resulted in periodic discharge of the ovulatory LH surge, and in turn, ovulation. In 1968 after studying over 50 normal menstrual

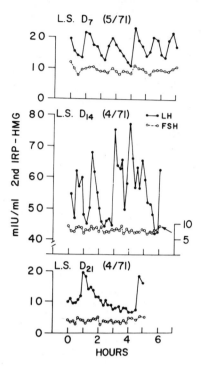

Fig.5 LH and FSH concentrations in a eugonadal woman on Day 7 (D_7), Day 14 (D_{14}), and Day 21 (D_{21}) of the menstrual cycle. Note the marked fluctuation in LH and the smaller fluctuations in FSH. (Reproduced from Yen *et al.*, 1972a.)

cycles in women, it became apparent that this might not be true. Three reasons for our conclusion existed: 1) Aberrant LH ovulatory-type surges never occurred during the normal menstrual cycle (Odell *et al.*, 1966; Odell *et al.*, 1967; Ross and Odell, 1967; Odell *et al.*, 1968). 2) When we suppressed castrate or postmenopausal women with generous doses of estrogens and continued such suppression for up to 60 days, aberrantly timed or sporadically occurring LH surges did not occur. This indicated that under these conditions in women an inherent rhythmicity did not exist. 3) If an engineer designed a control system for ovulation, he would ensure that a communication system existed to ensure that the ovulatory LH-FSH surge occurred only when a follicle was optimally ripe. For all of these reasons, we (Odell and Swerdloff, 1968) asked whether *an ovarian signal* might not be involved in timing of the ovulatory surge. By administering sequential estrogen plus progestogen medications to castrated or postmenopausal women, we produced LH-FSH surges which exactly mimicked the ovulatory surge, it was demonstrated that such a signal could occur. Such ovulatory type surges were produced by first suppressing the elevated LH and FSH concentrations with estrogen treatment, and then when a surge was desired, adding treatment with injected progesterone or oral medroxyprogesterone (Fig.7). In addition, studies from our laboratory first demonstrated that estrogens administered to premenopausal women stimulated LH, but not FSH secretion (Swerdloff and

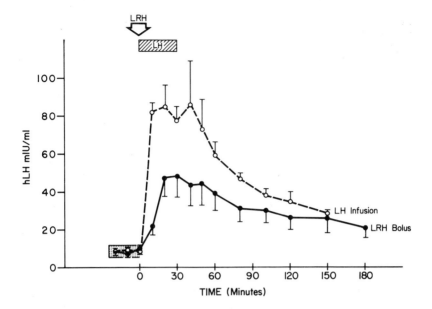

Fig. 6A LH response to LRH and LH administration. 100 μg of LRH given 0 min one day. On a separate day, 90 IU of LH given as bolus at 0 min followed by 22.5 IU over 30 min. Each point represents mean of four subjects and the bar represents ± SEM. Shaded area represents 95% confidence interval for control samples. (From Edmonds *et al.*, 1975a.)

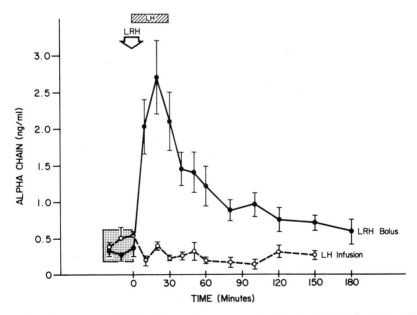

Fig. 6B Measurement of alpha chain on the same samples described in 6A. (Also from Edmonds *et al.*, 1975a.)

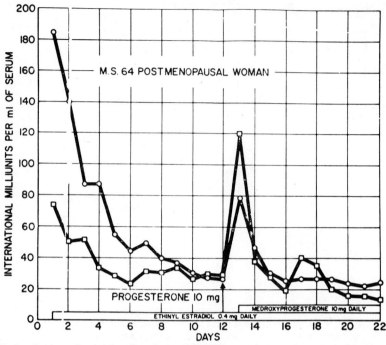

Fig. 7 Production of a simulated ovulatory LH-FSH surge in a postmenopausal woman. Changes in FSH and LH in a postmenopausal woman treated with ethinyl estradiol, days 1 to 22, with progesterone or medroxyprogesterone days 12 to 22. Note that on the control samples FSH is greater than LH (FSH:LH = > 1) and that with estrogen suppression FSH is ≈ LH. When progesterone was added, an LH-FSH surge occurred with FSH:LH < 1.0. In spite of continued exposure to E and P, the surge is self-terminating in 24 h, and LH and FSH are suppressed. (Reproduced from Odell and Swerdloff, 1968.)

Odell, 1969) (Fig.8A).

Furthermore, additional studies also demonstrated that progestogens administered alone to premenopausal women resulted in LH, but not FSH surge (Mishell and Odell, 1971) (Fig.8B). We were unable to demonstrate stimulation of LH and FSH by progestogens alone or estrogens alone administered to postmenopausal subject; this indicated that either the type of steroid acting in premenopausal women might be potentiated by endogenous basal ovarian steroid secretion. These studies were performed prior to those of Abraham and Klaiber (1970), which showed that estradiol concentrations in blood reached a peak prior to the LH-FSH ovulatory surge. The concentrations of LH, FSH, estradiol and progesterone are shown in Fig.9 from studies in our laboratory (Abraham *et al.*, 1972).

Subsequently, Yen and Tsai (1971) showed that estrogen administered to postmenopausal women by infusion would increase blood LH, but not FSH concentrations. Furthermore, studies in many animal species revealed that estrogens could stimulate LH secretion. These observations, coupled with those of Abraham and Klaiber (1970), Baird and Guevara (1969) and Tulchinsky and Korenman (1970), that estradiol reached a peak prior to the ovulatory surge, made estrogens

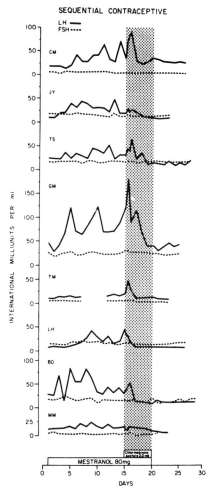

From Swerdloff & Odell
J Clin Endocr 29:157,
1969

Fig. 8A Daily LH and FSH concentrations in eight eugonadal women receiving a sequential contraceptive. In most women, LH secretion was stimulated during the mestranol treatment phase; note the very high irregular or bizarre pattern of LH secretion. When the progestogen (chlormadinone acetate) was added, most women had another LH peak and then LH concentrations fell to low concentrations. FSH concentrations were suppressed throughout estrogen alone and progestogen plus estrogen treatment. (Note the difference in response of these eugonadal women compared to the postmenopausal woman shown in Fig. 7. Reproduced from Swerdloff and Odell, 1969.)

a good candidate for the ovarian signal system. To support this hypothesis further, Ferin *et al.* (1969) demonstrated that antiestradiol antiserum administered to cycling rats prevented ovulation; antiprogesterone antibodies failed to block ovulation. We (Swerdloff *et al.*, 1972) restudied the phenomenon in rats and

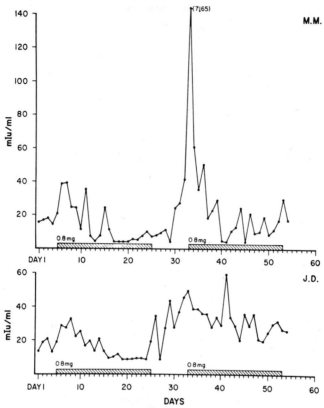

Fig.8B Serum LH concentrations in eugonadal women receiving only progestogen treatment. Note that high, spiking LH peaks were produced. The response in a large series of women studied may be summarized: high doses (2 mg) generally suppressed LH concentrations; progressively lower doses (to 0.1 mg) generally stimulated LH concentrations. (Reproduced from Mishell and Odell, 1971.)

demonstrated that large doses of estradiol benzoate administered to castrate rats would at first suppress LH, and then at day 4 (in rats that had 4-day cycles prior to castration) resulted in a typical LH-type ovulatory surge. Such estradiol dosages did not stimulate FSH during this LH surge, but (for FSH) showed only a suppressive action throughout the days of treatment. Lower doses of estradiol benzoate caused suppression of LH without stimulatory actions at day 4. However, if progesterone treatment were added at day 3 to the suppressive dose of estradiol benzoate, a typical ovulatory LH surge, *accompanied by an FSH surge* occurred. These findings are summarized in Figs 10ABC. Thus, we concluded that progesterone lowered the threshold for estrogen stimulation of LH, and was required for the concomitant FSH stimulation. Subsequently, Barraclough (1974) has shown that if all sources of progesterone are removed from the rat (bilateral adrenalectomy and oophorectomy) that estradiol alone will not result in ovulatory surges. In a recent workshop discussion (Odell, 1975) several animal species were discussed; review of existing data revealed that progesterone is probably a component of the ovulatory surge in all species with the exception of the Rhesus

Selected Aspects of Control of LH and FSH Secretion in Women

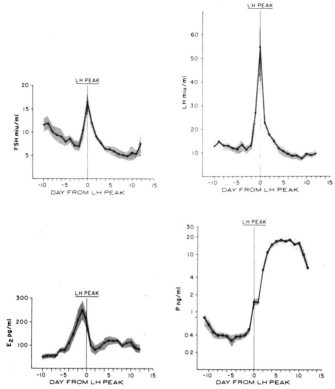

Fig. 9 Pattern of FSH, LH, estradiol (E_2) and progesterone (P) during the menstrual cycle of normal women. All data centered on the time of the LH peak for plotting purposes. Progesterone concentrations are shown in log scale to permit demonstration of changes in follicular phase concentrations. Dash line indicates average, shaded area SEM.

monkey. In this species, Knobil et al. (1974) have demonstrated that estradiol alone will result in LH-FSH ovulatory surges. Continued study of the interplay of progesterone and estradiol modifications of pituitary gonadotrophin secretion is warranted, but based on present data, one may conclude: 1) That an ovulatory surge is triggered by an ovarian signal which occurs at the time the follicle is fully developed; 2) that a major component of this signal is estradiol; 3) that such an estrogenic signal appears to be modified by a "fail-safe mechanism" consisting of progestogen secretion; and 4) that if the data of Barraclough apply to women, it is possible that some basal secretion of progestogens is necessary for the positive feedback of estrogens.

Although all of these data indicate that an ovarian signal controls the ovulatory LH-FSH surge, in general, the possibility that an inherent central-nervous-system rhythmicity exists has been discounted. However, Legan and Karsch (1975) published a most interesting paper which indicates that at least in rats a component of inherent rhythmicity does exist. Figure 11 depicts their data. These workers implanted estradiol containing silastic capsules into castrate female rats. Each afternoon the animals had an LH surge; such surges repetitively occurred for at least 9-10 days, even though estradiol concentrations were relatively constant throughout.

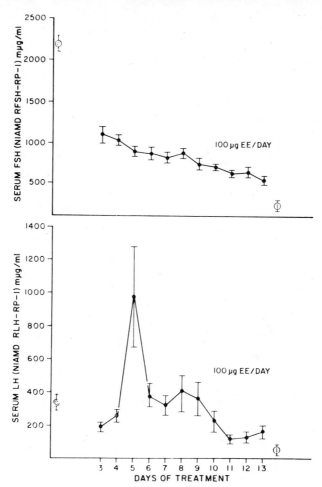

Fig.10A Serum FSH (above) and LH (below) during 13 days treatment with 100 μg/day of EE to oophorectomized rats. A marked rise in serum LH but not FSH was seen on day 5 of treatment. ⓒ = castrate controls: Ⓘ = intact controls. (From Swerdloff et al., 1972.)

Some data indicate that prolactin may modify LH and FSH secretion or modify the action of LH and FSH on the ovary. Thus, in women with persistent galactorrhea-amenorrheic syndromes, inhibition of prolactin secretion by ergot derivatives, results in resumption of menstrual cycles (Rolland et al., 1974; Zarate et al., 1972). A number of possible explanations exist. Zarate et al. (1972) demonstrated that gonadotrophins were ineffective if administered to lactating women in the postpartum period. They postulated prolactin had an antigonadotrophin action. It is also possible that prolactin *per se* inhibits LH and FSH secretion either at a hypothalamic level or pituitary level in so-called short-loop feedback.

A variety of indirect and direct studies in rodents have indicated that LH may feedback to suppress LH secretion and FSH feedback to suppress FSH secretion (Molitch et al., 1975). We (David et al., 1966; Corbin and Cobin, 1966;

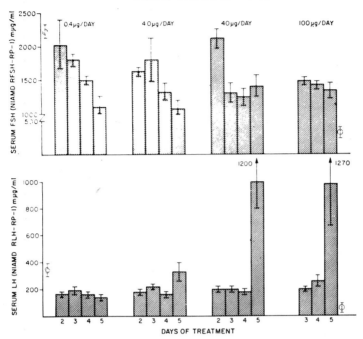

Fig. 10B Serum FSH (above) and LH (below) in oophorectomized rats treated daily with EE in dose range from 0.4 μg/day-100 μg/day. Serum LH and FSH levels fell with all doses of EE. A biphasic response of LH with an initial fall followed by a sharp rise was seen after the two highest doses (40 μg/day) of EE. Ⓒ = castrate concentrations; Ⓘ = intact concentrations. (From Swerdloff *et al.*, 1972.)

Corbin and Story, 1967; Molitch *et al.*, 1975) have recently directly studied this phenomenon in rabbits using a different approach. The experimental design was based on the fact that our rabbit LH RIA is specific for rabbit LH and shows no cross-reaction with human LH or hCG. Similarly, our human LH radioimmunoassay shows no reaction with rabbit LH. In short, when we infused purified human LH or hCG intravenously into male or female rabbits, castrated for two weeks or less, endogenous rabbit LH secretion was promptly suppressed. Interestingly, no suppression occurs if animals castrated for four weeks or longer are studied. These studies are illustrated by Fig.12, and indicate that in the short-term castrate (and presumably intact), but not the long-term castrate animal, a short-loop feedback for LH exists. It is attractive to think that this short-loop feedback is functionally important in the normal animal — possibly in assisting in termination of the mid-cycle ovulatory surge. Not only is it unknown if the short-loop is physiologically important, but also whether it acts at a pituitary or hypothalamic level.

V. INTEGRATION OF THE REPRODUCTIVE SYSTEM

The initiation of follicle growth at the beginning of a menstrual cycle appears to be the resultant of a rise in blood FSH concentrations and a fall in estrogen

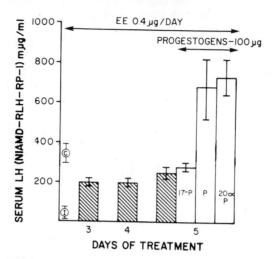

Fig. 10C Serum LH after EE + either 17P, P or 20∝P administration to oophorectomized rats. Hatched bars represent serum LH on days 3, 4, and 5 of 0.4 μg of EE given above. Open bars represent serum LH when 17∝P, 20∝P was added to EE treatment on day 5. Ⓒ = castrate control; Ⓘ = intact control. Serum LH after both P and 20P was significantly elevated above EE treatment alone ($p < 0.01$). (From Swerdloff *et al.*, 1972.)

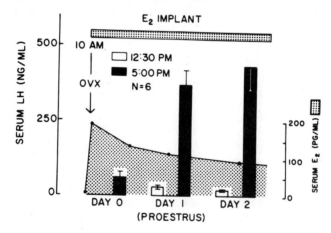

Fig. 11 Induction of ovulation in a hypophysectomized patient (SB) by treatment with HMG and hCG. (From Vande Wiele *et al.*, 1970.)

concentrations. As a result, a *group* of perhaps 6-12 follicles begins to develop. As local ovarian estrogen concentrations increase, one of the follicles continues to develop while all the remainder undergo atrophy. There is some uncertainty as to the need for continued increased FSH secretion to stimulate continued follicle development. Crooke *et al.* (1968) have shown that for some infertile women, only a short period of FSH treatment is required to initiate follicle growth. Afterwards, without additional FSH treatment, the follicle continues to develop to the preovulatory stage. Studies have not been performed in women totally devoid of FSH.

Selected Aspects of Control of LH and FSH Secretion in Women   105

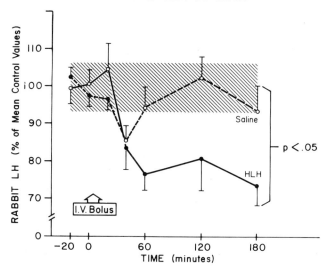

Fig. 12 Demonstration of short-loop feedback in castrated rabbits. Rabbit LH measured by specific RIA which does not cross-react with human LH; 10 μg of human LH given as bolus injection at time 0. The hatch area represents 95% confidence limits for the control values prior to human LH injection. The same animals were studied after saline injection as a control on another occasion, and all studies were performed on the unanesthetized animal using chronic indwelling catheterization. (Reproduced from Molitch et al., 1975.)

In normal women, it is presumed that the falling blood estrogen and progestogen concentrations, during the end of the luteal phase, are the cause of the increased FSH secretion initiating follicle growth. However, as discussed earlier, the hypothalamic-pituitary response to this fall in steroid concentrations may not act solely via increased GnRH secretion; the LH:FSH concentrations during this early follicular phase are < 1.0.

It is not certain what factors control the development of one follicle, while resulting in atresia of the remaining 6-12 that have initially begun development. One factor may be local ovarian estrogen concentrations. Harmon and Ross (1974) have shown that in rats, when estrogen effects are inhibited by antiestrogen treatment, FSH increases follicle atresia; when estrogens effects are increased follicle growth is promoted. Thus, local estrogen concentrations may determine whether FSH induces follicle maturation or atresia.

As the follicle develops to the preovulatory phase, increasing estradiol and 17-hydroxyprogesterone secretion occurs. Acting as an ovarian signal either alone, or modified by the increasing progesterone concentrations, which closely parallel the LH increase, an LH-FSH ovulatory surge occurs. This surge stimulates ovulation.

Subsequent to ovulation, the corpus luteum develops. This histological transformation follows ovulation and based upon *in vitro* studies requires at least small amounts of LH. Van Thiele *et al.* (1971) and Channing and Kammerman (1971) showed that culture porcine granulosa cells luteinize in the presence of small amounts of LH. In addition, monkey follicles induced by FSH (or hCG) *in vivo*, will luteinize *in vitro* if LH is present, but not if LH or hCG is absent. Pre-

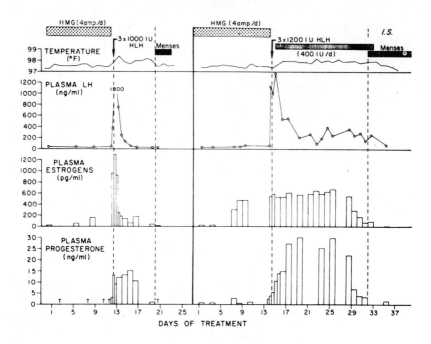

Fig.13 Demonstration of LH support of corpus luteum function in an infertile woman. In the left-half is depicted results of stimulation of follicular maturation with FSH rich gonadotropin (HMG) followed by ovulation induction with a single injection of 3,000 IU of human LH. As shown by progesterone concentrations and onset of menses, corpus function continued for only 4-6 days. In the right-half, the results show that if ovulation is induced with 3,600 IU of human LH, and additionally 400 IU daily of LH are continued, corpus luteum function continued for 15-19 days. (Reproduced from Vande Wiele *et al.*, 1970.)

sumably, LH or hCG is also a necessary component of luteinization in women.

Once the corpus luteum develops, its functional life span appears to be, in part, intrinsic and, in part, modifiable by the very low blood LH concentrations found after the LH-FSH surge. Vande Wiele *et al.* (1970) have shown that if ovulation is induced with LH* treatment in infertile women, the corpus luteum functions only a few days. In contrast, if small amounts of LH are administered daily, after ovulation is induced, corpus luteum functions persist for 12-18 days or longer. Thus, at least in part, corpus luteum life span is modifiable by the continual low blood concentrations found during the lutel phase of the normal menstrual cycle (Fig.13). Of course, if conception occurs, the rapidly rising human chorionic gonadotrophin concentrations in blood greatly prolong the corpus luteum life span.

During the corpus luteum, the combined effects of increased blood progesterone and estradiol suppresses blood LH and FSH to values below those observed in the follicular phase, when progesterone concentrations are not highly elevated (Odell *et al.*, 1967; Odell *et al.*, 1968). Similarly, estrogens, administered alone to castrate or postmenopausal women, do not suppress LH and FSH to as

---

* As opposed to the very long half-lived human chorionic gonadotrophin.

low concentrations as does administration of both estrogens and progestogens. It remains very poorly understood how progestogens can, at times, augment the stimulatory effects of estrogens and, at other times, augment their suppressive effects.

In summary, according to our present concepts, control of the reproductive process in women is a continued shuttling of information within the closed-loop feedback system, shown in Fig.1. The shuttling involved sequentially: 1) Falling blood steroid concentration → increased pituitary FSH and LH secretion; 2) Initiation of a group of follicles to growth; 3) Follicle atresia for most of the follicles and maturation development of one follicle; 4) Increased blood estradiol concentrations; 5) Progesterone induced modification of the central nervous system pituitary threshold for LH-FSH secretion; 6) Secretion of the LH-FSH ovulatory surge; 7) Ovulation; 8) Corpus luteum formation and function; 9) High estradiol and progesterone concentrations in blood; 10) Suppression of blood LH and FSH secretion (and GnRH secretion); 11) Maintenance of corpus luteum function for 12-14 days; 12) Functional death of the corpus luteum presumably partly as an intrinsic corpus luteum property; 13) Fall in blood estradiol and progesterone concentrations; and 14) Restart of a subsequent cycle.

Within this accumulated understanding of the normal cycle, a host of incompletely understood facets exist, amongst these are: 1) What are the hormones or hypothalamic factors other than GnRH which modify LH and FSH secretion by the pituitary? 2) What controls follicle atresia versus development? 3) What is the physiological importance of progestogen interplay during the ovarian signal induction of the LH-FSH ovulatory surge? 4) What are the factors making estrogen and progestogens at times stimulatory and at times inhibitory to LH-FSH secretion? 5) Is there a functional purpose to the mid-cycle FSH surge? 6) What is the physiological importance of the short-loop feedback system and does it act at a hypothalamic or pituitary level?

REFERENCES

Abraham, G.E., Odell, W.D., Swerdloff, R.S. and Hopper, K. (1972). *J. clin. Endocr. Metab.* **34**, 312.
Abraham, G.E. and Klaiber, E.L. (1970). *Am. J. Obstet. Gynec.* **108**, 528.
Amoss, M., Burgus, R., Blackwell, R., Vale, W., Fellows, R. and Guillemin, R. (1971). *Biochem. Biophys. Res. Comm.* **44**, 205-210.
Baird, D.T. and Guevara, A. (1969). *J. clin. Endocr. Metab.* **29**, 149-156.
Barraclough, C.A. (1974). *Program. of IV International Congress on Hormonal Steroids*, Mexico City.
Besser, G.M., McNeilly, A.S., Anderson, D.C., Marshall, J.C., Harsoulis, P., Hall, R., Ormston, B.J., Alexander, L. and Collins, W. (1972). *Brit. Med. J.* **3**, 267-271.
Channing, C.P. (1970). *Endocrinology* **87**, 156.
Channing, C.P. and Kammerman, S. (1973). *Endocrinology* **93**, 1035-1043.
Coble, Y.D., Kohler, P.O., Cargille, C.M. and Ross, G.T. (1969). *J. clin. Invest.* **48**, 359.
Corbin, A. and Cohen, A.I. (1966). *Endocrinology* **78**, 41-46.
Corbin, A. and Story, J.C. (1967). *Endocrinology* **80**, 1006-1012.
Crook, A.C., Morell, M. and Butt, W.R. (1968). *In* "Gonadotropins" (E. Rosenberg, ed) Geron-X, Inc., Los Altos, California.
David, M.A., Fraschini, F. and Martini, L. (1966). *Endocrinology* **78**, 55-60.
Edmonds, M., Molitch, M. and Odell, W.D. (1974). Unpublished observations.
Edmonds, M., Molitch, M. and Odell, W.D. (1974a). *Program of The Endocrine Society*, Abstract 26, p.A-68.
Edmonds, M., Molitch, M., Pierce, J. and Odell, W.D. (1975). Accepted *Clin. Endocrinology*.

Harmon, S. and Ross, G.T. (1974). *Program of The Endocrine Society*, **Abstract 221**.
Job, J.C., Garnier, P.E., Chaussain, J.L., Binet, E., Rivaille, P. and Milhaud, G. (1972). *Revues Europ. Etudés Clin. Biol.* 17, 411-414.
Job, J.C., Milhaud, G., Garnier, P.E., Chaussain, J.L., Binet, E., Canlorbe, P. and Rivaille, P. (1972a). *Annales D'endocrinologie (Paris)* 33, 537-538.
Kastin, A.J., Schally, A.V., Gual, C. and Arimura, A. (1972). *J. clin. Endocr. Metab.* 34, 753-756.
Knobil, E. (1974). *Program of IV International Congress on Hormonal Steroids*, Mexico City.
Kohler, P.O., Ross, G.T. and Odell, W.D. (1968). *J. clin. Invest.* 47, 38-47.
Laburthe, M.C., Dolais, J.R. and Rosselin, G.E. (1973). *J. clin. Endocr. Metab.* 37, 156.
Legan, S.J. and Karsch, F.J. (1975). *Endocrinology* 96, 57.
Matsuo, H., Baba, Y., Nair, R.N.G., Arimura, A. and Schally, A.V. (1971). *Biochem. Biophys. Res. Comm.* 43, 1334-1339.
McNeilly, A.S., Anderson, D.C., Besser, G., Marshall, J.C., Harsoulis, P., Alexander, L., Ormston, B.J., Hall, R. and Collins, W. (1972). *J. Endocrinol.* 55, 24-25.
Midgley, A.R., Jr. and Jaffe, R.B. (1971). *J. clin. Endocr. Metab.* 33, 962.
Milhaud, G., Rivaille, P., Garnier, P., Chaussain, J.L., Binet, E. and Job, J.C. (1971). *C.R. Acad. Sci. (Paris)* 273, 1858-1861.
Mishell, D.R., Jr. and Odell, W.D. (1971). *Am. J. Obstet. Gynec.* 109, 140-149.
Molitch, M., Edmonds, M., Jones, E. and Odell, W.D. (1975). *Am. J. Physiol. (in press)*.
Nakano, R., Kotsuji, F., Mizuno, T., Hashiba, N., Washio, M. and Tojo, S. (1973). *Acta Obstet. Gynec. Scand.* 52, 171-175.
Odell, W.D., Ross, G.T. and Rayford, P.L. (1966). *Metabolism* 15, 287-289.
Odell, W.D., Ross, G.T. and Rayford, P.L. (1967). *J. clin. Invest.* 46, 248-255.
Odell, W.D. and Swerdloff, R.S. (1968). *Proc. natn. Acad. Sci. U.S.A.* 61, 529-536.
Odell, W.D., Parlow, A.F., Cargille, C.M. and Ross, G.T. (1968). *J. clin. Invest.* 47, 2551-2562.
Odell, W.D. (1974). *Program of the IV International Congress on Hormonal Steroids*, Mexico City.
Papkoff, II. and Samy, T.S.A. (1967). *Biochim. biophys. Acta* 147, 175-177.
Parlow, A.F. and Shome, B. (1974). *J. clin. Endocr. Metab.* 39, 199-201.
Pierce, J.G., Liao, T.H., Howard, S.M., Shome, B. and Cornell, J.S. (1971). *Rec. Prog. Horm. Res.* 27, 165.
Rabinowitz, D., Benveniste, R., Frohman, L., Bell, J. and Spitz, I. (1973). *J. clin. Invest.* 52 67a **(Abstract)**.
Rolland, R., Schellekens, L.A. and Leguin, R. (1974). *Clin. Endocrinol.* 3, 155.
Ross, G.T. and Odell, W.D. (1967). *Science* 155, 1679-1680.
Schally, A.V., Arimura, A., Baba, Y., Nair, R.M.G., Matsuo, H., Redding, T.W., Dubeljuk, L. and White, W.F. (1971). *Biochem. Biophys. Res. Comm.* 43, 393-399.
Swerdloff, R.S. and Odell, W.D. (1969). *J. clin. Endocr. Metab.* 29, 63-71.
Swerdloff, R.S., Jacobs, H.S. and Odell, W.D. (1972). *Endocrinology* 90, 1529.
Tulchinsky, D. and Korenman, S.G. (1970). *J. clin. Endocr. Metab.* 31, 76-80.
Vande Wiele, R.L., Bogumil, J., Dyrenfurth, I., Ferin, M., Jewelewicz, F., Warren, M., Rizkallah, J. and Mikhail, G. (1970). *Rec. Prog. Horm. Res.* 26, 63-103.
Van Thiele, D.H., Bridson, W.E. and Kohler, P.O. (1971). *Endocrinology* 89, 622.
Von zur Muhlen, A. and Kobberling, J. (1972). *Acta Endocrinol.* **(Suppl.)** 159, 57.
Wollesen, F., Swerdloff, R. and Odell, W. (1975). Submitted to *J. clin. Invest.*
Wollesen, F., Swerdloff, R. and Odell, W. (1975a). Submitted to *Metabolism*.
Yen, S.S.C. and Tsai, C.C. (1971). *J. clin. Endocr. Metab.* 33, 882-887.
Yen, S.S.C., Rebar, R., Vanden Berg, G., Naftolin, F., Ehara, Y., Engblom, S., Ryan, K.J. and Bernishke, K. (1972). *J. clin. Endocr. Metab.* 34, 1108-1111.
Yen, S.S.C., Tsai, C.C., Naftolin, F., Vanden Berg, G. and Ajabor, L. (1972a). *J. clin. Endocr. Metab.* 34, 671.
Zarate, A., Canales, E.S., Soria, J., Ruiz, F. and MacGregor, C. (1972). *Am. J. Obstet. Gynec.* 112, 1130.

SOME ASPECTS OF HYPOTHALAMUS-HYPOPHYSEAL-OVARIAN

FEEDBACK MECHANISMS IN WOMEN

P. Franchimont, U. Gaspard, J.R. Van Cauwenberge and E. Depas

*Laboratory of Radioimmunology, Institute of Medicine, University of Liege
Department of Obstetrics and Gynaecology, Faculty of Medicine
University of Liege*

The pituitary gonadal axis is under the control of the hypothalamus via releasing hormones. The experiments of Harris *et al.* (1966), Martini *et al.* (1968) and McCann and Dhariwal (1966) in animals and the independent control of FSH and LH in man (see review, Franchimont and Burger, 1975; Naftolin *et al.*, 1972; Bowers *et al.*, 1973) suggest that there are two releasing hormones: LH-RH and FSH-RH. Porcine hypothalamic LH-RH has been isolated, purified, characterized and then synthesized by Schally and his group in 1971 whilst on the other hand, until now, no hypothalamic substance with the property of specifically releasing FSH has been isolated.

As LH-RH also leads to a release of FSH (though at doses clearly greater than those which release LH and with a different time course) in various animal species *in vivo* and *in vitro*, the concept has rapidly arisen that there is only a single hormone for these two gonadotrophins. The specificity of the gonadotrophin response would occur because of the specific synthesis and establishment of a releasable store of FSH or of LH under the influence of various gonadal hormones and of the endocrine equilibrium. In order to analyse hypothalamus-hypophyseal-ovarian feedback mechanisms, one may attempt to determine the influence of gonadal steroids on gonadotrophin basal levels and to study the response of the pituitary gonadotrophins to exogenous LH-RH injection in various physiological, experimental and pathological states.

Analysis of gonadal steroid effects on basal gonadotrophin levels and on response to LH-RH makes it possible to define the hypothalamic or hypophyseal levels of action. Hence, hypothalamic action may be accepted when a substance decreases basal FSH or LH levels without modifying the LH-RH response. A pitu-

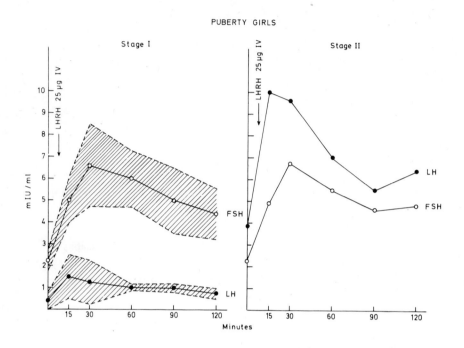

Fig. 1 Mean FSH and LH response to 25 µg of LH-RH in girls at Stage I (13 subjects) and II (9 subjects) of puberty.

itary inhibiting action is defined by decreased basal gonadotrophin levels and response to LH-RH.

We have investigated the influence of endocrine equilibrium on the basal FSH and LH levels and on the LH-RH response firstly in physiological conditions and secondly in normal eugonadal and postmenopausal women treated with several hormonal steroids.

I. CHANGES IN THE EFFECTS OF LH-RH IN PHYSIOLOGICAL CONDITIONS

Two conditions were studied: the response to LH-RH before and during puberty and the LH-RH induced gonadotrophin response during the menstrual cycle.

A. Changes of Pituitary Response to LH-RH Injection at Puberty

The occurrence of puberty leads to a significant change in the gonadotrophin response to the intravenous injection of LH-RH. Thus, prior to puberty, both in boys and girls at Stage I of pubertal development (Tanner, 1962) 26 µg of LH-RH leads to a small but significant LH release and a larger response of FSH. All investigators have noted this greater degree of FSH release as compared with LH before puberty in girls (Kastin *et al.*, 1972; Roth *et al.*, 1972; Job *et al.*, 1972;

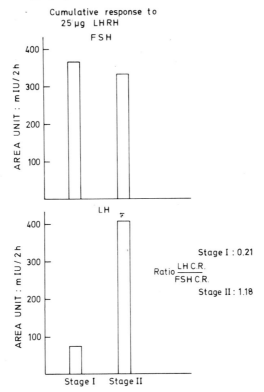

Fig. 2 Mean cumulative FSH and LH response to 25 µg of LH-RH in girls at Stage I (13 subjects) and II (9 subjects) of puberty. The ratio (cumulative LH response)/(cumulative FSH response) is 0.21 at Stage I and 1.18 at Stage II.

Franchimont et al., 1973a).

In contrast, when the child enters puberty (Stage II) and pubertal development continues (Stages III, IV, V) the LH response increases progressively whilst the FSH release does not differ statistically significantly or is even decreased from that seen before puberty. The increase in FSH is less than that in LH from that time onwards (Fig. 1). The ratio (LH cumulative response to LH-RH)/(FSH cumulative response to LH-RH) is less than 1 before puberty (Stage I) whereas it is always higher than 1 when puberty has started (Fig. 2).

When LH-RH is injected intravenously in doses of 25-100 µg daily for 4 days in pre-pubertal girls (Stage I), there is a considerable release of FSH which is always more marked than that of LH (Bourguignon et al., 1975) (Fig. 3).

Thus prior to puberty the pituitary essentially releases FSH under the influence of LH-RH, suggesting the likelihood of preferential synthesis and storage of FSH. When puberty occurs, LH-RH causes LH release which is greater the more advanced the stage of puberty.

The cause of the qualitative and quantitative change in gonadotrophic response to LH-RH injection at puberty is not yet clear. It can be postulated that gonadal steroids or other substances not yet identified appearing at puberty might stimulate the synthesis and storage of LH without changing, and perhaps even

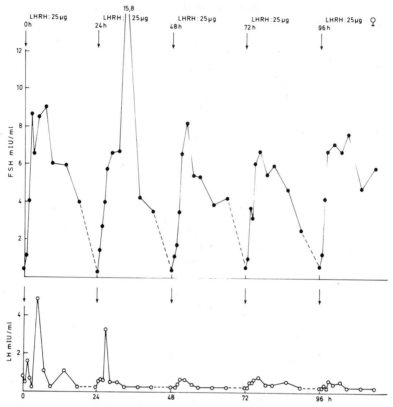

Fig.3 FSH and LH response to 25 µg of LH-RH intravenous injection repeated each morning for 5 days in a girl at Stage I of puberty. Times of sampling: 0-10-20-30-40-50-60-90-120 min.

diminishing, the synthesis and pituitary storage of FSH.

B. Effects of LH-RH in Normally Cycling Women

In the normal cycling female, the intravenous injection of exogenous LH-RH leads to an increase in LH, the degree of which changes as a function of the time during the cycle at which the test is carried out: the response is less during the early and mid-follicular phase and is generally highest at around the time of the spontaneous LH surge. Furthermore, the LH response is greater during the early and mid-luteal phase than during the follicular phase (Fig.4) (Nillius and Wide, 1972; Yen et al., 1972; Thomas et al., 1973; Franchimont et al., 1973a).

We have been able to show that during the luteal phase, basal levels of LH are statistically lower than during the follicular phase. Thus, with the exception of the responses seen in the hours which precede the ovulatory peak, the lower the basal levels of LH, the greater is the response (Franchimont et al., 1973a). An inverse relationship was found in the hypothalamic amenorrhea and in hypogonadotrophic hypogonadism; the LH response following LH-RH injection was greater the higher the baseline LH level.

FSH levels increase significantly following the injection of 50 µg or more of LH-RH. The increase in FSH concentration with the same amount of LH-RH

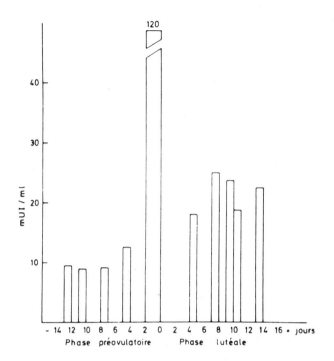

Fig.4 Maximum LH levels reached during LH-RH test (50 μg) in women at different stages of the cycle. Day 0 corresponds to the LH mid-cycle peak

appears to be greater during the pre-ovulatory phase than during the luteal phase (Franchimont *et al.*, 1973a; Le Marchand-Beraud *et al.*, 1973). During the first few days of the cycle the ratio (LH cumulative response)/(FSH cumulative response) is most often less than 1. When LH-RH is given in the hours prior to the LH peak, there is a marked release of FSH and LH which lasts for more than 120 minutes.

II. CHANGES IN THE GONADOTROPHIN BASAL LEVELS AND IN THE RESPONSE OF LH-RH UNDER THE INFLUENCE OF VARIOUS HORMONES

A. *In Normally Cycling Women*
1. *Gonadotrophin basal levels*

It is well known (see review of Franchimont and Burger, 1975) by the chronology of hormonal events that the preovulatory peak of oestrogens triggers LH mid-cycle peak most probably by a positive feedback mechanism.

The essential role of the mid-cycle oestradiol peak was also established by Vande Wiele *et al.* (1970). The administration of serum anti-17β-oestradiol to female rats inhibits the LH mid-cycle peak. However, LH surges occurred again after the administration of diethylstilboestrol, an oestrogen whose biological activity is not neutralised by antibodies to oestradiol.

Experiments involving acute oestrogen administration during the follicular

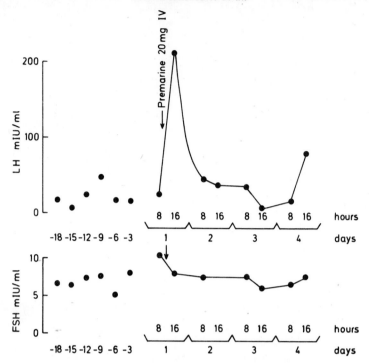

Fig.5 Effect of conjugate equine oestrogens (Premarin 20 mg) on FSH and LH levels in one case of functional amenorrhea. There is no obvious change in FSH concentration whereas a LH burst is induced.

phase of the cycle in eugonadal women have been conducted by Yen and Tsai (1972): their effect on gonadotrophin levels and ovarian function has been studied:

a) During the early follicular phase. The administration of ethinyl oestradiol (400 µg/day) during days 3-4-5 of the cycle abolishes the early follicular rise of gonadotrophin levels observed in control cycles. Approximately 36 hours after the cessation of oestrogen treatment a 'rebound' LH release appears with no FSH response and, after a latent period, the cycle goes on. The changes in gonadotrophin levels are associated with a disturbance of the normal sequence of follicular maturation: the follicular phase is prolonged and the mid-cycle peaks of oestradiol and LH are delayed.

b) During the mid-follicular *phase.* An infusion of oestradiol was given for 36 hours between day 7 and 9 of the cycle so as to mimic the preovulatory oestradiol peak. Twelve to 24 hours later, the usual FSH and LH suppressive effect is followed by a positive feedback action which is moderate for FSH but quite obvious for LH. Here again, the transient disruption of the normal sequence of gonadotrophin secretion is associated with the impairment of follicular maturation.

These data suggest that a rapid rise (or rather the acute fall) in serum oestradiol provides a major ovarian signal for triggering the gonadotrophin surge at mid-cycle.

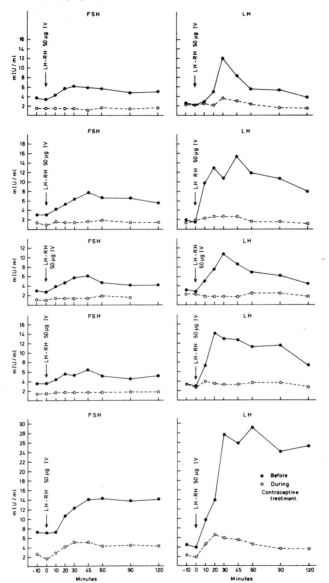

Fig. 6 Effect of 50 µg of LH-RH on FSH and LH levels in five young women, first without therapy (•) and subsequently given non-sequential contraceptive therapy (o). The LH-RH test was performed between the 22nd and 26th days of the control and treatment cycle.

These results, obtained during the cycle of eugonadal women may also be reinforced by evidence gathered from experiments in women suffering from functional amenorrhea (Millet *et al.*, 1973). These amenorrheic women had detectable secretion of both ovarian and pituitary hormones. Twenty µg of Premarin was administered to 15 such patients intravenously. In 6 cases there was no change in LH. In the remaining 9, however, significant rise in LH was found as compared

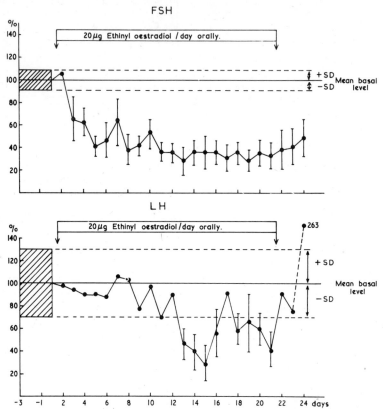

Fig. 7 Effect of ethinyloestradiol 20 µg/day on serum of FSH and LH in 4 postmenopausal women. The results are shown as percentage change from basal levels prior to treatment.

with the patients' baseline levels, which had been measured on at least three different occasions prior to injection. The rise in LH occurred after various latent periods. In 2, there was an early rise, i.e. 8 hours after the injection (Fig.5), and in 7 the rise was more delayed: 24 hours (2 cases), 32 hours (3) and 48 or more hours (2) after injection. FSH concentrations, in contrast, fell in 12 of the 15 and generally showed no rebound effect. Similar findings were obtained by Vande Wiele *et al.* (1970) in women with amenorrhea. It can thus be asserted that it is the preovulatory peak of oestrogens which triggers the LH surge but has no effect on the FSH peak.

2. *Response to LH-RH*

Recently, Jaffe and Keye (1974) have demonstrated that the intensity of gonadotrophin response to LH-RH depends not only on the amounts but also on the duration of oestrogen administration. Giving oestradiol benzoate to eugonadal women from the second day of the cycle to mimic endogenous 17β-oestradiol levels of late preovulatory phase, they observed that treatment for two days decreases the FSH and LH response. On the other hand, if oestradiol benzoate is given for 4-5 and 6 days the LH and FSH responses are increased compared with the control cycle without treatment.

In the first cycle of treatment, the administration of a non-sequential hormonal contraceptive (Norgestrel, 500 μg and EE 50 μg) considerably reduces the FSH and LH response following the intravenous injection of 50 μg of LH-RH (Fig.6) but does not alter the normal response to 100 μg (Franchimont et al., 1974a; Becker et al., 1973). Likewise, Taubert (1973) noted that therapy with contraceptives of the same type for 1-3 years led to a lowering of the basal levels of FSH and LH and of the response of these hormones to the intravenous injection of 12.5-25 μg of LH-RH. These results differ from those of Kastin et al. (1972) and Thomas and Ferin (1972) who carried out the test with bigger doses of LH-RH.

B. *In Menopausal Women*
   1. *Effect of oestrogens*
   a) *On basal gonadotrophin levels.* The unrestrained gonadotrophin release in the postmenopausal woman is an excellent model for the study of gonadotrophin-steroid feedback regulation.

On administration of oestrogens to postmenopausal women various changes occur in the secretory pattern of LH and FSH. Action of different types and doses of oestrogens was studied on gonadotrophin basal levels (Franchimont et al., 1972). When using low doses of ethinyl oestradiol, i.e. 20 μg daily for more than 20 days, an FSH decrease is seen whilst LH changes only slightly (Fig.7). On stopping treatment, a slow rise of FSH concentration towards normal values occurs while LH shows a marked rebound effect related to the disappearance of the oestrogens. With larger doses of ethinyl oestradiol (i.e. 50 μg daily) a greater inhibition of LH release, similar to that of FSH, can be seen and, in this experiment, no rebound effect occurs after stopping the drug. The same pattern of gonadotrophin release can be noted after administration of corresponding doses of mestranol or of l-hydroxy-ethinyl-oestradiol-1,3-diacetate. Finally, when 200 μg of 17-dioxanyl oestradiol is administered daily (Franchimont et al., 1972) there is virtually no depression of FSH and LH. Upon withdrawal of this therapy, there appears a clear-cut rebound effect for LH.

No positive feedback effect was found in our study when giving high doses of the oestrogens. However, in an experiment described by Yen (1971) the administration to postmenopausal women of 400 μg of ethinyl oestradiol daily for eight days produced a biphasic response. During the first three days there was a marked inhibition of FSH and LH. This was followed by the occurrence of an augmented phase of gonadotrophin release which consisted of wide oscillations: this can be interpreted as a positive feedback action of ethinyl oestradiol, quantitatively much greater for LH than for FSH.

Thus, in postmenopausal women, some points regarding the action of oestrogens on gonadotrophin basal levels can be emphasized:

   1. Oestrogens seem to exert a differential feedback control on FSH and LH: a more suppressive action on FSH release and a more stimulatory action on LH release. However, the positive feedback for LH is inconstant and may appear during or after cessation of oestrogen therapy.
   2. The three steroids having the same group (17α-ethinyl β-hydroxy) in the 17 position (ethinyl oestradiol, mestranol and 1-hydroxyethinyl-oestradiol-1,3-diacetate) lower serum levels of FSH and LH in doses of 50 μg/day. In contrast, the steroid, in which the hydroxyl group in the 17β position is substituted by 1,4-dioxane-2-oxy, namely 17-dioxanyloestradiol, fails to reduce FSH and LH

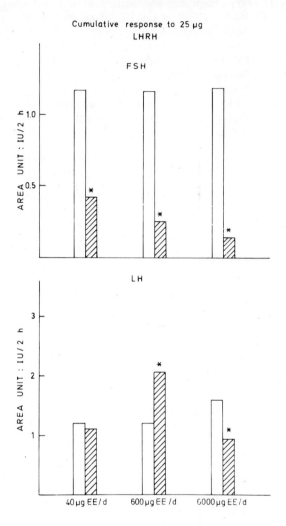

Fig. 8 Mean cumulative response of LH and FSH to 25 µg of LH-RH before and after 12 days of oral treatment with 40, 600 and 6,000 µg ethinyl oestradiol. Cumulative response corresponds to the areas circumscribed by the serum FSH and LH curves during the first 2 hours following LH-RH. * = $p < 0.05$. Each dose of EE was given to at least four normal postmenopausal women.

when administered at a dose level (200 µg/day) capable of exerting peripheral oestrogenic effects. Together, these findings suggest that the 17β-hydroxy group is essential for inhibition of gonadotrophin secretion.

  b) Gonadotrophin response to LH-RH. We have investigated the gonadotrophic response to the intravenous injection of LH-RH in menopausal women treated with several doses of ethinyl oestradiol (EE): 20, 40, 600 and 6,000 µg/day. Treatment was carried out for 12 days. Each dose of EE was given to at least four normal postmenopausal women. Twenty-five µg of LH-RH was injected intravenously prior to and on the last day of treatment.

All doses of EE significantly decreased the basal levels of FSH and LH (except small doses of 20 μg) whereas they increased prolactin levels. For all the doses of EE, the FSH response to LH-RH is significantly decreased. On the other hand, the response of LH to LH-RH injection was not changed after treatment with 40 μg of EE and increased after 600 μg of EE daily administration as compared with that seen in the same subjects prior to treatment. At dose levels of 6,000 μg of EE daily, the basal levels of LH were decreased after 12 days as was the amount of LH released following LH-RH (Fig.8).

Taubert (1973) gave 60 μg/day EE for one week to four menopausal women and gave an injection of 25 μg of LH-RH before and on the last day of treatment. He noted a reduction in the basal levels of LH in three of the four cases. The LH responses to LH-RH were identical before and after treatment in two cases, whilst in the other two the response was increased following treatment. In contrast, EE administration led to a significant decrease in the basal levels of FSH and in the release of FSH following LH-RH injection. In amenorrheic women several investigators have shown that increase in the level of circulating oestrogens by the administration of exogenous oestrogen (Nillius and Wide, 1972; Schneider et al., 1973; Valcke et al., 1974) or of exogenous gonadotrophins (Lunenfeld, 1973) leads to a reduction in the FSH response and an increase or no change in the LH response as compared with the pre-treatment observation. Our results in menopausal women are in agreement with other experiments in animals. Thus Arimura and Schally (1971) showed that the administration of small doses of oestrogen enhances the response of female rats to LH-RH; Reeves et al. (1971) made the same observation in the ewe.

2. *Effect of progestogens*

The intramuscular injection of 150 μg of medroxyprogesterone acetate leads to a statistically significant decrease in basal FSH and LH levels. However, the increase in LH and FSH following 25 μg of LH-RH is the same before and 12 days after the intramuscular injection of medroxyprogesterone.

The intramuscular injection of 100 mg of norgestrel undecyclate (NU) leads to a significant reduction in FSH and LH levels for six weeks. The intravenous injection of 25 μg of LH-RH on the day before and on the 42nd day after NU administration, leads to the same degree of LH and FSH release.

The oral administration of norethisterone acetate 10 mg/daily in four doses for four days reduces baseline levels of FSH and LH. However, the FSH responses to the injection of 25 μg of LH-RH is the same and the LH release is even increased after these four days of treatment compared with results obtained before therapy.

These three derivatives decreased basal FSH and LH levels but they do not modify the cumulative response to LH-RH. One may accept that they act at the hypothalamic level inhibiting the endogenous LH-RH secretion. On the other hand, they do not affect the synthesis and the storage of gonadotrophins at the pituitary levels.

III. MODIFICATION OF THE GONADOTROPHIN RESPONSE TO LH-RH IN PATHOLOGICAL CONDITIONS

LH-RH test is of only moderate use to localize a disease process in the hypothalamic-pituitary axis. Thus, all authors agree that in some patients with organic

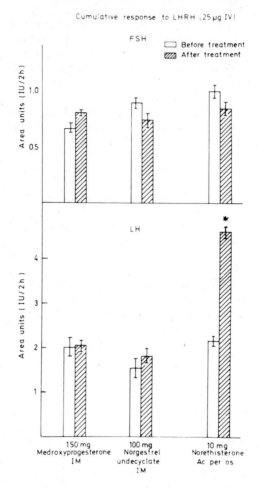

Fig. 9 Cumulative response to 25 μg LH-RH before and after treatment with medroxyprogesterone, norgestrel undecyclate and norethisterone acetate. Each of these progestogens was administered to 5 normal postmenopausal women. * = p < 0.05.

pituitary lesions LH-RH can provoke a normal or impaired FSH and LH response whereas in hypothalamic diseases LH-RH may be ineffective in releasing gonadotrophins (Nillius and Wide, 1972; Besser and Mortimer, 1974; Valcke *et al.*, 1974; Franchimont *et al.*, 1974b). This latter absence of response seems to be due to a chronic insufficiency in endogenous LH-RH.

On the other hand, qualitative aspects of the response are of great interest. Normally after puberty, LH response to LH-RH is consistently higher than that of FSH. Nevertheless, one may observe responses to LH-RH identical to that found in prepuberal girls, i.e. a marked increase of FSH and a weak, if any, increase in LH.

This kind of response could be found in hypothalamic or psychogenic amenorrhea of long duration, particularly in anorexia nervosa. A classification of functional amenorrhea can be attemped on the basis of baseline gonadotrophin levels, the degree of spontaneous fluctuation of LH, the responses seen to the administration of clomiphene, oestrogens and intravenous LH-RH (Franchimont et al., 1974b). Some cases of amenorrhea are characterized by normal levels of FSH and normal or slightly increased levels of LH. The LH levels fluctuated from one day to the next but there are none of the cyclical fluctuations which are characteristic of the menstrual cycle. Clomiphene stimulates an increase in the levels of FSH and LH and the intravenous injection of oestrogen leads to an increase in LH levels. The concentration of plasma 17β-oestradiol is always greater than 50 pg/ml. Twenty-five or 50 µg LH-RH intravenously causes normal or increased release of LH and normal release of FSH. On the basis of the sequential levels of the gonadotrophins and the results of dynamic tests, it can be assumed that there is a disturbance of hypothalamic regulation of gonadotrophin secretion without primary involvement of the pituitary.

Cases of functional secondary amenorrhea described by Zarate et al. (1973) could be classified in this first group on the basis of the basal levels of FSH and LH and the response to LH-RH injection. In contrast, another group of patients with amenorrhea are characterized by low and stable levels of LH always less than those of FSH, by lack of response to the administration of clomiphene and the intravenous injection of oestrogen. In these cases, oestrogen levels are very low and always less than 60 pg/ml. The administration of 25 µg of LH-RH causes abnormal responses of various types: absent or diminished release of FSH and LH, normal or increased release of FSH and virtually no LH. In these cases it is postulated that there is a prolonged depression of the hypothalamic regulation of LH and FSH secretion induced by oestrogens. Between these two well-defined groups of amenorrhea several situations are possible and are found.

A large number of patients with anorexia nervosa belong to this second group of amenorrhea; their LH and oestrogen levels are low and stable and LH-RH injection leads to a very poor LH response and a normal or increased FSH response (Franchimont et al., 1973b; Valcke et al., 1974). This response, which is similar to that found in girls prior to puberty, suggests a return of the cells to the prepubertal state because of the lack of certain stimuli such as oestrogens (Franchimont et al., 1973b, 1974b).

Comparing the basal levels of LH and 17β-oestradiol with those seen in normal women during the follicular phase of the cycle, Yen et al. (1973) picked out a category of amenorrhea characterized by low levels of LH and 17β-oestradiol, lack of fluctuation of LH levels, normal concentrations of FSH and a positive response to clomiphene. These subjects have amenorrhea of a type intermediate between those of our two groups. The injection of 150 µg of LH-RH led to FSH and LH responses greater than those seen in normal women during the follicular phase. For these authors this association of low normal levels of LH and an excessive response to LH-RH injection suggests that the deficit is due to a lack of endogenous LH-RH stimulation.

Numerous investigators have established that there is a correlation between the levels of circulating oestrogens, the basal level of LH and the degree of LH response to LH-RH injection (Nillius and Wide, 1972; Taubert, 1973; Vandekerckhove et al., 1975; Valcke et al., 1974; Bettendorf, 1974; Zarate et al., 1973). However, Valcke et al. (1974) showed that there is a correlation between the basal levels of FSH and the degree of FSH response to LH-RH injection, but that there was no correlation between the type of response to LH-RH on the one hand and the duration of the amenorrhea, the degree and duration of loss or gain in weight and the age of the patients on the other.

IV. DISCUSSION AND CONCLUSIONS

It could logically be postulated that there is only a single gonadotrophin releasing hormone and that the specificity of the response is at the level of the synthesis or reserve capacity or selective release of one or other gonadotrophin under the influence of various gonadal hormones. This effect of the endocrine equilibrium is well shown by the gonadotrophin responses seen before and after puberty. Before puberty, single or repeated injections of LH-RH lead particularly to an increase in FSH and little or no increase in LH. Likewise, foetal pituitaries in organotypic tissue culture produce increased quantities of FSH under the influence of LH-RH as compared with controls, whilst the release of LH is virtually unchanged (Pasteels and Franchimont, 1975). During and after puberty and in adulthood, the response to LH-RH becomes greater for LH than for FSH.

In women, when given at adequate doses and for a long enough prolonged period to eugonadal or menopausal or amenorrheic women oestrogens decrease FSH release but could increase that of LH. In contrast, when there is a chronic oestrogen deficiency as in anorexia nervosa or in hypogonadotrophic hypogonadism, FSH release is frequently greater than that of LH following LH-RH. Oestrogens could thus be one of the factors which alter the LH-RH response in women before and after puberty, decreasing the synthesis and/or storage of FSH whilst they have an inverse effect on LH.

The effect of gonadal steroid hormones, of synthetic analogues or of their antagonists may occur at the level of the hypothalamus or pituitary as suggested by the pattern of the basal levels of the gonadotrophins and their response to LH-RH injection before and after some days of administration of these substances. Two situations may occur. During the initial administration of the substance studied, there may be a significant decrease in basal gonadotrophin levels but no changes or even an increase in the response to LH-RH. This decrease in basal gonadotrophin levels is not related to a diminished pituitary gonadotrophin responsiveness to LH-RH, or to storage, but rather to an inability to release due to insufficient endogenous LH-RH stimulation. The substances, therefore, act primarily at the hypothalamic level. This hypothalamic effect is seen in women with various progestogens which have been examined: medroxyprogesterone, norgestrel, undecyclate and norethisterone acetate.

The second type of effect is characterized by a decrease in basal levels and gonadotrophin responses to LH-RH under the influence of small doses of a hormone given for a short time. This is the situation for the oestrogens which at all the doses used consistently act on FSH in women. Oestrogens thus appear to act directly at the level of the pituitary on the synthesis and storage of FSH.

REFERENCES

Arimura, A. and Schally, A.V. (1971). *Proc. Soc. exp. Biol.* L36, 290-293.
Becker, H., Kleibel, H.P., Reuter, A. and Franchimont, P. (1973). *Klin. Wschr.* 52, 759-761.
Besser, G.M. and Mortimer, C.H. (1974). *J. clin. Path.* 27, 173-184.
Bettendorf, G. (1973). Discussion in the Round Table on "Clinical Investigation with LH-RH",

International Symposium on Hypothalamic Hormones, Vienna, June 1973. Schattauer Verlag *(in press)*.
Bourguignon, J.P., Ernould, Ch., Brichant, J. and Franchimont, P. (1975). *(In preparation)*.
Bowers, C.Y., Currie, B.L., Johansoon, K.N.G. and Folkers, K. (1973). *Biophys. Res. Comm.* **50**, 20-26.
Franchimont, P., Legros, J.J. and Meurice, J. (1972). *Horm. Metab. Res.* **4**, 288-292.
Franchimont, P., Becker, H., Ernould, Ch., Thys, Ch., Demoulin, A., Bourguignon, J.P., Legros, J.J. and Valcke, J.C. (1973a). *Ann. d'Endocr.* **34**, 477-490.
Franchimont, P., Valcke, J.C., Schellens, A.M.C;M., Demoulin, A. and Legros, J.J. (1973b). *Ann. d'Endocr.* **34**, 491-501.
Franchimont, P., Becker, H., Ernould, Ch., Thys, Ch., Demoulin, A., Bourguignon, J.P., Legros, J.J. and Valcke, J.C. (1974a). *Clin. Endocr.* **3**, 27-39.
Franchimont, P., Valcke, J.C. and Lambotte, R. (1974b). *In* "Clinics in Endocrinology and Metabolism" Vol.III, pp.536-559. W.B. Saunders & Co., Ltd., London.
Franchimont, P. and Burger, H.G. (eds) (1975). "Human Growth Hormone and Gonadotrophin in Health and Disease". Excerpta Medica, North Holland.
Harris, G.W., Reed, M. and Fawcett, C.P. (1966). *Brit. Med. Bull.* **22**, 266.
Jaffe, R.B. and Keye, W.R. (1974). *J. clin. Endocr.* **39**, 850-855.
Job, J.C., Garnier, P., Chaussain, J.L., Binet, E., Rivaille, P. and Milhaud, G. (1972). *J. clin. Endocr.* **35**, 473-476.
Kastin, A.J., Gual, C. and Schally, A. (1972). *Rec. Prog. Horm. Res.* **28**, 201-227.
Le Marchand-Beraud, T., Ruedi, B. and Magrinni, G. (1973). *Acta endocr. (Copenh.)* **Suppl. 177**, 291.
Lunenfeld, B. (1973). Discussion in the Round Table on "Clinical Investigation with LH-RH", *International Symposium on Hypothalamic Hormones*, Vienna, June 1973. Schattauer Verlag *(in press)*.
Martini, L., Fraschini, F. and Motta, M. (1968). *Rec. Prog. Horm. Res.* **24**, 439-485.
McCann, S.M. and Dhariwal, A.P.S. (1966). *In* "Neuroendocrinology" (L. Martini and W.F. Ganong, eds) Vol.I, pp.261-296. Academic Press, New York and London.
Millet, D., Franchimont, P., Hayeck, G. and Netter, A. (1973). "Actualités Gynecologiques" (A. Netter and A. Gorins, eds) pp.83-89. Masson et Cie, Paris.
Naftolin, F., Yen, S.S.C. and Tsai, C.C. (1972). *Nature* **236**, 92-96.
Nillius, S.J. and Wide, L. (1972a). *J. Obstet. Gynec. Brit. Commwlth.* **79**, 874-882.
Nillius, S.J. and Wide, L. (1972b). *J. Obstet. Gynec. Brit. Commwlth.* **79**, 865-873.
Pasteels, J.L. and Franchimont, P. (1975). *(In preparation)*.
Reeves, J., Arimura, A. and Schally, A.V. (1971). *Biol. Reprod.* **4**, 88-92.
Roth, J.C., Kelch, R.P., Kaplan, S.L. and Grumbach, M.M. (1972). *J. clin. Endocr.* **35**, 926-930.
Schally, A.V., Arimura, A., Kastin, A.J., Matsuo, H., Baba, Y., Redding, T.W., Nair, R.M.G. and Debeljuk, L. (1971). *Science* **173**, 1036-1037.
Schneider, W.H.G., Matt, K. and Spona, J. (1973). Discussion in the Round Table on "Clinical Investigation with LH-RH", *International Symposium on Hypothalamic Hormones*, Vienna, June 1973. Schattauer Verlag *(in press)*.
Tanner, J.M. (1962). *In* "Growth and Adolescence", pp.433-469. Blackwell Scientific Publications.
Taubert, H.D. (1973). Discussion in the Round Table on "Clinical Investigation with LH-RH", *International Symposium on Hypothalamic Hormones*, Vienna, June 1973. Schattauer Verlag *(in press)*.
Thomas, K. and Ferin, J. (1972). *Contraception* **6**, 1-16.
Thomas, K., Cardon, M., Donnez, J. and Ferin, J. (1973). *Contraception* **7**, 289-297.
Valcke, J.C., Mahoudeau, J.A., Thieblot, Ph., Pique, L., Luton, J.P., Franchimont, P., Moreau, L. and Bricaire, H. (1974). *Ann. d'Endocr.* **35**, 423-444.
Vandekerckhove, D., d'Hont, M. and Van Eyck, J. (1975). *Acta endocr. (Copenh.)* **78**, 625-633.
Vande Wiele, R.L., Bougmil, J., Deyrenfurth, J., Ferin, M., Jewelewicz, R., Warren, M., Rizkallah, T. and Mikhail, G. (1970). *Rec. Prog. Horm. Res.* **26**, 63-95.
Yen, S.S.C. (1971). *J. clin. Endocr.* **33**, 882-887.
Yen, S.S.C. and Tsai, C.C. (1971). *J. clin. Endocr.* **34**, 298-305.
Yen, S.S.C., Rebar, R., Vandenberg, G. and Judd, H. (1973). *J. clin. Endocr.* **36**, 811-816.

Yen, S.S.C., Vandenberg, G., Rebar, R. and Ehara, Y. (1972). *J. clin. Endocr.* 35, 931-934.
Zarate, A., Canales, E.S., de la Cruz, A., Soria, J. and Schally, A.V. (1973). *Obstet and Gynec.* 41, 803-808.

OVARIAN STEROID SECRETION AND METABOLISM IN WOMEN

D.T. Baird

M.R.C. Unit of Reproductive Biology
Department of Obstetrics and Gynaecology, University of Edinburgh
39 Chalmers Street, Edinburgh EH3 9ER, Scotland

SUMMARY

Ovarian steroid secretion and metabolism can be studied by a) direct sampling of ovarian venous effluent b) measurement of steroid metabolites in urine c) measurement of hormone in peripheral plasma d) isotope dilution techniques. The advantages, limitations and disadvantages of each method are discussed. Using a combination of these techniques it has been demonstrated that the human ovary secretes progesterone, pregnenolone, 17α hydroxyprogesterone, dehydroepiandrosterone, androstenedione, testosterone, oestrone and oestradiol. Although all cell types in the ovary probably have the biosynthetic capacity to synthesize these steroids, each functional compartment (follicle, corpus luteum and stroma) has its own characteristic pattern of secretion. Thus androstenedione is secreted by all three compartments; oestradiol by the follicle and the corpus luteum and progesterone exclusively by the corpus luteum.

The secretion of steroid hormones is controlled by a negative feedback loop involving the hypothalamus and anterior pituitary. FSH and LH are secreted in a pulsatile fashion with peaks of secretion occurring every 60-120 minutes. Short term fluctuations in the concentration of oestradiol in plasma have also been recorded although they appear to bear little relationship to the fluctuation in gonadotrophin secretion. Using the autotransplanted ovary of the sheep it has been demonstrated that pulses of LH secretion are followed 5-10 minutes later by an increase in the secretion of oestradiol and androstenedione which in this species arises exclusively from the follicle. The secretion of progesterone from the corpus luteum remains relatively steady indicating that the control of steroid secretion by gonadotrophins from these two ovarian compartments is probably different.

INTRODUCTION

Study of ovarian secretion is complicated by the fact that the ovary contains a variety of cell types (e.g. theca, granulosa, interstitial cells) which are organized into different functional compartments (i.e. Graafian follicle, corpus luteum, stroma). Because the structure and function of these compartments is always changing, it is necessary to make serial measurements throughout at least one complete ovarian cycle in order to obtain an accurate assessment of ovarian activity. The situation is further complicated by the fact that the secretion of the ovary (like that of the testis and adrenal) fluctuates from minute to minute, and single measurements of the concentration of a steroid in peripheral venous blood may be unrepresentative of the amount secreted. It is the purpose of this chapter to consider the steroids secreted by the human ovary and to review the methods available for measuring their secretion and metabolism. The proportion of the total production rate of a steroid which is secreted by the ovary and where possible the individual cellular compartment, will be identified. Lastly, mention will be made of the mechanisms involved in the control of steroid secretion by the ovary.

I. METHODS OF STUDY

a) Direct proof of the secretion of a steroid by the ovary can be obtained by demonstrating that the concentration in ovarian venous effluent is higher than that in afferent arterial (or peripheral venous) blood. However because of the surgical interference necessary serial studies over a long period are precluded using this method. Using techniques of catheterization of the ovarian vein retrogradely via the femoral vein it is possible to collect several samples without the stress of major surgery (Kirschner and Jacobs, 1971) although clearly there are limitations as to the type of subject who can be studied in this way. Chronic preparations in sheep involving autotransplantation of the ovary to a superficial site in the neck have been used to make repeated measurements of secretion rates in the conscious unstressed animal (McCracken *et al.*, 1971).

b) The measurement of the excretion of urinary metabolites of steroids have been widely used in the study of ovarian activity (Brown and Matthew, 1962). However, urinary excretion only reflects ovarian secretion accurately if the steroid chosen is secreted exclusively by the ovary and is metabolized to a unique metabolite. Fortunately this is so for oestradiol and progesterone in women during their reproductive life so that the measurement of the excretion of an oestrogen metabolite (or indeed "total oestrogen") and pregnanediol is usually a reliable index of ovarian activity (Brown, 1971).

c) Measurements of the concentration of the steroid in peripheral venous blood have the advantage that it is possible to collect serial samples easily, so that rapid changes in secretion can be studied. How closely the concentration in plasma reflects ovarian secretion depends on whether its rate of metabolism or clearance from the blood changes significantly from hour to hour or day to day.

d) The techniques of isotope dilution were introduced in an attempt to study these variables. A tracer amount of radioactively labelled steroid is introduced into the blood stream either by single injection or by continuous infusion and its fate followed by collecting serial samples of blood and/or urine. The rate of metabolism can then be expressed as half-life time (T 1/2) or Metabolic Clear-

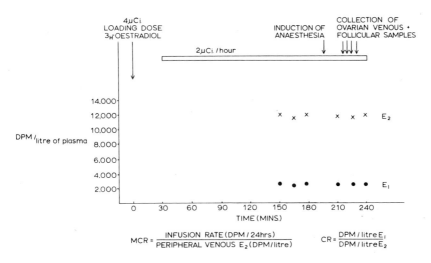

Fig. 1 Measurement of MCR by Constant Infusion Technique. After a priming dose, 6,7 ^{3}H oestradiol is infused at a constant rate into a vein in the antecubital fossa. When the steady state has been reached the concentration of radioactively labelled oestradiol (E_2) and oestrone (E_1) is measured. If as in this case the infusion is continued during operation, samples of ovarian venous blood and follicular fluid can be collected to measure ovarian secretion rate directly.

ance Rate (M.C.R.). The advantages and disadvantages of each technique have been discussed in various reviews (Tait, 1963; Vande Wiele et al., 1968; Baird et al., 1969).

Most of the dynamic results discussed in this chapter were obtained using the constant infusion technique first developed by Tait (1963) and applied to the different ovarian steroids by different members of his group - progesterone (Little and Billiar, 1969), androgens (Tait and Horton, 1966) and oestrogens (Baird et al., 1969). The technique illustrated in Fig.1 for oestradiol, involves the constant infusion of labelled steroid and the sampling of radioactivity in peripheral plasma when the steady state has been reached. The MCR and Blood Production Rate (P_B) can be calculated from this data. By isolating any radioactively labelled conversion product (e.g. oestrone) at the steady state the "conversion ratio" (or the extent to which the precursor contributes to the plasma concentration of product) can be calculated.

Doubt has recently been expressed as to the validity of two of the assumptions on which isotope dilution techniques involving single injections are based. Tritium labelled cortisol injected at 8 a.m. is metabolized slightly differently from ^{14}C cortisol injected in the evening (Kelly, 1970). After injection of labelled oestrone the specific activity of oestriol, oestrone and oestradiol in urine differs significantly indicating that it is unlikely that the tracer has mixed completely with endogenous oestrone (Rizkallah et al., 1975). These limitations probably do not apply to the constant infusion technique although some authors have claimed difficulty in achieving the steady state following the infusion of oestradiol

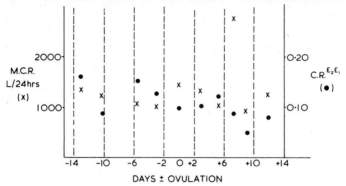

Fig. 2 The Metabolic Clearance Rate (MCR) and Conversion Ratio ($C_{BB}^{E_2 E_1}$) of oestradiol in 10 women throughtout the menstrual cycle. The subjects have been grouped around the day of estimated ovulation. The MCR and $C_{BB}^{E_2 E_1}$ were measured at the steady state after the infusion of ^{3}H oestradiol. There is no significant difference in the values at different stages of the cycle.

Table I. The Blood Production Rate (P_B), Adrenal (S_{ad}) and Ovarian (S_{ov}) Secretion Rates of seven steroids in women at three different stages of the menstrual cycle. The values, which are means from published series, are expressed in mgm/24 hours.

	Early Follicular			Pre-ovulatory			Mid-luteal		
	P_B	S_{ad}	S_{ov}	P_B	S_{ad}	S_{ov}	P_B	S_{ad}	S_{ov}
Progesterone	1.00	0.20	0.10	4.0	0.20	2.40	2.50	0.20	24.0
17α hydroxyprogesterone	0.50	0.10	0.350	4.0	0.10	3.7	4.0	0.10	3.7
Dehydroepiandrosterone	7.0	6.8	0.2	7.0	6.8	0.2	7.0	6.8	0.2
Androstenedione	2.6	1.2	0.9	4.7	1.2	3.1	3.4	1.2	2.8
Testosterone	0.200	0.040	0.040	0.236	0.040	0.080	0.174	0.040	0.050
Oestrone	0.050	0.01	0.010	0.350	0.01	0.035	0.250	0.01	0.025
Oestradiol 17β	0.036	0.000	0.032	0.400	0.000	0.380	0.250	0.000	0.240

(Hembree et al., 1969). Nevertheless the MCR as measured by constant infusion appears to be a fairly stable parameter (Baird and Fraser, 1974). The MCR of oestradiol for example does not change significantly throughout the menstrual cycle although there are large differences in endogenous oestrogen secretion (Fig. 2). Thus for oestradiol the concentration in plasma does represent production fairly accurately.

II. STEROIDS SECRETED BY THE OVARY

Using a combination of the above technique it has been established that the human ovary secretes at least eight steroid hormones; progesterone, pregnenolone, 17α hydroxyprogesterone, dehydroepiandrosterone, testosterone, androstenedione, oestrone and oestradiol (Baird, 1974). How well does the concentration in peripheral venous plasma of these steroids, all of which are secreted by the adrenal glands, reflect ovarian secretion? The ovarian secretion of pregnenolone and dehydroepiandrosterone is tiny compared to the adrenal secretion of these steroids which will not be discussed further. Table I summarises the blood production rate of seven steroids at different stages of the menstrual cycle. Measurements of the concentration of steroids in glandular venous effluents as well as dynamic studies have been used to calculate the secretion from the ovaries and adrenals.

In the normal ovarian cycle virtually all the oestradiol and progesterone is secreted by the ovary. By Day 7 of the cycle the preovulatory follicle is secreting over 90 per cent of the oestradiol while in the luteal phase almost all the oestradiol and progesterone originate from the corpus luteum. In the early follicular phase of the cycle the adrenal secretion of progesterone constitutes a significant proportion of the P_B so that the measurement of the metabolite pregnanediol in urine is not indicative of ovarian secretion (Little and Billiar, 1969). In the luteal phase, however, the secretion of progesterone from the corpus luteum is so much greater than the amount produced from other sources, that plasma progesterone or urinary pregnanediol reflects the function of the corpus luteum.

17α hydroxyprogesterone is secreted by the corpus luteum as well as the mature Graafian follicle. It was suggested at one time that its measurement in peripheral plasma would provide a useful monitor of follicular and luteal function (Strott et al., 1969). However, since the development of relatively simple methods for the measurement of oestradiol in plasma, the estimation of 17α hydroxyprogesterone has been largely restricted to early pregnancy when it reflects the activity of the corpus luteum as opposed to the trophoblast (Holmdahl and Johansson, 1972).

Oestrone, testosterone and androstenedione are all secreted in variable amounts by both ovaries and adrenal glands (Baird, 1971). In addition a significant proportion of the total P_B of these steroids is produced by the conversion of other precursor steroids of "pre-hormones" in tissues outside the endocrine glands e.g. skin, fat, liver, etc.. This extraglandular production of steroids may be a mechanism of achieving a high concentration of hormone locally in a peripheral tissue without the necessity of undesirably high concentrations in peripheral blood. For example over half the P_B of the potent androgens, testosterone and dihydrotestosterone are produced by extraglandular conversion of the biologically weaker androgens, androstenedione and dehydroepiandrosterone (Ito and Horton, 1970). A proportion of androstenedione is aromatized to oestrone in peripheral tissues and in some circumstances, e.g. polycystic ovarian disease and after the menopause, almost all the oestrogen is produced in this way (Siiteri and MacDonald, 1973).

The proportion of androstenedione secreted from the ovary and adrenal appears to vary depending on the time of day and the stage of the ovarian cycle (Abrahams, 1974). In the early follicular phase of the cycle, the daily adrenal secretion (1.2 mgm) exceeds that from the two ovaries (0.8 mgm). The adrenal secretion is very responsive to ACTH (Baird et al., 1969) and exhibits a marked diurnal rhythm similar to that of cortisol (Tunbridge et al., 1973). Thus in the early morning the adrenal secretion may account for over 80 per cent of the total P_B. The mature Graafian follicle secretes increasing amounts of androstenedione as indicated by the fact that the concentration and P_B^A rise nearly two-fold by mid-cycle (to 4.7 mgm) (Baird, 1973). The increased P_B^A is maintained during the second half of the cycle by the secretion from the corpus luteum. The fluctuations in ovarian secretion are more apparent when the adrenal secretion is absent (as in

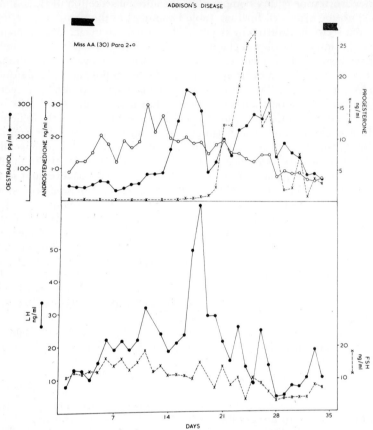

Fig.3 The concentration of oestradiol, progesterone, androstenedione, LH and FSH in women with Addison's Disease. The patient was maintained on 37.5 mgm cortisone acetate together with 0.1 mgm fluodro-hydro cortisone per day. The biphasic pattern of oestradiol, the mid-cycle peak of LH and the rise in the concentration of progesterone in the second half are characteristic of an ovulatory cycle. Note that the concentration of androstenedione rises from less than 1 ng/ml on Day 1 to greater than 3 ng/ml at mid-cycle.

adrenalectomized subjects) (Abrahams and Chakmakjian, 1973) or reduced due to suppression with dexamethasone (Abrahams, 1974) or in Addison's disease (Fig.3). In this subject the pattern of oestradiol, progesterone, LH and FSH indicate a normal ovulatory cycle. The concentration of androstenedione in the early follicular phase of the cycle was below the normal range for women, but rose four-fold toward mid-cycle.

Recent evidence in sheep suggests that androstenedione may play a role in controlling the secretion of FSH and LH. The secretion of LH is raised in ewes in which the biological activity of androstenedione has been neutralized by active immunization against 11α hydroxyandrostenedione hemisuccinate (Martensz *et al.*, 1976). In castrate ewes the sensitivity to the negative and positive feedback effects of oestrogen are enhanced by implants of androstenedione (Van Look and Scaramuzzi, unpublished observations). From these preliminary data it is not possible to determine whether the effect on secretion of gonadotrophins is due to a direct effect of androstenedione on the hypothalamic-pituitary system or due to its conversion to oestrone. When the secretion of androstenedione is elevated (as in polycystic ovarian disease) the pattern of secretion of gonadotrophins is disturbed. It has been suggested that the constant relatively high circulating levels of oestrone produced by extraglandular aromatization of androstenedione, result in the suppressed levels of FSH and elevated basal levels of LH which are characteristic of this condition (Baird, 1975).

III. OVARIAN COMPARTMENT RESPONSIBLE FOR THE SECRETION OF STEROIDS

The ovary can roughly be divided into three structural compartments - the Graafian follicle, the corpus luteum and the stroma. Each component synthesizes and secretes its own characteristic pattern of steroids. By comparing the secretion of steroids from the ovary containing the pre-ovulatory follicle or the corpus luteum with the secretion from the contralateral ovary, it is possible to obtain information of the secretory activity of the cellular compartments of the ovary *in vivo*.

The follicle secretes oestradiol, 17α hydroxyprogesterone and androstenedione although small quantities of progesterone are secreted by the mature follicle prior to ovulation at the time of the pre-ovulatory surge of LH (Johansson and Wide, 1969; Baird *et al.*, 1975). Steroids are secreted into the follicular cavity as well as into the ovarian vein so that the concentration of oestradiol (Baird and Fraser, 1975b) and progesterone (McNatty *et al.*, 1975) rises sharply in the pre-ovulatory follicle. However not all the steroids secreted into the ovarian vein are reflected by the concentration in follicular fluid. The concentration of androstenedione and testosterone remains constant within the follicle although there is a very large increase in the ovarian vein (De Jong *et al.*, 1974).

The corpus luteum secretes a range of steroids similar to that secreted by the pre-ovulatory follicle, i.e. oestradiol, 17α hydroxyprogesterone, androstenedione and progesterone. The secretion of progesterone increases markedly and the mature CL secretes over 20 mgm of progesterone per day.

The stroma or interstitial tissue secretes mainly androgens - androstenedione and testosterone. The stroma is particularly sensitive to stimulation by LH and the increased secretion of androstenedione in polycystic ovarian disease derives from this source.

IV. CONTROL OF OVARIAN STEROID SECRETION

The secretion of ovarian steroids is controlled by a negative feedback loop involving the ovary, hypothalamus and anterior pituitary. FSH is thought to be mainly concerned with the development and growth of the Graafian follicle although it potentiates the steroidogenic and ovulatory effects of LH. Administration of LH (or HCG) stimulates steroid synthesis and secretion *in vitro* (Savard *et al.*, 1965) and *in vivo* (Strott *et al.*, 1969). Specific receptors for LH are found on the membrane of the theca and granulosa lutein cell. A basal level of LH is probably necessary for the maintenance of steroid synthesis. Whether these changes in gonadotrophin secretion are due to oscillations of the negative feedback loop involving LH and oestradiol are unknown. Preliminary data in the rhesus monkey showing that the pulses of LH secretion persist after section of the pituitary stalk, suggest that they may reflect the inherent mechanism by which the pituitary gonadotrop releases its secretion (Vande Wiele and Ferin, 1974).

Short-term fluctuations in the concentration of oestradiol in plasma have also been recorded in women, although they appear to bear little relationship to the changes in secretion of gonadotrophins (Korenman and Sherman, 1973). We have recently studied the relationship between the secretion of ovarian steroids and pituitary LH in the ewe. In this species oestradiol and progesterone are secreted exclusively by the Graafian follicle and the corpus luteum respectively. The secretion of oestradiol, androstenedione and progesterone was measured serially at 10 minute intervals in 6 ewes in which the ovary and uterus had been autotransplanted to the neck (Harrison et al., 1968). During the luteal phase of the cycle pulses of LH, which occurred about every 2 hours, were followed within five minutes by a marked rise in the secretion of oestradiol (and androstenedione). The secretion reached a peak about 30 minutes after the pulse of LH and then declined gradually to control levels. In contrast over the same time period there was little, if any, stimulation in the secretion of progesterone. These findings suggest that the short term changes in the concentration of oestradiol in peripheral plasma probably reflect changes in ovarian secretion rather than alterations in the rate of metabolism. The follicle appears to be extremely sensitive to minor changes in the concentration of LH, confirming that the negative feedback loop operates mainly through follicular oestradiol (Baird et al., 1975).

In contrast, although LH is a necessary component of the luteotrophic complex in the ewe (McCracken et al., 1971), the secretion of progesterone from the corpus luteum is not under the same gonadotrophic control. Infusion of LH in large amounts through the ovary on Day 10 of the cycle fails to stimulate an increase in progesterone secretion (Land et al., 1974). Progesterone alone has very little negative feedback effect on secretion of LH (Goding et al., 1970) and the amount of progesterone secreted by the corpus luteum of the cycle is probably limited only by the biosynthetic capacity of the luteal cells.

It is unlikely, however, that the human corpus luteum is controlled by a similar mechanism. In women most of the oestradiol secreted during luteal phase arises from the corpus luteum unlike in the sheep. Thus the primate corpus luteum has direct access to the inhibitory loop of the negative feedback system controlling the secretion of gonadotrophins. For this reason it is possible that regression of the corpus luteum in women occurs because of lack of luteotrophic support in contrast to the sheep in which it is destroyed by a uterine luteolysin (prostaglandin $F_2\alpha$).

REFERENCES

Abraham, G.E. (1974). *J. clin. Endocr. Metab.* 39, 340-346.
Abrahams, G.E. and Chakmakjian, Z.H. (1973). *J. clin. Endocr. Metab.* 37, 581-587.
Baird, D.T. (1971). *In* "Control of Gonadal Secretion" (D.T. Baird and J.A. Strong, eds) pp.176-189. Edinburgh University Press, Edinburgh.
Baird, D.T. (1974). *Europ. J. Obstet. Gynec. Reprod. Biol.* 4, 31-39.
Baird, D.T. (1975). This volume, p.349.
Baird, D.T., Uno, A. and Melby, J.C. (1969). *J. Endocrinol.* 45, 135-136.
Baird, D.T., Horton, R., Longcope, C. and Tait, J.F. (1969). *Rec. Progr. Horm. Res.* 25, 611-656.

Baird, D.T. and Fraser, I.S. (1974a). *J. clin. Endocr. Metab.* **38**, 779-787.
Baird, D.T. and Fraser, I.S. (1974b). *Clin. Endocr.* **4**, 259-266.
Baird, D.T., Baker, T.G., McNatty, K.P. and Neal, P. (1975). *J. Reprod. Fertil.* **45**, 611-619.
Brown, J.B. (1971). *In* "Control of Gonadal Secretion" (D.T. Baird and J.A. Strong, eds) pp. 127-141. Edinburgh University Press, Edinburgh.
Brown, J.B. and Matthew, G.D. (1962). *Rec. Prog. Horm. Res.* **18**, 337-373.
Goding, J.R., Blockey, M.A. de B., Brown, J.M., Catt, K.J. and Cumming, I.A. (1970). *J. Reprod. Fertil.* **21**, 368-369.
Harrison, F.A., Heap, R.B. and Linzell, J.L. (1968). *J. Endocrinol.* **40**, viii.
Hembree, W.C., Bardin, C.W. and Lipsett, M.B. (1969). *J. clin. Invest.* **48**, 1809-1819.
Holmdahl, T.H. and Johansson, E.D.B. (1972). *Acta endocr. (Copenh.)* **71**, 765-772.
Ito, T. and Horton, R. (1970). *J. clin. Endocr.* **31**, 362-368.
Johansson, E.D.B, and Wide, L. (1969). *Acta endocr. (Copenh.)* **62**, 82-88.
De Jong, F.H., Baird, D.T. and Van der Molen, H.J. (1974). *Acta endocr. (Copenh.)* **77**, 575-587.
Kelly, W.G. (1970). *Steroids* **16**, 579-602.
Kirschner, M.A. and Jacobs, J.B. (1971). *J. clin. Endocr.* **33**, 199-209.
Korenman, S.G. and Sherman, B.R. (1973). *J. clin. Endocr. Metab.* **36**, 1205-1209.
Land, R.B., Collett, R.A. and Baird, D.T. (1974). *J. Endocrinol.* **62**, 165-166.
Little, B. and Billiar, R.B. (1969). *In* "Progress in Endocrinology" (C. Gaul, ed) pp.871-879. Excerpta Medica Foundation, Amsterdam.
McNatty, K.P., Hunter, W.M., McNeilly, A.S. and Sawers, R.S. (1975). *J. Endocrinol.* **64**, 555-571.
Martensz, N.J., Baird, D.T., Scaramuzzi, R.J. and Van Look, P.F.A. (1976). *J. Reprod. Fertil.* (submitted for publication).
McCracken, J.A., Baird, D.T. and Goding, J.R. (1971). *Rec. Prog. Horm. Res.* **27**, 537-582.
Rizkallah, T.H., Tovell, H.M.M. and Kelly, W.G. (1975). *J. clin. Endocr. Metab.* **40**, 1045-1056.
Savard, K., Marsh, J.M. and Rice, B.F. (1965). *Rec. Prog. Horm. Res.* **21**, 285-365.
Siiteri, P.K. and MacDonald, P.C. (1973). *In* "Handbook of Physiology", II. Female Reproductive System (R.O. Greep, ed) pp.615-629. American Physiological Society, Washington D.C.
Strott, C.A., Yoshimi, T., Ross, G.T. and Lipsett, M.B. (1969). *J. clin. Endocr.* **29**, 1157-1167.
Tait, J.F. (1963). *J. clin. Endocr. Metab.* **23**, 1285-1297.
Tait, J.F. and Horton, R. (1966). *In* "Steroid Dynamics", *Proc. Symposium on Dynamics of Steroid Hormones*, Tokyo (G. Pincus, T. Nakao and J.F. Tait, eds) pp.393-424. Academic Press, New York and London.
Tunbridge, D., Rippon, A.E. and James, V.H.T. (1973). *J. Endocrinol.* **59**, xxi.
Vande Wiele, R.L., Dyrenfurth, I. and Gurpide, E. (1968). *In* "Clinical Endocrinology II" (E.B. Astwood and C.E. Cassidy, eds) pp.603-620. Grune & Stratton, New York and London.
Vande Wiele, R.L. and Ferin, M. (1974). *In* "Chronobiological Aspects of Endocrinology" (F. Ceresa and F. Halberg, eds) pp.203-211. Schattauer Verlag, Stuttgart.

THE ROLE OF EXTRAGLANDULAR ESTROGEN IN WOMEN IN HEALTH AND DISEASE

C.D. Edman and P.C. MacDonald

*Department of Obstetrics and Gynecology
The University of Texas Southwestern Medical School
5323 Harry Hines Boulevard, Dallas, Texas 75235, USA*

The classic concepts of the physiologic impact of estrogens have evolved mainly around the secretion of estradiol by the follicle of the ovary, the growth promoting effects of estradiol on specific tissues, and the feedback mechanisms of estradiol on the brain, pituitary and ovary itself. The emphasis on these stimulating effects of estradiol, especially on protein synthesis in a few tissues, have tended to mask interest in the actions of other estrogens until recently. Only since we have begun to investigate the action and mechanisms of production of other estrogens, have we begun to escape this intellectual yoke. We now know that the commonly believed weak estrogen, estrone, and the weak androgen, androstenedione, are very active metabolically in both men and women. The demonstration that estrogen production in the human placenta was totally dependent on C_{19} precursors sharply focused attention on another mechanism of hormone production (Siiteri and MacDonald, 1963). From these studies, it became apparent that the peripheral or extraglandular conversion of androgens to estrogens in nonpregnant women and men was an important determinant in the hormonal milieu in both healthy and diseased states. Since our main theme is the ovary, we shall confine our discussion to the effects of extraglandular estrogen production in women.

The possible sources of estrogen production are 1) by direct glandular secretion and/or 2) by extraglandular formation. When discussing total estrogen production in women we need to identify clearly whether the patient is premenopausal or postmenopausal. Direct glandular secretion refers mainly to estradiol secretion, although small amounts of estrone may be secreted directly into blood by the ovary. In anovulatory premenopausal women, the extraglandular pathway(s) accounts for almost all of the estrogen production (Siiteri and MacDonald, 1973a). In postmenopausal women, total estrogen production occurs almost exclusively by the extraglandular mechanism. The adrenal and postmenopausal ovary secrete little if any estrogen. Extraglandular estrogen(s) arises principally, if not exclusively, by the aromatization of plasma and androstenedione to yield estrone (Grodin et al., 1973). In the female the conversion of testosterone to estradiol is very inefficient and only small amounts of testosterone are available for extraglandular metabolism, thus almost no estradiol is produced by this mechanism in normal women. Therefore, when we talk about extraglandular estrogens, we are speaking mainly of estrone.

The contribution of total estrone production made by extraglandular pathways, i.e. the conversion of circulating plasma androstenedione to estrone, in a nonpregnant normal ovulatory female was measured during various phases of the ovulatory cycle. The values were found to vary from 50 µg in early proliferative phase to 80 µg in midproliferative phase, 180 µg at ovulation, 140 µg in midluteal phase, and 80 µg in late luteal phase (Siiteri and MacDonald, 1973b). The amount of estrone produced by the extraglandular conversion of androstenedione is relatively constant, averaging about 40 µg/day, despite the fluctuating secretion of estradiol by the developing follicles or corpus luteum. Thus, 10 to 50% of the total estrone production arose from the extraglandular aromatization of androstenedione. The remainder of the estrone production is derived principally from the ovarian secretion of estradiol, which is metabolized peripherally to estrone and thereby contributes to the total estrone production rate. Several investigators have reported a plasma production rate of androstenedione in ovulatory women to be about 3000 µg/day (Baird et al., 1969). The average conversion of androstenedione to estrone in 6 normal ovulatory women was 1.3% (MacDonald et al., 1967a). Thus, the conversion of androstenedione to estrone normally contributes about 40 µg/day to total estrogen production in healthy, ovulatory premenopausal women.

The increased production of estrone by extraglandular mechanisms may occur as the result of one or both of two metabolic events. First, an increase in the extent of conversion of plasma androstenedione will increase the amount of estrone formed. Second, an increase in the production rate of the androgen, androstenedione, will also increase the extraglandular production of estrone.

The increase in the extent of conversion of plasma androstenedione to estrone in extraglandular sites has been investigated in many clinical situations (MacDonald et al., 1967b). The extent of conversion was computed for 38 premenopausal women who included normal, castrate, and adrenalectomized individuals as well as patients with arrhenoblastomas, polycystic ovarian disease, and gonadal dysgenesis. The extent of conversion expressed in percent varied from 1.2 to 1.4% in patients with pathologic problems. These values are nearly the same as the extent of conversion observed in normal ovulatory women.

When the estrone production in normal women 1 to 5 years following menopause was investigated, it was found that the extent of conversion was increasing. In this group of normal postmenopausal women, the extent of conversion of plasma and androstenedione to estrone was 2.7% compared to 1.3% observed in normal premenopausal women. However, the production rate of androstenedione was only 1,700 µg/day or approximately half that of premenopausal women. Androstenedione production, at this time in life, is normally from adrenal secretion. Judd and co-workers (1974) demonstrated that the postmenopausal ovary secretes little if any androstenedione. Thus, the daily production rate of estrone was 35-40 μg in these normal postmenopausal women. Furthermore, that amount of estrone produced by the extraglandular aromatization of plasma androstenedione is nearly the same as the total production of estrone that was secreted or was formed from extraglandular mechanisms as measured by the specific activity of urinary metabolites of estrone.

After observing a 2-fold increase in the extent of conversion in women in early menopause, an investigation of the estrone production in women more than 10 years beyond the menopause was begun. The amount of precursor, androstenedione, was unchanged from the early postmenopausal group. But the extent of conversion was continuing to increase, 4.5%. The resultant estrone production was also increased to 85 to 90 μg/day. Again, all the estrone production could be accounted for by extraglandular aromatization.

The next logical step was to see whether age was an independent variable influencing the observation of the increased conversion of androstenedione to estrone. When 23 women, ages 18-75 and controlled for weight, 120-150 lbs, were compared, a direct linear correlation between the conversion of androstenedione to estrone and age was found. The same correlation between the conversion and age was found for men of similar body weight. By the seventh decade this conversion had increased 3-4 fold over the younger counterpart of similar body weight. Although adrenal secretion of androstenedione tends to decrease in the elderly, nevertheless, a mechanism for increased estrone production is apparent and can account for the clinical findings of the well-estrogenized vaginal mucosa often observed clinically in elderly postmenopausal women without other pathology.

The group of women 10 years beyond the menopause had an average weight of 205 lbs as a group. Was weight also a variable which increased the extent of conversion of androstenedione to estrone? In a study of 88 women, a direct correlation was found between the conversion of androstenedione to estrone and total body weight in postmenopausal women with and without endometrial cancer as well as in premenopausal women. Not only did the extent of conversion rise with increasing body weight, but there was a two-fold difference between the extent of conversion in postmenopausal and premenopausal women. Furthermore, the conversion of androstenedione to estrone was nearly identical in postmenopausal women with and without endometrial cancer. We conclude, therefore, that this increased conversion is the consequence of the additive effects of obesity and aging and is not a unique finding in women with endometrial carcinoma.

The extent of conversion of androstenedione to estrone may be altered by hepatic disease. Liver failure has been implicated as a cause of hyperestrogenism in both men and women. In women of similar age and with severe hepatic disease (mainly cirrhosis), there is a 2-7 fold increase in the extent of conversion of androstenedione to estrone which resulted in an average production rate of 140 μg of estrone per day for these subjects. Likewise, the extent of conversion was also altered by increasing body weight in hepatic disease. Again, there was a direct correlation between the extent of conversion of androstenedione to estrone and body weight in 13 postmenopausal women with and 33 postmenopausal women without cirrhosis. The non-cirrhotic group included normal women, those with postmenopausal bleeding, with adenomatous atypical hyperplasia of the endometrium, and with adenocarcinoma of the endometrium. The extent of conversion observed in the group with cirrhosis was increased two-fold at any given weight over that observed in non-cirrhotic postmenopausal women irrespective of their endometrial histology. Thus, compared to subjects of similar age and body weight, the increased extent of conversion represents as much as a combined 10-fold increase in the extent of extraglandular aromatization of plasma androstenedione in hepatic disease. Again, the same correlation exists between the extent of conversion of androstenedione to estrone and body weight in the non-cirrhotic groups as was found in a previous study.

The extraglandular production of estrogen, principally estrone, was increased in all cirrhotic subjects. In this group of 13 patients with hepatic disease, 10 had endometrial neoplasia, 2 had uterine bleeding with proliferative endometrium; and a single subject with a daily estrone production of 56 μg, did not have uterine bleeding. This increased extraglandular formation of estrone appears, in part, to be the consequences of decreased hepatic extraction of androstenedione. In one postmenopausal patient with cirrhosis of the liver and adenomatous hyperplasia of the endometrium, we computed the hepatic extraction of androstenedione to be 9.6% compared to the 80-90% hepatic extraction observed in normal subjects reported by Rivarola *et al.* (1967). Furthermore, there was little transplanchnic or hepatic aromatization of plasma androstenedione observed in this subject. Likewise, the hepatic removal of estrone from plasma was low, 7%. These data suggest that extensive extrahepatic metabolism of androstenedione is occurring and that a principal site of aromatization might be adipose tissue. Studies are currently underway to investigate this hypothesis. Thus, to recapitulate, aging, obesity, and hepatic disease all may act in concert to increase the efficiency of plasma androstenedione to estrone to as much as a 20-fold or greater increase in the older obese woman. Furthermore, the extent of conversion of plasma androstenedione to estrone is not altered by oophorectomy, adrenalectomy, or both.

Pathologically, the increased production of the substrate, androstenedione, is tantamount to the increased efficiency of conversion in many clinical problems. The potential for the ovarian contribution of plasma prehormones exist in both the pre- and postmenopausal woman. As we indicated earlier, under normal circumstances, the postmenopausal ovary does not constitute a significant source of plasma androstenedione. However, clearly the postmenopausal ovary can synthesize androstenedione *in vitro* (Plotz *et al.*, 1967) and higher concentrations of androstenedione have been found in the ovarian venous blood than in the peripheral blood in postmenopausal women. The magnitude of the gradient between the concentrations of androstenedione in the ovarian venous blood is exceedingly small (Mikhail, 1970), therefore, while the capacity for ovarian contribution of androstenedione exists, it is unlikely that the ovary contributes a significant source of the plasma precursor under normal conditions.

In premenopausal women with chronic anovulation, but continued gonadotrophin production, a state analogous to chronic estrus may obtain. Moreover, the source of estrogen in these cases is extraovarian (MacDonald and Siiteri, 1973c). The maintenance of uterine and breast development, plus the predictable withdrawal uterine bleeding to progesterone indicate active estrogen production. A defect in this disorder, commonly known as polycystic ovarian disease, is the failure of follicular maturation and the hyperplasia of thecal and stromal elements which result in an increased secretion of androstenedione by the ovary. The estrogen production in a group of six anovulatory women with polycystic ovarian disease was predominantly from the aromatization of androstenedione. The mean age of this group was 24 years, and the mean body weight was 158 lbs. The secretion of androstenedione was 10.6 mg/day or 3-5 times that of normal ovulatory women. The extent of conversion was normal, 1.3%, but the production of estrone was definitely elevated 3-4 fold, 140-160 µg/day. Extraglandular aromatization of androstenedione accounted for almost all of the estrone production in this study. It is tempting to speculate that this large source of constant extraglandular estrogen is etiologically related to the acyclicity of gonadotrophin release and ultimately to the failure of adequate follicular maturation. This mechanism would readily explain the therapeutic effects of the ovarian wedge resection and adrenal suppression in some patients since these maneuvers are known to decrease androstenedione production. It is important to note that the hormonal milieu of polycystic ovarian disease is the extraglandular production of estrone rather than the follicular production of estradiol. This is the same hormonal milieu in which endometrial neoplasia exist and fully 80% of women under age 40 who develop endometrial cancer have co-existing polycystic ovarian disease.

In certain pathologic circumstances, androstenedione may be produced in large amounts by the abnormal postmenopausal ovary. For example, when the ovarian stromal hyperplasia is found in association with such nonsecretory ovarian tumors as the Krukenberg tumors, mucinous or serous cystadenoma or carcinoma, and Brenner tumors, there is direct secretion of massive amounts of androstenedione, 20 mg/day. The tumor cell *per se* does not secrete the steroidal hormones, only the surrounding theca-stromal cells secrete possibly under the influence of a glycoprotein secreted by the tumor, such as human chorionic gonadotrophin. A case in point is a 57 y.o. subject who presented with postmenopausal bleeding and was found to have an ovarian mucinous cystadenoma and adenomatous hyperplasia of the endometrium. In this case, there was marked hyperplasia of the theca-stromal elements. The androstenedione production rate was 11.7 mg/day. The extent of conversion was low, 1.5% for this age group but the estrone production was massive, 170 µg/day. Again, all of the estrone was formed by extraglandular aromatization. Moreover, these levels of estrone are commonly associated with endometrial neoplasia.

In some secretory tumors, the arrhenoblastoma, hilar cell tumor, and lipoid cell tumor of the ovary, massive amounts of androstenedione are formed. In an anovulatory 37 y.o. subject with a virilizing lipoid cell tumor of the ovary and ovarian stromal hyperplasia, androstenedione production was massive, 31 mg/day. These levels of androstenedione produced 340 µg of estrone each day by the extraglandular routes.

These data clearly establish that the mechanism for the increased estrone production in those young women at risk to develop endometrial carcinoma is a marked increase in androstenedione production, principally the consequences of secretion by abnormal ovaries. However, irrespective of the origin of plasma androstenedione, i.e. whether adrenal or ovarian, androstenedione entering the plasma represents the principal if not exclusive precursor for extraglandular production in women.

Our principal interest in extraglandular estrogen formation was the increased incidence of endometrial neoplasia found in women with increased extraglandular estrone. In a study of 46 postmenopausal women, estrone production ranged from 45 µg/day in 8 normal women to 155 µg/day in 8 women with non-neoplastic endometrium, but postmenopausal bleeding; 165 µg/day in 7 women with adenomatous atypical hyperplasia of the endometrium; and 122 µg/day for 23 women with endometrial cancer. The cirrhotic group was producing 140-180 µg/day and was discussed earlier. In the nonbleeding group the mean daily production of estrone, 45 µg, was the same by both the blood and urinary method. This finding provides direct evidence that all estrone produced at extraglandular sites entered the plasma as estrone. In the four groups with uterine bleeding, the average mean daily production of estrone was 150 µg. These levels are more than 3 times greater than that of the normal postmenopausal woman. There is no statistically significant difference among the four groups with uterine bleeding, but a highly significant difference between the four bleeding groups and the nonbleeding group. Additionally, in this study, all levels of estrone production above 70 µg/day was associated with uterine bleeding in all women studied.

In summary, estrone is the near exclusive estrogen produced in postmenopausal women. Aging, obesity, and hepatic disease increase estrone production and by the additive effect of each, the extent of conversion of plasma androstenedione to estrone increases independent of any neoplasia. The availability of the substrate, androstenedione, has been found to be massively increased in some subjects with hepatic disease and abnormal ovarian states. Estrone production was shown to be increased at least 3-fold in subjects with neoplastic changes of the endometrium. Moreover, in every patient in whom the estrone production exceeded 70 µg/day, uterine bleeding occurred. These often associated findings of endometrial aberrations suggest that the extraglandular production of estrone is involved in some yet undefined mechanism in the development of endometrial neoplasia. Moreover, those conditions which give rise to the increased production of extraglandular estrogens are precisely the same conditions that favor the increased occurrence of endometrial neoplasia, namely, menopause and aging, obesity, liver disease, polycystic ovarian disease, and some ovarian tumors (MacDonald and Siiteri, 1974).

REFERENCES

Baird, D.T., Horton, R., Longcope, C. and Tait, J.F. (1969). *Rec. Prog. Horm. Res.* **25**, 611-656.
Grodin, J.M., Siiteri, P.K. and MacDonald, P.C. (1973). *J. clin. Endocr. Metab.* **36**, 207-214.
Judd, H.L., Lucas, W.E. and Yen, S.S.C. (1974). *Am. J. Obstet. Gynec.* **118**, 793-798.
MacDonald, P.C., Rombaut, R.P. and Siiteri, P.K. (1967a,b). *J. clin. Endocr. Metab.* **27**, 1103-1111.
MacDonald, P.C. and Siiteri, P.K. (1974). *Gynec. Oncology* **2**, 259-263.
Mikhail, G. (1970). *Gynec. Invest.* **1**, 5-20.
Plotz, E.J., Weiner, M., Stein, A.A. and Hahn, B.D. (1967). *Am. J. Obstet. Gynec.* **99**, 182-196.
Rivarola, M.A., Singleton, R.T. and Migeon, C.J. (1967). *J. clin. Invest.* **46**, 2095-2100.
Siiteri, P.K. and MacDonald, P.C. (1963). *Steroids* **2**, 713-730.
Siiteri, P.K. and MacDonald, P.C. (1973a,b). *In* "Handbook of Physiology", Section 7.2 (S.R. Geiger, E.B. Astwood and R.O. Greep, eds) pp.615-629. The American Physiologic Society, Washington, D.C.

THE ROLE OF SEX HORMONE BINDING GLOBULIN IN HEALTH AND DISEASE

D.C. Anderson

Senior Lecturer in Medicine, University of Manchester and Honorary Consultant Physician, Manchester Royal Infirmary Manchester M13, England

Sex hormone binding globulin (SHBG), whose existence in the human has been known for ten years now (Mercier *et al.*, 1966), belongs to a family of chemically — and functionally — related plasma glyco-proteins which, apart from thyroxine-binding globulin, bind steroid hormones. Particularly when we consider those proteins that bind oestrogens, androgens or progestogens (the sex-hormone-binding globulins) we find a great diversity in binding characteristics of these proteins between species (Westphal, 1971; Raynaud *et al.*, 1971; Mahoudeau and Corvol, 1973; Gauthier-Wright *et al.*, 1973). There are also wide variations in concentration within the same species under different physiological and pathological states (Anderson, 1974).

This chapter concentrates principally on the human SHBG and will examine its possible role in health and disease in the human. After a brief review of the physical and chemical properties, the influence of SHBG upon the unbound concentrations of steroids, and upon their metabolic clearance rates will be discussed. Then some of the hormonal influences on SHBG will be considered. An attempt will be made to relate the known physiological changes in SHBG concentration to these hormonal influences. Some of the imperfections of our present understanding will be illustrated with a comparative physiological study of human and non-human primate SHBG. Finally, the possible role of SHBG in disease states will be raised and in particular whether SHBG has a truly hormonal role or is merely a hormonal barometer, which reflects but does not influence the balance between different hormones such as oestrogens and androgens. To be provocative I shall finally try to put SHBG into the wider context of sex hormone interaction and antagonism by proposing that SHBG may be one link in a

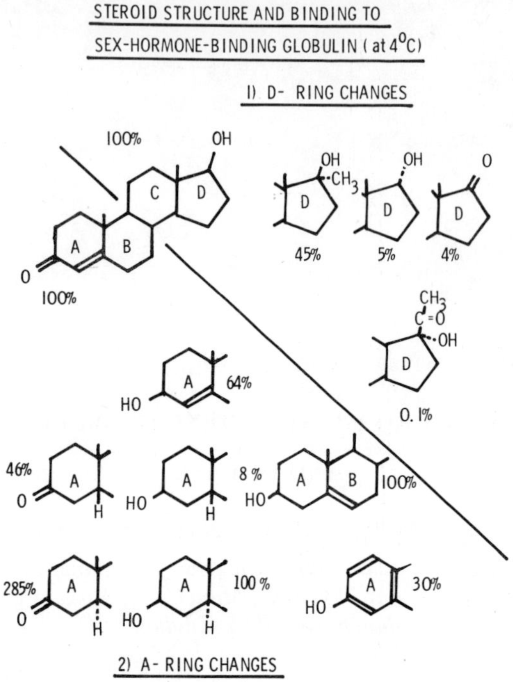

Fig.1 Binding characteristics of sex hormone binding globulin (from Anderson and Galvão-Teles, 1975). Observations made in diluted plasma at 37°C, using 3H-testosterone as ligand and florisil adsorption.

cascade system by which oestrogens and other hormones can oppose androgen action and *vice versa*.

I. PHYSICAL AND CHEMICAL PROPERTIES OF SHBG

SHBG, whose source is generally assumed to be the liver, has a molecular weight of about 100,000 (see Anderson, 1974; Rosner and Smith, 1975) and is relatively stable in plasma, but very labile in the pure state. The binding constant of the protein for testosterone at 37°C is in the region 10^9 M^{-1} (Burke and Anderson, 1972) which is one or two orders of magnitude less than that for the intracellular receptor proteins and 30,000 times greater than that for albumin. Heyns and De Moor (1971a) have shown association and dissociation rates to be very rapid; at 37°C the half-time of dissociation is about ten seconds.

The binding characteristics of SHBG are summarized in Fig.1. Changes in either the A or the D rings profoundly affect the capacity of the steroid molecule to bind to SHBG. If we assign to testosterone a potency of binding of 100%, 5α-dihydrotestosterone (DHT) binds nearly three times as well, and oestradiol about one third as well. Another steroid whose binding to SHBG may possibly be important is Δ5-androstenediol, which binds as well as does testosterone. We would therefore predict that the SBHG concentration will influence the percentage of unbound dihydrotesterone in plasma more than that of testosterone and androstenediol, which in turn will be affected more than that of oestradiol.

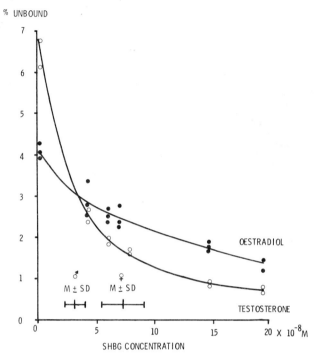

Fig.2 Differential effect of sex hormone binding globulin on testosterone and oestradiol binding in pools of steroid-free plasma *in vitro* at 37°C, pH 7.4. Percentage unbound T and E_2 measured by steady-state gel filtration, and SHBG by competitive adsorption (Anderson *et al.*, 1972). From Burke and Anderson (1972).

Androstenedione and dehydroepiandrosterone hardly bind to the SHBG at all.

From a practical point of view the high affinity of DHT, and its great specificity now make this steroid the preferred ligand in assays for determining the SHBG binding-site concentration (Corvol *et al.*, 1971; Heyns and De Moor, 1971b; Rosner, 1972; Anderson *et al.*, 1972; Tulchinsky and Chopra, 1973; Murray *et al.*, 1975).

II. EFFECTS OF SHBG ON THE UNBOUND CONCENTRATIONS

Figure 2 shows the results of an *in vitro* study carried out by Burke and Anderson (1972) in steroid-free plasma pools, in which the effect of changes in SHBG concentration was related to the percentage unbound T and E_2 measured by steady-state gel filtration. There is a marked effect for testosterone and relatively little effect for oestradiol. This is partly because of the higher affinity of T than E_2 for SHBG and partly because of the greater affinity of E_2 for albumin (Fig.3). These observations also hold when the comparison is made in plasma from a wide variety of different patients. As shown in Fig.4, there is a reasonable

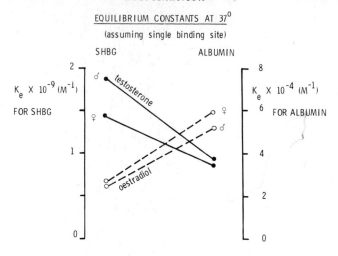

Fig.3 Equilibrium constants at 37°C for binding testosterone and oestradiol, to SHBG and Albumin. From Burke and Anderson (1972).

linear correlation between the logarithm of SHBG concentration and the percentage unbound T, at least over the range of $2 - 25 \times 10^{-8}$ mole/litre (Fisher and Anderson, 1975). On the other hand there is a very poor correlation, as shown in Fig.5, between percent unbound oestradiol and log SHBG. If comparisons of the latter are made in diluted plasma a better relationship is seen (Tulchinsky and Chopra, 1973) which is probably artifactual due to diluting out the relatively weak albumin binding.

A central question regarding the role of plasma steroid binding-proteins concerns whether the unbound steroid in plasma alone is biologically active. The question can be phrased more accurately: "Does a steroid in plasma influence the end-organ response solely as a function of its unbound concentration?" There is not time to discuss this in detail but extensive data reviewed previously (Anderson, 1974) suggests at least that the *bound* steroid is inactive. Only non protein-bound steroid can diffuse out of the plasma compartment into the interstitial space. We would expect the concentration here in turn to determine either the free steroid concentration within the cell, or the flux of steroid into the cell. It seems inherently unlikely that the target tissue can be directly influenced by steroid bound to a protein or proteins of molecular weight 100,000 in a fluid (plasma) which (except in the case of liver cells) is not even bathing the cells directly. Clearly other factors within the cell will influence the biological activity; much will be said at this meeting on this subject by others, but I shall at the end of this chapter try to put the free plasma concentration into the wider context, clearly of relevance in disorders of sex-hormone balance.

The unbound concentration is a complex function of the total concentration of the steroid, and the concentration and affinities of the proteins to which it binds. It is also influenced to a small extent by the concentration of steroids competing for binding sites. However, at physiological levels of T, E_2 and SHBG the percentage unbound steroid is relatively little affected by the total concentration of that steroid, or of other competing steroids.

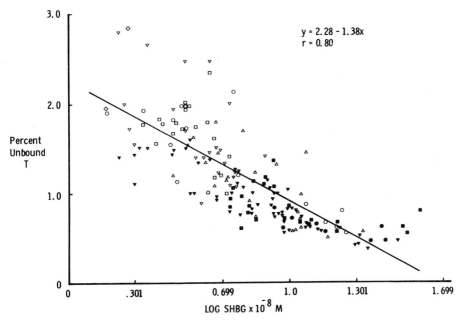

Fig.4 Correlation between percent unbound T and log SHBG in plasma samples from subjects under a wide variety of conditions. From Fisher and Anderson (1975).

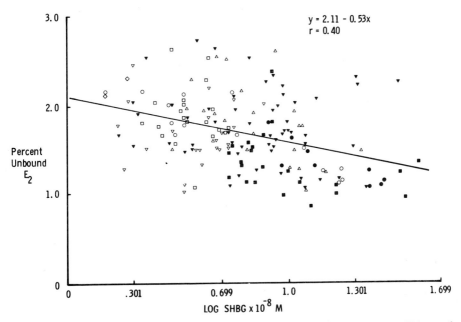

Fig.5 Much poorer correlation between percent unbound E_2 and log SHBG. From Fisher and Anderson (1975).

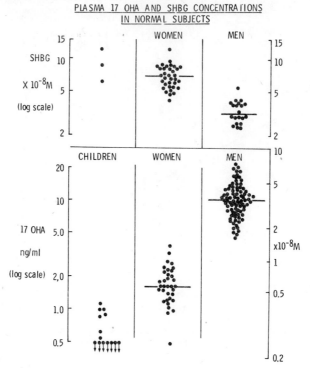

Fig.6 Plasma 17β-hydroxy androgen (17OHA) and SHBG concentrations in normal subjects. From Anderson (1974). Note SHBG concentrations in adult men are approximately half those in women.

Table I. Relation between metabolic clearance rates and SHBG binding.

Steroid	MCR (L/m² /24 h) Adult ♂	♀	Relative[1] SHBG Binding
5α-Dihydrotestosterone[2]	340	150	285
Testosterone[3]	635	370	100
Oestradiol[4]	990	790	30
Androstenedione[3]	1270	1130	4

1 Own data; by competitive adsorption with florisil in dilute plasma at 4°C. 2 Mahoudeau et al., 1971. 3 Southren et al., 1968. 4 Longcope et al., 1968 (all mean values).

As first suggested by Daughaday (1959), in the case of CBG, protein-binding will damp out the effect of acute variations of steroid secretion upon the unbound concentrations of the steroid. It will do this by increasing the size of the circulating steroid pool (and of the total steroid level) at any given unbound level. Such a dampening effect might be of great physiological importance in feed-back regulation, although little is known about this.

Plasma protein-binding undoubtedly influences the metabolic clearance rate of steroids. Table I shows published metabolic clearance rates in men and women

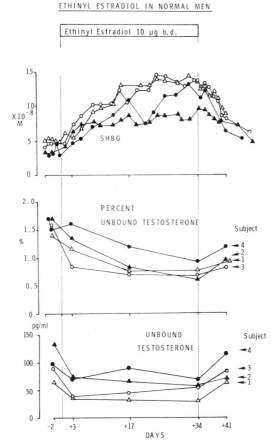

Fig. 7 Effect of administration of Ethinyl Oestradiol 20 g twice daily in four normal men, on SHBG (top), percent unbound T (middle) and actual unbound testosterone (bottom). Note slow, progressive 3-fold rise in SHBG, and complementary fall in percent unbound T.

for 4 steroids, DHT, T, E_2 and androstenedione and their relative binding to SHBG measured *in vitro*. It is evident that there is a marked decrease in clearance rate of the steroid as binding of the steroid to SHBG increases. The sex-difference in clearance rates can be accounted for by higher SHBG levels in women than men, and is not seen for instance with such steroids as androstenedione which do not bind well to this or another high-affinity plasma protein.

III. HORMONAL INFLUENCES ON SHBG: POSSIBLE AMPLIFIER FUNCTION

We now turn briefly to reviewing hormonal influences on SHBG, with some examples from physiological and pharmacological experiments.

A   Oestrogens and Androgens
As shown in Fig.6, SHBG levels are about two-fold higher in adult women

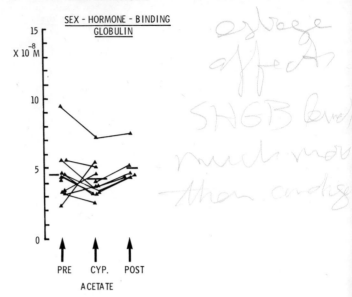

Fig.8 a) Effect of Cyproterone Acetate (100 mg/day for 5 weeks) on SHBG in adult male sexual offenders. No significant change. b) Same subjects, given Ethinyl Oestradiol 20 µg/day for 5 weeks, to show marked effect on SHBG (as seen also in normal men).

than in men. This difference can be simply accounted for by the sensitivity of SHBG production to oestrogens, and the antagonistic effect of androgens. For example, Fig.7 shows the results of administering 20 µg/day of ethinyloestradiol for five weeks to four healthy men. The total T levels changed relatively little during treatment, although there was a significant fall in the unbound levels of T. There was a progressive three-fold rise in SHBG, due to the small dose of oestrogen, which clearly occurred in the face of only a slight fall in production of T. A significant rise in SHBG in men can be demonstrated with as little as 5 micrograms per day of EE_2 (unpublished data). This effect is much more dramatic than that of oestrogens on other human plasma binding proteins (Kley *et al.*, 1973). As shown by Vermeulen *et al.* (1969), and illustrated later in this chapter, when we discuss hirsutism, androgens antagonise this effect. However, studies with cyproterone acetate suggest that this antagonism may not be mediated directly through the specific androgen receptor. We had the opportunity to study the SHBG levels in a group of adult male sexual offenders, who had been treated with cyproterone acetate therapeutically for 5 weeks. The results are shown in Fig.8; there is clearly no effect of the anti-androgens on the low levels of SHBG found in men, in contrast to the effect of ethinyloestradiol.

Before considering other hormonal influences it is appropriate to consider briefly the so-called "amplifier hypothesis", which we proposed (Burke and Anderson, 1971, 1972) as the result of these observations. To recapitulate, the effect of changes in SHBG concentration on the unbound oestradiol fraction is relatively small, and that on testosterone much greater; on the other hand, oestrogens have a marked effect in stimulating SHBG production, while testosterone and other androgens have the opposite effect. In Fig.9, taken from that publication, we depict the unbound testosterone and oestradiol concentrations

Role of Sex Hormone Binding Globulin in Health and Disease

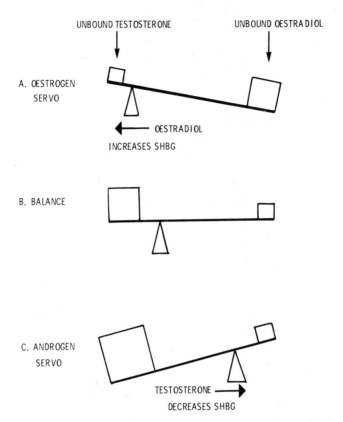

Fig. 9 Postulated role of SHBG as an amplifier in testosterone-oestradiol balance (from Burke and Anderson, 1972). See text for discussion.

as boxes on either end of a see-saw. The fulcrum is SHBG, and its concentration is depicted by the position of the fulcrum. An increase in oestradiol secretion will of course lead to an increase in unbound oestradiol level. In the long-term it will also stimulate SHBG production, and thereby selectively reduce the unbound testosterone fraction. Conversely an increase in androgen production will increase the unbound testosterone and, by lowering SHBG, amplify its own effect relative to oestradiol. SHBG might thereby itself amplify any change in oestrogen-androgen balance. Furthermore, any third factor that modifies SHBG production, might exert an influence on the balance between oestrogens and androgens. The principal attractions of this hypothesis are as follows:

    a) It gives a physiological role to the physico-chemical characteristics of SHBG, and the marked and opposing effects of oestrogens and androgens on its level.

    b) In general, states with elevated SHBG levels, including those such as thyrotoxicosis where the elevation may be a direct effect of thyroid hormones, are associated with feminization and those with low SHBG with virilization.

Table II. Metabolic clearance rates and percent unbound T and E_2 compared.

Steroid			MCR L/m²/24 h	% Unbound[3]	Ratio $\frac{MCR}{\% \text{ Unb}}$
Testosterone	Normal	♂	635[1]	1.71	371
(T)	adult	♀	370[1]	1.02	363
Oestradiol	Normal	♂	990[2]	1.79	553
(E_2)	adult	♀	790[2]	1.49	530

1 Southren et al. (1968). 2 Longcope et al. (1968). 3 Fisher et al. (1974).

However, there are several objections to this hypothesis. It would be irrelevant, for instance, if oestrogen and androgen production were regulated independently; however, in the male at least they appear to be closely interlocked; and it is unlikely that androgen levels are closely monitored and regulated in the female. The most serious objection is that it does not take account of the effects of SHBG upon metabolic clearance rates. This can be seen if we consider the relationships as shown below.

$$\text{Unbound (S)} = \text{Total (S)} \times \% \text{ Unbound (S)}$$

$$\text{Total (S)} = \frac{\text{Production rate (S)}}{\text{MCR (S)}}$$

$$\therefore \text{Unbound (S)} = \frac{\text{Production rate (S)} \times \% \text{ Unbound (S)}}{\text{MCR (S)}}$$

Now, as already discussed, an increase in protein-binding of the steroid lowers both the unbound fraction of the steroid (% unbound (S)) and its metabolic clearance rate (MCR (S)). It is evident therefore that the increase in SHBG alone will only lower the unbound *concentration* of testosterone as opposed to its unbound *fraction* if the fall in percentage unbound is proportionately greater than the fall in its clearance rate. The available evidence suggests that the unbound fraction and the metabolic clearance rate of testosterone, in some circumstances at least, do fall in proportion with increased SHBG. Certainly if we compare our own direct measurements of the % unb T and % unb E_2 to published clearance rates in men and women (Table II) the differences in unbound fractions for T and E_2 in men and women are reflected quite closely in differences in their metabolic clearance rates. This would minimise any amplifier function for SHBG on unbound T:E_2 balance *in vivo*.

Nevertheless it remains true that the binding characteristics of SHBG are such that changes in its concentration differentially affect the percentage unbound T to a greater extent than that of E_2. In any situation in which this is not cancelled out by a simultaneous effect on the metabolic clearance rates of the steroids then we predict that the change in SHBG will alter the balance between unbound T and E_2 concentrations.

B   *Other Hormonal Influences*

Progestational agents (Forest and Bertrand, 1972), glucocorticoids (Vermeulen et al., 1969), growth hormone (De Moor et al., 1972) and thyroid hormones (Crépy et al., 1967; Ruder et al., 1971; Anderson, 1974; Akande and Anderson, 1975) may all exert an action on SHBG *in vivo*. Of these, only thyroid

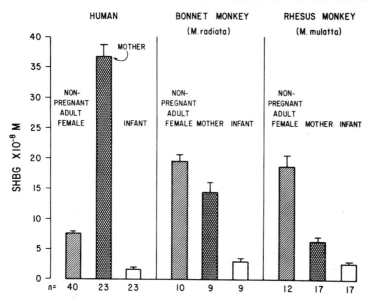

Fig. 10 Sex hormone binding globulin levels in humans and monkeys compared (from Anderson *et al.*, 1975b).

hormones have been found to stimulate SHBG production. It is not known whether these effects are independent of one another, or mediated by interfering with oestrogen or androgen action on SHBG synthesis, or indeed on its half-life. Because of the complexities of hormonal interactions, it seems unlikely that the mechanisms of regulation of SHBG production will be solved by *in vivo* experiments. It will certainly remain an open question, until an *in vitro* model system is developed, whether SHBG production is directly modulated by thyroid hormones or whether all effects are due to influences on oestrogen and androgen action.

IV. PHYSIOLOGICAL AND PATHOLOGICAL CHANGES IN SHBG

To date I know of no really convincing evidence for a structural abnormality of SHBG in the human, analagous to the well-described defects in CBG or TBG.

Time does not permit an exhaustive review, and so I shall concentrate on a few specific aspects, not considering for instance the increase in SHBG in men with advancing age (Vermeulen *et al.*, 1972) and only briefly referring to the changes in hepatic cirrhosis (Rosenbaum *et al.*, 1966; Vermeulen *et al.*, 1969; Galvão-Teles *et al.*, 1973).

A Physiological Changes in SHBG in Pregnancy and the New Born

In collaboration with Fisher, Lasley and Hendrick (Anderson *et al.*, 1975b) I have recently compared the SHBG levels in three primate species, in mother and foetus at term. The high maternal and low foetal levels in the human at term

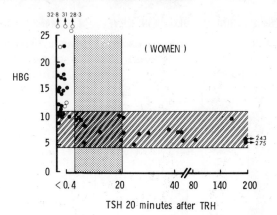

Fig.11 Relation between SHBG levels and the TSH response to TRH, in women with thyrotoxicosis on carbimazole therapy. Open circles untreated, closed circles on therapy. Note that TRH response returns only when SHBG levels are normal.

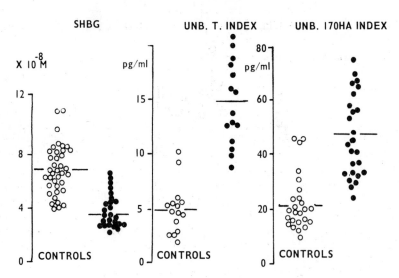

Fig.12 Basal levels of SHBG, unbound testosterone index, and unbound 17OHA index, in hirsute women (from Anderson *et al.*, 1975a).

have been known since the studies of Rivarola *et al.* (1968). We confirmed that there is a 20-fold gradient from mother to foetus which is independent of foetal sex. Tulchinsky and Chopra (1973) have shown in the human neonate that SHBG rises in the first few days after birth. In this situation at least, oestrogens can hardly be exercising an important role, since E_2 levels are high and fall abruptly after birth.

To examine this further we looked at the Rhesus (*M. Mulatta*) and the Bonnet monkey (*M. Radiata*), to see if these species might provide suitable animal models for studying this question. We found evidence, in keeping with that of

Gauthier-Wright et al. (1973) in the stump-tailed macaque, for a very similar protein. However, the non-pregnant levels were much higher in these species than in the human (Fig.10). In contrast to changes in the human, SHBG levels *fell* modestly during pregnancy in the Bonnet monkey mother, and markedly in the Rhesus. Foetal levels were similar to those in the human, in both species. Furthermore, in the Bonnet monkey, we studied the changes post-partum, and found a very marked and brisk rise in SHBG in the newborn period. This may in part be accounted for by a post-natal surge in thyroid hormones, which we found, and has been well documented in the human previously (Abuid et al., 1974).

Recently, Resko et al. (1975) have reported oestrogen levels in the Rhesus mother and foetus. It is well known that oestriol production is much lower in the macaque than in the human during pregnancy. However, there is a marked rise in E_1 and E_2 production and yet maternal SHBG levels fall markedly. Clearly, either oestrogens influence SHBG production in the macaque monkey and the human in different ways; or other hormonal influences are negating the oestrogen effect. Interestingly, a similar fall has been shown to occur in maternal CBG levels in Rhesus pregnancy (Beamer et al., 1973).

In the human, the magnitude of the transplacental SHBG gradient is such that quite possibly it has a physiological role as a magnet, drawing steroids that bind to it, such as dihydrotestosterone, 5α-androstenediol and testosterone, from the foetal circulation across the placenta into the maternal circulation for degradation.

B Pathological Changes in SHBG Concentration

SHBG levels are high in men with cirrhosis of the liver, in whom recent work has demonstrated little change in oestradiol, but increased plasma levels of estrone and oestriol, and a marked fall in unbound testosterone (Galvão-Teles et al., 1973; Kley et al., 1975; Thiyssen et al., 1975; Green et al., 1975). SHBG levels are even higher in active thyrotoxicosis in both sexes; interestingly, in this disorder the sex difference in SHBG levels is preserved. In this situation, SHBG is an excellent thyroid function test, of comparable value to the TRH test. Figure 11 shows the relation between the SHBG concentration on the ordinate, and the response to TRH on the abscissa, in 7 women with thyrotoxicosis before and during carbimazole therapy. In the untreated thyrotoxic state (open symbols) there is a four- to five-fold elevation in SHBG, and no response to TRH. Lagging a week or so behind the fall in free T_3 and T_4 levels the SHBG level falls to normal and there is excellent agreement between normalisation of the SHBG and the return of the TRH response.

Low levels of SHBG are found in acromegaly, and De Moor et al. (1972) have shown that the protein appears to be inhibited by administration of growth hormone in G.H. deficiency. It is indeed tempting to speculate that it is high G.H. levels *in utero* combined with low tridothyronine levels which are inhibiting SHBG production before birth.

The most frequent disorder in which SHBG plasma levels are reduced is hirsutism. The magnitude of this fall is demonstrated in Fig.12. This is from a study of the effects of long-term therapy in 29 hirsute women (Anderson et al., 1975a). Mean basal SHBG levels are reduced to about half normal in hirsute women. Unbound testosterone levels were, in these subjects, quite consistently raised, because of a modest rise in total testosterone, combined with the low SHBG. Similarly, unbound 17β-hydroxyandrogen levels (which include andro-

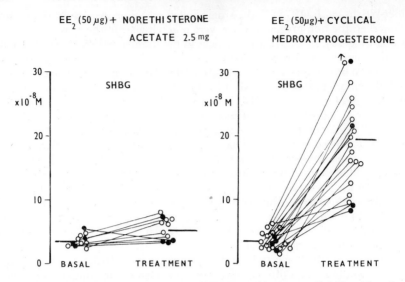

Fig.13 Comparison of long-term response of SHBG to ethinyl oestradiol 50 µg/day plus norethisterone acetate 2.5 mg (left) or combined with cyclical medroxyprogesterone acetate (5 mg), in hirsute women. Open symbols - also on prednisolone. EE_2 and norethisterone acetate given as "Norlestrin" 21 days out of 28. EE_2 + medroxyprogesterone; EE_2 given continuously, and medroxyprogesterone for 7 days of each month. (From Anderson *et al.*, 1975a).

stenediol, the androstenediols and dihydrotesterone) are markedly raised.

Others in this symposium are going to be speaking more authoritatively than I can on the question of the basic nature of the androgen disorder in hirsutism. The most difficult problem is which of the abnormalities is primary and which is secondary. In the case of SHBG we can now be sure that it is secondary. Figure 13 shows two things. On the right are shown the levels of SHBG basally and on treatment with ethinyl oestradiol 50 µg daily, given continuously, combined with cyclical medroxyprogesterone acetate. There is a very marked oestrogenic effect on SHBG production, demonstrating conclusively that the low levels found in women with hirsutism are a secondary phenomenon — secondary either to oestrogen lack or androgen excess. Possibly, in view of the recent studies of Poortman *et al.* (1975) showing competition of androstenediol for oestrogen receptors, it is in fact due to this steroid acting as an anti-oestrogen. However, in this case we would expect SHBG to rise as androstenediol is suppressed on glucocorticoid therapy, and this does not occur.

On the left of Fig.13 is shown the result when the same dose of ethinyl oestradiol was combined with the 19-nor testosterone progestational agent, norethisterone acetate (2.5 mg/day). Clearly, the effect of the oestrogen on SHBG production is virtually abolished on this preparation. This, finally, emphasises the importance we must attach to the kind of progestogen or antiandrogen used in combination with oestrogen to suppress ovarian androgen production in these women. It also brings us full circle, in demonstrating the place of SHBG as, among other things, a plasma marker for oestrogen-androgen interaction.

In conclusion, I should like to be provocative and suggest a very simplified

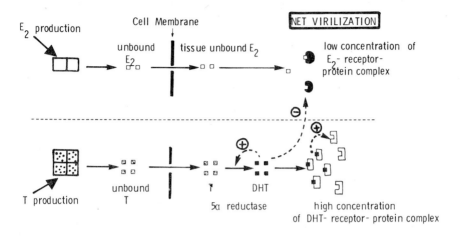

Fig.14 Simplified hypothesis to account for oestrogen-androgen antagonism. Above dashed line represents oestrogens and their target tissues, below line represents androgens. E_2 production, which is small, determines the unbound E_2 in plasma, which determines tissue unbound E_2, which leads to a low E_2-receptor concentration in oestrogen target tissues. T production which is high determines plasma unbound T, and so tissue unbound T. In androgen target tissues 5α-reductase is stimulated, so DHT levels are high, and receptor-protein synthesis stimulated. DHT-receptor complex levels are increased and androgen action stimulated. The postulated inhibition of E_2 receptor production by androgens is speculative; alternative rôle of 5-androstenediol in binding to E_2 receptor is suggested by the work of Poortman *et al.* (1975). *Net effect is virilization.* The situation is envisaged in the normal male, and virilized female.

scheme and the place of SHBG within it, for the mutual antagonism of oestrogens and androgens. Figure 14 shows how we might envisage the situation in the virilized individual such as a normal man. E_2 production determines the unbound E_2, which determines the tissue unbound E_2. Oestrogen action on oestrogen target-tissue is minimal, because of the low concentration of oestrogen receptors, and therefore of E_2-receptor complexes. The testosterone production rate determines the unbound testosterone in plasma, which determines the tissue unbound T. T is converted to DHT by 5α-reductase, the action of which (Adachi and Kano, 1972; Moore and Wilson, 1973) is stimulated by androgens. Androgens also stimulate the synthesis of DHT-receptor, and, in oestrogen target tissues might possibly inhibit oestrogen receptor formation. The balance is firmly in favour of androgen action, and androgen target tissues, and against oestrogen action.

Change the situation in the same individual, by modestly increasing oestradiol production, as shown in Fig.15. The plasma unbound E_2 is increased. SHBG synthesis increases and this leads to a fall in the unbound testosterone fraction and (with the reservation that I have already discussed) its absolute level. The tissue unbound E_2 is elevated in oestrogen target tissues, and this in turn stimulates the production of more oestrogen receptor-protein, and leads to feminization. The tissue unbound testosterone will fall, and together with the rise in oestrogen this may reduce the activity of 5α-reductase leading to a fall in DHT, and a reduction in DHT receptor-protein synthesis (Mauvais-Jarvis and Bercovici, 1967; Mauvais-Jarvis *et al.*, 1968; Shimazah *et al.*, 1969). The net result is one of

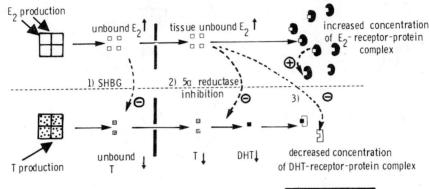

Fig.15 Same as Fig.14, but postulated doubling of E_2 production leading to *Net Feminization*. It is suggested that several steps may then amplify this effect both to increase oestrogen action, and to antagonise androgen action in respective target tissues. Rise in plasma unbound E_2 and therefore tissue unbound E_2: this may induce synthesis of increased amounts of E_2-receptor-protein complex. Meanwhile, androgen action may be antagonised first by the stimuation of SHBG production (see text); unbound T falls, and this in turn leads to a fall in tissue unbound T and DHT. 5α-reductase is inhibited by this fall in androgens, and possibly by the rise in E_2; receptor-protein production falls and there is decreased concentration of DHT-receptor complex and therefore decreased androgen effect. This scheme ignores feedback effects on gonadotrophin production, and possible anti-androgenic action of progesterone, and parts are speculative or imperfectly documented.

decreased concentration of DHT-receptor protein complex, and decreased androgen action. There are many factors totally neglected in this proposal including the anti-androgenic action of progesterone (Mauvais-Jarvis *et al.*, 1974) and feedback effects modulating gonadotrophin production. However, I put this forward simply to suggest that we have in oestrogen-androgen balance some of the features of a cascade system. A defect at any step in the cascade may lead to other secondary changes further down the line.

In the higher primates the importance of SHBG in oestrogen-androgen balance no doubt varies greatly under different physiological and pathological conditions. Even if further work shows that its active role is trivial, SHBG remains an intriguing, and clinically useful pointer both to oestrogen-androgen balance, and to other hormonal interactions. It certainly merits further study.

ACKNOWLEDGEMENTS

This work was supported by a research grant from the Medical Research Council. Part of the work was done when in receipt of a Goldsmith's Travelling Fellowship.

The help of Dr. C.W. Burke, Mr. R. A. Fisher, Dr. A. Galvão-Teles, Mrs. H. L. Goble, Dr. B.L. Lasley, Dr. M.A.F. Murray, Dr. M.O. Thorner, Mrs. J.P. Woodham, and many others who collaborated in these studies is gratefully acknowledged.

REFERENCES

Abuid, J., Klein, A.H., Foley, T.P. (Jr.) and Larsen, P.R. (1974). *J. clin. Endocr. Metab.* **39**, 263.
Adachi, K. and Kano, M. (1972). *Steroids* **19**, 567.
Akande, E.O. and Anderson, D.C. (1975). *Brit. J. Obstet. Gynec.* **82**, 557.
Anderson, D.C., Peppiatt, R., Schuster, L. and Fisher, R. (1972). *J. Endocrinol.* **55**, xi *(Proceedings)*.
Anderson, D.C. (1974). *Clin. Endocrinol.* **3**, 69.
Anderson, D.C., Thorner, M.O., Fisher, R.A., Woodham, J.P., Goble, H.L. and Besser, G.M. (1975a). *Acta endocr. (Copenh.)* **Suppl. 199**, 224.
Anderson, D.C., Lasley, B.L., Fisher, R.A., Shepherd, A., Newman, L. and Hendricx, A.G. (1975b). *Submitted for publication.*
Anderson, D.C. and Galvão-Teles, A. (1975). *(In press)*.
Beamer, N., Hagemenas, F. and Kittinger, G.W. (1972). *Endocrinology* **90**, 325.
Burke, C.W. and Anderson, D.C. (1971). *J. Endocrinol. (Proceedings)* **53**, xxvi.
Burke, C.W. and Anderson, D.C. (1972). *Nature* **240**, 38.
Corvol, P.L., Chrambach, A., Rodbard, D. and Bardin, C.W. (1971). *J. biol. Chem.* **246**, 3435.
Crépy, O., Dray, F. and Sebaoun, J. (1967). *Compt. Rend. Acad. Sci.* **264**, 2651.
Daughaday, W.H. (1959). *Physiological Reviews* **39**, 885.
De Moor, P., Heyns, W. and Bouillon, R. (1972). *J. Steroid Biochem.* **3**, 593.
Fisher, R.A., Anderson, D.C. and Burke, C.W. (1974). *Steroids* **24**, 809.
Fisher, R.A. and Anderson, D.C. (1975). *Submitted for publication.*
Forest, M.G. and Bertrand, J. (1972). *Steroids* **19**, 197.
Galvão-Teles, A., Anderson, D.C., Burke, C.W., Marshall, J.C., Corker, C.S., Bown, R.L. and Clark, M.L. (1973). *Lancet* **1**, 173.
Gauthier-Wright, F., Baudot, N. and Mauvais-Jarvis, P. (1973). *Endocrinology* **93**, 1277.
Green, J.R.B., Mowat, N.A.G., Fisher, R.A., Anderson, D.C. and Dawson, A.M. (1975). *In preparation.*
Heyns, W. and De Moor, P. (1971a). *J. clin. Endocr. Metab.* **32**, 147.
Heyns, W. and De Moor, P. (1971b). *Steroids* **10**, 245.
Kley, H.K., Herrmann, J., Morgner, K.D. and Krüskemper, H.L. (1973). *Horm. Metab. Res.* **5**, 21.
Kley, H.K., Nieschlag, E., Wiegelmann, W., Solbach, M.G. and Krüskemper, H.L. (1975). *Acta endocr. (Copenh.)* **79**, 275.
Longcope, C., Layne, D.S. and Tait, J.F. (1968). *J. clin. Invest.* **47**, 93.
Mahoudeau, J.A., Bardin, C.W. and Lipsett, M.B. (1971). *J. clin. Invest.* **50**, 1338.
Mahoudeau, J.A. and Corvol, P. (1973). *Endocrinology* **92**, 1113.
Mauvais-Jarvis, P. and Bercovici, J.P. (1967). *Compt. Rend. Acad. Sci.* **264**, 2825.
Mauvais-Jarvis, P., Floch, H.H. and Bercovici, J-P. (1968). *J. clin. Endocr. Metab.* **28**, 460.
Mauvais-Jarvis, P., Kuttenn, F. and Baudot, N. (1974). *J. clin. Endocr. Metab.* **38**, 142.
Mercier, C., Alfsen, A. and Baulieu, E-E. (1966). *Proceedings of the Second Symposium on Steroid Hormones*, Ghent, 1965. Excerpta Medica International Congress Series **101**, 212.
Moore, R.J. and Wilson, J.D. (1973). *Endocrinology* **93**, 1277.
Murray, M.A.F., Bancroft, J.H.J., Anderson, D.C., Tennent, T.G. and Carr, P.J. (1975). *J. Endocrinol. (in press).*
Olivo, J., Vittek, J., Southren, A.L., Gordon, G.G. and Rafii, F. (1973). *J. clin. Endocr. Metab.* **36**, 153.
Poortman, J., Prenen, J.A.C., Schwarz, F. and Thijssen, J.H.H. (1975). *J. clin. Endocr. Metab.* **40**, 373.
Raynaud, J.P., Mercier-Bodard, C. and Baulieu, E-E. (1971). *Steroids* **18**, 767.
Resko, J.A., Ploem, J.E. and Stadelman, H.L. (1975). *Endocrinology* **97**, 425.
Rivarola, M.A., Forest, M.G. and Migeon, C.J. (1968). *J. clin. Endocr. Metab.* **28**, 34.
Rosenbaum, W., Christy, N.P. and Kelly, W.G. (1966). *J. clin. Endocr. Metab.* **26**, 1399.
Rosner, W. (1972). *J. clin. Endocr. Metab.* **34**, 983.
Rosner, W. and Smith, R. (1975). *Methods Enzymol.* **36** (00), 109.
Ruder, H., Corvol, P., Mahoudeau, J.A., Ross, G.T. and Lipsett, M.B. (1971). *J. clin. Endocr. Metab.* **33**, 382.
Shimazaki, J., Matushita, I., Furiya, N., Yamanaka, H. and Shida, K. (1969). *Endocr. Jap.* **16**, 453.

Southren, A.L., Gordon, G.G. and Tochimoto, A. (1968). *J. clin. Endocr. Metab.* **28**, 1105.
Thijssen, J.H.H., Lourens, J., Donker, G.H. and Schwarz, F. (1975). *Acta endocr. (Copenh.)* **Suppl. 199**, 234.
Tulchinsky, D. and Chopra, I.J. (1973). *J. clin. Endocr. Metab.* **37**, 873.
Tulchinsky, D. and Chopra, I.J. (1974). *J. clin. Endocr. Metab.* **39**, 164.
Vermeulen, A., Verdonck, L., Van der Straeten, M. and Orie, N. (1969). *J. clin. Endocr. Metab.* **29**, 1470.
Vermeulen, A., Rubens, R. and Verdonck, L. (1972). *J. clin. Endocr. Metab.* **34**, 730.
Westphal, U. (1971). *Proceedings of the Third International Congress of Hormonal Steroids.* Exerpta Medica International Congress Series **219**, 410.

DIVERSITY IN ANDROGENIC EFFECT AS STUDIED IN RAT AND MOUSE

P. De Moor, G. Verhoeven, W. Heyns and H. Van Baelen

Rega Instituut, Laboratorium voor Experimentele Geneeskunde Departement voor Ontwikkelingsbiologie, K.U. Leuven, Belgium

The various cells of a mammalian organism do not respond in the same way to androgenic stimulation. Drawing from the literature as well as from personal data, four possible causes for this diversity will be discussed.

I. DIVERSITY IN BINDING OF ANDROGENS BY NON-RECEPTOR PROTEINS

In this chapter we will discuss *diversity in binding by non-receptor proteins* as a possible cause of diversity in androgen effect. Tissues contain two classes of high affinity binders for androgenic steroids: receptor proteins and non-receptor proteins. The former transfer steroid hormones to the nucleus and so induce androgen activity, the latter do not. The concentration of receptor protein does not seem to vary widely: we will show later that rat uterus, mouse kidney and rat prostate contain about the same number of receptor sites for androgen although the responsiveness of these three tissues for androgens differs rather widely. Non-receptor binding, on the contrary, seems to vary from tissue to tissue and can be located either in the interstitial fluid surrounding certain target cells or in certain types of target cells themselves. The presence of these proteins probably explains why the androgen level in target tissues often exceeds the levels in blood and why target tissues differ as to the kind of androgen they concentrate.

Androgens are concentrated in the germinal epithelium of the testes through specific binding in the interstitial fluid surrounding this epithelium. The protein involved has been called androgen binding protein or ABP and is not found in other tissues. It seems to play an important role in delivering concentrated amounts of testicular androgen to sites of action in the germinal epithelium (Ritzen *et al.*, 1972a; Hansson, 1972; Hansson *et al.*, 1973a; Ritzen *et al.*, 1973;

Hansson et al., 1973b; Ritzen et al., 1974; French and Ritzen, 1973; Ritzen et al., 1974).

The target cells themselves contain variable amounts of binding proteins distinct from the receptor proteins that transfer the steroid hormones into the nucleus. Not only the amount of androgen that is bound, but also the ratio of bound DHT over bound testosterone varies from tissue to tissue. This is exemplified in the following experiment. In cytosol of rat uterus, rat prostate or mouse kidney equilibrium dialysis in the presence of 10^{-6} M of testosterone or DHT gave bound over unbound androgen ratios that were always higher for DHT than for testosterone. The B/U ratios for DHT were respectively 7.3 in prostate, 5.7 in uterus and 2.7 in kidney, while for testosterone these ratios were respectively 2.1, 1.1 and 1.4. When, however, the receptor binding capacity in the prostate was measured specifically by ammonium sulfate precipitation (Verhoeven et al., 1975c), the receptor capacity for DHT was the same as the receptor capacity for testosterone.

Non-receptor binding of steroid hormones in the cytosol of target cells has been referred to in the literature either as non-specific, non-saturable binding or as non-transferable binding. It could be that the macromolecules described as binding subunits of the genuine receptor represent the same entity. *Non-specific, non-saturable binding* has been described in rat and human prostate i.a. by Mainwaring (1969), Baulieu and Jung (1970), Mainwaring and Peterken (1971) and Grant and Giorgi (1972). *Non-transferable binding* was first mentioned by Liao et al. (1971) and by Fang and Liao (1971). They described in prostate cytosol an α-protein with a sedimentation coefficient of 3.5 S to be distinguished from the genuine receptor not only by its sedimentation coefficient (8 S) but also by ammonium sulfate precipitation, gel filtration and binding characteristics. Two bits of evidence seem to indicate that the cytosolic steroid receptor is made up of at least two subunits: a steroid binding subunit and a transfer subunit that binds to chromatin. As already stated, it could be that the steroid subunit accounts, at least in part, for the non-saturable, non-transferable binding described above. The evidence for the existence of steroid binding subunits is as follows. Sherman and co-workers (1974) by treating partially purified progesterone receptor with Ca^{++} or other divalent cations obtained a steroid binding macromolecule that could, however, not bind to nuclear chromatin. Whereas the intact receptor had a molecular weight of about 100,000 and was an asymmetric acidic protein, the steroid binding moiety had a lower molecular weight and was a compact basic protein. On the other hand, Sibley and co-workers (cited by Edelman, 1975) have studied genetic variants of lymphoma cells in tissue culture and could isolate and identify three variants. In the first one there was no high affinity binding in the cytosol, in the second there was glucocorticoid binding but no transfer of the formed complexes into the nucleus and in the third variant there was binding and transfer but no physiological response.

In conclusion, differences in non-receptor binding probably explain why certain cells take up more androgens and so respond better.

II. DIVERSITY THROUGH TARGET CELL METABOLISM OF ANDROGEN: WHY DO CERTAIN CELLS USE DHT AND CERTAIN OTHERS TESTOSTERONE?

Certain tissues such as prostate build up tissue plasma gradients for DHT (e.g. Robel et al., 1973) while others such as levator ani muscle, hypothalamus,

Table I. Metabolism of androgens in cytosol prepations at 4°C.

Origin of Cytosol	Androgen	Per Cent Untransformed Androgen	
		In Total Cytosol	In Receptor Bound Fraction
Rat prostate	testosterone	76	87
	dihydrotestosterone	56	92
Mouse kidney	testosterone	81	88
	dihydrotestosterone	20	77
Rat uterus	testosterone	90	96
	dihydrotestosterone	10	89

pituitary and kidney concentrate testosterone (Robel et al., 1973). Moreover, certain tissues respond to androgens while retaining testosterone in their nuclei. Nuclear retention of testosterone has been reported in rat and mouse kidney (Bullock et al., 1971; Ritzen et al., 1972b), mouse submandibular gland (Goldstein and Wilson, 1972), immature rat uterus (Giannopoulos, 1973), rat anterior hypophysis (Jouan et al., 1973), rat testis (Hansson et al., 1974), rat bone marrow (Valladares and Minguell, 1975), erythropoietic mouse spleen (Hadjan et al., 1974) and Shionogii carcinoma (Bruchovsky, 1972). On the contrary, in most cells from male accessory organs of reproduction androgenic stimulation seems only to imply nuclear uptake of DHT (Wilson and Gloyna, 1970). Selective uptake around or in certain target cells could explain part of the observed differences. Could other parts of the receptor machinery of the target cells also be involved? A review of the literature as well as personal data sustain the following conclusions.

A *A Common Cytosolic Receptor for DHT and Testosterone*

The physiochemical behaviour, the number of binding sites and the ligand specificity of the cytosolic androgen receptors were compared first between ligands in the same tissue then between tissues with the same ligand. Due to extensive DHT metabolism *in vitro* at 4°C in rat uterus and mouse kidney cytosol (Table I), part of the planned experiments could not be done.

1. Comparison of DHT and testosterone binding in prostatic tissue

Using ammonium sulfate precipitation (Verhoeven et al., 1975c) instead of density gradient centrifugation so as to avoid dissociation of the formed testosterone receptor complexes (Verhoeven et al., 1975b) it was shown that the *number of binding sites* for DHT (55 fmol/mg protein) compared very well with the value (60 fmol/mg protein) found in prostate for testosterone. *Ligand specificity* was studied by competition with either labelled DHT or labelled testosterone. Five different steroids, including DHT and testosterone, competed equally well for the DHT and testosterone binding sites in prostatic tissue (Table II).

2. Comparison of testosterone binding in tissues with DHT retention in vivo *(rat prostate) and in tissues with testosterone retention* in vivo *(rat uterus and mouse kidney).*

The receptor proteins in these three tissues displayed *identical behaviour* on density gradient centrifugation and ammonium sulfate precipitation. In all three

Table II. Ligand specificity of the androgen receptor protein in rat prostate, mouse kidney and rat uterus cytosol.[a]

Origin of Cytosol Labelled Ligand	PROSTATE		KIDNEY	UTERUS
	testosterone	dihydrotestosterone	testosterone	testosterone
COMPETITOR	APPARENT K_i VALUE			
	nM	nM	nM	nM
testosterone	4.0	3.3	1.3	2.0
dihydrotestosterone	2.0	2.0	4.1	4.3
epitestosterone	74	33	20	33
5α-androstane-3α,17β-diol	–	35	26	39
5α-androstane-3β,17β-diol	–	9.9	10	7.2
17β-estradiol	–	12	5.3	6.1
17α-methyl 19-nor-testosterone	1.5	0.9	0.2	0.4
methandrostanolone	2.2	2.0	4.8	3.3
cyproterone acetate	–	4.1	5.4	2.6

[a] The concentration of the labelled ligands was 1.25×10^{10} M. The concentration of competitors was varied from 10^{-11} to 10^{-7} M. Values were calculated from logit-log plots.

tissues low capacity high affinity binding proteins were precipitated at 30-40 per cent of saturation. The *number of binding sites* for testosterone was very similar, amounting respectively to 60, 66 and 22 fmol/mg protein in prostate, uterus and kidney. *Ligand specificity* was studied by competition with testosterone in the three tissues. Similar K_i values were found for 26 competing steroids. Part of these results are shown in Table II. It should be noted that not only potent androgens compete efficiently for the binding site, but that the apparent K_i values for antiandrogens such as cyproterone acetate is in fact very close to the corresponding value for testosterone. This proves unambiguously that receptor binding does not necessarily parallel androgenic activity.

Methodological problems may have obscured the issue in previous studies (Heyns *et al.*, 1975; Verhoeven *et al.*, 1975b). The experiments presented here make it, however, likely that the same receptor protein is involved in the binding of DHT and testosterone in tissues which *in vivo* retain DHT as well as in tissues which *in vivo* retain testosterone.

B   *Both the Testosterone-Receptor Complex and the DHT-Receptor Complex Bind Equally Well* In Vitro *to Nuclei of Prostate or Kidney.*

No specific nuclear acceptor site seems to be involved. Early attempts to demonstrate *in vitro* binding of testosterone-receptor complexes in prostatic nuclei failed (Jung and Baulieu, 1971). Recently, however, one of us (Verhoeven, unpublished results) was able to demonstrate that DHT and testosterone receptor complexes bind equally well to highly purified rat prostate and mouse kidney nuclei (Fig.1). Thus, this part of the receptor machinery for androgens too seems not to discriminate between DHT and testosterone.

C   *Testosterone as a Second Choice Messenger: Does the Higher Affinity of DHT for the Androgen Receptor and Thus Its Preferential Binding Explain the Discrepancies between Tissues with Nuclear Uptake of DHT and Testosterone?*

When measuring B/U ratios at different loads of androgen one gets the impression that both DHT and testosterone have the same affinity (K_d 1.3 and 1.5

NUCLEAR TRANSPORT OF ANDROGEN-RECEPTOR COMPLEXES IN HOMOLOGOUS SYSTEMS

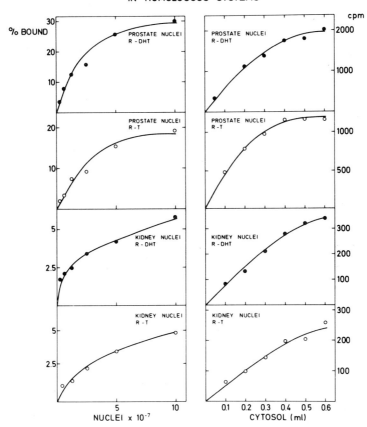

Fig.1 Binding of androgen-receptor complexes to nuclei *in vitro*. Receptor preparations were obtained by passing cytosol, equilibrated with 10^{-9} M of H^3-DHT (●—●) or H^3-testosterone (○—○), over a Sephadex G50 column and collecting the void volume. Purified rat prostate or mouse kidney nuclei were incubated with the homologous cytosol-receptor preparations for 30 min at 18°C. In the left panels 0.5 ml of a receptor preparation (containing 2×10^4 cpm/ml) was added to increasing amounts of nuclei. In the right panels varying amounts of receptor were added to 2×10^7 nuclei and the final volume was made up to 0.6 ml with buffer. The incubation was stopped by placing the tubes in melting ice for 5 min and centrifugation at 3000 rpm for 10 min at 4°C. The nuclear pellets were washed three times with buffer containing tris-HCl, pH 7.4 (1=0.01); NaCl, 50 mM; $MgCl_2$, 5 mM and EDTA, 5×10^{-5} M and the radioactivity in the nuclei was quantitated by liquid scintillation counting. Results represent the amount or the fraction of the incubated radioactivity associated with the washed nuclear pellets (Verhoeven, unpublished results).

nM) for the cytosolic androgen receptor. When, however, dissociation rates are measured, one observes (Fig.2) that the DHT-receptor complex dissociates much more slowly than the testosterone-receptor complex. Since association rates do not vary much in biological systems the two above findings are contradictory. By closer scrutiny, the K_d for DHT seems to be underestimated by non-receptor binding included in the free fraction.

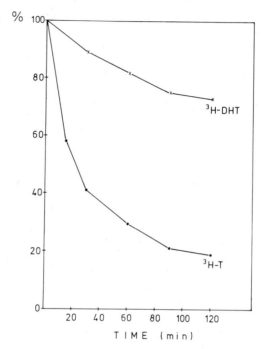

Fig. 2 Dissociation of testosterone-receptor or dihydrotestosterone-receptor complexes in rat uterus cytosol. Chase experiments at 25°C.

If DHT has, indeed, a higher affinity for the androgen receptor, it will, if available, bind preferentially to this receptor. One could thus surmise that testosterone will get a chance of forming a complex with the latter receptor and of mediating androgen action, only when not enough DHT is available. Whether DHT will be available in a given target cell depends on the balance between DHT formation and DHT catabolism. The factors regulating DHT catabolism are not well documented. On the other hand, DHT is formed in the target cells not only by the microsomal 5α-reductase (for review see e.g. Verhoeven, 1975a), but apparently also by other enzymes, i.a. by a NADH-dependent, microsomal 3α-hydroxysteroid dehydrogenase (3α-HSD). The latter enzyme was found to be distinct from the well-known and ubiquitous NADPH-dependent soluble 3α-HSD and from a NADPH-dependent microsomal 3α-HSD described by one of us (Verhoeven, 1975a) and found only in the kidney of male rats (Table III). The NADH-linked 3α-HSD was detected in several tissues including the prostate. Its activity in these various tissues paralleled roughly the formation of DHT from 3α-androstanediol by tissue slices *in vitro* and with the responsiveness to 3α-androstanediol of these tissues (Verhoeven, 1975a).

However, non-availability of DHT does not automatically imply androgen action through testosterone as will be shown by the following findings. Testo-

Table III. A comparison of several properties of the 3α-hydroxysteroid dehydrogenases in rat kidney.

	ENZYME		
	NADPH-soluble	NADPH-particulate	NADH-particulate
sex difference	♀ > ♂	♀ <<< ♂	♀ < ♂
(NH$_4$)$_2$SO$_4$ precipitation	60 - 80%		
Sephadex G 100	MW ≃ 30 000	Vo*	Vo*
optimum activity at pH	5.5 — 7	8	4.5
influence of ionic strength	—	↑ ↑	—
influence of phosphate	—	↓ ↓	—
50% denaturation after 10 min incubation at:	52°C	41°C	58°C
app Km for dihydrotestosterone	0.3 nM	> 100 nM	2.4 nM
app Km for 3α-androstanediol	> 100 nM	3.14 nM	0.6 nM

* trition-solubilized enzyme or trace of this enzyme found in the cytosol fraction. Vo = void volume.

sterone is responsible for the development of the internal organs of male reproduction whereas DHT mediates the development of the external genitalia. In line with this, Wolffian duct has an undetectable 5α-reductase activity (Wilson and Lasnitzki, 1971; Kelch et al., 1971; Siiteri and Wilson, 1974) and concentrated testosterone (Wilson, 1973). But, recently Walsh et al. (1974) and Imperato-McGinley et al. (1974) have found patients with normal internal organs of male reproduction in which the external genitalia such as the prostate fail to develop. These patients have no 5α-reductase activity and no DHT available. Testosterone apparently can not take over the role of the missing DHT in these organs.

D Conclusion

It is as yet not clear why certain tissues use testosterone to mediate androgenic messages. Apparently, these tissues do not have a special androgen receptor in their cytosol nor a special acceptor site in their chromatin. Tissues which can not build up DHT gradients do not automatically use testosterone as a second choice androgenic messenger.

III. PITUITARY HORMONES FAVOUR OR HAMPER THE ACTIVITY OF TESTOSTERONE IN CERTAIN TARGET CELLS POSSIBLY BY ALTERING TESTOSTERONE METABOLISM IN THESE TISSUES

As has been shown in the previous section, kidney microsomes contain at least two 3α-hydroxysteroid dehydrogenases (3α-HSD), one of which is NADPH-linked, and the other depends on NADH as cofactor. Both dehydrogenases are more active in male than in female rats. Their activity decreases in castrated males and increases when the latter animals or (castrated) females are treated with testosterone propionate (Verhoeven, 1975a). We compared the effects of testosterone propionate (TP) treatment on these two enzymes both in gonadectomized rats and in hypophysectomized-gonadectomized animals (De Moor et al., accepted for publication). TP had an effect on both 3α-HSD in the hypophysectomized

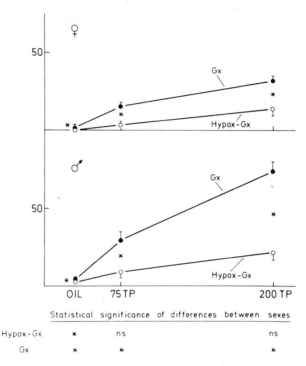

Fig.3 Renal NADPH-dependent 3α-hydroxysteroid dehydrogenase (μmol/g protein/h ± SD) in gonadectomized (●) and hypophysectomized-gonadectomized (○) rats of both sexes treated for 14 consecutive days with oil, or with respectively 75 and 200 μg testosterone propionate (TP) per 100 g body weight. The significance of the difference *within-sex* between hypophysectomized-gonadectomized rats and gonadectomized rats is indicated by the symbols NS (non-significant) or by an asterisk (P < 0.01) printed between both curves. The same symbols are also used at the bottom of the figure to indicate the significance of the *between-sex* differences in rats pre-treated similarly.

animals. But, whereas hypophysectomy decreased the effect of testosterone on the NADPH-linked enzyme, it increased its effect on the NADH-linked enzyme. The pituitary apparently helped testosterone to act on the NADPH-linked enzyme but antagonized the testosterone effect on the NADH-linked enzyme (Figs 3,4). On closer scrutiny it appears that there are similar examples in the literature of such a dual response on hypophysectomy (Table IV).

How can the pituitary modulate certain androgenic responses? The most promising hypothesis in this respect seems to be that pituitary hormones alter the metabolism of testosterone in certain target cells and not in others and so affect androgen response. Some data in the literature sustain this contention. Indeed, Ebling and collaborators (1971) could show that whereas in intact animals testosterone was clearly more active on sebum production than DHT or androstanedione, the opposite was true in hypophysectomized-gonadectomized animals. Similarly Nikkari and Valavaara (1970) found that 3β-androstanediol stimulates sebum secretion in hypophysectomized rat skin. A possible interpretation of these

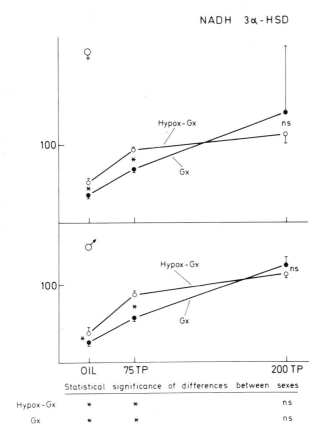

Fig.4 Renal NADH-dependent 3α-hydroxysteroid dehydrogenase (μmol/g protein/h ± SD) in gonadectomized (●) and hypophysectomized-gonadectomized (○) rats of both sexes treated for 14 consecutive days with oil, or with respectively 75 and 200 μg testosterone propionate (TP) per 100 g body weight. The significance of the different *within-sex* between hypophysectomized-gonadectomized rats and gonadectomized rats is indicated by the symbols NS (non-significant) or by an asterisk ($P < 0.01$) printed between both curves. The same symbols are also used at the bottom of the figure to indicate the significance of the *between-sex* differences in rats pretreated similarly.

results according to Ebling and collaborators (1971) is "that within the sebaceous glands either the release of testosterone from its conjugates or its conversion to an active metabolite is under pituitary control". In accordance with this hypothesis the same authors (Ebling *et al.*, 1969) could show that a preparation of porcine growth hormone and a preparation of prolactin which had no growth hormone activity, were each capable of fully restoring the response of the sebaceous glands to testosterone in hypophysectomized male rats, though these pituitary hormones had no effect by themselves.

Later on, Thody and Shuster (1973, 1975) reported that synthetic α-MSH increased sebum production in intact or gonadectomized-hypophysectomized rats of both sexes and corrected the decrease in sebum production resulting from

Table IV. Effects of the pituitary on the expression of maleness in the rat.

	— Hypophysectomy "feminizes" these parameters — The pituitary favours testosterone effects	— Hypophysectomy "masculinizes" these parameters especially in females The pituitary hampers testosterone effects
Testosterone augments protein synthesis, enzyme activity or growth Male > Female	— a_{2U} sex protein (Kumar et al., 1969; Roy, 1973) — Hepatic steroid sulphatase (Burstein, 1968) — Hepatic hexobarbital hydroxylase (Nair and Bau, 1967; Nair et al., 1970; Jam and Du Bois, 1967) — N-hydro-2-acetyl aminofluorene sulfotransferase in liver (De Baun et al., 1970) — 32 P incorporation in hepatic phospholipid (Tate and Williams-Ashman, 1967) — Polyploidy in hepatic nuclei (Hoffman and Schwartz, 1962) — Carcinogen binding proteins (Weisburger et al., 1968) — Renal NADPH 3a-hydroxysteroid dehydrogenase (this chapter)	— Hepatic 3a, 3β, 5β and 17a hydroxysteroid dehydrogenase (Denef, 1974; Gustafsson and Stenberg, 1974a) — Hepatic 2a, 2β, 6β, 16a and 18 steroid hydroxylase (Gustafsson and Stenberg, 1974a) — Hepatic and renal catalase (Gaebler and Mathies, 1951; Utsugi, 1960) — Renal NADH 3a-hydroxysteroid dehydrogenase (this chapter)

the removal of the neurointermediate lobe of the pituitary in male rats. Much larger responses were obtained in hypophysectomized rats that had received treatment with either TP or progesterone. Similar results were obtained as far as weight *of* (Thody and Shuster, 1975) or lipid synthesis *in* (Cooper et al., 1975) the preputial glands were concerned. It is clear from these experiments that a-MSH reinforces the activity of testosterone on sebum and preputial glands. But whether, as suggested by these authors, a-MSH also has a *direct* influence on these parameters remains to be proven. Indeed, the animals they studied were hypophysectomized and gonadectomized but were not adrenalectomized and it is known that a-MSH has an adrenocorticotrophic activity and thus could stimulate adrenal androgen production.

In conclusion, pituitary polypeptides, such as a-MSH, act synergistically with testosterone on certain target cells possibly by altering the metabolism of testosterone in these target cells. The cells possessing the membrane receptors for these polypeptide hormones can thus answer better or differently to androgenic signals.

IV. PARADOXICAL EFFECTS OF ANDROGENS ARE MEDIATED BY THE HYPOTHALAMO-HYPOPHYSIAL AXIS AND PARTLY IMPRINTED AT BIRTH

As a rule androgens promote growth and increase protein concentrations or enzyme activities. There are however a few exceptions which we would like to call the *paradoxical* effects of androgens. Indeed, androgens involute the thymus (Dougherty, 1952; Senelar et al., 1973) and probably also the adrenals, since the relative weight of the latter organs is greater in females than in males (Kitay, 1963; Kubota and Suzuki, 1974; Toh, 1973). Androgens, also depress the concentrations of the activity of liver 5α-reductase (Yates et al., 1958; Forchielli et al., 1958; Troop, 1958), serum transcortin (Westphal, 1971; De Moor et al., accepted for publication), liver 7α-steroid hydroxylase (Berg and Gustafsson, 1973) and pituitary 5α-reductase (Massa et al., 1975). As far as these parameters are concerned, intact adult female rats have higher values than intact males.

Hershko and collaborators (1971) have put forward the hypothesis that, except for a few so-called specific effects on enzyme induction, the various biochemical events regulated by glucocorticoids are related to the inhibitory effects of these hormones on cell growth. They call the biochemical processes related to cell growth the *pleiotypic effect* of the steroid hormone and suppose that this effect is mediated by one or a few secondary signals. Besides a genuine pleiotypic effect of inhibition of cell growth, glucocorticoids seem to possess a paradoxical pleiotypic program of cell growth in the liver. In analogy, the atrophy of the thymus and the lower adrenal weight implemented by androgens could be called the paradoxical pleiotypic effect of androgens. According to the Hershko hypothesis the various paradoxical biochemical events, described above, should be part of this negative pleiotypic program and should therefore contribute, somehow, to the involution of thymus or adrenals.

A   No Paradoxical Effects of Testosterone After Hypophysectomy

The parallelism between the mechanism of glucocorticoid and androgen action can be drawn a step further. Rousseau, in a stimulating review (1975) suggests that the paradoxical pleiotypic effects of glucocorticoids are mediated by other hormones. Indeed, whereas lymphoid cells or fibroblasts can be inhibited *in vitro* by glucocorticoids, some anabolic effects (paradoxical pleiotypic effect) described in the liver, such as the increased synthesis of protein, RNA and glycogen appear to require administration of the hormone to the whole animal and are difficult or impossible to reproduce in the perfused organ or on liver slices. Rousseau concludes: it could be that insulin or other hormones released into the blood following glucocorticoid administration participate in this paradoxical pleiotypic response of the liver. Apparently a similar situation prevails as far as the paradoxical effects of androgens are concerned. Indeed, testosterone does not depress any further the liver 5α-reductase activity (Denef, 1974) or the hepatic microsomal 7α-hydroxylase (Gustafsson and Stenberg, 1974b) and the serum transcortin concentration (De Moor et al., accepted for publication) in hypophysectomized rats. All these paradoxical androgen effects are completely mediated by pituitary hormones.

B Experiments with Uncoupled Pituitaries Suggest that Testosterone Does Not Act Directly on the Pituitary but on the Production of Hypothalamic Inhibiting Factors

First, Denef (1974) has shown that implantation of a pituitary under the kidney capsule makes hypophysectomized male rats react as females as far as hepatic microsomal 5α-reductase (and 3α-hydroxysteroid reductase) activity is concerned. We obtained similar results from serum transcortin. Hypophysectomized rats of both sexes (6 males and 4 females) and low (masculine) transcortin levels (De Moor, accepted for publication). These levels increased (feminized) almost 100 per cent after implantation of a pituitary under the capsule of the kidney of these hypophysectomized rats. Six male rats increased their transcortin levels from 54.7 ± 9.6 (SD) mg/l to 85.5 ± 21.3 mg/l and in 4 female rats the levels went up from 56.2 ± 8.8 to 98.9 ± 13.9 mg/l. The sex of either the donor or the receptor was of no importance: female rats implanted with a male pituitary reacted as male rats which had received a female pituitary graft.

Secondly, in a certain number of hypophysectomized rats the initial low transcortin levels did increase spontaneously while in most others they remained low. This secondary increase in serum transcortin levels in hypophysectomized rats seems to be due to regeneration of uncoupled pituitary tissue in the sella turcica.

Rats with uncoupled pituitaries do not respond to testosterone propionate (0.2 mg per rat per day for 7 days). The transcortin levels in 6 male rats were 8.5 ± 21.3 (SD) mg/l before TP treatment and 87.2 ± 20.7 afterwards; in 4 female rats the figures were respectively 98.9 ± 13.9 and 101.7 ± 14.5 These findings show clearly that androgens influence the pituitary through the hypothalamus. Furthermore, the fact that uncoupling the pituitary of the hypothalamus has the opposite effect to hypophysectomy, at least for the three parameters (5α-reductase, 3β-reductase and transcortin) studied up till now, implies that hypothalamic *inhibiting* factors are involved. If *releasing* factors were involved uncoupling would have the same result as hypophysectomy. Indeed, it is admitted that uncoupled pituitaries produce only hormones whose secretion under physiological conditions is preponderantly regulated by hypothalamic inhibition. The two best known examples of pituitary polypeptides in this group are prolactin and MSH.

C Serum Transcortin Levels Increase in Hypophysectomized Rats Treated with Synthetic β-MSH

Two groups of 6 hypophysectomized rats (3 males and 3 females in each group) were treated for 7 consecutive days either with 0.1 mg synthetic β-MSH (a gift from Ciba-Geigy, Basle) in 50 μl carboxy methylcellulose gel or with gel alone. Transcortin was measured by radial immunodiffusion (Van Baelen and De Moor, 1974) using a rat antitranscortin. After 7 days the transcortin levels were significantly higher in the MSH treated group than in the control group (49.1 ± 4.2 v. 36.4 ± 5.4 mg/l $\pm$ SD; $p < 0.001$). Experiments with hypophysectomized, gonadectomized, *adrenalectomized* animals are now in order.

D Neonatal Imprinting of All Paradoxical Testosterone Effects Studied Up Till Now

In a series of experiments published in 1968 (De Moor and Denef, 1968; Denef and De Moor, 1968) it was demonstrated that the lower hepatic 5α-reductase activity found in males was partly the immediate but transient result of testosterone secreted by the adult testes but partly also the delayed irreversible

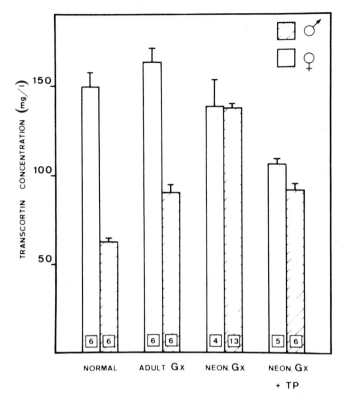

Fig.5 Effect of gonadectomy (gx) and testosterone propionate (TP) administration on the serum transcortin level of adult rats. Blood was collected on day 62 except for blood of adult gonadectomized rats which was collected on day 86. The amount of transcortin measured by single radial immunodiffusion is expressed in mg/l (mean ± SEM). The number of animals in each group is given at the bottom of the bars.

result of testosterone secreted at birth. These results were later confirmed by Gustafsson and collaborators (Einarsson *et al.*, 1973; Gustafsson and Stenberg, 1974a) and extended to another paradoxical effect of testosterone, i.e. the 7α- hydroxylase active on 5α-androstane-$3\alpha,17\beta$ diol. Pituitary 5α-reductase activity is also depressed by neonatal testosterone (Massa *et al.*, 1975). Recently, we showed that the same was true for transcortin (Van Baelen *et al.*, submitted for publication). Male rats castrated at birth develop in adult life transcortin levels similar to those of female animals spayed at birth or to those of adult intact female animals. Animals of both sexes gonadectomized at birth and treated neonatally with a single injection of testosterone propionate show transcortin concentrations significantly lower than those of intact female animals (Fig.5).

E   *Conclusions*

In a few instances androgens depress and involute instead of inducing protein synthesis or promoting organ growth. It is hypothesized that these seemingly unrelated paradoxical androgenic effects are part of a coordinated negative pleiotypic program. As for glucocorticoids these paradoxical androgen effects appear to be indirect. Indeed, all of these biochemical events seem to be due to suppression in the hypothalamus of the production of certain hypophysial hormones. Although all paradoxical androgen effects seem to be indirect and mediated, all mediated androgen effects are not paradoxical: certain genuine androgen effects (e.g. 3β-hydroxysteroid reductase) are also mediated by the hypothalamo-hypophysial axis, The indirect (mediated) effects of androgens are partly imprinted at birth by testosterone.

REFERENCES

Baulieu, E.E. and Jung, J. (1970). *Biochem. Biophys. Res. Comm.* **38**, 599.
Berg, A. and Gustafsson, J.A. (1973). *J. Biol. Chem.* **248**, 6559.
Brushovski, N. (1972). *Biochem. J.* **127**, 561.
Bullock, L.P., Bardin, C.W. and Ohno, S. (1971). *Biochem. Biophys. Res. Comm.* **44**, 1537.
De Moor, P. and Denef, C. (1968). *Endocrinology* **82**, 480.
De Moor, P., Adam-Heylen, M., Van Baelen, H. and Verhoeven, G. (1975). *J. Endocrinol. (accepted for publication).*
Denef, C. and De Moor, P. (1968). *Endocrinology* **83**, 791.
Denef, C. (1974). *Endocrinology* **94**, 1577.
Dougherty, T.F. (1952). *Physiol. Rev.* **32**, 379.
Ebling, F.J., Ebling, E. and Skinner, J. (1969). *J. Endocrinol.* **45**, 245.
Ebling, F.J., Ebling, E., McCaffery, V. and Skinner, J. (1971). *J. Endocrinol.* **51**, 181.
Einarsson, K., Gustafsson, J.A. and Stenberg, A. (1973). *J. biol. Chem.* **248**, 4987.
Forchielli, E., Brown-Grant, K. and Dorfman, R.I. (1958). *Proc. Soc. exp. Biol. Med.* **99**, 594.
French, F.S. and Ritzen, E.M. (1973). *Endocrinology* **93**, 88.
Giannopoulos, G. (1973). *J. biol. Chem.* **248**, 1004.
Goldstein, J.L. and Wilson, J.D. (1972). *J. clin. Invest.* **51**, 1647.
Grant, J.K. and Giorgi, E.P. (1972). *J. Steroid Biochem.* **3**, 471.
Gustafsson, J.A. and Stenberg, A. (1974a). *J. biol. Chem.* **249**, 711.
Gustafsson, J.A. and Stenberg, A. (1974b). *Endocrinology* **95**, 891.
Hadjian, A.J., Kowarski, A., Dickerman, H.W. and Migeon, C.J. (1974). *J. Steroid Biochem.* **5**, 346 (Abs.).
Hansson, V. (1972). *Steroids* **20**, 575.
Hansson, V., Djoseland, O., Reusch, E., Attramadal, A. and Torgerson, O. (1973a). *Steroids* **21**, 457.
Hansson, V., Reusch, E., Trugstad, O., Torgerson, O., Ritzen, E.M. and French, F.S. (1973b). *Nature New Biol.* **246**, 56.
Hansson, V., McLean, W.S., Smith, A.A., Tindall, D.J., Weddington, S.C., Nayfeh, S.N., French, F.S. and Ritzen, E.M. (1974). *Steroids* **23**, 823.
Hershko, A., Mamont, P., Shields, R. and Tomkins, G.M. (1971). *Nature New Biol.* **232**, 206.
Imperato-McGinley, J., Guerrero, L., Gautier, I.T. and Peterson, R.E. (1974). *Science* **186**, 1213.
Jouan, P., Samperez, S. and Thieulant, M.L. (1973). *J. Steroid Biochem.* **4**, 65.
Jung, I. and Baulieu, E.E. (1971). *Biochimie* **53**, 807.
Kelch, R.P., Lindholm, U.B. and Jaffe, R.B. (1971). *J. clin. Endocr. Metab.* **32**, 449.
Kitay, J.I. (1963). *Acta endocr. (Copenh.)* **43**, 601.
Kubota, K. and Suzuki, M. (1974). *Endocr. Japon.* **21**, 167.
Kumar, M., Roy, A.K. and Axelrod, A.E. (1969). *Nature* **223**, 399.
Mainwaring, W.I.P. and Peterken, B.M. (1971). *Biochem. J.* **125**, 285.
Mainwaring, W.I.P. (1969). *J. Endocrinol.* **45**, 531.
Nikkari, T. and Valavaara, M. (1970). *J. Endocrinol.* **48**, 373.
Ritzen, E.M., Dobbins, M.C., French, F.S. and Nayfeh, S.N. (1972a). Excerpta Medica Int. Congress Series **256**, 79.
Ritzen, E.M., Nayfeh, S.N., French, F.S. and Aronin, P.A. (1972b). *Endocrinology* **91**, 116.
Ritzen, E.M., Dobbins, M.C., Tindall, D.J., French, F.S. and Nayfeh, S.N. (1973). *Steroids* **21**, 593.
Ritzen, E.M., Hansson, V., French, F.S. and Nayfeh, S.N. (1974). *J. Steroid Biochem.* **5**, 334 (Abs.).
Robel, P., Corpechot, C. and Baulieu, E.E. (1973). *FEBS Letters* **33**, 218.

Rousseau, G.G. (1975). *J. Steroid Biochem.* **6**, 75.
Roy, A.K. and Neuhaus, O.W. (1967). *Nature* **214**, 618.
Roy, A.K. (1973). *J. Endocrinol.* **56**, 295.
Rubin, B.L. (1957). *J. biol. Chem.* **227**, 917.
Senelar, R., Catayee, G., Escola, R., Escola, M.J. and Bureau, J.P. (1973). *Path. Biol.* **21**, 937.
Siiteri, P.K. and Wilson, J.D. (1974). *J. clin. Endocr. Metab.* **38**, 113.
Thody, A.J. and Shuster, S. (1971). *J. Endocrinol.* **51**, vi-vii.
Thody, A.J. and Shuster, S. (1973). *Nature* **245**, 207.
Toh, Y.C. (1973). *Advances in Cancer Research* **18**, 155.
Valladares, L. and Minguell, J. (1975). *J. Steroids* **25**, 13.
Van Baelen, H. and De Moor, P. (1974). *J. clin. Endocr. Metab.* **39**, 160.
Van Baelen, H., Adam-Heylen, M. and De Moor, P. (1975). *Endocrinology (submitted for publication).*
Verhoeven, G., unpublished results.
Verhoeven, G. (1975a). "A comparative study of the androgen receptor apparatus in rodents", Acco, Leuven, Belgium (Thesis).
Verhoeven, G., Heyns, W. and De Moor, P. (1975b). *Vitamins and Hormones* **33** *(in press).*
Verhoeven, G., Heyns, W. and De Moor, P. (1975c). *Steroids (in press).*
Walsh, P.C., Madden, J.D., Harrod, M.J., Goldstein, J.L., McDonald, P.C. and Wilson, J.D. (1974). *New Engl. J. Med.* **291**, 944.
Westphal, U. (1971). "Steroid Protein Interactions" Chapter IX, pp.237-293. Springer Verlag, New York.
Wilson, J.D. and Gloyna, R.E. (1970). *Rec. Progr. Horm. Res.* **26**, 309.
Wilson, J.D. and Lasnitzki, I. (1971). *Endocrinology* **89**, 659.
Wilson, J.D. (1973). *Endocrinolgoy* **92**, 1192.
Yates, F.E., Herbst, A.L. and Urquhart, J. (1958). *Endocrinology* **63**, 887.

HORMONAL CONTROL OF PROGESTERONE RECEPTOR IN THE GUINEA PIG UTERUS

M.T. Vu Hai Luu Thi and E. Milgrom

*Groupe de Recherches sur la Biochimie Endocrinienne
et la Reproduction (INSERM U.135)
Faculté de Médecine Paris-Sud, 94270 Bicêtre, France*

In estrogen primed castrated guinea pigs the uterine cytosol progesterone receptor is a 6.7 S species which can be dissociated into 4.5 S subunit(s) by high ionic strength. It binds progesterone with an association constant $K_A = 1.10^9$ M^{-1} at 0°. Various steroids compete with ^3H progesterone for the binding site according to their progestational activity. When ^3H progesterone has been administered to the animals or if uteri have been incubated with the radioactive hormone a 4S binding species can be extracted from the nuclei. Similar receptors have been found in the cervix, vagina and mammary gland, whereas no specific binding was detected in the hypothalamus or the pituitary (Atger *et al.*, 1974).

I. VARIATIONS DURING THE OESTROUS CYCLE

A Quantitative Changes

Using an exchange technique (Milgrom *et al.*, 1972a) the total concentration (occupied or not by the hormone) of the cytosol receptor was studied during the oestrous cycle (Milgrom *et al.*, 1972b). The largest amount of receptor was found in prooestrus (about 40 000 binding sites/cell). There was then a fast decrease during oestrus and postoestrus to 16-fold lower values at dioestrus (Fig.1).

B Variations in the Sedimentation Coefficient of the Radioactive Hormone Receptor Complexes Were Also Observed

In prooestrus, a 6.7 S peak was observed with a shoulder in the 4-5 S region. In oestrus, the sucrose gradient showed 2 peaks of about equal height, one in the 4-5 S region and another one at 6.7 S. The latter was completely destroyed by

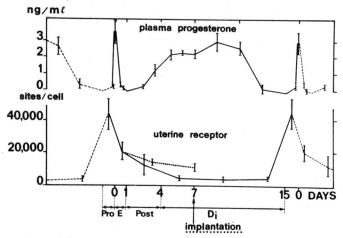

Fig.1 Comparison of plasma progesterone levels and uterine receptor sites concentrations during the oestrous cycle and beginning of pregnancy in the guinea pig. Pro: proestrus, E: oestrus, Post: postoestrus, Di: dioestrus. Plasma progesterone concentrations are from Feder et al. (1968). Dotted line: pregnancy.

p-hydroxymercuribenzoate (PHMB) whereas only a part of the former disappeared. In postoestrus only a shoulder was seen in the 6.7 S region. In dioestrus, the heavier binding had nearly completely disappeared and the only observed binder had a sedimentation coefficient of about 4.5 S.

Thus there appeared to be a progressive decay of the 6.7 S binder from prooestrus to dioestrus. The heavier (6.7 S) binding seems to be composed only of receptor protein sensitive to PHMB, and it is found only in target organs and binds specifically progestagens. In the 4.5 S region of the cytosol ultracentrifugation pattern, there seems to be a mixture of receptor, which adheres to the above-mentioned criteria, and of other uncharacterized binding macromolecules which do not.

II. THE BEGINNING OF PREGNANCY (Fig.1)

The receptor concentration is very similar to that of non-pregnant animals. At day 7 (implantation takes place at day 6.5) the concentration of receptor is very low. This points to the possible physiological importance of progesterone secreted some days before implantation at a moment where the receptor concentration is highest.

III. MECHANISMS OF THE HORMONAL CONTROL (Milgrom et al., 1973)

The variations over the oestrous cycle imply the probable existence of hormonal mechanisms controlling progesterone receptor concentration in the uterus. The rise of receptor after oestrogen injection, observed in preliminary experiments, pointed to the existence of a positive control. But also the fact that oestrogen deprived animals (castrated at various periods of the cycle or prepuberal guinea pigs) had higher levels of receptor than dioestrous animals, indicated the possible existence of another, probably negative, control.

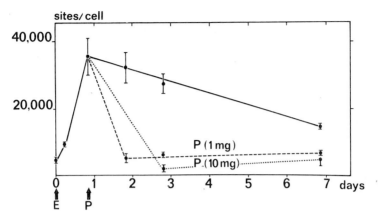

Fig.2 Oestrogen and progesterone effect on uterine cytosol progesterone receptor concentration. E: oestradiol injection; P: progesterone injection.

A Effect of Oestrogen on Progesterone Receptor

Castrated guinea pigs received a single injection of oestradiol. Uteri were analysed at various times after the injection (Fig.2). The concentration of receptor (per cell) had already doubled 6 h after oestrogen administration. A peak was observed at 20.5 h with an 8-fold increase over the basal level. At subsequent times, a slow decrease was observed: 7 days after oestradiol, receptor concentration was still 300% of the basal level.

B RNA and Protein Synthesis and Oestrogen Effects on Progesterone Receptor

When cycloheximide or actinomycin D were injected 15 min before oestrogen, there was a marked inhibition of increase in progesterone receptor measured 20.5 h after oestrogen injection. On the contrary, when these compounds were injected 20.5 h after oestradiol (at the moment of the peak of receptor concentration), there was no effect on receptor concentration measured 67 h after oestrogen. In this case, however, a single injection of a moderate amount of cycloheximide might not have given a sufficiently prolonged inhibition of protein synthesis capable of modifying receptor concentration measured 47 h later. In order to control this point, the amount of cycloheximide necessary to block 86% of protein synthesis during 6 h was determined. This large amount of cycloheximide was then injected twice at 20.5 h and 26.5 h after the oestradiol injection and the progesterone receptor concentration was measured at 32.5 h. Even in these conditions where protein synthesis was inhibited for 12 h, there was no effect on receptor concentration.

These experiments suggested that after injection of oestrogen, two different time periods could be observed. During the first day there was a rise in the concentration of progesterone receptor, depending on RNA and protein synthesis. During a second period (up to the 7th day in these experiments), there was a slow decrease in this concentration, RNA and protein synthesis inhibitors being without effect. It is probable that this decrease reflected mainly the inactivation of the receptor since the rate of synthesis as judged by inhibitor experiments was very slow. If this assumption is correct, then the half life of the receptor mole-

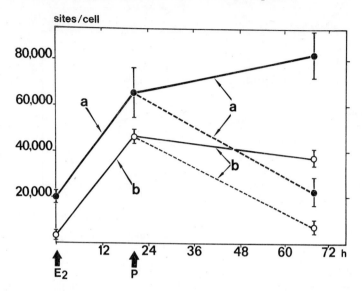

Fig.3 Hormonal control of progesterone receptors in the endometrium and the myometrium. The concentration of progesterone receptors in the endometrium (a) and the myometrium (b) was measured 20.5 h and 67 h after oestradiol injection (full line). In other animals a similar experiment was performed but progesterone was injected 20.5 h after oestradiol and receptor concentration was measured at 67 h (broken line).

cule would be about 5 days as may be seen in Fig.2.

C   *Effect of Progesterone on Progesterone Receptor*

Progesterone was injected in physiological amounts (1-10 mg) 20.5 h after oestrogen, and the uteri were examined 24, 46.5 and 147.5 h later. The concentration of progesterone receptor fell to very low values 24 h after the injection of progesterone and slowly returned to basal levels (as found in castrated non-hormone injected animals) at 147.5 h (Fig.2). Six days after progesterone injection, the concentration of receptors was 1/3 that of controls (oestrogen but not progesterone treated).

This effect of progesterone was not due to masking of sites by unlabelled progesterone since the concentration of the latter was very low in the uteri, 1, 2 or 6 days after injection, and since furthermore this "endogenous" progesterone was taken into account when measuring receptor concentration by the exchange technique previously described. This fall in progesterone receptors was not due to an accumulation in the nuclei of steroid-receptor complexes.

The variation of the receptor concentration may be due to a modification either of the rate of synthesis or of the rate of inactivation. Since it has been already indicated that, at the moment of progesterone injection, the rate of synthesis is very slow, the effect of progesterone can probably be explained by a rise in the inactivation rate (whatever the molecular mechanism of this inactivation might be).

D   *Protein Synthesis and Progesterone Effect on Progesterone Receptor*

The loss of progesterone receptors after progesterone injection was also observed when protein synthesis was blocked by administration of cycloheximide

15 min before progesterone.

Since the effect of progesterone was not mediated by protein synthesis, ruling out the induction of an inactivating enzyme, a further possibility was that it could be eventually related to the transfer of the receptor from the cytosol to the nucleus under the influence of progesterone. To test this hypothesis, the stability of receptor was studied, when present in the cytosol and in the nucleus, respectively.

E   Stability of Cytosol and Nuclear Progesterone Receptors

Cytosol and nuclei were prepared from the uteri of oestradiol treated castrated guinea pigs injected with ^{3}H progesterone. The effect of heating for various periods of time at 25°C was studied by measuring the residual cytosol and nuclear receptors in the corresponding extracts. The inactivation of the cytosol receptor was very limited whereas the nuclear receptor disappeared with a half life of about 20 min.

The interpretation of these experiments is very difficult since it is possible that proteases might have been released during the homogenization. It is then impossible to say if the nuclear inactivating process observed *in vitro* is the same which works *in vivo* or if it is simply an artefact.

IV. COMPARISON OF THE CHARACTERISTICS AND OF THE HORMONAL CONTROL OF ENDOMETRIAL AND MYOMETRIAL PROGESTERONE RECEPTORS (Fig.3)

The characteristics of endometrial and myometrial progesterone receptors are very similar (Luu Thi *et al.*, 1975). Affinity for progesterone, hormone specificity, sedimentation properties (in oestrogen primed animals), inhibition of binding by p-hydroxymercuribenzoate were found to be identical in both tissues. Differences were however observed in the hormonal control of the concentration of receptor. In all situations which were studied, the concentration of receptors was higher in the endometrium than in the myometrium. In guinea pigs castrated at dioestrus the concentration was 3,500 binding sites per diploid genome in the myometrium and 20,300 in the endometrium. One to three days after oestradiol injection, this concentration was raised to 46,000-38,000 and 65,000-83,000 binding sites respectively. Thus the absolute rise was similar in both tissues but the relative increase was about 15-fold in the myometrium and only 3-4-fold in the endometrium. After injection of 2 mg progesterone, the concentration of receptor previously induced by oestrogen returned to very low values similar to those observed in non-hormally treated controls.

This difference between endometrial and myometrial receptors could be due either to a faster turnover of the latter or to the existence of a stable non-hormally controlled population of receptors, present only in the endometrium.

V. DISCUSSION

The progesterone receptor in the guinea pig uterus is under a double hormonal control mechanism. Its induction by oestrogen probably explains the priming effect of these hormones on progesterone action. Its inactivation is enhanced by progesterone administration and thus progesterone has probably a self-limiting action on its own hormonal activity. The interplay between these ovarian hormones yields cyclic variations of receptor concentration. The pre-ovulatory peak of oestrogen is followed shortly by a rise in progesterone receptor concentrations. Low concentrations are observed during the maximal secretory activity of the corpus luteum. If progesterone action is quantitatively related to the concentration of progesterone-receptor complexes in the nucleus, high hormonal activities should be obtained with low hormone concentrations during the periovulatory period whereas very high progesterone concentrations would be necessary to obtain effects of the same magnitude during diestrus.

REFERENCES

Atger, M., Baulieu, E.E. and Milgrom, E. (1974). *Endocrinology* **94**, 161-167.
Feder, H.H., Resko, J.A. and Goy, R.W. (1968). *J. Endocrinol.* **40**, 505.
Luu Thi, M.T., Baulieu, E.E. and Milgrom, E. (1975). *J. Endocrinol.* **66**, 349.
Milgrom, E., Perrot, M., Atger, M. and Baulieu, E.E. (1972a). *Endocrinology* **90**, 1064.
Milgrom, E., Atger, M., Perrot, M. and Baulieu, E.E. (1972b). *Endocrinology* **90**, 1071.
Milgrom, E., Thi Luu, M., Atger, M. and Baulieu, E.E. (1973). *J. biol. Chem.* **248**, 18, 6366.

SPECIFIC BINDING OF HUMAN LUTEINIZING HORMONE AND HUMAN CHORIONIC GONADOTROPIN TO OVARIAN RECEPTORS

C.Y. Lee and R.J. Ryan

Department of Molecular Medicine, Mayo Clinic and Mayo Foundation Rochester, Minnesota 55901, USA

SUMMARY

Studies on the interaction of hLH-hCG with ovarian receptors have given new insight into the basic mechanism of gonadotropin action.

The binding of hLH or hCG to receptors is a highly specific, saturable and reversible process which is dependent upon time, temperature, pH and salt concentrations. Equilibrium studies revealed high affinity and a finite number of binding sites for ovarian receptors. Association and dissociation rate constants have been derived from kinetic analyses.

Gonadotropin receptors from luteinized rat ovaries have been solubilized with Triton X-100 and partially purified by affinity chromatography. Purified receptors retained specific binding activity and showed only a single band in disc-gel electrophoresis.

Attempts to correlate hCG binding to receptors and biological effects have been made. HCG-hLH receptors of rat ovaries can be induced by pregnant mare's serum gonadotropin and ovine LH (or hCG). The marked increase in hCG binding following receptor induction and the decline in hCG binding during luteolysis correlate well with changes of ovarian progesterone concentration and hCG-mediated adenylate cyclase activity. Estrogen treatment which prevented luteolysis, as evidenced by sustained high levels of progesterone in both the blood and ovary, also prevented the decline in hCG binding and hCG activation of adenylate cyclase.

The gonadotropin-receptor interaction has been used as a competitive binding assay for serum and urinary hCG from pregnant women. This radio-receptor assay provides a rapid and sensitive method for measurement of biologically active hLH or hCG.

The existence of hLH receptors in human ovaries has been demonstrated. The properties of hLH-human ovarian receptor interaction have been shown to be similar to those found in the rat ovary as described above. Further studies of hormone-receptor interaction will undoubtedly yield greater insight about the role of gonadotropin receptors in normal and pathological endocrine function of the human ovary.

INTRODUCTION

The primary event in gonadotropin action appears to be the binding of the hormone to its specific receptors in target tissue. Receptors specific for hCG-LH* have been demonstrated *in vitro* in human (Lee et al., 1973a) and rat (Lee and Ryan, 1971a; Danzo et al., 1972) ovaries and rat testes (Catt et al., 1971). Such an *in vitro* system has proved extremely useful for the direct study of gonadotropin-receptor interaction and the elucidation of mechanism of gonadotropin action. The purpose of this chapter is to review our studies on gonadotropin binding to ovarian receptors.

METHODS

A Labeling of hLH and hCG

The procedure for iodination was a modification (Lee and Ryan, 1973b) of the method of Greenwood et al. to minimize the damage of hormones during labeling. Hormone (10 µg in 10 µl) was added to 25 µl of 0.5 M phosphate buffer (pH 7.5) containing 1.0 mCi of $Na^{125}I$. The reaction was initiated by addition of 10 µl of chloramine-T (12.5 mg/10 ml of 0.5 M phosphate buffer). After 40-60 seconds in an ice bath, 250 µl of sodium metabisulfite (10 mg/100 ml) of 0.5 M phosphate buffer) was added, followed by 100 µl of 1% KI. Labeled hormone was purified by gel filtration on Bio-Gel P10. The mole ratio of hormone: ^{125}INa:chloramine-T in the reaction vial was 1:1:100. The labeled gonadotropin had specific activity of approximately 60-70 µCi/µg which corresponds to 1.3-1.5 atoms of ^{125}I per molecule of hCG and 1 atom of ^{125}I per molecule of hLH.

B Preparation of Receptor

Female Holtzman rats were primed with 50 IU of PMSG at 25 days of age and 50 IU of hCG (or 125 µg of ovine LH equivalent to NIHLH S18) 56 h later (Lee and Ryan, 1971a). PMSG-hCG priming causes great increases in both ovarian receptor sites and ovarian weight. Animals were used between eight and fifteen days after PMSG injection when the ovaries were heavily luteinized.

Particulate fraction (2,000 × g pellets) was prepared from luteinized rat ovaries as described previously (Lee and Ryan, 1973b). Fresh particulate fraction was prepared for most of our studies. Otherwise, the 2,000 × g pellets can be resuspended in 10% sucrose and stored at −60°C for months without loss of binding activity.

Soluble receptor was prepared from the particulate fraction by extraction with 0.25% Triton X-100 as reported before (Lee and Ryan, 1974a). The extract was centrifuged for 1 h at 110,000 × g in a Beckman L2-65 centrifuge. The supernatant appeared clear and contained soluble receptors.

---

*Abbreviations used are: hCG, human chorionic gonadotropin; hLH/oLH, human/ovine luteinizing hormone; PMSG, pregnant mare's serum gonadotropin.

C Assay of Binding Activity
   1. *Ovarian slices*
   Two slices (30 to 40 mg) made from rat ovaries with a Stadie-Riggs microtome were incubated with [^{125}I] hCG in modified Krebs-Ringer phosphate buffer according to the method published before (Lee and Ryan, 1974b).
   2. *Particulate fraction*
   The typical assay of binding activity was performed at 25°C for 5 to 16 h by which time equilibrium was reached. The procedure in detail has been described before (Lee and Ryan, 1973b). In essence, approximately 2 ng of [^{125}I] hCG was incubated with 2 to 5 mg equivalent wt of 2,000 X g pellets of ovarian homogenate in a final volume of 1 ml of 40 mM Tris buffer containing 0.1% bovine serum albumin, pH 7.4. Millipore membrane filtration was employed to separate bound from free hormone. Nonspecific binding was measured in the presence of a large excess of unlabeled hCG (200 IU/ml) and represented only 1-2% of total binding. Specific binding was obtained by subtraction of the nonspecific component from the total binding.
   3. *Solubilized receptors*
   The condition for assay of soluble and purified receptor binding activity has been reported previously (Lee and Ryan, 1974a). Bound and free hormone were separated using polyethylene glycol (MW 6,000-7,500) which precipitates bound hormone. Nonspecific binding assessed in the presence of large amounts of unlabeled hCG represented about 5% of total binding which could be further reduced by solution and reprecipitation of the hormone receptor complex.

D Adenylate Cyclase Assay
   Adenylate cyclase activity was determined according to the method of Saloman *et al.* (1974) using Dowex and alumina chromatography for isolation of cyclic AMP. The radioactivity present in eluate containing [^{32}P] cyclic AMP and [^{3}H] cyclic AMP was determined in a Beckman LS-100C liquid scintillation counter.

RESULTS AND DISCUSSION

A *General Properties of hCG-LH Binding to Receptors*
   Labeling of hLH or hCG with substantial retention of biologic activity has proved to be a critical step for studies of gonadotropin binding to receptors. The modified iodination procedure described under Methods using a low ratio of chloramine-T to hormone at low temperature, minimized damage to gonadotropin molecules. Under such conditions, hLH and hCG can be labeled to approximately 1 atom of ^{125}I per gonadotropin molecule with retention of biologic activity. The estimated bioactivity of labeled gonadotropins will be given in the section B.
   The gonadotropin-receptor interaction is dependent on pH, time, temperature, and salt concentration (Lee and Ryan, 1972). Maximal binding appeared to occur at neutral pH. Binding of LH to receptor *in vitro* reaches an equilibrium state in 30 min at 37°C (Fig.1), approximately 4 h at 25°C and 48 h at 4°C. All salts (MgSO$_4$, CaCl$_2$, NaCl and KCl) at concentration of 150 mM significantly inhibited binding as compared with 40 mM Tris buffer. A nonlinear relationship was observed between the concentrations of receptor prepararion used and the amounts of labeled hLH bound. The reduced binding at the higher concentrations

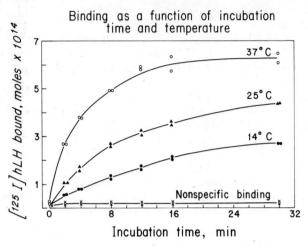

Fig.1 Binding of human [^{125}I] hLH to ovarian homogenates as a function of incubation time and temperature. Nonspecific binding at 37°C (x—x) was determined by the addition of a high concentration of hCG.

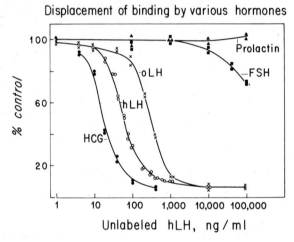

Fig.2 Displacement of [^{125}I] hLH binding by unlabeled hormones. (●) hCG, (○) hLH, (x) oLH, (■) FSH, (▲) prolactin.

of ovarian tissue is largely due to inactivation of hormone during incubation. Assay conditions can be modified to minimize hormone damage by utilization of small amounts of the receptor preparation, low temperature of incubation, and prior washing of the tissue preparation.

Binding of hCG-LH to ovarian receptors is highly specific for both hormone and tissue. Unlabeled hLH, hCG and ovine LH readily displace [^{125}I] hLH binding to its receptor (Fig.2). FSH and prolactin are without effect (Lee and Ryan, 1972). The subunits of hCG and oLH have little or no binding activity and are considerably less potent than native hormone in competing with labeled hLH for binding to receptor (Lee and Ryan, 1971b).

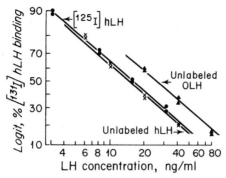

Fig.3 Radioreceptor assay of [^{125}I] hLH using [^{131}I] hLH. Biological (receptor-binding) activity of [^{125}I] hLH was assayed using unlabeled hLH and oLH as references. The tracer was [^{131}I] hLH. The concentration of [^{125}I] hLH was determined by radioimmunoassay.

Target tissue specificity and subcellular localization of binding have been studied. Labeled hLH binds significantly only to ovary and not muscle, liver, spleen and kidney (Lee and Ryan, 1971a). Autoradiographic studies revealed a high binding of [^{125}I] hCG to corpora lutea (luteal cells) and slight binding to interstitial and theca cells (Han et al., 1974). Centrifugal fractionation studies (Rajaniemi et al., 1974) and EM autoradiography (Han et al., 1974) of rat ovarian tissue suggest that LH receptors are primarily located in the plasma membrane.

Binding is a function of hormone concentration. Scatchard analysis revealed high affinity and a finite number of binding sites in the ovary. Quantitative aspects of gonadotropin-receptor interactions will be described in Section B.

Chemical nature of the gonadotropin receptor has been studied by its susceptibility to degradation by a variety of enzymes. A lipoportein nature of the LH receptor is suggested by the evidence that binding was reduced by the treatment of the receptor with proteolytic enzymes, phospholipase C and phospholipase D (Lee and Ryan, 1972).

B Equilibrium and Kinetic Studies of Gonadotropin-Receptor Interactions

   1. *Assessment of concentration of biologically active [^{125}I] labeled hCG and hLH*

Because of iodination damage of hormones and inaccurate estimation of the recovery of labeled hormone following the iodination procedure, the concentration of biologically active [^{125}I] hCG or hLH is often overestimated. The overestimated concentration of hormone results in erroneously high receptor sites and low association constant (or high dissociation constant). In order to measure the biological activity of small amounts of labeled gonadotropin, a sensitive radioreceptor was employed, since conventional bioassay becomes impractical for the assay of small amounts of labeled hormones. The radioreceptor assay is utilized for estimation of biological activity on the basis of our finding that the receptor assay is in good agreement with the ovarian ascorbic acid depletion assay for LH potency estimation of pituitary extracts (Lee and Ryan, 1971a).

When the biologic activity of [^{125}I] hLH was measured by the radioreceptor assay, in which the tracer was [^{131}I] hLH, parallel dose-response curves were obtained for [^{125}I] hLH, unlabeled hLH and oLH (Fig.3). The biologic activity of [^{125}I] hLH could, therefore, be estimated using unlabeled purified hLH prep-

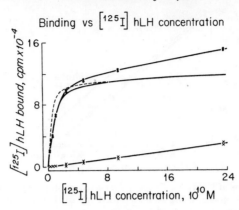

Fig. 4 Binding v. [^{125}I] hLH concentration. Nonspecific binding (x) was determined in the presence of excess hCG (200 IU/ml). Specific binding (—) was obtained by subtraction of the nonspecific component from total binding (●). Specific binding corrected for inactivation (---).

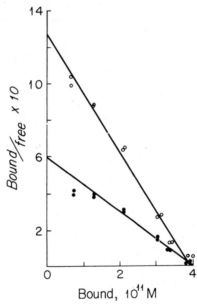

Fig. 5 Scatchard plot of specific binding data of [^{125}I] hLH illustrated in Fig. 4. (○) corrected for inactivation during incubation, (●) uncorrected.

aration as a reference. The concentration of [^{125}I] hLH was determined by the radioimmunoassay and unlabeled hLH was used as a reference. The biologic activity of [^{125}I] hLH was estimated to be 80 to 90% of the activity of unlabeled hLH, and labeled hCG retained 90 to 100% of its activity (Lee and Ryan, 1973b). In all equilibrium and kinetic studies, the concentration of labeled hormone was calculated on the basis of mass of biologically active hLH or hCG.

2. *Assessment of gonadotropin inactivation*

Labeled gonadotropin was degraded by ovarian receptor preparations during

Table I. Dissociation constants for hCG-LH binding to ovarian receptors.

		Dissociation constant	Temp.	Reference
rat luteinized ovary	hCG	5.0×10^{-10} M 6.2×10^{-10} M	4°C 37°C	Danzo (1973)
rat luteinized ovary	hCG hLH	3.8×10^{-11} M 4.9×10^{-11} M	25°C 25°C	Lee and Ryan (1973b)
bovine corpus luteum	oLH	3.0×10^{-9} M	23°C	Gospodarowicz (1973)
bovine corpus luteum	hCG hLH	1.5×10^{-10} M 1.4×10^{-10} M	37°C 37°C	Haour and Saxena (1974)
bovine corpus	hCG	1.2×10^{-10} M	38°C	Rao (1974)

incubation. Attempt to correct the inactivation of free hormone was made for equilibrium studies of hormone binding. The inactivation of labeled free gonadotropin incubated with receptor preparation and subsequently separated from bound hormone was assessed for its ability to bind to fresh receptor (Lee and Ryan, 1973b). For the control, labeled gonadotropin in assay buffer and receptor preparation were incubated separately and mixed immediately before Millipore filtration.

3. Equilibrium dissociation constant and number of binding sites

The quantitative studies of gonadotropin-receptor interaction were carried out at 25°C, in 40 mM Tris-HCl, pH 7.4. Fresh preparations of thrice washed 2000 X g pellets of luteinized rat ovaries were used for each experiment. Figure 4 illustrates that binding is a function of [^{125}I] hLH concentration. A saturable curve was obtained when the nonspecific binding was subtracted from the total binding. When correction was made for the inactivation of free hLH, a small shift in the specific binding curve occurred as shown by the dashed line in Fig.4.

The specific binding data shown in Fig.4 can be analyzed by a Scatchard plot (Scatchard, 1949), which is illustrated in Fig.5. The slope of the lines give the reciprocal of the equilibrium dissociation constant (K_d) and the intercept at the abscissa yields the number of binding sites. The K_d value was calculated to be 6.7×10^{-11} M (Lee and Ryan, 1973b). Correction for the inactivation of free hLH decreased the value of K_d (to 3.1×10^{-11} M) by approximately a factor of 2 (Fig.5). Since free hCG is less affected than hLH by the inactivation process, only a small decrease of K_d was observed (Lee and Ryan, 1973b). Correction for the inactivation did not alter the number of binding sites for either hCG and hLH.

Table I lists the dissociation constants of gonadotropin-receptor interaction reported from different laboratories. The discrepancies could result from the degree of iodination damage, the receptor preparations, the assay conditions (for example, salt concentration used and the extent of inactivation) and the gonadotropin preparations used as tracer. Nevertheless, the values of dissociation constant reported are consistent with the general concept of high affinity of ovarian receptors for gonadotropins.

A Hill plot for gonadotropin binding was made from specific binding data.

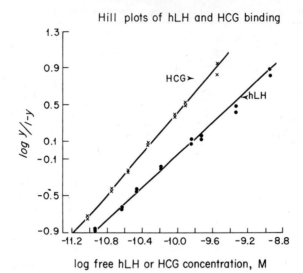

Fig.6 Hill plots of hLH (•) and hCG (x) binding. Correction was made for inactivation during incubation.

A value of 0.9 was obtained for the Hill coefficient of hLH binding. The result indicates the absence of cooperativity for hLH-receptor interaction. A similar result was found for hCG binding with a Hill coefficient of 1.1 (Fig.6).

4. Rate constants for hLH and hCG binding

Various concentrations of receptors and gonadotropins were used to obtain the association rate constant. Essentially similar results were obtained. The association rate constant for hCG at 25°C was found to have a mean value of 5.7×10^6 M^{-1} sec^{-1} (Lee and Ryan, 1973b). The rate constant for hLH under the same conditions ranged from 2.5 to 3.8×10^6 M^{-1} sec^{-1}.

The dissociation of the gonadotropin-receptor complex was biphasic when plotted as a first-order reaction. The dissociation rate constants of hLH-receptor complex for the fast (k_{-1}) and slow (k_{-2}) phases were estimated to be 1.4×10^{-4} sec^{-1} and 6.7×10^{-6} sec^{-1} respectively (Lee and Ryan, 1973b). The dissociation reaction is temperature dependent.

The equilibrium dissociation constant (K_d) can be calculated from the ratio of dissociation rate constants (k_{-1} or k_{-2}) and association rate constant (k_{+1}). The dissociation constant (K_d) calculated from the ratio k_{-1}/k_{+1} compares favorably with the constant obtained from equilibrium data.

C Correlation of Binding with Biological Effects

1. HCG binding activity, ovarian progesterone and adenylate cyclase.

Binding of [^{125}I] hCG to rat ovarian slices was followed daily after PMSG-oLH injection. Ovarian progesterone concentration was measured simultaneously by radioimmunoassay (Lee et al., 1973a). A 7-fold increase in ovarian binding activity (tissue/medium ratio) was observed following luteinization of the ovary. Increased binding activity was accompanied by an increase in ovarian progesterone content and preceded by ovarian weight increase as illustrated in Fig.7 (Lee et al., 1975b). Binding was maintained at maximum for 9 days and subsequently followed by a sharp decline. The decrease in binding activity was also temporarily

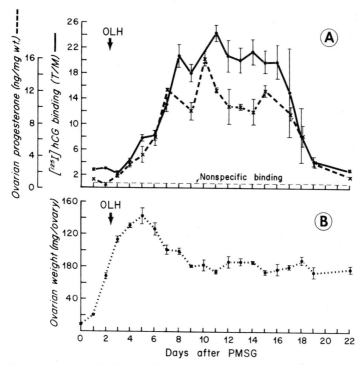

Fig. 7 Daily changes in hCG binding (A), progesterone concentration (A) and weight of rat ovaries (B) following PMSG-oLH priming. Binding of hCG (●—●) by ovarian slices is expressed as a tissue/medium ratio.

correlated with luteolysis as evidenced by diminished ovarian progesterone. This study demonstrates that ovarian binding of hCG is closely related to the corpora luteal function of the ovary.

The correlation of hCG-sensitive adenylate cyclase activity with hCG binding activity has recently been examined following induction of hCG-LH receptors by PMSG-oLH priming. The close correlation of binding and adenylate cyclase activities strongly suggest the degree of activation of adenylate cyclase hCG is dependent upon the number of hCG receptor sites available (Lee, 1975c).

2. *Effect of estrogen treatment*

The effects of estrogen injection on hCG binding, ovarian progesterone concentration and hCG-sensitive adenylate cyclase were investigated. The PMSG-oLH primed rats were injected daily (10 μg/rat) for 5 days or with a single injection (50 μg/rat) of polyestradiol phosphate (long-acting estrogen) at day 10 after PMSG. Two effects on hCG binding and ovarian progesterone level were observed. First, both binding activity and progesterone concentration in ovary and serum were enhanced 5 days after estrogen treatment (Lee and Ryan, 1974b). Secondly, estrogen treatment maintained hCG binding and ovarian progesterone concentrations at high levels for a longer period of time than the control group (Fig. 8).

Similar results were found for hCG-sensitive adenylate cyclase activity following estrogen injection. Estrogen treatment rescued the luteinized ovary from regression and thus increased and maintained the enzyme activity for a longer

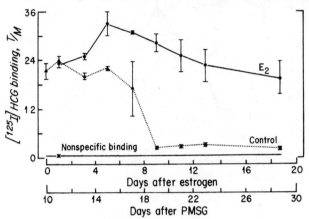

Fig.8 Effect of estrogen treatment on [^{125}I] hCG binding by ovarian slices. Binding is expressed as a tissue/medium ratio. Nonspecific binding was determined in the presence of a high concentration of hCG (200 IU/ml).

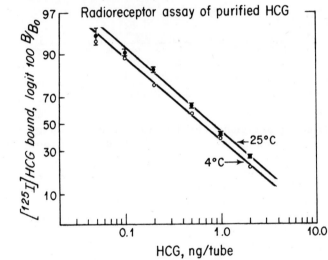

Fig.9 Dose-response curves for the radioreceptor assay of purified hCG at 4°C and 25°C.

period of time (Lee, 1975c) when corpora lutea regressed in the control group, the hCG receptor sites and hCG-activated adenylate cyclase also decreased.

These results demonstrate that estrogen has stimulatory effects on hCG binding, hCG-sensitive adenylate cyclase and progesterone synthesis by luteinized rat ovary and that increased hCG binding following estrogen treatment is associated with enhanced biological effects (e.g. adenylate cyclase activity and progesterone synthesis).

D *Radioreceptor Assay of hCG-LH*

Because of its high affinity and specificity, gonadotropin binding to receptors can be used as a competitive protein-binding assay for hCG or hLH (Lee and Ryan,

Specific Binding of hLH and hCG to Ovarian Receptors

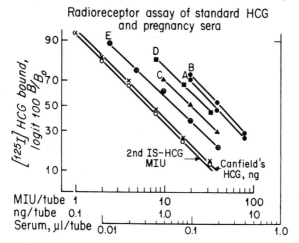

Fig.10 Dose-response curves for the radioreceptor assay (high sensitivity) of 2nd IS-hCG and pregnancy sera. Each point represents the mean of 3 determinations.

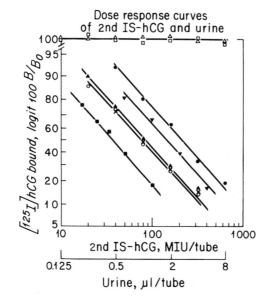

Fig.11 Dose-response curves for the radioreceptor assay (low sensitivity) of 2nd IS-hCG and pregnancy urine. Urine from nonpregnant women (△) and men (□) produced no response. (■) 2nd IS-hCG, (○ × ▲ ▼ ●) 5 pooled pregnancy urines.

1975a). A sensitive radioreceptor assay capable of detecting 100 pg of hCG has been developed (Fig.9). HCG concentrations in urine and sera of pregnant women have been determined by receptor assay. The dose-response curves of a series of pregnancy urines and sera were shown to be parallel to those of the standard hCG reference preparations (Figs 10 and 11) measurement of hCG concentrations in the urine of pregnant and nonpregnant women, using a receptor assay, showed good correlation with the result obtained from the standard agglu-

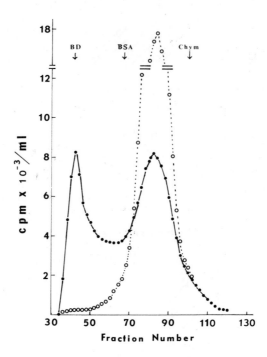

Fig. 12 Gel filtration on Sephadex G-150 of purified ovarian receptor after incubation with [^{125}I] hLH alone (●—●) or [^{125}I] hLH plus 200 IU/ml of hCG (o—o). Standard markers are BD (blue dextran) BSA (bovine serum albumin) and Chym (chymotrypsinogen). The peak eluted at the void volume indicated by BD represents [^{125}I] hLH-receptor complex. Free [^{125}I] hLH was eluted between BSA and Chym.

tination pregnancy test.

The major advantage of receptor assay is the measurement of biologically active forms of the hormone rather than immunologic determinants as is the basis for radioimmunoassay. Using this technique, it is possible to detect heterogeneous forms of gonadotropin in biological samples and potential inhibitors of receptor binding. The radioreceptor assay should also provide a more precise and rapid method of defining structure-functional relationships of hCG-LH molecule.

E   *Solubilization and Purification of LH-hCG Receptors*

Attempts to solubilize and purify the membrane-bound LH-hCG receptors from luteinized rat ovaries have been made. The particulate fraction (2000 X g pellets) of ovaries were solubilized in cold 0.25% Triton x-100. The soluble receptor was shown capable of binding hCG as demonstrated by Sephadex G-100 chromatography or by the polyethylene glycol separation method. Binding of labeled hCG to soluble receptor can be displaced by unlabeled hCG (Lee and Ryan, 1974a).

The soluble receptor was further purified by affinity chromatography using hCG-Sepharose columns. Approximately 60 to 100-fold purification was achieved from solubilized receptors. When purified receptor was subjected to SDS gel electrophoresis, a single band was observed, whereas, multiple bands were found for crude Triton x-100 extracts of ovary. Purified receptors still retain specific binding activity. Binding of [^{125}I] hLH to purified receptors was inhibited by unlabeled hCG (Fig.12) but not by FSH (Lee and Ryan, 1974a).

Considerable difficulty was encountered during purification of soluble receptor using affinity chromatography. The receptor absorbed to affinity column could not be eluted without drastic treatment using 3M guanidine. It is possible that eluted receptor was altered since the equilibrium dissociation constant of [^{125}I] hCG binding to purified receptor was found to be 7.6×10^{-10} M compared to 4.1×10^{-11} M for hCG binding to membrane-bound receptor. It appears that utilization of gonadotropin with low affinity toward receptors (e.g. chemically modified or LH from nonhuman sources) for making affinity columns may improve the purification procedure. The alternate approach is to improve the elution conditions. Elution of receptors at low pH (3.2) from the affinity column has been utilized (Dufau et al., 1975).

Solubilization of hCG receptor from the rat testis (Dufau et al., 1973) and [^{125}I] hCG-receptor complex from the rat ovary (Bellisario and Bahl, 1975) have also been reported. The Stokes radius of hCG-receptor complex was estimated to be 64A and the molecular weight of receptor was calculated to be 200,000. The Stokes radius (65A) of soluble hCG-receptor complex extracted *in vitro* was found identical with the complex formed *in vivo* (Bellisario and Bahl, 1975).

F   Luteinizing Hormone Receptors in Human Ovary

The existence of hLH receptors has been demonstrated in the human ovary using homogenates of luteal tissue (Lee et al., 1973a). [^{125}I] hLH binding to human ovarian receptors has similar characteristics to those found in luteinized rat ovary. Binding of [^{125}I] hLH to human luteal tissue is a time-dependent and saturable process that can be competitively inhibited by unlabeled hLH. FSH is without effect. The equilibrium dissociation constant at 37°C for hLH binding to human ovarian homogenates was estimated to be 1.1×10^{-9} M at 37°C. The constant is in reasonable agreement with the value of 7.9×10^{-10} M found in rat luteal tissue (Lee and Ryan, 1972) using similar experimental conditions.

The binding activity of human ovarian tissue showed considerable variability which appeared to depend on the physiologic state. However, vigorous equilibrium and kinetic studies as described in rat ovarian binding are needed to define the quantitative aspects of hLH-human ovarian receptor interaction. Further studies of gonadotropin-receptor interaction will certainly offer an opportunity for the determination of the role of gonadotropin receptors in both normal and pathological endocrine functions of the human ovary.

ACKNOWLEDGEMENTS

Figures 1 and 2 are reproduced from *Proc. natn. Acad. Sci. U.S.A.* 69, 3520 (1972): Figures 4-6 from *Biochemistry* 12, 4609 (1973) (copyright by the American Chemical Society): Figure 7 from *Proc. Soc. Exp. Biol. Med.* 148, 505 (1975) (copyright by the Society for Experimental Biology and Medicine):

Figure 8 from *Endocrinology* **95**, 1691 (1974): Figures 9-11 from *J. clin. Endocr. Metab.* **40**, 228 (1975) (copyright by The Endocrine Society): Figure 12 from "Gonadotropins and Gonadal Function" (Moudgal, N.R., ed), Academic Press, New York, p.444 (1974) (copyright by Academic Press). Figures are reproduced by permission of the publishers.

The work was supported by Grants NIH-HD-08099 and HD-09140 from USPHS and the Mayo Foundation.

REFERENCES

Bellisario, R. and Bahl Om P. (1975). *J. biol. Chem.* **250**, 3837.
Catt, K.J., Dufau, M.L. and Tsuruhara, T. (1971). *J. clin. Endocr. Metab.* **32**, 860.
Danzo, B.J., Midgley, A.R. and Kleinsmith, L.J. (1972). *Proc. Soc. exp. Biol. Med.* **129**, 88.
Danzo, B.J. (1973). *Biochim. Biophys. Acta* **304**, 560.
Dufau, M.L., Charreau, E.H. and Catt, K.J. (1973). *J. biol. Chem.* **248**, 6973.
Dufau, M.L., Ryan, D.W., Baukal, A.J. and Catt, K.J. (1975). *J. biol. Chem.* **250**, 4822.
Gospodarowicz, D. (1973). *J. biol. Chem.* **248**, 5042.
Han, S.S., Rajaniemi, H.J., Cho, M.I., Hirschfield, A.N. and Midgley, A.R., Jr. (1974). *Endocrinology* **95**, 589.
Haour, F. and Saxena, B.B. (1974). *J. biol. Chem.* **249**, 2195.
Lee, C.Y. and Ryan, R.J. (1971a). *Endocrinology* **89**, 1515.
Lee, C.Y. and Ryan, R.J. (1971b). *In* "Protein and Polypeptide Hormones" (M. Margoulies and F.C. Greenwood, eds) pp.332-333. Excerpta Medica, Amsterdam.
Lee, C.Y. and Ryan, R.J. (1972). *Proc. natn. Acad. Sci. U.S.A.* **69**, 3520.
Lee, C.Y., Coulam, C.B., Jiang, N.S. and Ryan, R.J. (1973a). *J. clin. Endocr. Metab.* **36**, 148.
Lee, C.Y. and Ryan, R.J. (1973b). *Biochemistry* **12**, 4609.
Lee, C.Y. and Ryan, R.J. (1974a). *In* "Gonadotropins and Gonadal Function" (N.R. Moudgal, ed) pp.444-459. Academic Press, New York.
Lee, C.Y. and Ryan, R.J. (1974b). *Endocrinology* **95**, 1691.
Lee, C.Y. and Ryan, R.J. (1975a). *J. clin. Endocr. Metab.* **40**, 228.
Lee, C.Y., Tateishi Kayoko, Ryan, R.J. and Jiang, N.S. (1975b). *Proc. Soc. exp. Biol. Med.* **148**, 505.
Lee, C.Y. (1975c). Unpublished data.
Rajaniemi, H.J., Hirshfield, A.N. and Midgley, A.R., Jr. (1974). *Endocrinology* **95**, 579.
Rao, Ch.V. (1974). *J. biol. Chem.* **249**, 59.
Scatchard, G. (1949). *Ann. N.Y. Acad. Sci.* **51**, 660.
Salomon, Y., Londas, C. and Rodbell, M. (1974). *Analytical Biochem.* **58**, 541.

SOME ASPECTS OF THE ACTIVITY OF THE HYPOTHALAMO-PITUITARY GONADAL SYSTEM IN CHILDREN

J. Girard and P.W. Nars

Endocrine Department, University Childrens Hospital, Basel, Switzerland

INTRODUCTION

The long process of sexual maturation is completed by a rapid evolution with a variety of very obvious events such as growth-spurt, development of secondary sexual characteristics, growth of genital organs, etc.. The pattern of hormone-secretion, which is concomitant with this period of 'puberty' shows dramatic changes, which are well recognized.

It is only recently, however, that it has been possible to demonstrate a definitive activity of the hypothalamo-pituitary gonadal system in newborns, infants and prepubertal children (Forest *et al.*, 1973; Angsusingha *et al.*, 1974; Bidlingmaier *et al.*, 1973; Kenny *et al.*, 1973; Grumbach *et al.*, 1974; Faiman and Winter, 1974).

Sex differences in hormone-secretion are present in the perinatal period, but are no longer detectable during later infancy and childhood. These differences reappear with the final period of sexual maturation.

Studies elucidating endocrine 'profiles' in pediatrics rely on the development of specific sensitive assay-methods, which require plasma-volumes in the 100 μl range. For a cautious interpretation of results, the technical problems of the assay-procedures have to be well understood. Interlaboratory comparisons are impossible if the technical details of the estimations are not given. Furthermore, the reagents (i.e. antibodies) used may differ from one laboratory to another and thus lead to different results in spite of an identical methodological procedure (see Fig.1).

As random plasma values may only represent instantaneous hormone-secretion, an evaluation of endocrine activities requires stimulation and suppression tests. For ethical reasons, however, these cannot be performed in normal children.

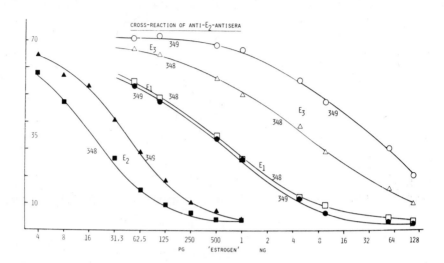

Fig.1 Different standard curves obtained with antiserum 348 and 349 in spite of practically identical cross-reactions (given for one antiserum in Table II). Antiserum 348 is not 'more sensitive', but gives absolute plasma values with an 80% difference to antiserum 349. Note the difficulty of interpreting absolute plasma concentrations.

Interpretation of urinary excretion has to take into account hormone metabolism and renal function. Hepatic metabolism of sex-steroids for instance is known to change with age (Blunck, 1958). On the other hand plasma-concentrations can represent glandular activity or peripheral interconversion (Van de Wiele *et al.*, 1963).

The activity of circulating sex-steroids depends on the unbound fraction of the hormone in circulation. It has been shown that the percentage of total plasma-sex-steroids bound to carrier proteins changes with age (Forest *et al.*, 1973; Anderson, 1974).

The investigations performed in our laboratory do not include the perinatal period or early infancy. Testosterone, estradiol, androstenedione and estrone-concentrations in plasma and urine are presented in relation to bone-age and/or secondary sexual characteristics rather than to chronological age. For this symposium on ovarian function the main interest will be focused on hormonal activity in girls, but values in boys are given for comparison. Random LH and FSH values and their response to LH-RH will only be briefly mentioned.

PATIENTS AND METHODS

Table I includes data of patients investigated in the present chapter.

LH. For iodination with the chloramine-T-method the LH-preparation from NAP was used. Purification over Sephadex G25 followed by adsorption to cellulose (3 ml), wash with phosphate buffer and elution with 20% acetone, 5% BSA. Reference preparation LH-LER 960. Antiserum raised with HCG. Sensitivity 1.5 mU/ml plasma. Mean coefficient of variation 17%.

Table I.

BOYS (N = 45)				
$\bar{X}$ Testes volume (ml)	$\leqslant 2$	$2 \leqslant 5$	$5 \leqslant 12$	> 12
N =	15	15	6	9
Chronological age (Years)	9.1 (6.7–11.4)	11.1 (5.9–14.5)	13.2 (8.5–16.5)	14.6 (13.1–15.8)
Bone-age (Years)	8.7 (4.3–10.4)	11.0 (6.0–13.6)	13.2 (8.5–15.8)	14.8 (13.1–16.1)
GIRLS (N = 47)				
Breast-stage	B I	B II	B III-IV	B V
N =	14	11	10	12
Chronological age (Years)	8.2 (2.3–10.4)	10.8 (7.8–14.3)	13.6 (12.7–15.6)	15.0 (13.1–17.2)
Bone-age (Years)	8.1 (2.5–12.0)	11.0 (7.5–14.2)	13.5 (12.0–15.5)	15.2 (13.0–17.3)

Table II. Cross-reaction of anti-estradiol-antiserum 349/3 for 50% inhibition of positive controls.

Estradiol-17β	100
Estradiol-17β-glucoronide	0.12
Estradiol-17β-sulfate	2.5
Estrone	0.6
Estriol	0.1
Testosterone	0.08
Androsterone	0.02
Etiocholanolone	0.003
Dehydroepiandrosterone	0.05
Cortisol	0.0005
Progesterone	0.002
Pregnanediol	0.0008
Cholesterol	< 0.0004

Table III. Cross-reaction of anti-testosterone-antiserum for 50% inhibition of positive controls.

Testosterone	100
Androstenedione	0.8
Androsterone	0.02
Dehydroepiandrosterone	0.02
Etiocholanolone	0.004
Estrone	0.007
Estradiol-17β	0.007
Estriol	0.004

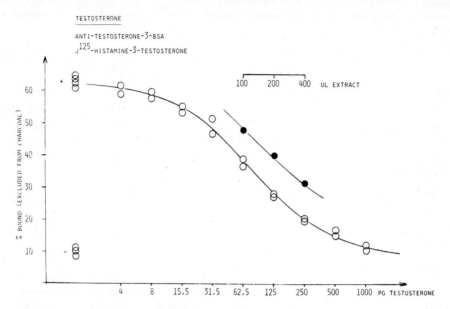

Fig. 2 Typical testosterone calibration curve showing parallelism of doubling dilutions of plasma extracts. (Cross-reactions of antiserum see Table III.)

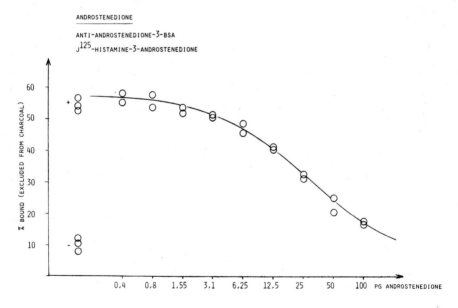

Fig. 3 Androstenedione calibration curve showing sensitivity of the assay. (Cross-reactions of antiserum see Table IV.)

FSH. For iodination fractions CPDS6 (gift from Dr. W. Butt, Birmingham). Purification like LH. Reference preparations MRC 69-104. Antiserum gift from

Table IV. Cross-reaction of anti-androstenedione-antiserum for 50% inhibition of positive controls.

Androstenedione	100
Testosterone	2.0
Dehydroepiandrosterone	1.7
Estrone	0.02
Deoxycortisol	0.02
Cortisol	0.02
Progesterone	0.03

Dr. W. Butt. Sensitivity 1 mU/ml plasma. Mean coefficient of variation 26.9%.

Estradiol 17β. (Method for all steroid-assays based on the method described by Hunter et al., 1973). Tracer ^{3}H-estradiol (E_2). Antiserum raised in rabbits using estradiol-6-BSA (Dighe and Rutherford). Cross-reactions given in Table II. Extraction of E_2 from 1 ml plasma with diethyl-ether. Dry organic phase. Redissolve in petrol-ether. Phase separation 0.1 M NaOH. 0.4 ml of water phase for RIA. Assay in acid buffer, containing gelatine, separation with charcoal dextran. Recovery from extraction 91.2 ± 2.5%. Recovery after phase separation. 90.5 ± 5.3%. Sensitivity (using 1 ml of plasma) 2.6 pg/ml plasma. Mean coefficient of variation 10.7%.

Figure 1 gives an example of estradiol calibration curves obtained with 2 antisera, which have very similar characteristics, but give a different sensitivity and widely differing plasma values.

Estrone. Assay procedure as for the estradiol-assay. Antiserum raised against estrone-6-BSA (Dr. W.M. Hunter, Edinburgh). Sensitivity 7 pg/ml plasma. Mean coefficient of variation 14%.

Testosterone. Antiserum raised against testosterone-3-BSA (Dr. A. Ismail, Edinburgh). Cross-reaction given in Table III. Tracer: I^{125}-histamine-3-testosterone (Nars and Hunter, 1972). Separation charcoal dextran. 0.5 ml plasma, extracted with diethyl-ether, 3 extracts, redissolve in assay diluent. Recovery 98.5%, sensitivity 50 pg/ml plasma. Coefficient of variation 10.9-16.1%. A typical standard curve is given in Fig.2.

Androstenedione. Antiserum raised against androstenedione-3-BSA (Dr. A. Ismail, Edinburgh). Cross-reaction given in Table IV. Tracer I^{125}-histamine-3-androstenedione (Nars and Hunter, 1972). Extraction according to testosterone-assay. Recovery 98.5%. Sensitivity 40 pg/ml plasma. Mean coefficient of variation 13.3%. A typical standard curve is given in Fig.3.

PERINATAL/POSTNATAL PERIOD UP TO 6 MONTHS OF AGE

Gonadotropins (LH, FSH)

FSH and LH are certainly detectable in neonates of both sexes (Kulin and Reiter, 1973). Faiman and Winter (1971) found higher plasma FSH and LH levels in girls than in boys during the first 2 years of life. Considerably higher FSH-levels with a wide range have been described in female infants below 2 years of age (Faiman and Winter, 1974). Increased mean LH-concentrations during the first 2 years have been found in boys (Lee, 1973). Buckler and Clayton (1970) described no sex differences in 24-hours-gonadotropin-excretion, but higher LH-

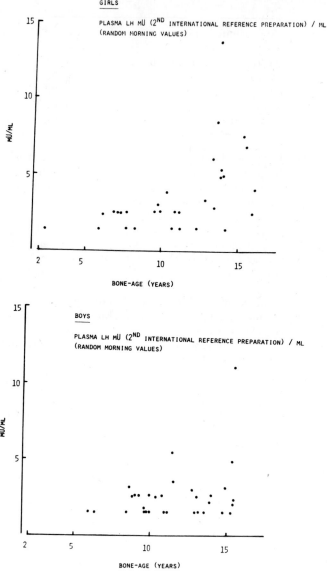

Fig. 4 Random plasma LH concentrations in boys and girls related to bone-age. No obvious sex difference.

excretions for both sexes during the first 6 months as compared to later.

These findings support the evidence given below for an increased activity of the hypothalamo-pituitary gonadal axis during the first 6 months.

Androgens

Testosterone

Forest *et al.* (1973) have recently clearly shown a significantly higher testo-

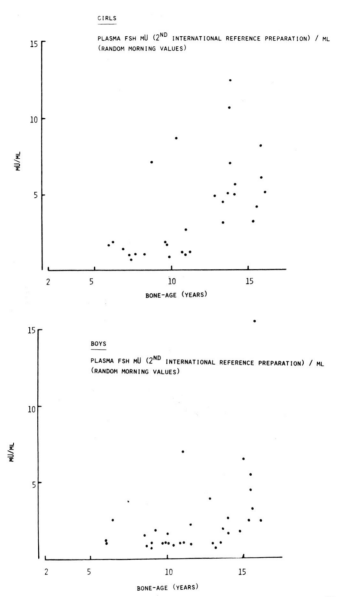

Fig. 5 Random plasma FSH concentrations in boys and girls related to bone-age. No obvious sex difference.

sterone concentration in cord-blood of males as compared to females. In male infants, testosterone concentrations show a significant increase during the first 3 months of life. About 50% of the normal adult male values are reached. Testosterone concentrations subsequently decrease at the age of 7 months to prepubertal levels. In the female neonate testosterone concentrations are in the range of 2.6 ng/ml and decrease thereafter to prepubertal levels from 2 weeks onwards (Forest

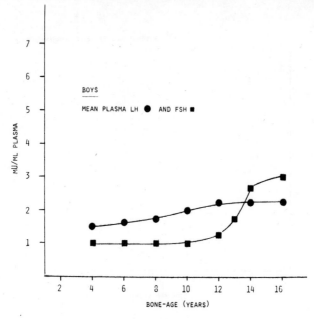

Fig. 6A

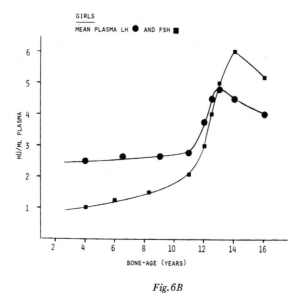

Fig. 6B

Figs 6A,B Mean plasma LH and FSH concentrations in boys and girls related to bone-age. Somewhat higher mean plasma LH concentrations in girls as opposed to boys and distinct earlier FSH increase.

et al., 1973).

Estrogens
   Estradiol
   Estradiol concentrations in cord-blood are about a thousand times higher than those found later (Bidlingmaier et al., 1973; Kenny et al., 1973). A sex difference has not been described and estradiol values decline rapidly.
   Estrone concentrations show a similar pattern (Bidlingmaier et al., 1973).

In the neonatal period an activity of the gonadal system is thus definitely proved in boys and is very probable in girls. The physiological significance of this activity remains to be elucidated.

FROM 6 MONTHS TO 'PUBERTY'

Gonadotropins
   During this relatively slow development of sexual maturation, dramatic changes in endocrine secretion are not apparent. A definitive sex difference in random plasma LH and FSH concentration does not exist (Figs 4,5). Our mean results (Figs 6A,B) can however be interpreted as showing somewhat higher LH-values in girls and an earlier increase in FSH concentration in girls. Lee (1973) described a higher plasma-FSH-concentration in girls than in boys in the first 6 years of life. This finding could correspond to the earlier increase in FSH-concentration found in our series. Random plasma values show a considerable overlap between prepubertal and adult stages of development. The plasma levels in adults are only 2-3 times higher than in children.
   Kulin and Reiter (1973) in reviewing the subject of gonadotropins in children found a steadily increasing FSH and LH plasma concentration and urinary gonadotropin excretion. Our results (individual points are given in Figs 7A,B and mean curves in Figs 8A,B) demonstrate a urinary gonadotropin excretion which shows no sex difference during the period from a bone-age of 4 to 12 years. The pattern of hormone excretion is, however, completely different from the results obtained in random plasma samples.
   Mean urinary LH excretion is about 1/12 in prepubertal children as compared to adults. The FSH excretion must increase 6 times from the prepubertal stage to reach adult levels (Kulin and Reiter, 1973).
   As gonadotropin concentrations in this period are often at the lower limit of sensitivity of the assay, it is difficult to demonstrate a clear sex difference.
   The responses of LH and FSH to LH-RH are, however, in this period of development, different between girls and boys (Grumbach et al., 1974) (Fig.9).

Androgens
   Testosterone
   Figure 10 shows plasma testosterone concentrations in girls and boys related to bone-age. A very slight increase from the prepubertal values of 0.4 ng/ml is noted in girls. In boys at a bone-age of 12 years a clear increase is evident. These data as well as those given for testosterone concentrations in relation to breast development and testes size (Table V) correspond well to the figures published by Forest et al.,(1973) and Faiman and Winter (1974).
   An adrenal origin of plasma androgen activity during this period is indicated by the increase in total ketosteroids and the increasing concentrations of DHA, DHAS, androsterone and etiocholanolone in urine. Furthermore plasma testo-

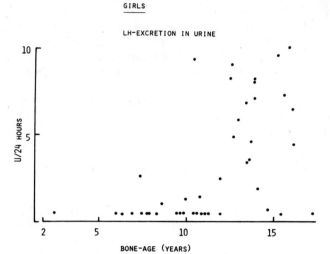

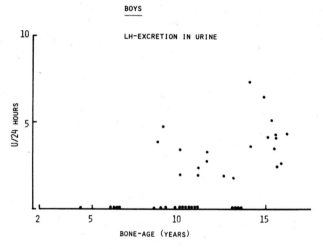

Fig. 7A Urinary LH excretion in girls and boys related to bone-age. In girls increase to higher LH levels.

sterone concentrations can be stimulated by ACTH and partially suppressed by dexamethasone (Forest et al., 1973).

Androstenedione

Androstenedione concentrations (Fig.11) are very similar in boys and girls during this prepubertal period. The mean concentration of 0.5 ng/ml starts to increase slightly at a bone-age of 6 years in both sexes. Mean androstenedione concentrations are somewhat higher in girls than in boys.

The ratio of testosterone to androstenedione is around 0.5 in girls and 0.8 in boys without a noticeable change until puberty (Fig.12). Similarly androstenedione-excretion and plasma concentrations in boys and girls related to the puberty

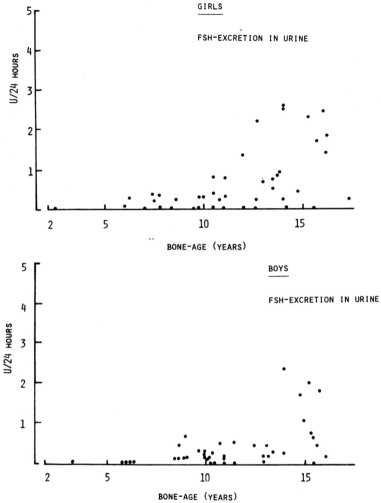

Fig. 7B Urinary FSH excretion in boys and girls related to bone-age. No obvious sex difference.

ratings show no sex difference.

Estrogens
    Estradiol

Although plasma estradiol concentrations are very similar in boys and girls (individual data Fig.13, mean values Fig.14), it is noteworthy that in prepubertal girls (B I) 15.4% of plasma estradiol concentrations were found to be below the assay sensitivity as compared to 57% unmeasurable values in boys with a testes size below 2 ml. The plasma estradiol concentrations in relation to puberty ratings are given in Table VI. Plasma estradiol concentrations described here compare well with those given by Forest *et al.* (1973), Bidlingmaier *et al.* (1973) and Kenny *et al.* (1973).

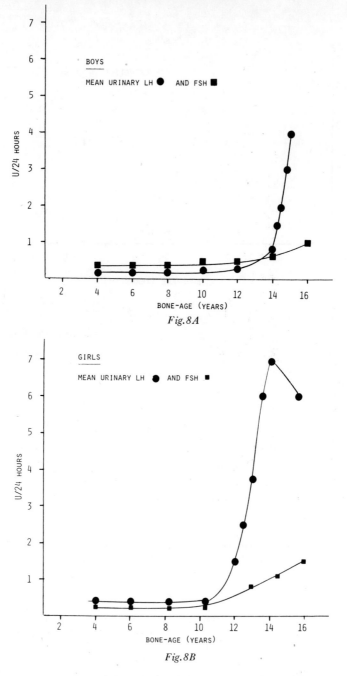

Figs 8A,B Mean LH and FSH excretion in boys and girls related to bone-age; during the prepubertal period from a bone-age of 4-12 years no sex-difference and a pattern thereafter which is different from that observed in plasma samples.

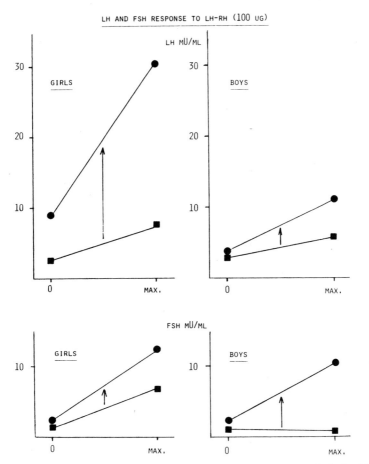

Fig. 9 Mean basal LH and FSH concentrations and maximum values reached after LH-RH in prepubertal and pubertal boys and girls. Note: nearly absent FSH response in prepubertal boys as opposed to a definite response in girls. LH response in boys and girls. In the prepubertal period very similar, in the pubertal period very much higher response in girls.

Urinary estradiol concentrations could be measured in all 24-hour-urines and again showed no sex-difference during this period of development. Individual data are given in Fig. 15 and mean values for estrogen output in 24-hour-urine in Fig. 16. The data in relation to puberty ratings are given in Table VII.

Estrone

Plasma estrone values were found to be around 5-10 pg/ml in boys and girls (Fig. 17). In both sexes 9% of the values were undetectable. An increase in plasma estrone concentration becomes obvious at a bone-age of 8 years in girls (Fig. 17). At this stage boys show only a clear increase in urinary estrone output (Fig. 18). The data for estrone plasma concentrations and excretion related to puberty ratings are given in Tables VIII and IX.

Similar to the androgens, the ratio of estradiol to estrone changes only in girls and only at the final stage of maturation (Fig. 19).

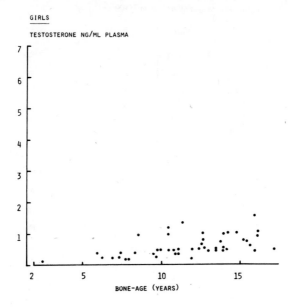

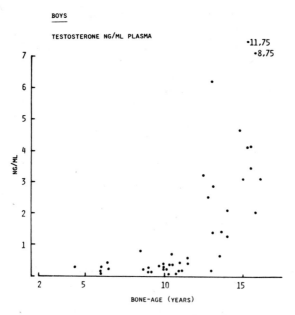

Fig.10 Plasma-testosterone concentration in girls and boys related to bone-age.

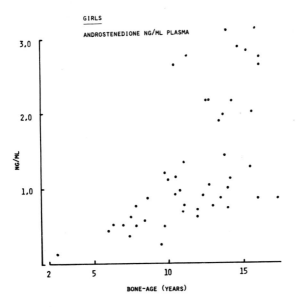

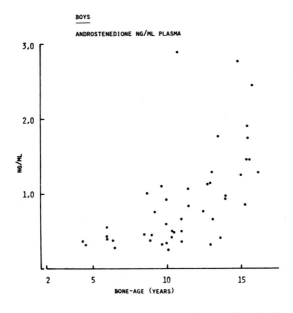

Fig. 11 Androstenedione concentrations in girls and boys related to bone-age.

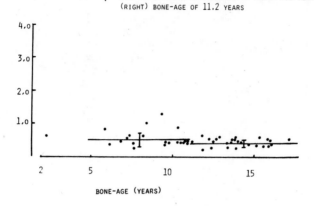
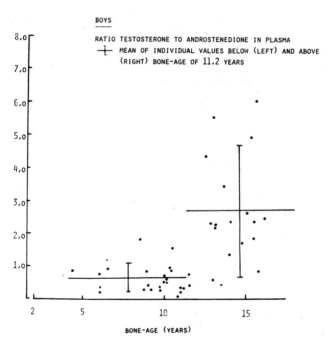

Fig. 12 Ratio of plasma testosterone to androstenedione concentrations in girls and boys related to bone-age.

The Hypothalamo-Pituitary Gonadal System in Children

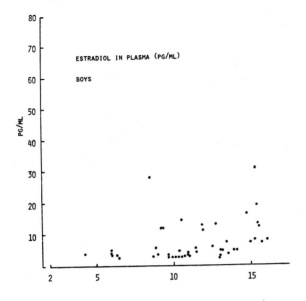

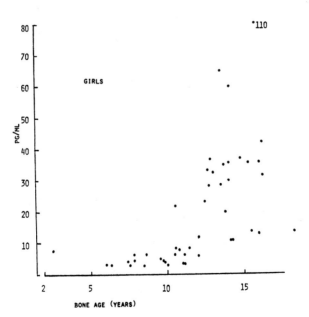

Fig.13 Individual plasma estradiol concentrations in boys and girls related to bone-age. In prepubertal values no obvious sex difference.

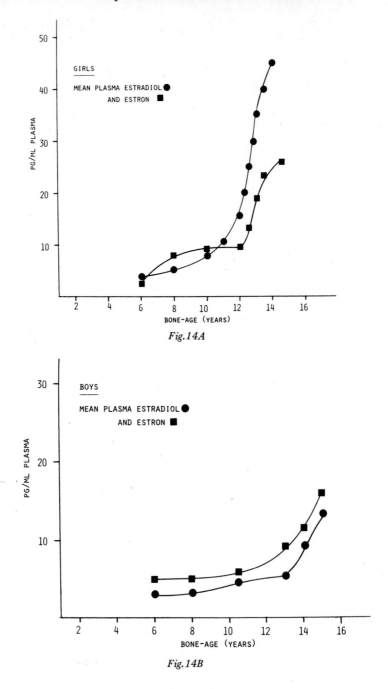

Figs 14A,B Mean plasma estradiol and estrone concentrations in boys and girls related to bone-age.

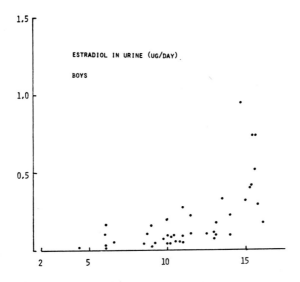
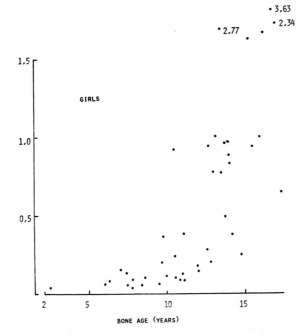

Fig. 15 Estradiol excretion in urine in boys and girls related to bone-age.

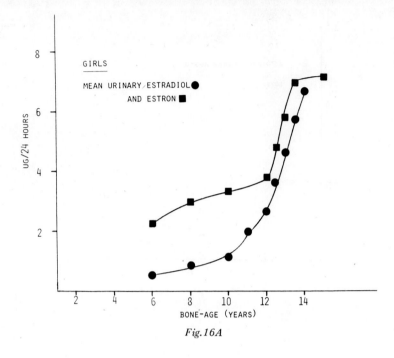

Fig. 16A

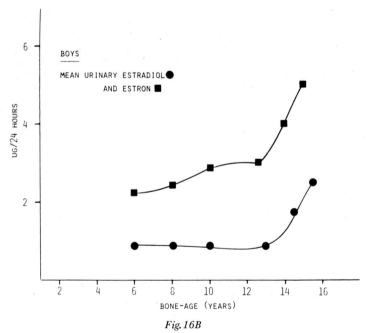

Fig. 16B

Figs 16A,B Mean estradiol and estrone excretion in 24-hour-urine in girls and boys.

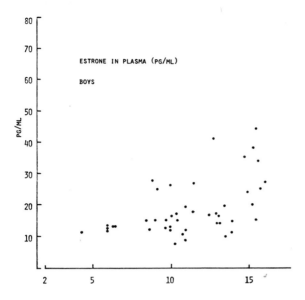
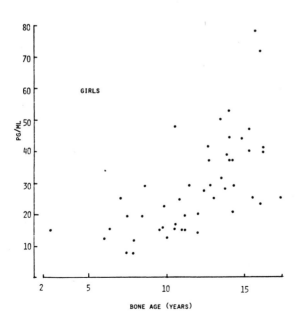

Fig. 17 Plasma estrone concentrations in boys and girls in relation to bone-age.

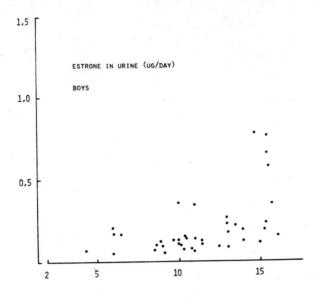

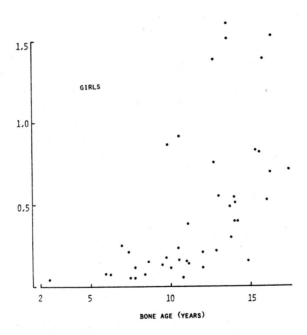

Fig.18 Urinary estrone excretion in boys and girls in relation to bone-age.

The Hypothalamo-Pituitary Gonadal System in Children   217

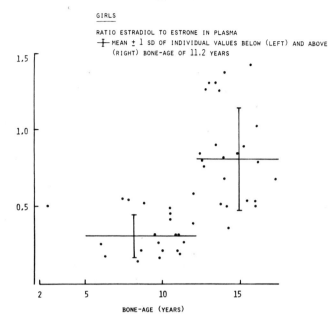

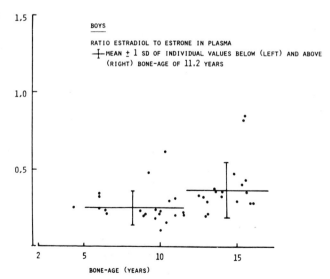

Fig.19 Ratio of plasma estradiol to estrone concentrations in girls and boys related to bone-age. No significant sex difference in the prepubertal ratios for girls and boys and no appreciable increase of the ratio in pubertal boys as opposed to a dramatic increase in girls.

Table V.
Girls: Testosterone ng/ml plasma related to breast-development.

Breast-stage	X̄	S.D.	S.E.M.	Range
B I	0.349	0.262	0.070	0.075—0.947
B II	0.539	0.333	0.100	0.312—1.280
B III—IV	0.625	0.262	0.083	0.132—0.990
B V	0.747	0.351	0.101	0.398—1.525

Boys: Testosterone ng/ml plasma related to testes volume.

Testes volume	X̄	S.D.	S.E.M.	Range
2 ml	0.279	0.109	0.029	0.074—0.454
2 ≤ 5 ml	1.217	1.705	0.440	0.098—6.250
5 ≤ 12 ml	2.994	2.982	1.222	0.622—8.750
> 12 ml	3.951	3.217	1.072	0.663—11.750

Table VI.
Girls: Estradiol pg/ml plasma related to breast-development.

Breast-stage	X̄	S.D.	S.E.M.	Range
B I	5.2	2.6	0.7	2.6—11.5
B II	8.5	7.2	2.2	2.6—21.3
B III—IV	30.1	16.1	5.1	5.4—65.0
B V	37.7	26.4	7.6	10.5—110.0

Boys: Estradiol pg/ml plasma related to testest volume.

Testes volume	X̄	S.D.	S.E.M.	Range
≤ 2 ml	3.1	0.8	0.2	2.6—5.3
2 ≤ 5 ml	4.3	2.3	0.6	2.6—11.7
5 ≤ 12 ml	11.7	8.9	3.6	3.8—28.1
> 12 ml	12.2	8.9	3.0	3.0—31.3

FINAL SEXUAL MATURATION

Gonadotropins

A large number of publications (references see Grumbach et al., 1974) confirm that plasma gonadotropin secretion rapidly increases from the time of development of the secondary sexual characteristics to adulthood. FSH concentration starts to increase before LH. From our own results this is certainly true for girls (Fig.6). In general it is thought that LH shows a more impressive change. In our series this is true for urinary gonadotropin excretion (Fig.8), whereas in

Table VII.

Girls: Estradiol in urine (µg/24 h) related to breast-development.

Breast-stage	X̄	S.D.	S.E.M.	Range
B I	0.124	0.096	0.026	0.029–0.372
B II	0.161	0.151	0.050	0.034–0.462
B III–IV	0.773	0.776	0.245	0.135–2.772
B V	1.439	0.933	0.295	0.651–3.626

Boys: Estradiol in urine (µg/24 h) related to testes volume.

Testes volume	X̄	S.D.	S.E.M.	Range
⩽ 2 ml	0.079	0.053	0.014	0.013–0.196
2 ⩽ 5 ml	0.090	0.072	0.020	0.027–0.266
5 ⩽ 12 ml	0.249	0.186	0.084	0.042–0.510
> 12 ml	0.454	0.280	0.093	0.170–0.937

Table VIII.

Girls: Estrone pg/ml plasma related to breast-development.

Breast-stage	X̄	S.D.	S.E.M.	Range
B I	16.5	6.2	1.6	7.5–28.7
B II	22.2	9.9	3.0	12.5–47.5
B III–IV	33.7	10.5	3.3	13.8–50.0
B V	42.1	17.5	5.1	23.1–77.5

Boys: Estrone pg/ml plasma related to testes volume.

Testes volume	X̄	S.D.	S.E;M.	Range
⩽ 2 ml	14.5	5.6	1.4	7.5–27.5
2 ⩽ 5 ml	15.5	4.9	1.3	10.0–26.3
5 ⩽ 12 ml	26.0	9.9	4.0	15.0–41.0
> 12 ml	25.2	11.3	3.8	14.1–43.8

plasma a convincing increase is obvious for FSH in girls (Fig.6). Most of the results on plasma gonadotropins rely, however, on random plasma determinations and represent cross-sectional data.

A pulsatile FSH and LH secretion with secretory peaks during sleep develop as an important indication of advanced maturation (Grumbach et al., 1974; Boyar et al., 1973; Boyar et al., 1972a; Boyar et al., 1972b; Kelch et al., 1973). The response of LH and FSH to LH-RH changes to an adult pattern (Fig.9) (Grumbach et al., 1974) and is further modified in girls by the menstrual cycle.

Table IX.

Girls: Estrone in urine (µg/24 h) related to breast-development.

Breast-stage	X̄	S.D.	S.E.M.	Range
B I	0.189	0.208	0.056	0.038−0.868
B II	0.262	0.279	0.093	0.048−0.924
B III−IV	0.799	0.577	0.217	0.114−1.980
B V	0.848	0.577	0.183	0.397−2.280

Boys: Estrone in urine (µg/24 h) related to testes volume.

Testes volume	X̄	S.D.	S.E.M.	Range
≤ 2 ml	0.124	0.079	0.021	0.041−0.349
2 ≤ 5 ml	0.138	0.086	0.024	0.041−0.338
5 ≤ 12 ml	0.241	0.219	0.098	0.059−0.578
> 12 ml	0.373	0.278	0.093	0.147−0.661

Androgens
   Testosterone

Mean testosterone values (Table V) demonstrate the rapid increase in plasma testosterone concentration in boys in relation to testes size. The increased secretions depending on bone-age are given in Fig.10. The characteristic sex differences of plasma testosterone concentrations are easily recognizable during this period.

   Androstenedione

As is evident from Fig.11 androstenedione concentrations show a similar increase in girls and boys.

Ratios of testosterone to androstenedione are stable in girls and thus confirm the adrenal origin of these steroids. As evidence of the different origin in boys the ratio of testosterone to androstenedione dramatically changes during 'puberty' (Fig.12).

Estrogens
   Estradiol

Female sex steroids develop in a similar way to androgens rising to adult levels. A gradual increase of plasma estradiol concentrations with advancing breast development in girls and only a slight increase in boys can be seen (Table VI and Fig.13). In girls from breast stage III onwards estradiol concentrations are detectable in all plasma samples (Table VI). Starting at a bone-age of 12 years, the mean concentration increases from 5 to 45 pg/ml (Fig.14). In boys an increase to 12 pg/ml starts at a bone-age of 14 years (Nars and Hunter, 1972). At a testes size of 2 to 5 ml 21.4% of plasma samples show undetectable estradiol concentrations in boys (Table VI). These changes in plasma estradiol concentrations in boys and girls are in a similar way reflected in urinary estradiol output (Fig.15 and Table VII).

   Estrone

Estrone plasma concentrations in girls increase 2.3 times from prepubertal to pubertal levels. In boys estrone concentrations increase by a factor of 1.7

(Fig.14 for mean values and Fig.17 for individual values). A significant sex difference is obvious from a bone-age of 12 years on.

Table VIII shows the increase in plasma estrone concentrations in relation to puberty ratings. Although estrone concentration increases in boys and girls, a sex difference is obvious as soon as secondary sex characteristics start to develop. The same is evident from the urinary estrone output (Table IX). Urinary estrone excretions related to bone-age show again an increasing output in girls above a bone-age of 12 years (Fig.18). The mean increase in estrone excretion in 24-hour-urine starts earlier than the increase of estradiol (Fig.16).

The ratio of estradiol to estrone demonstrates the predominantly adrenal origin of these steroids in boys, where the ratio estradiol to estrone increases only slightly whereas in girls a dramatic change is evident, which is similar to the testosterone-androstenedione-ratio in boys (Fig.19).

SUMMARY

Extensive studies using sensitive immunoassays provided evidence of changing activities of the hypothalamo-pituitary-gonadal axis during the development from the neonate to the adult state. An early phase of increased activity during the first 6 months of life is followed by an apparently quiescent period with relatively constant plasma and urinary hormone concentrations. These concentrations require however an increasing hormone production to keep up with the steadily increasing body-mass.

The final period of rapid sexual maturation appears to be initiated by the change in adrenal enzyme activity, which leads to an increased secretion of sex steroid precursors. At the same time a sleep correlated gonadotropin secretion can be observed. The hypothalamic gonadostat changes its setpoint and a dramatic increase in the activity of the hypothalamo-pituitary gonadal axis can be observed with the development of secondary sexual characteristics until the adult stage is reached.

The number of patients investigated in Tables V—IX is given in Table I.

ACKNOWLEDGEMENTS

We wish to thank Dr. W. Butt for the kind gift of FSH and FSH-antiserum, the MRC Mill Hill for reference preparations and the National Pituitary Agency, University of Maryland, for LH-preparations.

We wish to thank Dr. W.M. Hunter and collaborators in Edinburgh for the various steroid-antibodies. The steroid-assays have been performed by P.W. Nars after his return from a sabbatical in Dr. W.M. Hunter's laboratories.

This work was partially supported by Schweiz. Nationalfonds zur Förderung der wissenschaftlichen Forschung, Bern (Credit-Nr. 3.690.73).

REFERENCES

Angsusingha, K., Kenny, F.M., Nankin, H.R. and Taylor, F.H. (1974). *J. clin. Endocr. Metab.* 39, 63.
Anderson, D.C. (1974). *Clin. Endocr.* 3, 69.
Bidlingmaier, F., Wagner-Barnack, M., Butenandt, O. and Knorr, D. (1973). *Pediat. Res.* 7, 901.
Blunck, W. (1958). *Acta endocr. (Copenh.)* **Suppl.** 59, 154.

Boyar, R.M., Finkelstein, J.W. and Roffwarg, H. (1972a). *J. clin. Invest.* 51, 13.
Boyar, R.M., Finkelstein, J.W.,and Roffwarg, H. (1972b). *N. Engl. J. Med.* 287, 582.
Boyar, R.M., Finkelstein, J.W., David, R., Roffwarg, H., Kapen, S. and Weitzman, E.D. (1973). *N. Engl. J. Med.* 289, 282.
Buckler, J.M.H. and Clayton, B.E. (1970). *Arch. Dis. Child.* 45, 478.
Faiman, C. and Winter, J.S.D. (1971). *Nature* 232, 130.
Faiman, C. and Winter, J.S.D. (1974). *In* "The Control of the Onset of Puberty" (M.M. Grumbach, G.D. Grave and F.E. Mayer, eds) pp.32-61. John Wiley & Sons, New York - London - Sydney - Toronto.
Forest, M.G., Saez, J.M. and Bertrand, J. (1973). *Paediatrician* 2, 102.
Grumbach, M.M., Roth, J.C., Kaplan, S.L. and Kelch, R.P. (1974). *In* "The Control of the Onset of Puberty" (M.M. Grumbach, G.D. Grave and F.E. Mayer, eds) pp.115-181. John Wiley & Sons, New York - London - Sydney - Toronto.
Hunter, W.M., Bolton, A.B. and Nars, P.W. (1973). *J. Endocrinol.* 58, XXI.
Kelch, R.P., Conte, F.A., Kaplan, S.L. and Grumbach, M.M. (1973). *J. clin. Endocr. Metab.* 36, 424.
Kenny, F.M., Angsusingha, K., Stinson, D. and Hotchkiss, J. (1973). *Pediat. Res.* 7, 826.
Kulin, E.H. and Reiter, E.O. (1973). *Pediatrics* 51, 260.
Lee, P.A. (1973). *Clin. Endocr.* 2, 255.
Nars, P.W. and Hunter, W.M. (1972). *J. Endocrinol.* 57, XLVII.
Van de Wiele, R., McDonald, P., Gurpide, E. and Lieberman, S. (1963). *Horm. Res.* 19, 275.

THE EFFECT OF HYPOTHALAMIC AND PITUITARY TUMOURS ON THE DEVELOPMENT OF PUBERTY

J.S. Jenkins

Department of Medicine, St. George's Hospital, London SW1
England

Animal experiments suggest that the medial basal (hypophysiotrophi) area of the hypothalamus is responsible for controlling the tonic secretion of gonadotrophin, and, in the female, the anterior hypothalamic and preoptic area is responsible for the phasic control necessary for ovulation. Somewhat paradoxically, there is evidence that lesions made in various parts of the hypothalamus, particularly in the anterior area, may actually advance sexual development (Donovan and Van der Werff ten Bosch, 1965). The picture is further complicated by the action of the pineal gland as an inhibitory influence on gonadotrophin secretion, at least in rodents, where pinealectomy leads to the advancement of puberty. It is therefore to be expected that tumours involving the pituitary, the stalk, and various sites in the hypothalamus or pineal would profoundly affect the development of puberty, providing that the tumours arise at an age before sexual maturation of the child is complete.

PITUITARY ADENOMA

Adenomas of the pituitary are a fairly common cause of hypogonadism in the adult but they are rarely responsible for complete failure of pubertal development, simply because such tumours are very rare in children of prepubertal age. In a series of 260 patients with pituitary adenomas reviewed by our neurosurgical department (Elkington and McKissock, 1967), the youngest patient was 14 years old. Occasionally, however, the tumour does present early enough in life to interfere with the full development of puberty, and primary amenorrhoea occurs. This may result simply from a deficiency of gonadotrophin secretion, but there may be another reason, as the following case illustrates.

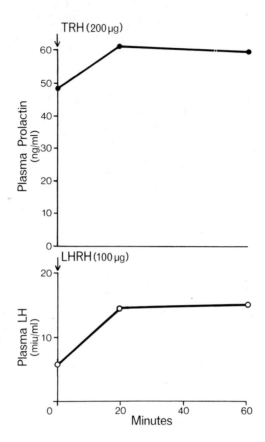

Fig.1 Plasma LH response to LHRH and plasma prolactin response to TRH in patients with pituitary chromophobe adenoma and primary amenorrhoea.

The patient was first seen at the age of 18 with primary amenorrhoea and failing vision due to a chromophobe pituitary adenoma. Body hair was somewhat scanty although there was normal breast development. Following treatment by surgery and radiotherapy amenorrhoea persisted.

Figure 1 shows that the amenorrhoea was not associated with low levels of gonadotrophin, the LH being within the normal range, and there was a response to the administration of LHRH in spite of the pituitary lesion. Also shown here is the plasma level of prolactin which was above the upper limit of our normal range (6-15 ng/ml) and which increased further after the administration of TRH. It is now known that many chromophobe adenomas are far from being "functionless" and recently excessive prolactin secretion has been incriminated as an important factor in interfering with normal ovulation in patients with pituitary adenomas even without the clinical features of galactorrhoea (Child *et al.*, 1975). The elevated levels of prolactin appear to affect the gonad directly, causing inhibition of steroidogenesis (McNatty *et al.*, 1974). It is important to recognize excessive prolactin secretion in these patients since it can now be effectively reduced by the

Effect of Hypothalamic and Pituitary Tumours on Development of Puberty

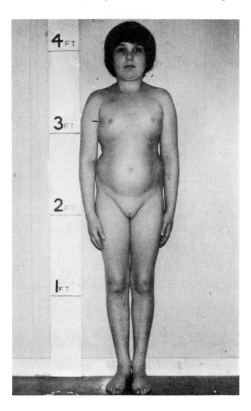

Fig. 2 Girl aged 15 with a craniopharyngioma showing failure of pubertal development and retardation of growth.

ergot alkaloid bromoergocryptine, which in a high proportion of cases will induce normal ovulation.

CRANIOPHARYNGIOMA

The commonest tumour to affect hypothalamic-pituitary function in children is a craniopharyngioma. This tumour most often arises in the pars tuberalis at the front of the upper end of the stalk, but by the time that it is usually seen it has involved the floor of the 3rd ventricle and ultimately results in destruction of the stalk. During the last ten years we have seen a total of 41 such patients of whom 19 presented for the first time in childhood or adolescence. There was a strong male predominance in this series. Three were still of prepubertal age and in all but one of the remainder there was some impairment of pubertal development. The presenting feature in all instances was either impairment of vision or symptoms of raised intracranial pressure. In the younger children retardation of growth was the dominant endocrine disturbance and this was always associated with failure of puberty.

The familiar clinical picture of retardation of growth with sexual infantilism is seen in this 15-year-old girl with a large calcified craniopharyngioma (Figs 2

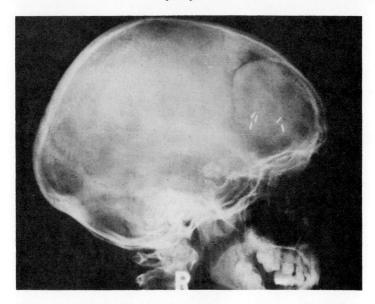

Fig.3 X-ray of skull showing large calcified craniopharyngioma from patient shown in Fig.2.

and 3). In spite of the predominantly suprasellar lesion, the plasma gonadotrophin response to the administration of LHRH was completely absent (Fig.4). On the other hand a typically hypothalamic delayed type of TSH response to TRH was obtained, and there was no growth hormone response to insulin-induced hypoglycaemia. ACTH secretion was normal as demonstrated by the cortisol response to hypoglycaemia. Table I shows the plasma gonadotrophin response to LHRH in 10 cases of craniopharyngioma, together with other aspects of pituitary function. The first three patients were untreated and the remainder had received radiotherapy before investigation.

All patients showed a similar pattern, the basal levels of gonadotrophin being below normal, sometimes very low, and the response to LHRH was grossly deficient considering that the ages of all the patients, apart from the last, were such that a pubertal or an adult type of response would be expected.

The response to LHRH is therefore of no value in determining the site of the lesion and these results are broadly similar to those reported by Mortimer *et al.* (1973). The failure of response implies that when the pituitary is deprived of endogenous gonadotrophin releasing hormone in early life the synthesis of gonadotrophin, quite apart from its release, is prevented. Indeed, Mortimer *et al.* (1974) have shown that by administering large doses of LHRH three times a day, the responsiveness of patients with hypothalamic lesions can be restored, and in this way LHRH can even be used therapeutically. It is significant that growth hormone was always found to be deficient in these patients, so that in prepubertal children lack of growth hormone will almost invariably presage failure of puberty. Hypopituitarism is generally more severe in these patients with craniopharyngiomas than in those with pituitary adenomas.

Attention should be called to the patient R.P. (Table I). Although he was hypogonadal when seen 12 years after treatment, with scanty pubic hair, small

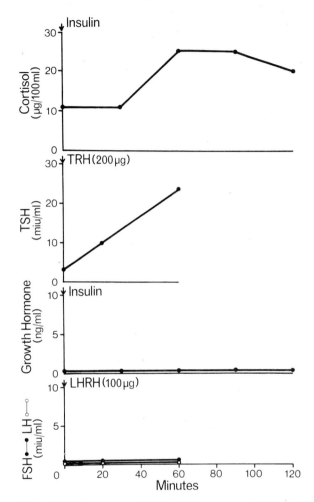

Fig. 4 Plasma LH and FSH response to LHRH, TSH response to TRH, and growth hormone and cortisol response to insulin-induced hypoglycaemia (same patient as Fig. 2).

testes and no facial hair, the paediatrician who saw him when he first presented at the age of 6 with neurological symptoms reported on an abnormally large penis and testes for his age, together with some pubic hair. Very rarely, craniopharyngiomas have been reported as a cause of precocious puberty, but in this case advanced sexual maturation was later followed by hypogonadism.

TUMOURS OF THE 3RD VENTRICLE

Any tumour which arises in the middle part of the hypothalamus in the floor of the 3rd ventricle may be expected to damage neurosecretory tissue, and gliomas of various degrees of malignancy occur in this area which interfere with the development of puberty.

Table I. Plasma gonadotrophin response to LHRH and other pituitary function tests in patients with craniopharyngiomas.

Patient	Age	LH (m.i.u./ml) Basal	LH (m.i.u./ml) Max. after LHRH	FSH (m.i.u./ml) Basal	FSH (m.i.u./ml) Max. after LHRH	G.H.	T.S.H.	A.C.T.H.
C.J.	14	1.1	1.1	1.2	1.9	D	D	D
A.E.	17	2.1	2.4	1.7	3.4	D	D	D
R.R.	17	0.5	0.5	—	—	D	D	D
A.O.	15	0.5	0.6	0.6	1.5	D	I	N
G.T.	14	1.8	1.8	0.6	0.6	D	D	I
A.Y.	18	2.2	2.7	1.9	1.6	D	N	N
P.P.	16	1.6	2.1	1.7	2.1	D	D	I
R.B.	19	2.2	3.9	1.5	2.2	D	D	D
R.P.*	18	2.7	4.0	2.9	3.6	D	D	D
M.E.	9	0.6	0.6	2.1	2.0	D	N	N
Normal range		5.2-15.2	10.1-34.0	3.1-12.0	7.4-18.0			

* Initially sexual precocity.

Table II. Optic nerve glioma. G.A. Female aged 16.

Plasma	L.H.	3.4 (normal 5.2-15.2 m.i.u./ml)
	F.S.H.	1.5 (normal 3.1-12.0 m.i.u./ml)
	Thyroxine	0.8 (normal 3.0-6.6 µg/100 ml)
	Cortisol	3.4 µg/100 ml (no response to hypoglycaemia)
	G.H.	1.7 ng/ml (no response to hypoglycaemia)

Teratoma of the 3rd Ventricle (Ectopic Pinealoma)

A rare tumour which requires special mention is a teratoma arising in the floor of the 3rd ventricle. Because the histological appearance is similar to tumours found in the pineal gland, it is sometimes referred to as an ectopic pinealoma, but there seems little to favour this term, since there is no evidence of truly ectopic pineal tissue. The following case is an example.

This girl presented at the age of 13 with signs of raised intracranial pressure. A ventriculogram revealed a large tumour in the floor of the 3rd ventricle extending forwards to the optic chiasma and the supra-optic area. She was treated by radiotherapy and our investigations were carried out 2 years later, by which time puberty had not occurred and there was amenorrhoea. Basal gonadotrophin levels were very low, there was no response to LHRH, and there were also deficiencies of growth hormone, TSH and ACTH. Histologically, the tumour was a malignant teratoma, very similar to those found in the pineal.

Tumours of the Anterior Hypothalamus

Experimentally, section of fibres in the anterior hypothalamic area of rats results in failure of ovulation. Conversely, electrolytic lesions in this area may lead to the advancement of puberty, although the exact anatomical specificity of the lesion is not well defined. In man, tumours arising from the optic chiasma affecting the hypothalamus may influence puberty. These tumours are uncommon but in the last 10 years we have seen nine such patients of whom 7 were females, presenting in early childhood with visual impairment, and this is the usual age and sex incidence reported for this type of tumour. Two of the 5 who have reached the age of puberty are hypogonadal, of whom one is of particular interest.

This patient presented at the age of 9 with visual impairment and signs of raised intracranial pressure, but of importance was the fact that the menarche had already occurred and breast development was well advanced. She was investigated 6 years later following treatment by surgery and radiotherapy, by which time there were signs of hypogonadism, with amenorrhoea and lack of body hair.

Investigation at this time showed low plasma gonadotrophins and the other pituitary function tests revealed generalised hypopituitarism (Table II). Obesity was also a feature as in other cases where the anterior hypothalamus is involved. Precocious puberty has been recorded as a complication of optic nerve glioma, and it is tempting to make the parallel with anterior hypothalamic lesions and precocious ovulation in animals. In fact, precocious puberty in children with organic disease of the central nervous system has more often been ascribed to posterior hypothalamic lesions, but it must be remembered that the localisation of tumours cannot usually be very precise by the time that they cause clinical signs.

PRECOCIOUS PUBERTY

Iso-sexual precocity is found far less commonly than impaired puberty in association with a hypothalamic tumour. In reviews of large numbers of cases (Jolly, 1955; Wilkins, 1965; Sigurjonsdottir and Hayles, 1968) there is no doubt that precocious puberty is much commoner in females in whom the so-called idiopathic variety comprises the greatest number, whereas in boys a greater proportion is associated with a cerebral lesion.

Pineal Tumours

For many years an association between tumours of the pineal gland and sexual precocity has been recorded. Kitay and Altschule (1954) reviewed 160 cases of children with this tumour of whom 43, curiously all boys, developed precocious puberty. Support for the relationship has been obtained from animal experiments where pineal extracts containing melatonin and other substances inhibit gonadotrophin secretion, and pinealectomy advances puberty. In fact pineal tumours are uncommon in children and in 65 cases reported by Ringertz (1954) there were only 7 patients less than 11 years of age.

During the last 10 years we have seen 17 patients with tumours of the pineal, of whom 5 were children. The only girl died at a prepubertal age, and the effect of the tumour in the remainder was variable; two were endocrinologically normal, one was hypogonadal and one showed precocious puberty. The latter patient (Fig.5) presented at the age of 7 with symptoms of raised intracranial pressure and the typical inability to perform conjugate upward eye movements associated with tumours of the pineal area. It was noted that he was taller than his fellows and already there was facial hair, marked acne, and increased size of penis and testes. A ventriculogram showed a large tumour occupying the pineal and posterior part of the 3rd ventricle (Fig.6). He was seen before radioimmunoassays of gonado-

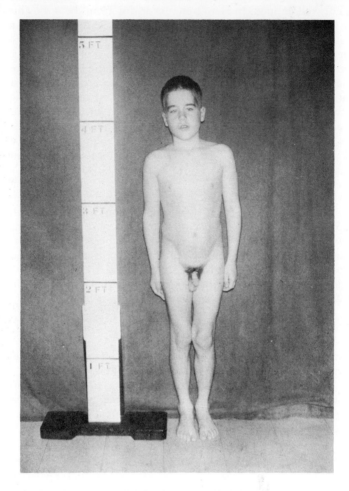

Fig. 5 Boy aged 7 with pinealoma showing precocious puberty.

trophin were readily available. Urinary gonadotrophins measured by bioassay (Table III) were not detected and this is a frequent finding even in cases of neurogenic sexual precocity (Sigurjonsdottir and Hayles, 1968) and may reflect the insensitivity of the method. Urinary testosterone was also not elevated but the level of 17-oxosteroids was increased for the age of the patient.

The patient died soon afterwards and the histological appearance of the tumour was that of a malignant teratoma very similar to the tumour of the 3rd ventricle described previously in a girl with hypogonadism.

In contrast, a second patient was seen for the first time at the age of 16 with similar neurological symptoms including paralysis of upward eye movements. Again, a ventriculogram showed a pineal tumour but in this case pubertal development was impaired. The histology of this tumour is unknown but the gonadotrophin levels together with those of a third patient with pinealoma and normal puberty are shown in Table IV.

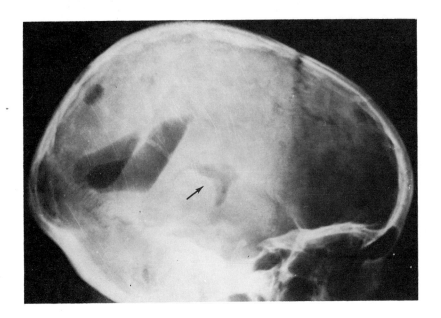

Fig.6 Air ventriculogram of same patient showing large pineal tumour (arrowed).

Plasma gonadotrophins were below normal and there was no response to LHRH in the hypogonadal subject, whereas normal basal values and a normal response was seen in the third patient.

The problem which requires consideration is why tumours of the pineal should sometimes be associated with precocious puberty and sometimes with hypogonadism or normal sexual development. According to one view the histological nature of the tumour is important in determining the effect on puberty, sexual precocity being associated with non-parenchymatous tumours, most often teratomas, which destroy the inhibitory influence of the gland. Very rarely these teratomas may actually secrete gonadotrophins directly. True parenchymatous pinealomas may elaborate the inhibitory substances and cause hypogonadism (Wurtman, 1968). Unfortunately, as in our cases, it is not always possible to determine the histological nature of a pineal tumour to substantiate this correlation. The other possibility is that pinealomas influence puberty by causing hypothal-

Table III. Pineal teratoma with precocious puberty. P.G. Male aged 7.

> Urinary gonadotrophins (bioassay)
> None detected
> Urinary testosterone
> 7.0 µg/24 h (normal adult male 30-150)
> Urinary 17-oxosteroids
> 7.1 mg/24 h

Table IV. Plasma gonadotrophin response to LHRH in patients with pinealoma.

Patient	Age	Puberty	LH (m.i.u./ml)		FSH (m.i.u./ml)	
			Basal	Max. after LHRH	Basal	Max. after LHRH
R.W.	19	Impaired	1.8	3.0	2.2	2.5
M.P.	14	Normal	5.8	23.0	8.0	33.4

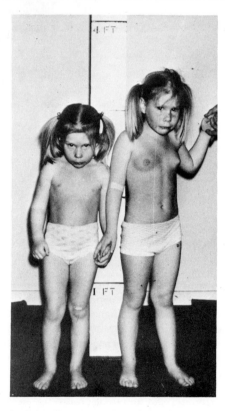

Fig. 7 Girl aged 4 showing premature breast development and growth spurt compared with normal twin.

Table V. Precocious puberty in girls.

Patient	Age	Bone age	E.E.G.	Plasma LH (m.i.u./ml) Basal	Max. after LHRH	Plasma FSH (m.i.u./ml) Basal	Max. after LHRH
R.T.	7	11	Abnormal	*5.0	–	*5.2	–
D.P.	6	10	Abnormal	6.2	10.0	9.8	11.8

* Medroxyprogesterone 20 mg q.i.d.

amic damage and in this respect are no different from tumours arising in various parts of the hypothalamus which can produce a similar effect.

Hamartoma

In girls, iso-sexual precocity is not commonly associated with any demonstrable brain lesion but in both sexes a hamartoma may occasionally be present. This is a small pedunculated tumour about 0.5-1.0 cm in diameter which is really a congenital malformation of nerve cells and nerve fibres, some of which may be myelinated, attached to the posterior part of the tuber cinereum just anterior to the mamillary bodies. It is usually too small to cause increased intracranial pressure and hangs down into the interpeduncular fossa without compressing the hypothalamus (Schmidt et al., 1958). Very small tumours are found not uncommonly at routine autopsy (Sherwin et al., 1962) but in those patients with precocious puberty the presence of fibres connecting the tumour to the hypothalamus has been emphasized, and may contain neurosecretory material (Wolman and Balmforth, 1963).

We have seen two cases of precocious puberty in girls in whom there was no clinical evidence of an intracranial lesion and they would ordinarily be classed as idiopathic sexual precocity. Both children, however, had a markedly abnormal EEG pattern suggesting a brain stem disorder. Such abnormalities of the EEG have been reported previously in children with precocious puberty (Liu et al., 1965). The first patient was aged 7 when breast development was first noticed and then was soon followed by pubic hair, and within a year she began to menstruate. She also had a congenital scoliosis which masked any growth spurt. The second girl presented at the age of 4 with enlarging breasts followed by pubic hair and menstruation within 3 months. This 'telescoping' of the events of puberty is often observed when the onset is premature. She was one of non-identical twins, and it is apparent in the picture (Fig.7) how the development of puberty is associated with a premature growth spurt when compared with her unaffected twin.

The endocrine data are shown in Table V. Bone age was advanced in both cases. Basal gonadotrophin levels are within the normal range expected for the pubertal status and not the chronological age, as reported by other authors (Blizzard et al., 1972) and an LHRH response was obtained in the second case which was compatible with the pubertal state. Treatment with medoxyprogesterone in the dosage shown here did not reduce gonadotrophin levels below normal in case R.T. although menstruation was effectively suppressed. In view of the abnormal EEG pattern a neurological lesion cannot be excluded and prolonged surveillance of these patients is essential. While many of the "idiopathic" cases apparently remain well, Wolamn and Balmforth (1963) reported a patient who presented with the menarche at the age of 7, but a hamartoma connecting with the

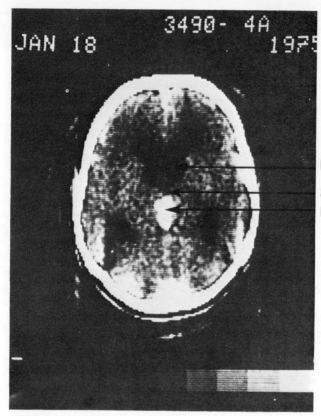

Fig. 8 Computerised scanning axial tomography of brain showing pinealoma.

hypothalamus was not found until death at the age of 62.

Computerised scanning axial tomography. There is understandable reluctance to embark upon unpleasant and possibly risky radiological investigations in children without neurological symptoms but with the advent, during the last two years, of the new radiological technique of computerized scanning axial tomography of the brain, it is possible to obtain clear pictures of the hypothalamic area without invasive methods. At present both girls described above have a normal brain, but, for comparison, Fig. 8 is a scan showing a pinealoma. This method has opened up new possibilities for obtaining detailed information on intracranial tumours relatively easily.

SUMMARY

Tumours of the pituitary rarely arise at an age early enough to interfere with puberty but when a chromophobe adenoma does occur in adolescence in association with primary amenorrhoea, excessive prolactin secretion must be considered in addition to a deficiency of gonadotrophin secretion.

Craniopharyngiomas and tumours in various parts of the hypothalamus often cause a failure of puberty, which can be explained by destruction of the area responsible for LHRH secretion. On the other hand an exact correlation between the localisation of the tumour and its effects is not always easy because the tumour has often spread beyond its original site by the time that clinical signs are apparent.

The association of hypothalamic and pineal tumours with sexual precocity is more difficult to explain in terms of the current views on the control of the onset of puberty. The most favoured hypothesis is that, with increasing maturity of the child, the hypothalamus becomes progressively more insensitive to negative feedback control by gonadal hormones, and this is followed in the female by the development of a positive feedback mechanism to produce a cyclical release of gonadotrophin.

It seems that, on occasions, hypothalamic tumours of various kinds, particularly when associated with internal hydrocephalus, may prematurely desensitise the hypothalamus to negative feedback control without actually destroying the hypophysiotropic area. In the female, the increase in ovarian secretion which follows, may then be responsible for inducing the positive feedback or phasic control.

Pineal tumours probably act in a similar manner although the possibility of a specific inhibitory action of the pineal on gonadotrophin secretion being present in man cannot be excluded. The relationship between a hamartoma and puberty may well be different, in view of its site and connections. It could advance puberty either by influencing nervous pathways within the hypothalamus causing abnormal secretion of gonadotrophin releasing hormone or by itself elaborating a gonadotrophin releasing substance.

These reflections are somewhat speculative, and some of these tumours are admittedly rare, but their investigation by modern techniques of radiology and endocrinology may yield further clues as to the mechanisms involved in the control of puberty in man.

REFERENCES

Blizzard, R.M., Penny, R., Foley, T.P., Baghdassarian, A., Johanson, A. and Yen, S.S.C. (1972). *In* "Gonadotrophins" (B.B. Saxena, C.G. Beling and H.M. Gandy, eds) pp.502-523. Wiley-Interscience, New York.
Child, D.F., Nader, S., Mashiter, K., Kjeld, M., Banks, L. and Russell Fraser, T. (1975). *Brit. Med. J.* **1**, 604.
Donovan, B.T. and van der Werff ten Bosch, J.J. (1965). "Physiology of Puberty". Williams and Wilkins, Baltimore.
Elkington, S.G. and McKissock, W. (1967). *Brit. Med. J.* **1**, 263.
Jolly, H. (1955). "Sexual Precocity". Blackwell Scientific Publications, Oxford.
Kenny, F.M., Midgley, A.R., Jaffe, R.B., Garces, L.Y., Vazquez, A. and Taylor, F.H. (1969). *J. clin. Endocr.* **29**, 1272.
Kitay, J.I. and Altschule, M.D. (1954). "The Pineal Gland". Harvard University Press, Cambridge, Mass.
Liu, N., Grumbach, M.M., de Napoli, R.A. and Morishima, A. (1965). *J. clin. Endocr.* **25**, 1296.
McNatty, K.F., Sawers, R.S. and McNeilly, A.S. (1974). *Nature* **250**, 653.
Mortimer, C.H., Besser, G.M., McNeilly, A.S., Marshall, J.C., Harsoulis, P., Tunbridge, W.M.G., Gomez-Pan, A. and Hall, R. (1973). *Brit. Med. J.* **4**, 73.
Mortimer, C.H., McNeilly, A.S., Fisher, R.A., Murray, M.A.F. and Besser, G.M. (1974). *Brit. Med. J.* **4**, 617.
Ringertz, N., Nordestam, H. and Flyger, G. (1954). *J. Neuropath.* **13**, 540.
Schmidt, E., Hallervorden, J. and Spatz, H. (1958). *Dtsch. Z. Neurenheilk* **177**, 235.

Sherwin, R.P., Grassi, J.E. and Sommers, S.C. (1962). *Lab. Invest.* **11**, 89.
Sigurjonsdottir, T.J. and Hayles, A.B. (1968). *Amer. J. Dis. Child.* **115**, 309.
Wilkins, L. (1965). "The Diagnosis and Treatment of Endocrine Disorders in Childhood and Adolescence". Thomas, Springfield, Ill.
Wolman, L. and Balmforth, G.V. (1963). *J. Neurol. Neurosurg. Psychiat.* **26**, 275.
Wurtman, R.J. (1968). *In* "Endocrine Pathology" (J.M.B. Bloodworth, ed) pp.117-132. Williams and Wilkins, Baltimore.

POSTMENOPAUSAL OVARIAN FUNCTION

A. Vermeulen

Department of Endocrinology, University of Ghent
Ghent, Belgium

Relatively few data are available concerning the secretory activity of the postmenopausal ovary. It is generally accepted that during reproductive life the follicular cells secrete estradiol, androstenedione and 17αOH-progesterone, the luteal cells secrete progesterone, 17αOH-progesterone, androstenedione and estradiol, whereas the stroma (interstitial tissue) secretes mainly androstenedione. In accordance with this view, steroids secreted by the cycling ovary show cyclical variations: this concerns estrogens, progesterone, 17αOH-progesterone, androstenedione, testosterone and dihydrotestosterone.

As the postmenopausal ovary contains only atretic follicles, it is not surprising that plasma estrogen levels decrease markedly after cessation of ovulatory cycles (Baird, 1969a; Korenman, 1969; Longcope, 1971; Judd, 1974a, 1974b). Nevertheless there is some evidence that the postmenopausal ovary continues to secrete significant amounts of hormones, as oophorectomy in postmenopausal women results in a significant decrease of testosterone and androstenedione plasma levels (Judd, 1974a). In order to gain some insight into the contribution of the postmenopausal ovary to plasma steroid levels as well as into the origin of plasma steroids in these women, we studied plasma levels of estrone (E1), estradiol (E2), progesterone (P), 17αOH-progesterone (17OHP), testosterone (T), dihydrotestosterone (DHT), androstenedione (Δ4) and dehydroepiandrosterone (DHEA) in postmenopausal women under basal conditions, after ACTH stimulation, under dexamethasone treatment and under HCG stimulation during dexamethasone treatment; moreover the same steroids were studied in a small group of ovariectomized women. All women were postmenopausal for a minimum of 4 and a maximum of 10 years and were between 52 and 65 years old; they were in general good health and upon routine gynecological examination a normal postmenopausal status was found.

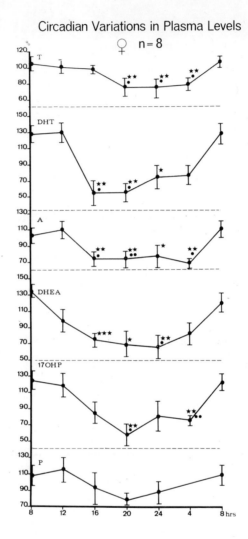

Fig. 1

Steroids were determined by radioimmunoassay methods as previously described (Vermeulen, 1975).

Basal steroid levels were determined in plasma samples taken between 8 and 10 a.m.

Circadian rhythm was studied by taking plasma levels in the recumbent subject every 4 h for 24 h starting at 8 a.m.

ACTH stimulation tests were performed in the fasting subject by injecting intravenously 0.25 mg of tetracosactide (Synacthen®), blood samaples were taken -5, 0, 30, 60 and 120 min after injection.

Dexamethasone suppression tests were performed by giving 1 mg orally every 8 h for 5 days; blood samples were taken at 8 a.m.

Table I. Plasma steroid levels in normal women (ng/100 ml).

	Mean value during cycle (15 complete cycles) age 18-25 yrs	Postmenopausal women age 51-65 yrs		Ovariectomized women (n = 8) age 51-62 yrs
E1	—	4.9 ± 0.5		4.8 ± 0.6
E2	—	2.0 ± 0.1*	(n = 20)	1.8 ± 0.4
T	44 ± 2.8	29.7 ± 4.0	(n = 19)	12 ± 2
DHT	33 ± 3.8	9.7 ± 2.1	(n = 16)	< 10
Δ4	184 ± 16	99 ± 13	(n = 17)	64 ± 9
DHEA	450 ± 43	197 ± 43	(n = 13)	126 ± 36
17OHP	—	40 ± 5	(n = 19)	44 ± 8
P	—	19 ± 3	(n = 22)	18 ± 1

* SEM

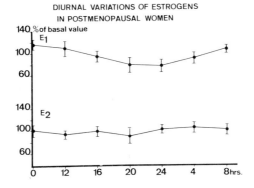

Fig.2

HCG tests were performed by administering during 4 days 5000 I.U. of Pregnyl® at 8 p.m., plasma samples being taken at 8.a.m., while continuing dexamethasone treatment.

RESULTS

Basal plasma levels of different steroids studied are indicated on Table I. Basals leves of E1 and E2 are significantly lower than corresponding values during the early follicular phase of the cycle. Similarly progesterone values are lower than during the follicular phase whereas 17OHP levels are similar. T, DHT, Δ4 and DHEA levels are all significantly lower than the mean value observed during the cycle.

In *ovariectomized women* (Table I), estradiol and estrone levels are similar to values observed in normal postmenopausal women. T, DHT, and Δ4 levels were significantly lower, but 17OHP and P levels are similar to normal postmenopausal women. DHEA levels are slightly but not significantly lower.

All steroids studied, with the exception of E2 show a *significant nyctohemeral* variation (Figs 1 and 2) with, for all steroids studied, highest levels at 8 a.m. and a nadir either at 8 p.m. or midnight. The amplitude of these variations is most important for DHEA (70 ± 17%), 17OHP (67 ± 18%), and DHT (70 ± 12%), E1 and P variations showed a comparable amplitude of about 40%. As far

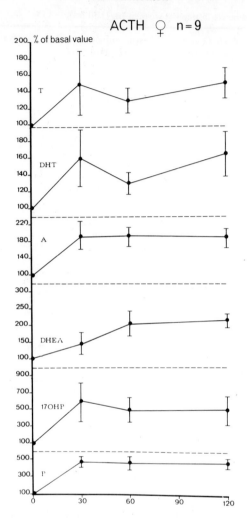

Fig.3

as data in the literature are concerned, nyctohemeral variations of T and DHT levels have been observed by Givens et al. (1974) in normal menstruating females whereas Lobotsky and Lloyd (1973) reported 5 p.m. plasma levels of Δ4 during the menstrual cycle to be 29% lower than the 9 a.m. levels. Runnebaum et al. (1972) found no circadian rhythm in P levels during the luteal phase in contrast to the P levels during the 3rd trimester of pregnancy. Finally, Baird and Guevara (1969a) reported a nyctohemeral rhythm for E1 but not for E2 levels.

ACTH stimulation (Figs 3 and 4) causes a significant increase of all steroids studied with the single exception of estradiol. For T, DHT, Δ4 and DHEA, the increase varies between 60 and 100%; for 17OHP and P however this increase varies between 350 and 500%.

Dexamethasone suppression (Table II) on the other hand causes a decrease of all steroids studied: E1, T, Δ4, 17OHP, and DHEA levels are approximately

Table II. Dexamethasone suppression test in postmenopausal women.

	Basal	During DX (ng/100 ml)	P
T	33 ± 7	20 ± 6 (58 ± 12%)	< 0.01
DHT	7 ± 1	< 5	< 0.001
Δ4	92 ± 29	27 ± 4 (47 ± 14%)	< 0.01
DHEA	163 ± 41	57 ± 16 (44 ± 16%)	< 0.01
17OHP	36 ± 5	17 ± 3 (50 ± 7%)	< 0.01
P	20 ± 6	< 5	< 0.01
E2	2.0 ± 0.2	1.2 ± 0.2	< 0.01
E1	4.9 ± 0.5	1.1 ± 0.2	< 0.001

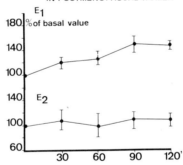

Fig. 4

halved, whereas DHT and P levels become undistinguishable from zero; E2 levels are often at the limit of the sensitivity of the method used. Dexamethasone administered to *ovariectomized postmenopausal* women resulted in a decrease of all steroids studied to levels below detectability.

HCG stimulation performed while continuing dexamethasone suppression causes a slight increase of E1, T, Δ4, 17OHP and P levels, but this difference is only statistically significant for T, DHEA, and 17OHP; E2 levels remain unchanged after HCG stimulation (Table III).

DISCUSSION

This study was performed with the purpose of gaining some insight into the origin of plasma steroid levels in postmenopausal women. The results obtained permit in our opinion the following conclusions:

Table III. HCG stimulation test in postmenopausal women.

	Basal value* ng/100 ml	During HCG	P
T	20 ± 5	35 ± 9 (122 ± 9%)	< 0.05
DHT	–	–	–
Δ4	25 ± 5	30 ± 6 (126 ± 22%)	ns
DHEA	57 ± 16	95 ± 30 (147 ± 23%)	< 0.05
17OHP	19 ± 2	27 ± 1 (167 ± 33%)	< 0.05
P	10 ± 2	14 ± 4 (131 ± 18%)	ns
E2	1.1 ± 0.2	1.2 ± 0.3 (104 ± 26%)	ns
E1	4.9 ± 0.5	5.4 ± 0.5 (116 ± 12%)	ns

* Basal values after 3 days of DX (3 mg/day) suppression, HCG stimulation performed during Δ4.

1. *Progesterone and 17OHP* levels in postmenopausal women have an almost exclusive adrenal origin: indeed after dexamethasone treatment, P and 17OHP levels become almost undetectable; after ACTH they increase about 500% whereas HCG causes no, or an insignificant increase in plasma levels, and levels in ovariectomized women are similar to values in normal postmenopausal women.
2. The major fraction of *DHEA* in postmenopausal women appears to have an adrenal origin, but on the basis of the lower levels in ovariectomized women and the slight increase upon HCG stimulation we conclude that the ovary secretes small but significant amounts of DHEA. It is known that during the menstrual cycle DHEA does not show any cyclical variations, and Abraham and Chakmakjian (1973) observed a decrease of DHEA levels to 25% of basal values after dexamethasone administration. Nevertheless McDonald *et al.* (1963) demonstrated the secretion of DHEA by the normal ovary. As in males, DHEA levels are significantly lower in old age than in young women: in view of the essentially adrenal origin of this steroid, this suggests a change in the adrenal secretory activity with age.
3. *T, DHT and Δ4 levels* in postmenopausal women are significantly lower than the mean value observed during the cycle, but comparable to values in the early follicular phase. After dexamethasone, levels of these steroids are approximately halved, except for DHT, the values of which fall below the limit of detection; after ACTH levels are doubled, as in ovariectomized women T, DHT and Δ4 are significantly lower than in normal postmenopausal women and as HCG increases T levels this suggests that the postmenopausal ovaries contribute about 50% of T and about 1/3 of Δ4 levels. This confirms studies by Judd *et al.* (1974a)

who suggested that in postmenopausal women with endometrial cancer, the ovaries contribute about 50% of T and secrete moderate amounts of androstenedione. Barberia and Thorneycroft (1974) reported a similar contribution of the ovaries to T levels wherea; others (Gandy, 1968; Lobotsky, 1964; Abraham, 1969) did not observe any difference in T levels after ovariectomy.

As far as the origin of *estrogens* is concerned, in accordance with Barlow et al. (1969) and Saez et al. (1972) we found similar values in ovariectomized women as in normal postmenopausal women, whereas dexamethasone decreased significantly both E1 and E2 levels, while ACTH increased significantly E1 but not the E2 levels. The latter may be a consequence of the very short stimulation of the adrenal cortex and it is not impossible that longer stimulation might result in an increase of E2 levels as well. Whereas we observed a significant decrease of E1 and E2 levels after dexamethasone, Saez et al. (1972) found an important decrease only for E1, the decrease in E2 levels being very slight. HCG in the dosage used did not increase significantly estrogen levels. It seems therefore that the adrenal is the main source of these estrogens. Of course this does not indicate that the adrenal cortex secretes estrogens, as other experiments (Siiteri, 1973; Grodin, 1973) have shown estrogens in postmenopausal women to arise mainly by peripheral conversion of androgens, mainly androstenedione. Nevertheless Baird et al. (1969b) reported direct estrone secretion by the adrenal cortex after stimulation. McDonald (1963) suggests a positive correlation to exist between the extent of conversion of $\Delta 4$ to E1 and body weight, tending to implicate adipose tissue in this conversion.

In summary, using a different approach, we arrive at the same conclusion as Judd et al. (1974b) that the postmenopausal ovary still secretes appreciable amounts of T and $\Delta 4$, and although to a lesser extent, also of DHEA. The main source of these steroids appears however to be the adrenal cortex, whereas for 17OHP and P, the latter seems to be the almost unique source. As estrogens originate essentially from peripheral conversion of $\Delta 4$, it is logical that the adrenal cortex is the main primary source of these steroids. Although the postmenopausal ovaries appear to have some secretory activity, this secretion is hardly influenced by a four days stimulation with 5000 IU. This should be kept in mind when dealing with problems of androgen sources in pathological conditions in elderly women.

REFERENCES

Abraham, G.E., Lobotsky, J. and Lloyd, C.W. (1969). *J. clin. Invest.* 48, 696.
Abraham, G.E. and Chakmakjian, Z.H. (1973). *J. clin. Endocr. Metab.* 37, 581.
Baird, B.T. and Guevara, A. (1969a). *J. clin. Endocr. Metab.* 29, 149.
Baird, B.T., Uno, A. and Melby, J.C. (1969b). *J. Endocrinol.* 45, 138.
Barberia, J.M. and Thorneycroft, I.H. (1974). *Steroids* 23, 757.
Barlow, J.J., Emerson, K. and Saxena, B.M. (1969). *N. Engl. J. Med.* 280, 633.
Gandy, A.M. and Peterson, R.A. (1968). *J. clin. Endocr. Metab.* 28, 949.
Givens, J.R., Andersen, R.M., Ragland, J.B. and Fish, S.A. (1974). *Abstr. 56th Meeting Endocr. Soc.* 124, 117.
Grodin, J.M., Siiteri, P.K. and McDonald, P.C. (1973). *J. clin. Endocr. Metab.* 36, 207.
Judd, H.L., Lucas, W.E. and Yen, S.S.C. (1974a). *Am. J. Obstet. Gynec.* 38, 793.
Judd, H.L., Judd, G.E., Lucas, W.E. and Yen, S.S.C. (1974b). *J. clin. Endocr. Metab.* 39, 1020.
Korenman, S.G., Perrin, L.E. and McCallum, T.P. (1969). *J. clin. Endocr. Metab.* 29, 879.
Lobotsky, G., Wyss, H.I., Segre, J. and Lloyd, C.W. (1964). *J. clin. Endocr. Metab.* 24, 1261.
Lobotsky, G. and Lloyd, C.W. (1973). *Steroids* 22, 133.

Longcope, C. (1971). *Am. J. Obstet. Gynec.* 111, 778.
McDonald, P.C., Vande Wiele, R.L. and Lieberman, L.S. (1963). *Am. J. Obstet. Gynec.* 86, 1.
Runnebaum, B., Rieben, W., Bierwirth, A.M., Münstermann, V. and Zander, J. (1972). *Acta endocr. (Copenh.)* 69, 731.
Saez, J.M., Morrera, A.M., Dazord, A. and Bertrand, J. (1972). *J. Endocrinol.* 55, 41.
Siiteri, P.K. and McDonald, P.C. (1973). *In* "Handbook of Physiology, Section 7, Endocrinology II, Female Reproductive System, Part I" (R. Greep, ed) pp.615-629. American Physiological Society, Washington.
Vermeulen, A. and Verdonck, L. (1975). *J. Ster. Biochem. (in press).*

FEMALE HYPOGONADISM – THERAPY ORIENTATED DIAGNOSIS OF SECONDARY AMENORRHOEA

M.G.R. Hull, M.A.F. Murray, S. Franks*, B.A. Lieberman
and H.S. Jacobs

*Department of Obstetrics and Gynaecology
St. Mary's Hospital Medical School
London, W2*

*The Middlesex Hospital, London, W1

Secondary amenorrhoea is a common symptom which accounts for the majority of referrals to a gynaecological endocrine unit. Authorities differ on the definition of this symptom, which of course occurs physiologically during puerperal lactation and after the menopause. If pregnancy is regarded as a valid cause, definitions that require an interval between periods exceeding nine months are inappropriate. Since the length of the normal menstrual cycle is so variable (Treloar *et al.*, 1967; Chiazze *et al.*, 1968) and occasional missed periods are almost a feature of the normal menstrual history, we have restricted our definition to an interval between episodes of vaginal bleeding of at least 120 days. Like all others this definition is arbitrary, but so far, at least, it has excluded all undiagnosed pregnancies referred to us as cases of amenorrhoea.

The causes of secondary amenorrhoea vary from the most trivial to life-threatening conditions. Although the latter are rare they make a certain level of diagnostic inquisitiveness obligatory. Nevertheless, it is accepted that the length to which one should pursue investigations is to a large extent determined by the patient's own wishes – the problem in a 29 year old woman desperate for her first pregnancy is very different from that presented by an older woman requiring advice about contraception, or an adolescent whose health is seriously imperilled by anorexia nervosa. We have attempted to evolve a protocol of simple diagnostic manoeuvres (Table I) which can be adapted to the needs of the majority of patients. The steps in this protocol are presented in a sequential way, but in practice the

Table I. Protocol of basic diagnostic manoeuvres for patients with secondary amenorrhoea.

1. Measure FSH
2. X-ray pituitary region
3. Measure prolactin
4. Assess oestrogen production
5. Clomiphene Test

Table II. Functional classification of secondary amenorrhoea.

1. Primary ovarian disease
2. Hyperprolactinaemia
3. Defects of gonadotrophin secretion
   A. Deficiency
   B. Disordered regulation
      Defect of: cyclic centre*
      tonic centre*
      cycle initiation

* The use of the word "centre" does not imply a discrete anatomic area but rather a physiological control mechanism.

patients all have the same tests and, as will become evident, the whole cycle of investigation takes about six weeks.

Table II depicts the forms of amenorrhoea that we recognise. Conditions characterised by hyperandrogenisation are excluded, not because amenorrhoea is an unimportant feature but because in these patients we consider a different approach is required (Jacobs, 1974). The definitions and clinical correlates of the various forms of anovulatory amenorrhoea indicated in Table II will be described below, but here it is necessary to emphasise that the nosology is based upon a functional classification in which the therapeutic response dominates. It is recognised that this approach ignores the traditional medical view of the precedence of aetiology and pathology. To this criticism we would only reply that it makes little difference to the woman with primary ovarian failure what particular condition has caused her premature menopause — the essential point is that if she is oestrogen deficient she will require oestrogen replacement. It is intended therefore that each manoeuvre elicits either a therapeutic response or a further diagnostic enquiry.

1. SERUM GONADOTROPHIN MEASUREMENTS

The first diagnostic manoeuvre in the assessment of patients with secondary amenorrhoea concerns the exclusion of primary ovarian failure. Although some help may be obtained from the history (for instance vaso-motor instability) and examination (atrophy of the skin, breasts and lower genital tract) the condition is more precisely characterised by measurement of serum gonadotrophin concentrations. In common with other workers (Goldenberg et al., 1973b) we have found a four to five fold elevation of serum FSH concentrations in patients with proven primary ovarian failure (Jacobs, 1975). As shown in Fig.1, serum LH con-

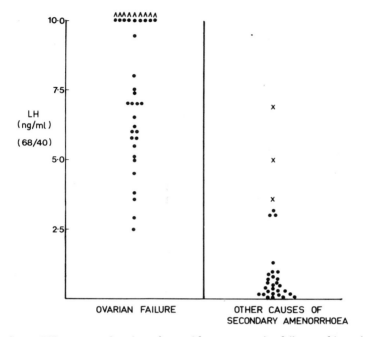

Fig.1 Serum LH concentrations in patients with proven ovarian failure and in patients with other forms of secondary amenorrhoea. Note the overlap of LH concentrations in the two groups (x indicates patients with polycystic ovaries).

centrations are also always elevated in patients with ovarian failure, but there is considerable overlap with the levels found in other forms of amenorrhoea. Figure 2 shows the results of measurements of FSH in the same samples and it can be seen that in this case the discrimination is easily made. Though there is some fluctuation of FSH concentrations when daily samples from the same patient are compared, the obvious distinction is preserved.

The finding of an elevated FSH concentration in the presence of amenorrhoea suggests activation of the hypothalamic tonic centre controlling gonadotrophin secretion — that is to say, recognition of low prevailing gonadal steroid levels by a mechanism sensitive to negative feedback. This association of increased trophic hormone secretion in the presence of target hormone deficiency occurs in other endocrine systems, such as the adrenal- and the thyroid-pituitary axes.

It is of interest that reports of elevated trophic hormone concentrations occurring with normal range thyroid hormone (Evered and Hall, 1972) and testosterone (Bramble et al., 1975) concentrations indicate the existence in some patients of a state of compensated target organ failure. That the same may be true of the ovary is suggested by the existence of cases of the "resistant ovary syndrome", in which increased gonadotrophin secretion is associated with numerous primordial follicles in the ovarian biopsy, normal or low oestrogen production and an impaired response to treatment with exogenous gonadotrophins (Jones and Moraes Ruehsen, 1969). The condition has mainly been recognised in patients with primary amenorrhoea but we have recently had the opportunity to study 3 patients with secondary amenorrhoea in whom this diagnosis has been made.

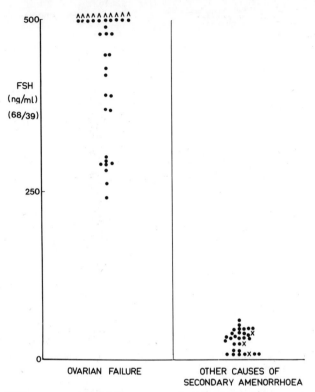

Fig.2 Serum FSH concentrations in the same samples whose LH results are given in Fig.1. Note the excellent discrimination between the two groups (data from Jequier et al., 1975).

The striking feature has been the association of serum gonadotrophin concentrations appropriate to the menopausal years with evidence of oestrogen production at mid-follicular phase levels. Ovarian biopsy in one patient revealed a histological appearance very similar to that described in the original reports, with large number of follicles in all stages up to and including antrum formation. In addition there were large numbers of follicles in various stages of atresia and numerous solid follicles containing Call-Exner bodies. There was also evidence of luteinisation. One patient has responded to HCG given five days after a course of clomiphene (to which when given alone she had not responded) with an increase of progesterone concentrations to 25 ng/ml. Whether this response represents ovulation or merely luteinisation of abnormal follicles we do not yet know.

The implication of cases of the resistant ovary syndrome is that a compensated ovarian failure may exist, is readily identifiable by assessing oestrogen production as well as gonadotrophin secretion, and may prove to be a remediable condition. We suspect it is not uncommon. Its existence limits the diagnostic value of elevated levels of FSH as the sole indicator of ovarian failure and as a consequence we consider an assessment of oestrogen production is also necessary to confirm the diagnosis of ovarian failure.

The aetiology of primary ovarian failure is often obscure, but there are some well-recognised causes (Table III). One form that is occurring with increasing fre-

Table III. Some causes of primary ovarian failure.

Chromosomal	—	e.g. Turner's Syndrome
Enzyme Defect	—	e.g. 17α-Hydroxylase deficiency
Iatrogenic	—	e.g. Cytotoxic drugs
		Pelvic irradiation
		Ovariectomy
Auto Immune	—	e.g. Associated with Addison's Disease and anti steroid-producing cell antibodies
Idiopathic	—	e.g. Premature menopause

This list is not exhaustive but is intended to indicate the heterogeneity of the causes of ovarian failure.

quency follows pelvic irradiation or anti-cancer chemotherapy in women with malignant disease (Thomas et al., 1976). We have been impressed that following inadvertent irradiation of the ovaries one of our patients experienced attacks of flushing, amenorrhoea with high gonadotrophins and low serum oestradiol concentrations, but a subsequent return of menstrual cycles. Though we cannot yet be certain she is ovulating it may be that the prognosis of this form of ovarian damage is better than in other conditions.

If a diagnosis of ovarian failure is made and there is evidence of oestrogen deficiency, we advise, as with failure of other endocrine target glands, hormone replacement: we give cyclical therapy with oestrogen, usually combining it with a progestogen in a single oral contraceptive tablet each day.

2. PITUITARY RADIOLOGY

The next series of investigations concerns the exclusion of an expanding lesion in the hypothalamic-pituitary region. A lateral x-ray of the skull is required to see calcification in a craniopharyngioma, which will already have been suspected in patients with primary amenorrhoea associated with short stature. Much more commonly the problem concerns the exclusion of a pituitary tumour. As discussed below, pituitary tumours which cause disorders of reproduction are almost without exception "prolactinomas". The important point to emphasise here is that, as graphically described by Hardy (1973), such tumours are usually small and since they arise in the lateral parts of the pituitary, they expand asymmetrically. The radiological appearances are quite subtle and are frequently missed: they include minor changes in size and shape of the sella and the characteristic double contour to the floor of the fossa (Kaufman et al., 1973). Coned views of the sella are routinely required and tomography in all hyperprolactinaemic patients is obligatory.

The cost of missing such lesions has been emphasised recently by Gemzell (1975) who described 4 patients with pituitary tumours in whom pregnancy following induction of ovulation was associated with expansion of the tumour. In one patient the symptoms were so severe that surgery during pregnancy was required. Similar cases have been reported by others (Burke et al., 1972). A vital point made by Gemzell is that the risk of expansion increases with succeeding

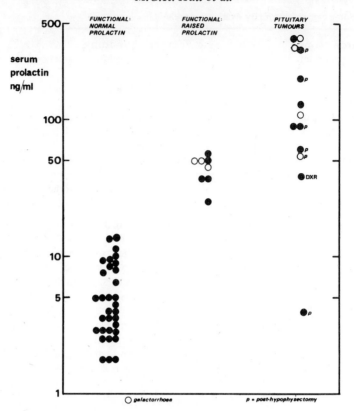

Fig.3 Serum prolactin (VLS, No.1) concentrations in 13 patients with pituitary tumour and in 40 patients with so-called functional secondary amenorrhoea. The scale is logarithmic. Note the rarity of galactorrhoea (data from Franks *et al.*, 1975, with permission).

pregnancies, a phenomenon presumably related to the increase in size of the normal pituitary during pregnancy (Erdheim and Stumme, 1909). Presumably a micro-tumour too expands during pregnancy and, most importantly, does not regress post partum. Consequently a history of normal pituitary radiology in connection with a previous pregnancy does not absolve one from the obligation to repeat the x-rays if further induction of ovulation is proposed. If a tumour is suspected, and particularly in the presence of hyperprolactinaemia, an air encephalogram is required to exclude a lesion of the hypothalamus or pituitary stalk, to exclude upward extension and to determine whether there is any air in the fossa. A very small tumour with plenty of room for expansion in a patient never pregnant before allows a less vigorous approach than is otherwise required. If there is evidence of a pituitary tumour delineation of the visual fields and full endocrine evaluation of the anterior pituitary is required (Jacobs and Nabarro, 1969). Upward extension exceeding 1 cm dictates a trans-frontal approach; otherwise we have been highly impressed with transsphenoidal hypophysectomy (Williams *et al.*, 1975). Internal (Burke *et al.*, 1972) or external (Kramer, 1973) irradiation may also be used.

3. SERUM PROLACTIN MEASUREMENTS

The next condition to be sought is hyperprolactinaemia. The relation of prolactin secretion to reproduction is described elsewhere in this volume and here we would only emphasise a few points. Galactorrhoea is an unreliable indicator of hyperprolactinaemia since it is present in only a third of patients with amenorrhoea who have aetiologically relevant hypersecretion of prolactin (Franks et al., 1975). In a series of over 100 women with amenorrhoea in whom prolactin secretion was assessed these authors found hyperprolactinaemia occurred in all but one of 13 patients with pituitary tumours (Fig.3). Surprisingly, the levels were just as high in the cases studied only after surgery as in those studied before treatment. One other patient, who had her galactorrhoea abolished several years earlier by external irradiation, still had amenorrhoea and hyperprolactinaemia. Treatments with clomiphene and Pergonal were unsuccessful but reduction of prolactin secretion by treatment with bromocriptine induced ovulation and led to a pregnancy within two months of starting treatment (Edington et al., in preparation).

Hyperprolactinaemia causing amenorrhoea may also occur in the absence of any evident pituitary tumour. Many factors, including a variety of drugs (Fluckiger, 1972), may cause hyperprolactinaemia and galactorrhoea. It is essential to exclude hypothyroidism as a cause of amenorrhoea with or without prolactinaemia (Bray and Jacobs, 1974); the diagnosis may be difficult to make clinically and as a cause of amenorrhoea is rare, but it is worth purusing with biochemical tests since its treatment is precise and specific. For reasons we do not yet understand we have on occasion had to add bromocriptine to thyroxine therapy to lower prolactin levels and promote ovulation in previously hypothyroid patients.

There remains a group of patients — in our view perhaps the most important group — in which amenorrhoea and hypersecretion of prolactin occur without any other evident abnormality. Being oestrogen deficient and unresponsive to clomiphene, such patients have often received treatment with exogenous gonadotrophins. If it is certain that the sella is radiologically normal, treatment with bromocriptine is safe and offers effective induction of ovulation, even in the absence of galactorrhoea. When prolactin levels are reduced the reproductive defect may be repaired astonishingly quickly, analysis of temperature charts indicating that patients usually enter a follicular phase within a day or two of starting treatment. It appears that as many as 20% of patients with so-called functional secondary amenorrhoea have aetiologically relevant hyperprolactinaemia (Franks et al., 1975; Hull et al., 1975). The cause in these patients is not clear, but we postulate that they have micro tumours of the pituitary (Jacobs and Franks, 1975). In this regard it is interesting to note the positive correlation of the logarithm of the prolactin concentration and the area of the sella noted by Child et al. (1975) in patients with pituitary tumours. In our radiologically normal cases the mean prolactin concentration (Fig.3) was 44 ng/ml, which is less than half of the levels found in most of our cases of pituitary tumour (Franks et al., 1975).

In our experience hyperprolactinaemic patients with secondary amenorrhoea are markedly oestrogen deficient, as assessed by the appearance of the lower genital tract, their failure to bleed after treatment with progestogen and by measurements of serum oestradiol-17β (Jacobs et al., 1976). It has generally been argued that the reproductive defect is caused primarily by a prolactin induced impairment of the ovarian response to gonadotrophins (Franks et al., 1975). However we have been impressed that although serum oestradiol levels are often in the

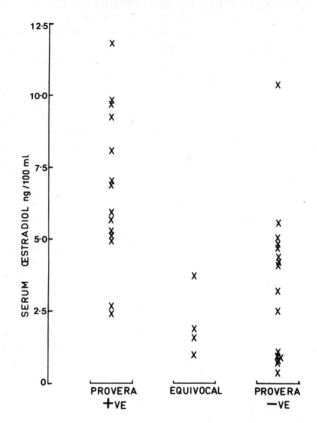

Fig.4 Serum oestradiol-17β concentrations in patients with secondary amenorrhoea, arranged according to their response to treatment with Provera. The equivocal group contains patients who experienced less than 2 days bleeding. In general a negative or an equivocal response to Provera was associated with serum oestradiol concentrations below 5 ng/100 ml; a positive response was associated with concentrations above 5 ng/100 ml (data from Hull *et al.*, 1975).

menopausal range, gonadotrophin concentrations were usually within or just above the normal range for the follicular phase. Thus for some reason serum LH and FSH concentrations do not rise as high as one would expect for the degree of oestrogen deficiency. While these findings could be explained by prolactin induced increased production of androstenedione (thence of oestrone) we consider a more probable explanation is that, in addition to its deleterious effect on the ovary, hyperprolactinaemia also inhibits gonadotrophin secretion (Jacobs *et al.*, 1976). Moreover this effect is presumably more upon release than on synthesis, since such patients typically show exaggerated responses of LH and FSH to injections of LHRH, despite near normal basal gonadotrophin concentrations. Finally it must be admitted that there are few data available on the gonadotrophin response to clomiphene or to exogenous oestrogen in hyperprolactinaemic women.

4. ASSESSMENT OF OESTROGEN PRODUCTION

We now turn to the analysis of patients with "hypothalamic amenorrhoea". Since these patients have already been defined as not having elevated gonadotrophin concentrations, the next point to establish is the patient's oestrogen status. This may be achieved by direct measurement of total urine oestrogen excretion or of serum oestradiol 17β concentrations, or indirectly by assessment of the lower genital tract, or by the uterine response to treatment with progestogen. An oestrogenised endometrium treated with progestogen is shed with bleeding after the drug is withdrawn. Failure to bleed indicates an oestrogen-deficient endometrium (Lunenfeld and Insler, 1974). As seen in Fig.4, which gives results in 35 consecutive patients, the response to progestogen correlates surprisingly well with the concentration of oestradiol-17β in serum (Hull et al., 1975).

The association of impaired oestrogen production with gonadotrophin concentrations that are not elevated suggests a defect of the tonic control of gonadotrophin secretion, since the normal response to impaired oestrogen production is hypersecretion of the trophic hormones. Negative responses to progestogen and subnormal concentrations of serum oestradiol occur, of course, in primary ovarian failure, hyperprolactinaemic amenorrhoea and hypogonadotrophic hypogonadism of any cause. When these conditions are excluded the most striking clinical correlate of the hormonal profile of low oestrogen and low gonadotrophin levels is anorexia nervosa. While this pattern is a familiar finding in a grossly underweight patient (Jacobs, 1975), we have been impressed by the large proportion of our patients who have a partially recovered form of the disease as the presumed cause of their amenorrhoea. In a study of 15 patients with this condition, though the mean lowest weight was 36 ± 6 kg (SD), the weight at presentation at our clinic was 52 ± 6 kg. The latter figure, whilst significantly lower than the computed ideal weight (56 ± 2 kg), was within 10% of it. Some of the lowest serum oestradiol concentrations that we have seen in young women in the reproductive phase of life have occurred in these patients. Nonetheless mean basal gonadotrophin concentrations were neither elevated nor subnormal but were in the upper part of the normal follicular phase range.

It is not clear why this severe reproductive defect should be so critically related to the patient's weight, though the findings in these patients are reminiscent of the observations of Frisch (1974) on the critical relation of body weight to normal puberty. It may of course be that the pre-morbid weight is of more significance than the ideal (normal population's) weight for the subject's height and age. If, as is not uncommon in anorexia nervosa, these patients had had an episode of pre-morbid obesity, the weight deficit might be greater than implied by the figures given. Since no explanation of this very stereotyped clinical picture appears in the literature, we feel free to speculate and postulate that the marked oestrogen deficiency occupies a critical role in the disorder. In the anovulatory woman a major proportion of oestrogen is derived from extra-gonadal sources (Siiteri and MacDonald, 1973; Baird et al., 1975), conversion of androstenedione to oestrone in fat tissue playing a major role in oestrogen economy. In our patients with partially recovered anorexia nervosa it may be that pre-morbid obesity led to an increase of the number of fat cells (Bray, 1970) and that the adipocytes in the partially recovered patients remain small and malfunctioning despite near-normal total body weight. Alternatively the starvation of severe anorexia nervosa may itself damage fat cells so that they are subsequently unable to aromatise oestrogen precursors to a normal extent. In any event oestrogen deficiency persists and since it is known from experiments in rats that treatment with oestrogen promotes the action of FSH on the ovary (Goldenberg et al., 1972), the deficiency of oestrogen in these patients may account for the failure of the ovaries to respond to endogenous gonadotrophins, which at this stage are in the high normal range (Jacobs, 1975; Hull et al., 1975). One endocrine feature that we have observed is however unexplained by this hypothesis. Basal serum prolactin and the prolactin response to TRH was subnormal in these patients and the levels did not increase in two subjects given cyclical oestrogen replacement in doses sufficient to cause withdrawal bleeding and suppression of the gonadotrophin response to LHRH. These data suggest an additional and persisting defect of hypothalamic-pituitary function.

Whatever the explanation offered, the syndrome is common and accounts for some 20% of patients with secondary amenorrhoea referred to our gynaecological endocrine clinic. These patients, because of their psychiatric condition, rarely request induction of ovulation. The question of replacement treatment with oestrogen is however more difficult. As mentioned, the oestrogen deficiency is severe but a decision about instituting cyclical hormone replacement treatment, with the inevitable return of (artificial) menstrual cycles, is often best taken in consultation with the patient's psychiatrist. When induction of ovulation is requested, as in other patients with defects of the tonic centre, treatment with exogenous gonadotrophins is required. These patients may in future be appropriate for treatment with LHRH.

5. ASSESSMENT OF THE RESPONSE TO CLOMIPHENE

If oestrogen production is not impaired, the probability of a successful response to treatment with clomiphene is greatly enhanced (Goldenberg et al., 1973b). The response to clomiphene that we aim for is induction of ovulation. The ultimate proof of ovulation is of course pregnancy, but this may not occur for various reasons and indeed many of our patients have no wish for it, being young women wanting to be reassured rather than treated. Ovulation is then inferred from the development of the corpus luteum, the characteristic endocrine feature of which is secretion of progesterone. The familiar progesterone-induced luteal phase rise of basal body temperature may be measured, but this index has serious drawbacks that include both recorder and observer error. The fundamental physiological problem is that the dose-response relationship of body temperature to progesterone is flat and soon levels off. As shown by Dhont et al. (1974), in normal women plateau levels of temperature are reached at progesterone concentrations of 4-5 ng/ml, concentrations that are not compatible with a normal luteal phase (Abraham et al., 1974). Conversely, the temperature occasionally fails to rise in the presence of normal luteal phase progesterone levels (Johansson et al., 1972).

Ovulation usually occurs about 5 days after completing a 5-day course of clomiphene (Insler et al., 1973). The chief value of recording the temperature is in timing the progesterone measurement, so to assess the response we measure serum progesterone concentrations about 12 days after stopping clomiphene, hopefully in the middle of the putative luteal phase. Concentrations of progesterone above 8 ng/ml are accepted as consistent with ovulation (Radwanska and Swyer, 1974), but when there is doubt several measurements are made (Abraham et al., 1974).

Despite the occurrence in some of a defective luteal phase, most patients who respond to clomiphene with an increase of serum progesterone concentrations of 8 ng/ml or more achieve a cycle in which conception readily occurs. However most patients seem unable to initiate further cycles without additional courses of treatment (Odell et al., 1972). It would appear that these patients can organise a perfectly normal single cycle but are unable to secure its repetition. We regard the underlying pathophysiological disturbance as constituting a "cycle initiation defect" (Hull et al., 1975). It is of interest that patients with this disorder often have a delayed menarche with a menstrual history characterised by episodes of oligo-amenorrhoea — features quite distinct from the stereotyped history usually obtained from our patients with anorexia nervosa, in whom the age of menarche is usually normal and the menstrual history without blemish until the abrupt onset of the weight disorder.

Patients with cycle initiation defects are usually very fertile once the anovulation is overcome. The dose of clomiphene is selected prospectively on the basis of the patient's oestrogen production, of which a contemporaneous assessment is readily provided by her response to the preceding treatment with progestogen. If the patient bleeds she takes clomiphene 50 mg per day for 5 days; if she does not she tries 100 mg per day. We have very rarely needed to advise more than 50 mg daily in patients who bleed after treatment with progestogen. The reason may be that the preceding treatment with progestogen causes an abrupt discharge of LH and FSH (Goldenberg et al., 1973b). This burst of gonadotrophin release may perhaps provide the patient with a stimulus to the ovary akin to that speculatively attributed to the mid-cycle peak of FSH secretion and so render the ovary more sensitive to subsequent clomiphene stimulated gonadotrophin secretion. Another possible advantage of prior treatment with progestogen is that it strips the endometrium, exposed for the duration of the amenorrhoea to the effects of oestrogen alone, in preparation for a normal start to a new cycle and for possible implantation of a blastocyst. A further point in this regard relates to what is being attempted by treatment with clomiphene. Contrary to its characterisation as a "fertility" drug, clomiphene should be viewed as an agent whose only favourable action is to induce ovulation. Indeed as the dose is raised, despite inducing ovulation clomiphene may actually impair fertility through its anti-oestrogenic action on the lower genital tract. As with all medicaments the optimum dose is the lowest one that works.

Patients with cycle initiation defects are seen in our clinic with perhaps half the frequency of those with tonic centre defects. Because of the characteristic history we presume the condition is often self-limiting. These patients may form a large proportion of those in whom the diagnosis of post-pill amenorrhoea has been made, since some reports of this condition (for instance, Steele et al., 1973) refer to a high incidence of preceding episodes of oligo-amenorrhoea.

Finally, there is the patient who is well-oestrogenised but who fails to make an ovulatory response to clomiphene. The response to this drug, like the normal ovulatory cycle, depends not only upon initiating gonadotrophin secretion at the beginning of the cycle to induce follicular development, but also upon the oestrogen stimulated mid-cycle surge of gonadotrophins, which is triggered by the rising late follicular concentrations of oestradiol (Ross et al., 1970). This action is possibly facilitated by progestogens (Swerdloff et al., 1972). Baird and his colleagues have postulated that in women with dysfunctional uterine bleeding, an-

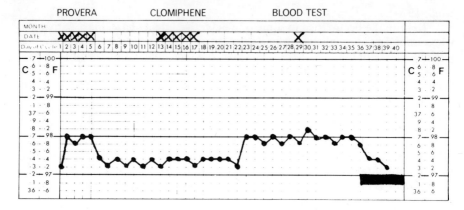

Fig.5 Idealised basal body temperature chart of a patient given Provera (note progestogen induced temperature rise) and clomiphene. The timing of the blood sample for progesterone measurement is indicated.

ovulatory cycles are caused by defects in oestrogen-mediated positive feedback, as a consequence of which there is no mid-cycle release of gonadotrophin despite marked fluctuations of oestradiol concentrations (Fraser *et al.*, 1973). In these patients the dysfunctional uterine bleeding results from the uterine response to the changing oestrogen environment; since ovulation does not occur progesterone levels do not rise, sloughing of the endometrium is incomplete and prolonged and irregular bleeding occurs.

A defect of oestrogen-mediated positive feedback in patients with dysfunctional uterine bleeding has been demonstrated by the failure of serum LH levels to rise in such patients after administration of ethinyl oestradiol (Van Look *et al.*, 1975). When this defect is combined with a defect of cycle initiation, the predicted response to clomiphene is failure to ovulate despite an increase of gonadotrophin levels during, and of oestradiol levels after, the treatment is administered. A group of patients with this pattern of endocrine events during attempted induction of ovulation with clomiphene has been reported by Marshall *et al.* (1973). These workers attributed their results to a defect of positive feedback, a conclusion that we would endorse with the one proviso that amenorrhoeic patients with this disorder must presumably have an additional cycle initiation defect.

Defects of oestrogen-mediated positive feedback are readily detected by determining serum LH concentrations after oral administration of ethinyl oestradiol (Van Look *et al.*, 1975) or after intramuscular injections of estradiol benzoate (Shaw *et al.*, 1975). An alternative method using exogenous progestogens is based upon observations in ovariectomised rats (Swerdloff *et al.*, 1972) and in post menopausal women (Odell and Swerdloff, 1968). In both species the addition of progestogens to treatment with oestrogen leads to an abrupt but temporary increase of serum gonadotrophin concentrations, the timing and levels of which simulate the preovulatory surge of LH and FSH. These results, which were only obtained when progestogen was added to oestrogen treatment, suggest that progestogens lower the threshold for the positive feedback of oestrogens on the hypothalamus. When anovulatory women were given an injection of progesterone, Goldenberg *et al.* (1973) reported an increase in gonadotrophin levels in those women who were well oestrogenised (as shown by a post-injection withdrawal bleed) and who were subsequently shown to ovulate in response to treatment with clomiphene. Although recognising that measuring serum LH concentrations after giving a progestogen is a less direct method of detecting a defect in oestrogen-mediated positive feedback than after giving an oestrogen, we have preferred the progestogen method, since the patient obtains all of the advantages of progestogen pretreatment referred to above. Accordingly in our clinic, the well-oestrogenised patient with normal basal gonadotrophin concentrations who does not respond to clomiphene has LH measurements made on the two days following an injection of intramuscular progesterone 100 mg. If the LH concentrations do not at least double (Goldenberg et al., 1972) a "cyclical centre" defect is diagnosed. Subsequent treatment is with clomiphene to initiate the cycle and HCG is added on the 5th day after stopping treatment to replace the absent midcycle surge of gonadotrophins.

SUMMARY AND CONCLUSIONS

The protocol of investigation described in this paper was designed with two major objectives. The first was an attempt to link diagnostic tests as directly as possible to therapy, with the intention of limiting the amount of time-consuming and expensive investigations that are now available to those with a defined therapeutic dividend. The initial protocol takes six weeks to complete (Fig.5). A blood sample for measurement of basal serum gonadotrophins, prolactin, thyrotrophin, thyroxine and free thyroxine index is drawn on the day the patient is seen and examined. An x-ray of the pituitary region is obtained then. Optimal additional investigations, such as measurements of serum oestradiol and of the response to a combined injection of thyrotrophin releasing hormone and gonadotrophin releasing hormone can be performed on the same morning. In our experience the latter investigations, while of great research interest, rarely alter a diagnosis made using the information provided by the combination of the basal measurements and the ensuing dynamic tests. The patient then takes medroxy progesterone acetate (Provera, Upjohn) 5 mg daily for 5 days, and one week later she takes a 5-day course of clomiphene, the dose of which depends on her response to the progestogen. The response to clomiphene is evaluated by measuring the serum progesterone concentration 12 days after stopping clomiphene, provided that the basal body temperature chart indicates that the sample was obtained in the middle of the putative luteal phase. If the timing is shown to be incorrect and the patient has not become pregnant, the clomiphene test is repeated and she is advised how to use the temperature chart to ensure that the sample is obtained on the optimum days(s). The patient returns for review six weeks after the tests were initiated and if required further investigations (pituitary tomography, intramuscular progesterone tests, insulin hypoglycaemia, etc.) are arranged then. As soon as the information is available she is informed of the diagnosis and, depending on that and her wishes, she is either reassured, referred to an appropriate specialist clinic (psychiatric, neuro-radiological, etc.) or placed upon definitive therapy. Analysis of over 100 consecutive patients with secondary amenorrhoea who have completed this protocol supports the validity of our approach (Hull et al., 1976).

The second major objective in designing this protocol has been our attempt to analyse reproductive disorders in purely functional terms. Thus diagnoses such as "hypothalamic amenorrhoea" or "post-pill amenorrhoea" do not feature in this system, not necessarily because we do not believe in their existence, but because such diagnoses do not imply any particular course of action. Nevertheless there remains the salutory thought that with the pace of modern research, the reproductive physiology upon which this approach is based will soon undergo revision and no doubt many of our diagnoses will appear as antiquated as those they were designed to replace.

ACKNOWLEDGEMENTS

The reagents used in the gonadotrophin assays were kindly provided by Dr W D Odell and the MRC of the UK; those for prolactin and FSH were provided by the Hormone Distribution Programme of the National Pituitary Agency of the NIAMDD, to which we express our appreciation. The antiserum used for the radioimmunoassay of oestradiol-17β was kindly supplied by Dr G Knaggs and was prepared in goats using a hapten (6-oxo-oestradiol) prepared by Dr D Exley.

Bromocriptine (CB 154, Sandoz) was provided by the late Dr E R Evans and to him and to Sandoz Products Limited we express our gratitude for financial support. LHRH was generously provided by Drs W Bogie and E Vogel of Hoechst Pharmaceuticals. During the time the work was being performed SF was in receipt of a Saltwell Fellowship from the Royal College of Physicians.

It is a pleasure to acknowledge the support and guidance of former colleagues and mentors, many of whose names appear in this paper. Particular gratitude is owed to Dr J D N Nabarro for encouragement and guidance over many years, and to Professor Bruno Lunenfeld with whom we have recently enjoyed a very stimulating and valuable association. Finally we are indebted to the physicians and gynaecologists of St. Mary's Hospital and the Middlesex Hospital for allowing us to investigate patients under their care.

REFERENCES

Abraham, G.E., Maronlis, G.B. and Marshall, J.R. (1974). *Obstetrics and Gynaecology* **44**, 522.
Baird, D. (1975). *J. Endocrinol.* **66**, 12p.
Bramble, F.J., Houghton, A.L., Eccles, S.S., Murray, M.A.F. and Jacobs, H.S. (1975). *J. Endocrinol.* **66**, 10p.
Bray, G.A. (1970). *Am. J. clin. Nutrition* **23**, 1141.
Bray, G.A. and Jacobs, H.S. (1974). *In* "Handbook of Physiology, Section 7: Endocrinology Vol.III: The Thyroid" (M.A. Greer and D.H. Solomon, eds) p.413. Amer. Physiol. Soc. Washington D.C.
Burke, C.W., Joplin, C.F. and Fraser, T.R. (1972). *Proc. Roy. Soc. Med.* **65**, 486.
Chiazze, L., Brayer, F.T., Macisco, J.J., Parker, M.P. and Duffy, B.J. (1968). *J. Amer. Med. Assoc.* **203**, 377.
Child, D.F., Nader, S., Mashiter, K., Kizeld, M., Banks, L. and Russel Fraser, T. (1975). *Brit. Med. J.* **1**, 604.
Dhont, M., Vande Kerkhove, A., Vermeulen, A. and Vandeweghe, M. (1974). *Europ. J. Obstet. Gynaec. Reprod. Biol.* **4**, **Supplement**, S 153.
Erdheim, J. and Stumme, E. (1909). *Zieglers Beitr. Pathol. Anat.* **46**, 1.
Evered, D.C. and Hall, R. (1972). *Brit. Med. J.* **1**, 290.
Franks, S., Murray, M.A.F., Jequier, A.M., Steele, S.J., Nabarro, J.D.N. and Jacobs, H.S. (1975). *Clin. Endocr.* **4**, 597.

Fraser, I.S., Michie, E.A., Wide, L. and Baird, D.T. (1973). *J. clin. Endocr. Metab.* 37, 407.
Fluckiger, E. (1972). *In* "Prolactin and Carcinogenesis" (A.R. Boyns and K. Griffiths) p.162. Alpha Omega Alpha Publishing, Cardiff, Wales.
Frisch, R.E. (1974). *In* "Control of the Onset of Puberty" (M.M. Grumbach, G.D. Grave and F.E. Mayer, eds) p.403. John Wiley & Sons, New York.
Gemzell, C. (1975). *Am. J. Obstet. Gynec.* 121, 311.
Goldenberg, R.L., Vaitukaitus, J.L. and Ross, G.T. (1972). *Endocrinology* 90, 1492.
Goldenberg, R.L., Grodin, J.M., Rodbard, D. and Ross, G.T. (1973a). *Am. J. Obstet. Gynec.* 116, 1003.
Goldenberg, R.L., Grodin, J.M., Vailukaitus, J.L. and Ross, G.T. (1973b). *Am. J. Obstet. Gynec.* 115, 193.
Hardy, J. (1973). *In* "Diagnosis and Treatment of Pituitary Tumours" (P.O. Kohler and G.T. Ross, eds) p.179. Exerpta Medica, Amsterdam, Int. Cong. Series No.303.
Hull, M.G.R., Franks, S., Murray, M.A.F. and Jacobs, H.S. (1976). *Brit. J. Obstet. Gynce. (in press).*
Insler, V., Zakut, H. and Serr, D.M. (1973). *Obstetrics and Gynaecology* 41, 602.
Jacobs, H.S. (1974). *In* "Clinical Physiology" (E.J.M. Campbell, C.J. Dickinson and J.D.H. Slater, eds) p.754, 4th ed. Blackwell Scientific Publications, Oxford.
Jacobs, H.S. (1975). *Post Graduate Med. J.* 51, 209.
Jacobs, H.S. and Franks, S. (1975). *Brit. Med. J.* 2, 141.
Jacobs, H.S. and Nabarro, J.D.N. (1969). *Quart. J. Medicine* 38, 475.
Jacobs, H.S., Franks, S., Murray, M.A.F., Hull, M.G.R., Steele, S.J. and Nabarro, J.D.N. (1976). *Clin. Endocr. (in press).*
Johansson, E.D.B., Larsson-Cohen, V. and Gemzell, C. (1972). *Am. J. Obstet. Gynec.* 113, 933.
Jones, G.S. and Moraes Ruehsen, M.De. (1969). *Am. J. Obstet. Gynec.* 104, 597.
Kaufman, B., Pearson, O.H. and Chamberlain, W.B. (1973). *In* "Diagnosis and Treatment of Pituitary Tumors" (P.O. Kohler and G.T. Ross, eds) p.100. Exerpta Medica, Amsterdam, Int. Cong. Series No. 303.
Kramer, S. (1973). *In* "Diagnosis and Treatment of Pituitary Tumors" (P.O. Kohler and G.T. Ross, eds) p.217. Exerpta Medica, Amsterdam, Int. Cong. Series No.303.
Lunenfeld, B. and Insler, V. (1974). *Clin. Endocr.* 3, 223.
Marshall, J.R., Morris, R., Court, J. and Butt, W.R. (1973). *J. Endocrinol.* 59, xxxiv.
Odell, W.D. and Swerdloff, R.S. (1968). *Proc. Nat. Acad. Sci. (USA)* 61, 529.
Odell, W.D., Swerdloff, R.S., Abraham, G.E., Jacobs, H.S. and Walsh, P.C. (1972). *In* "Gonadal Steroid Secretion" (J.A. Strong and D. Baird, eds) p.32. Pfizer Medical Monographs No.6. Edin. Univ. Press.
Radwanska, E. and Swyer, G.I.M. (1974). *J. Obstet. Gynaec. Brit. Comm.* 81, 107.
Ross, G.T., Cargille, C.M., Lipsett, M.B., Rayford, P.L., Marshall, J.R., Strott, C.A. and Rodbard, D. (1970). *Rec. Prog. Horm. Res.* 26, 1.
Ross, G.T. and Vande Wiele, R.L. (1974). *In* "Textbook of Endocrinology" (R.H. Williams, ed) p.368. W.B. Saunders Company, Philadelphia.
Shaw, R.W., Butt, W.R. and Thordon, D.R. (1975). *Brit. J. Obstet. Gynaec.* 82, 337.
Siiteri, P.K. and MacDonald, P.C. (1973). *In* "Handbook of Physiology, Section 7: Endocrinology, Vol.II: Female Reproductive System, Part 1" (R. Greep, ed). Amer. Physiol. Soc. Washington D.C.
Steele, S.J., Mason, B. and Brett, A. (1973). *Brit. Med. J.* 2, 343.
Swerdloff, R.S., Jacobs, H.S. and Odell, W.D. (1972). *Endocrinology* 90, 1529.
Thomas, P.R.M., Winstanly, D., Peckham, M.J., Austin, D.E., Murray, M.A.F. and Jacobs, H.S. (1976). *Brit. J. Cancer (in press).*
Treloar, A.E., Boynton, R.E., Borghild, G.B. and Brown, W.B. (1967). *Int. J. of Fert.* 12, 77.
Van Look, P.F.A., Fraser, I.S., Hunter, W.M., Michie, E.A. and Baird, D.T. (1975). *J. Endocrinol.* 64, 54-55p.
Williams, R.A., Jacobs, H.S., Kurtz, A., Millar, J.G.B., Oakley, N.W., Spathis, G.S., Sulway, M.J. and Nabarro, J.D.N. (1975). *Quart. J. Med. New Series* 46, 79.

AMENORRHOEA – ITS INVESTIGATION AND THE ASSESSMENT OF THE ROLE OF PROLACTIN

G.M. Besser

The Medical Professorial Unit, St. Bartholomew's Hospital London EC1A

For the purpose of this review I shall assume that the patient under consideration either has secondary amenorrhoea or if the amenorrhoea is primary that the patient is over the age of 16 years. In this group of patients it is essential to ascertain whether the failure to ovulate and menstruate is due to hypothalamic, pituitary or primary gonadal disease. I shall consider the place of investigations in the establishment of an accurate diagnosis.

PRIMARY AMENORRHOEA

Measurement of plasma luteinising and follicle stimulating hormone levels are of great value in patients expected to be past the age of puberty. Thus if their levels are low or in the range of the follicular phase of the adult menstruating woman and the breasts are poorly developed, then it is probable that the patient has simple delayed puberty, or less commonly isolated gonadotrophin deficiency. In the latter event other endocrine function tests such as thyroid hormone assays, and insulin-induced hypoglaemic tests for measurement of growth hormone and ACTH reserve must be performed to establish the monohormonal deficiency. Clearly delayed puberty and isolated gonadotrophin deficiency are identical in the early stages unless other associated stigmata such as hair lip, cleft palate or anosmia suggest that there is a true gonadotrophin secretory defect. Even in the absence of these features after the age of 19 years the patient should be treated as if she has permanent lack of LH and FSH since the primary amenorrhoea is unlikely to resolve spontaneously. Primary hypothyroidism must always be excluded by assay of thyroxine since such patients may present during their teens or early twenties, with either primary or secondary amenorrhoea without any other features, and it may be impossible to diagnose them clinically.

DEVELOPMENTAL DEFECTS

If the levels of FSH and LH are high, urinary estrogens low and there is little or no breast development, then the patient has clearly passed through the phase of hypothalamic-pituitary development normally associated with puberty, and primary gonadal failure may be present. Even if the patient does not show the typical clinical features of Turner's syndrome this diagnosis must be entertained and a chromosome analysis obtained from lymphocyte culture. It is not adequate to rely on a buccal smear since mosaicism is common and may give a normal chromatin pattern on the buccal smear. As an alternative true gonadal agenesis may be present, i.e., there has been failure of development of the gonads.
This could be in a patient with either a 46XX or 46XY karyotype. Each will look the same, with a small uterus and infantile external genitalia, since in the absence of the Müllerian duct inhibitor normally derived from the foetal testis, or in the female patient with undeveloped ovaries, the Müllerian system develops and the patient will be phenotypically female. Although both types of patient must be reared as females, and given long-term oestrogen-progestogen replacement therapy to promote and maintain normal female sexual characters and to protect against development of osteoporosis, the patient with a 46XY karyotype should be explored and the gonadal streaks removed since there is an increased incidence of the development of a dysgerminoma in the pelvis of phenotypic females with karyotypes containing Y chromosomes.

If the patient has adequate breast development yet has primary amenorrhoea and absence of secondary sexual hair then careful examination to exclude inguinal testes should be carried out since the testicular feminisation syndrome might be present. Such patients have a 46XY karyotype, male levels of plasma testosterone and oestrogen, and high LH levels. There is a short vagina and infantile vulva but no uterus. The patients are resistant to androgens, hence the lack of male external genital development, but sensitive to oestrogens. They cannot be virilised and should be reared as females. The testes should be removed after pubertal breast development has occurred because of the risk of malignant change.

THE POLYCYSTIC OVARY SYNDROME

Although more commonly secondary, occasionally primary amenorrhoea may be a feature of the polycystic ovary syndrome. Such patients usually have good breast development and oestrogen secretion in the range normally seen in the early follicular phase of the normal menstrual cycle. LH levels may be elevated, but without showing cyclicity, but FSH levels are normal. In these young patients hirsuties and obesity are often not present. Laparoscopy with ovarian biopsy, or gynaecography allow the diagnosis to be confirmed.

CLOMIPHENE TESTING

Clomiphene is an anti-oestrogen which combines with and occupies gonadal steroid receptors of the hypothalamus and pituitary, activating the negative feedback mechanisms to promote gonadotrophin secretion. These mechanisms are involved when this drug is used to promote ovulation. Clomiphene can be used to investigate the functional integrity of the hypothalamic-pituitary axis. 2 mg/kg/day of clomiphene, up to a maximum dose of 200 mg/day, is given in divided doses for 7 to 10 days and blood is sampled on days 0, 4, 7 and 10 for plasma LH and FSH levels. It is not possible to establish a true 'normal' range for the clomiphene test in amenorrhoeic patients, for the normally menstruating woman cannot be compared with amenorrhoeic women in this assay. Nevertheless a normal response is taken to be present when there is a clear rise in plasma LH, the values usually doubling to reach levels outside the range seen in the first week of normally cycling women. FSH responses are less consistent.

Absent responses are seen in patients with organic hypothalamic and pituitary diseases, isolated gonadotrophin deficiency and in severe anorexia nervosa. In the latter conditions as the patient improves clomiphene responsiveness returns long before spontaneous menstruation and may be taken to be a good prognostic sign (Marshall and Fraser, 1971). Patients with the polycystic ovary syndrome and unexplained secondary 'functional' amenorrhoea usually respond to clomiphene but patients with hyperprolactinaemia rarely respond. A good response indicates that the patient is likely to ovulate and become fertile on clomiphene alone or on clomiphene plus H.C.G.

Prepubertal patients, including those with delayed puberty, often show a reduction of basal gonadotrophins in response to clomiphene. Elevated levels in patients with gonadal failure do not alter.

TESTING WITH THE GONADOTROPHIN RELEASING HORMONE (Gn-RH)

By and large little diagnostic help has been obtained from the use of Gn-RH. The majority of patients with amenorrhoea whether associated with hypothalamic or pituitary disease, the polycystic ovary syndrome or idiopathic can be made to release LH and FSH from their pituitaries after 100 or 500 μg Gn–RH (Mortimer et al., 1973). Smaller doses show inconsistent results even in normal subjects. Patients with long-standing amenorrhoea of hypothalamic or pituitary origins may show an impaired response initially but repeated administration eventually results in secretion of hormone. Indeed ovulation can be induced in such patients with Gn-RH. The situation is similar with anorexia nervosa. In the latter patients, and in many others with unexplained 'functional' secondary amenorrhoea, a state of acquired Gn-RH deficiency exists. Since Gn-RH is required for LH and FSH synthesis as well as their release, the gonadotrophin stores may be low if hypogonadism has been present for some time. The situation can be restored by Gn-RH therapy. Although sometimes insensitive initially, patients with 'isolated gonadotrophin deficiency' can also be made to secrete gonadotrophin with Gn-RH despite their insensitivity to clomiphene, and this suggests that the primary defect is within the hypothalamus rather than the pituitary; this syndrome might be called more realistically the 'isolated gonadotrophin releasing hormone deficiency' (Marshall et al., 1972; Mortimer et al., 1973).

Patients with delayed puberty given Gn-RH show the pattern of LH and FSH secretion characteristic of the prepubertal state rather than that of the adult. After Gn-RH, adults show greater LH and FSH secretion, but prepubertal subjects show only slight LH release, whereas FSH secretion is similar to that of adults. As the patient passes through puberty, LH secretion increases until it is greater than that of FSH and the adult pattern is achieved (Bourguinon et al., 1974). When hypogonadotrophic patients are treated with long-term Gn-RH a similar sequence is seen as treatment progresses. Puberty can be induced with Gn-RH therapy (Mortimer et al., 1974).

PROLACTIN AND GONADAL FUNCTION

Elevated circulating prolactin levels are accompanied by disturbed gonadal function — women are either amenorrhoeic or have abnormal and anovular menstrual cycles, and men are virtually or absolutely impotent and may be oligospermic.

Hypogonadism may be commonly related to hyperprolactinaemia without the presence of clinically evident galactorrhoea. It is clearly mandatory, therefore, to measure plasma prolactin levels in all patients with amenorrhoea. The relationship between hypogonadism and hyperprolactinaemia appears to apply irrespective of the particular cause of the raised prolactin. Thus during the normal postpartum period if lactation is established resumption of menstruation is delayed and fertility is reduced as long as circulating prolactin levels are elevated. During phenothiazine therapy, metoclopramide, sulpuride, reserpine or methyl dopa administration, and often after the oral contraceptive, amenorrhoea and galactorrhoea are common, as they are when hypothalamic or pituitary diseases are present. Similarly when there is no obvious cause for the amenorrhoea hyperprolactinaemia is often present. Routine plasma prolactin assay should be obtained preferably on two or three separate days since prolactin secretion is pulsatile and single samples may not accurately reflect the true prolactin secretory status.

The explanation of the relationship between elevated prolactin levels and hypogonadism has in the past been held to be due to competition at the pituitary or hypothalamic level between the mechanisms controlling prolactin and gonadotrophin secretion. Thus it was held that when prolactin secretion was active, gonadotrophin secretion is decreased, and *vice versa*. Although this may be the case in animals, and even this seems doubtful, there is no evidence for this in man. Gonadotrophin levels basally and after Gn-RH in hyperprolactinaemic patients are almost always normal or even occasionally excessive (Mortimer et al., 1973; Thorner et al., 1974).

It seems clear now that prolactin blocks the actions of the gonadotrophins on the gonad and this, rather than impairment of gonadotrophin secretion, is the cause of the hypogonadism (Thorner et al., 1974; McNatty et al., 1974; Besser and Thorner, 1975).

Since the ergot alkaloid bromocriptine acts directly in the pituitary to suppress prolactin levels, hyperprolactinaemic patients may be effectively treated with this compound (Besser et al., 1972; Thorner et al., 1974) and menstruation, ovulation and fertility rapidly return without any other treatment (Besser and Thorner, 1975).

The availability of this simple therapy stresses the need for routine plasma prolactin measurements in amenorrhoeic patients, with or without galactorrhoea.

CONCLUSIONS

By combination of measurements of plasma LH, FSH, oestradiol, thyroxine and prolactin, together with clomiphene and Gn-RH testing, radiology of the skull and laparoscopy, an accurate diagnosis of the cause of the amenorrhoea may be made. Differentiation into delayed puberty, gonadal failure, premature menopause, hyperprolactinaemia, the polycystic ovary syndrome or hypogonadotrophic hypogonadism may be achieved and usually the precise cause determined. Clearly this is necessary before rational and effective therapy can be instituted.

REFERENCES

Besser, G.M., Parke, L., Edwards, C.R.W., Forsyth, I.A. and McNeilly, A.S. (1972). *Brit. Med. J.* 3, 669.

Besser, G.M. and Thorner, M.O. (1975). *Eleventh Symposium on Advanced Medicine*, Pitman Medical, London *(in press)*.

Bourguinon, J.P., Burger, H. and Franchimont, P. (1974). *Clin. Endocr.* 3, 437.

Marshall, J.C. and Fraser, T.R. (1971). *Brit. Med. J.* 4, 590.

Marshall, J.C., Harsoulis, P., Anderson, D.C., McNeilly, A.S., Besser, G.M. and Hall, R. (1972). *Brit. Med. J.* 4, 643.

McNatty, K.P., Sawers, R.S. and McNeilly, A.S. (1974). *Nature New Biol.* 250, 653.

Mortimer, C.H., Besser, G.M., McNeilly, A.S., Marshall, J.C., Harsoulis, P., Tunbridge, W.M.G., Gomez-Pan, A. and Hall, R. (1973). *Brit. Med. J.* 4, 73.

Mortimer, C.H., McNeilly, A.S., Murray, M.A.F., Fisher, R.A.F. and Besser, G.M. (1974). *Brit. Med. J.* 4, 617.

Thorner, M.O., McNeilly, A.S., Hagen, C. and Besser, G.M. (1974). *Brit. Med. J.* 2, 419.

CLASSIFICATION AND PROCEDURE FOR THE EVALUATION OF AMENORRHEA; THE CLINICIAN'S VIEW

G. Giusti, M.P. Piolanti, M. Pazzagli, F. Bassi and C. Vigiani

Endocrinology Unit, University of Florence, Italy

Disorders of ovarian function and, in particular, amenorrhea, depend on many factors. Some of these factors are already well-known, others are still undiscovered. Therefore, classifying amenorrhea is quite difficult if you take into consideration that for practical purposes the classification must be as simple as possible, in easy terminology and be readily adaptable to new developments.

We propose a classification in which primary and secondary amenorrhea are considered side by side (see Fig.1), since often the same factors may provoke either one or the other, even though in varying degrees. The classification is divided into two types of amenorrhea according to the possibility of demonstrating the presence of organic factors or not. The second type presents the greatest difficulties in diagnosis, but offers more interesting opportunities for research.

A protocol of diagnostic investigations valid for every case cannot be used because of quantity and diversity of causal factors. It is always necessary to take a careful history and make a clinical examination, since often this leads to a correct diagnosis.

In any case, the clinical examination is always a preliminary step towards choosing the right protocol of hormonal investigation. Hormonal investigations are of primary importance in certain cases, but we must distinguish the tests which, while of great research interest, rarely alter the clinical diagnosis.

I. AMENORRHEA CAUSED BY ORGANIC FACTORS (Fig.1)

Cases due to diseases of the genital tract (A) must be distinguished from those due to disorders of the hypothalamic-hypophyseal-ovarian system (B). In the latter case, further subdivisions must be made in cases where a primary ovarian

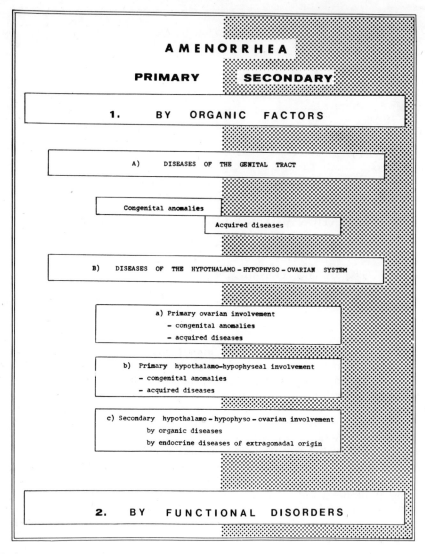

Fig.1

disorder is evident (a), where a primary disorder of the hypothalamic-hypophyseal system is present (b) and where the hypothalamic-hypophyseal-ovarian disorder is secondary either to endocrine disorders of extragonadal origin or to other types of disease (c).

Congenital anomalies of the genital tract always provoke primary amenorrhea. The presence of amenorrhea with normal sexual characteristics leads us to suspect the presence of a congenital anomaly which can easily be detected by gynaecological or pneumopelvigraphic examination. Hormonal investigations confirm the normal function of the hypothalamic-hypophyseal-ovarian axis (Fraser et al., 1973).

We consider that the "feminizing testis syndrome" should not be included in this group, since it should be classified in the group of male pseudohermaphroditism. The feminizing testis syndrome should only be taken into account from a differential diagnostic point of view. For this purpose we must point out that the feminizing testis syndrome is characterized by frequent absence of sexual hair, presence of inguinal hernias or their scars, negative chromatin and male karyotype and lastly, absence of ovaries on pneumopelvigraphic examination.

Even cases included in subgroup (c) usually have clinical signs which lead to a correct diagnosis. In fact it is often the primary endocrinological disease which gives the clinical signs in cases of amenorrhea due to extragonadal endocrine disorders with characteristics of adrenocortical pathology (Cushing's syndrome, Addison's disease, etc.) or with characteristics of thyroidal pathology (for instance hypothyroidism).

However, in some cases of hypothyroidism the clinical signs may be negligible or they may appear gradually (for instance ectopic thyroid).

Amenorrhea Caused by Primary Disorders of the Hypothalamo-Hypophyseal-Ovarian System

As shown in Fig.2 there are many factors which can provoke primary and secondary amenorrhea due to either primary ovarian disease or a primary lesion of the hypothalamic-hypophyseal system. It is evident that measurement of serum gonadotrophism, and FSH in particular, is one of the most important diagnostic investigations for distinguishing between the two groups (Goldenberg et al., 1973). Nevertheless a careful history and clinical examination often allows a correct diagnosis or offer valid support for functional investigations.

Primary Ovarian Involvement

In the group with primary ovarian damage, several cases can be easily recognized because of characteristic physical signs of Turner's syndrome. Genetic investigations which demonstrate often different anomalies of karyotype (the classic 45X, malformations of the X chromosome, mosaicism, etc.) explain the large variety of clinical pictures (Giusti et al., 1966).

On the other hand, some difficulty arises in the diagnosis of "pure gonadal dysgenesis" which is included in the "syndrome of rudimentary gonads".

In these patients stature is normal and somatic abnormalities are absent. The high gonadotrophin level constitutes the most evident sign of primary gonadal damage; the diagnosis may be confirmed by the presence of chromosome anomalies (XY or other mosaicism) and above all by histological examination of the ovary.

The presence of menstrual disorders and frequent association of hirsutism and/or obesity can lead to a diagnosis of polycystic ovary syndrome (PCOS) which can be confirmed by hormonal investigations (high values of LH with marked day-to-day variation and without ovulatory peak, low FSH values, increased production of extraglandular estrone with modification in the E_1/E_2 ratio, increased production of testosterone and androstenedione of glandular origin) and lastly by histological examination of the ovary. However, the precise classification of polycystic ovary syndrome is still under discussion (Borghi et al., 1972).

The existence of a normal gonadotrophin feedback mechanism (positive and negative) in patients affected by PCOS seems to suggest that the disorder of the hypothalamo-hypophyseal system has a functional origin (Yen et al., 1970).

On the other hand, the origin of the androgen hyperproduction in patients affected by PCOS is still uncertain; increased production of androgen by the ovary or the adrenal gland, or both, have been demonstrated.

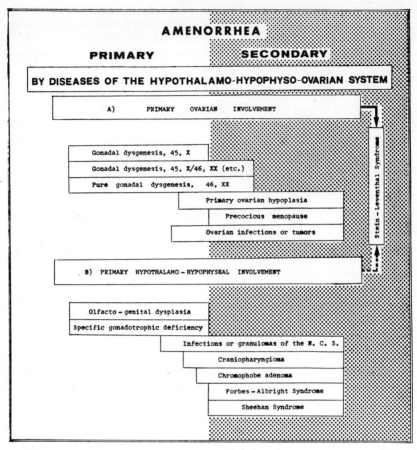

Fig. 2

Primary Hypothalamic-Hypophyseal Involvement
In the group of patients with amenorrhea caused by primary damage to the hypothalamic-hypophyseal system, the isolated gonadotrophin deficiencies are usually congenital and therefore provoke primary amenorrhea; for instance "olfacto-genital dysplasia" or "specific gonadotrophin deficiency".

"Olfacto-genital dysplasia" is already a well-known syndrome even in females; the anosmia (or hyposmia) is of course the guiding symptom. "Specific gonadotrophin deficiency" is still a syndrome to be precisely investigated and in practice may be recognised only where the other causes of hypogonadotrophic hypogonadism or delayed puberty have been excluded.

A variable degree of gonadotrophin deficiency may be associated with other congenital anomalies or in the "Lawrence-Moon-Biedl syndrome", "familiar cerebellar ataxia" or "Prader-Willie syndrome".

More frequently the gonadotrophin deficiency is connected with other hypothalamic or pituitary disorders (infective or vascular disease, tumors, etc.) which can damage this structure. The clinical picture is of course different in relation to age: before puberty we find growth deficiencies and genital infantilism (associated with obesity or hyperphagia in Frölich's syndrome). After puberty we can find secondary amenorrhea with changes of secondary sexual characteristics.

In all these cases it is necessary to carry out skull X-ray and other investigations in order to uncover organic lesions of the hypothalamic-hypophyseal structure of the brain.

The possibilities of recognizing whether gonadotrophin deficiency is due to either hypothalamic or hypophyseal lesions is still under investigation. Other authors in this book report research in this sector. We can only make the following two observations of general interest.

A) It is not always true that amenorrhea (even though long-standing) due to organic lesions of the pituitary gland, is the result of absolute failure of the gonadotrophin producing cells. For example we have observed two cases (one with amenorrhea of 3 years and the other of 13 months) of active acromegaly with evident enlargement of the sella turcica but without visual defects in whom regular menses reappeared after one month of bromocryptine treatment. In both cases galactorrhea with high prolactin values was present. We would mention also a case of amenorrhea with galactorrhea post-partum, in whom a large hypophyseal adenoma with extrasellar extension and hemianopia was discovered: in this patient after bromocryptine treatment galactorrhea disappeared and normal menstruation returned.

B) The meaning of high prolactin values in cases of partial or total pituitary deficiency is yet to be clarified. According to some authors, high prolactin values in patients with pituitary deficiency leads one to suspect damage at hypothalamic or portal vessels level (secondary hypopituitarism) without lesions of the pituitary gland (Woolf *et al.*, 1974). For instance we have observed one young girl, 22 years old, operated on 7 years before for craniopharingioma with classic Frölich's disease, who had short stature, primary amenorrhea with marked galactorrhea, and incomplete secondary sex characteristics. In this patient we found low values of cortisol, GH, LH and FSH without any response to insulin hypoglycemia, to arginine, to L-DOPA and to metyrapone. On the other hand we found high values of prolactin and normal values of TSH with normal response to TRH.

II. AMENORRHEA WITH FUNCTIONAL DISORDERS (Fig.3)

In clinical practice, the number of cases of secondary amenorrhea not due to organic lesions of the hypothalamic-hypophyseal structures (secondary functional amenorrhea) has increased in the last few years, especially in young women (Franchimont *et al.*, 1974). These patients with secondary functional amenorrhea are of scientific interest, but from a diagnostic point of view present great difficulties. These difficulties are often increased by previous pharmacological treatments of various types.

The clinical history frequently shows that the menstrual disorders appeared concomitantly with psychological problems (acute or chronic) often associated with gain or loss of weight.

Sometimes amenorrhea appeared after pharmacological treatments such as tranquilizing agents, contraceptive drugs, etc.. Several of these aspects are often associated and galactorrhea may even be present.

Several hormonal investigations have been performed to clarify causes and

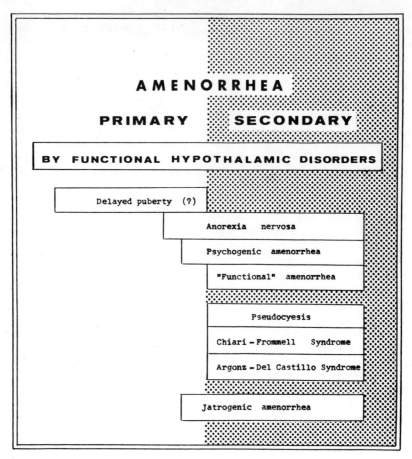

Fig. 3

mechanisms able to provoke secondary functional amenorrhea. These investigations have been made under basal and dynamic conditions (for instance GN-RH stimulation, clomiphene test, estrogen provocation test, etc.). Particular attention has been paid to the function of the feedback mechanism (positive and negative) and therefore of the so-called tonic and cyclic centers. These particular aspects are still under investigation and other authors report in this book some results and their possible explanation. However, for practical purposes, we wish to point to the increasing importance of the determination of prolactin in women affected by secondary functional amenorrhea with and without galactorrhea.

The presence of high prolactin levels in some patients affected by functional secondary amenorrhea has been reported (Franks *et al.*, 1975) and confirmed in our cases. Treatment with anti-prolactin drugs (bromocryptine) quickly induces normal cycles in these patients. These experimental observations raise important problems of the relationship between prolactin and gonadotrophins in human beings.

On the other hand, in the patients affected by amenorrhea with galactorrhea, excluding cases due to hypothyroidism or to treatment with particular drugs (tranquilizing and contraceptive agents) (Besser et al., 1972), studies on prolactin levels under basal conditions, and during inhibition and stimulation tests, associated with repeated skull X-rays, demonstrate that the classical distinction between functional and organic forms seems unrealistic (Zarate et al., 1974). The presence of hypophyseal microadenomas in several cases classified as functional demonstrates that careful observation is necessary of those cases which after treatment with bromocryptine ovulated and conceived. In two of our patients who became pregnant after bromocryptine treatment, amenorrhea with galactorrhea reappeared after delivery.

REFERENCES

Besser, G.M. and Edwards, C.R.W. (1972). *Brit. Med. J.* 2, 280.
Boczkowski, K. and Teter, J. (1966). *Acta endocr. (Copenh.)* 51, 497.
Borghi, A., Giusti, G. and Zampi, G. (1972). "La Sindrome dell'ovaio policistico iperandrogenico". Ed. Medico-Scientifiche, Turin.
Franchimont, P., Valke, J.C. and Lambotte, R. (1974). *In* "Clinics in Endocrinology and Metabolism". W.B. Saunders, London.
Franks, S., Murray, M.A.F., Jequier, A.M., Steele, S.J., Nabarro, J.D.R. and Jacobs, H.S. (1975). *Clin. Endocr.* 4, 597.
Fraser, I.S., Baird, D.T., Hobson, B.M., Michie, E.A. and Hunter, W. (1973). *J. clin. Endocr. Metab.* 36, 634.
Giusti, G., Borchi, A. and Bigozzi, U. (1966). *XI Congr. Soc. Ital. Endocrinol. Societa tipografica Piemontese*, Turin.
Goldenberg, R.L., Giodin, J.M., Rodbard, D. and Ross, G.T. (1973). *Am. J. Obstet. Gynec.* 116, 1003.
Sohval, A.R. (1951). *In* "Diseases of the Endocrine Glands" (J.L. Soffer, ed). Lea and Febigei, Philadelphia.
Woolf, P.D., Jacobs, L.S., Donotrio, R., Burgay, S.Z. and Schalch, D.S. (1974). *J. clin. Endocr. Metab.* 38, 71.
Yen, S.S.G., Vela, P. and Rankin, J. (1970). *J. clin. Endocr. Metab.* 30, 435.
Zarate, A., Canales, E.S., Soria, J., Garrido, J., Jacobs, L.S. and Schally, A.V. (1974). *Ann. Endocrin.* 35, 535.

MODIFICATION OF GONADOTROPIN RELEASE AFTER MULTIPLE OR PROLONGED STIMULATION WITH SYNTHETIC LH-RH

G. Lafuenti, G.C. Giraldi, S. Fazioli, P. Muscatello,
M. Signa, R. Caniglia and G.B. Serra

Department of Obstetrics and Gynecology
Catholic University of Sacred Heart, Rome, Italy

Previous studies conducted in a total of 85 patients with normo- and hypogonadotropic amenorrhea, excluding all cases with primary gonadal failure or utero-vaginal atresia, have shown that the rapid i.v. injection of 25 mcg of synthetic LH-RH (Relisorm, Institute Farmacologico Serono) may elicit different patterns of pituitary response in gonadotropin plasma levels. Comparing these profiles to the range of values obtained in normal subjects, it has been possible to differentiate essentially three types of responses for the release of LH (Fig.1):

— Absent or deficient response, without increase in LH levels after the administration of LH-RH, or with values well below the range of levels measured in normal subjects;
— Normal response, with LH levels within the range of normality;
— Exaggerated response, characterized by normal or low basal values, and by high levels after the injection of LH-RH.

As can be seen in Fig.2 'deficient' responses were found in 19 patients out of the 85 patients affected by amenorrhea, representing 22.3% of the total. 'Normal' responses were present in 40 patients (47.1%) and 'exaggerated' responses were found in 26 subjects (30.6%). Considering that all patients were affected by amenorrhea and hypogonadism, the finding of this large percentage of 'normal' responses seemed of interest. If this test were a true expression of pituitary function, it could be inferred that in all 40 cases the pituitary gland was normally active.

In order to correlate our responses to the clinical syndrome, all data have been distributed among the 3 groups in which the patients were classified; primary amenorrhea (P.A.), secondary amenorrhea (S.A.) and oligomenorrhea.

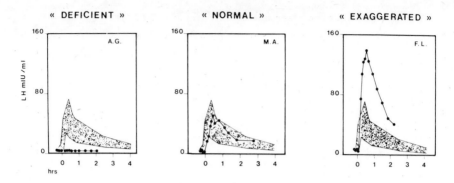

Fig. 1 Patterns of LH response to rapid i.v. injection of 25 mcg of LH-RH in patients with hypo- and normo-gonadotropic amenorrhea.

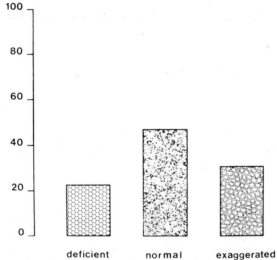

Fig. 2 Percent distribution of deficient, normal and exaggerated responses to 25 mcg of LH-RH in 85 patients with hypo- and normo-gonadotropic amenorrhea.

Table I shows the distribution of 'normal' and abnormal responses among our patients. As can be seen 'normal' responses were present in 7 out of 21 cases (33.3%) of P.A., 27 out of 52 (51.9%) of patients with S.A. and 6 out of 12 cases (50.0%) of subjects with oligomenorrhea. 'Deficient' responses were found with a different incidence in the group of P.A. and in patients with S.A. or oligomenorrhea: to 14 patients (66.7%) found with deficient or absent responses among the cases with P.A., corresponded only 4 patients (7.7%) with S.A. and 1 patient (8.3%) with oligomenorrhea. A similar reverse distribution between subjects affected by P.A. and those affected by S.A. or oligomenorrhea was also found for the 'exaggerated' responses: while 21 patients with S.A. and 5 cases with oligomenorrhea (40.4 and 41.7% respectively) exhibited 'exaggerated' responses, this type of response was not found in subjects affected by P.A.

On the basis of these results, the aim of this study was to investigate the sig-

Table I. LH responses to 25 mcg of LH-RH found in 85 patients with normo- and hypogonadotropic amenorrhea grouped as P.A., S.A. and oligomenorrhea.

RESPONSE	T	«Deficient»		«Normal»		«Exaggerated»	
	n	n	%	n	%	n	%
Primary Amenorrhea	21	14	66.7	7	33.3	—	—
Secondary Amenorrhea	52	4	7.7	27	51.9	21	40.4
Oligo-menorrhea	12	1	8.3	6	50.0	5	41.7

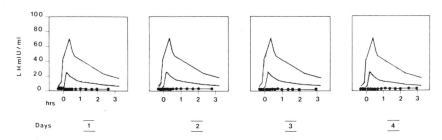

Fig.3 Deficient LH responses after daily administration of 25 mcg of LH-RH in a patient affected by P.A.

nificance of the pituitary responses in order to understand better the mechanisms regulating the LH release in normal pathological conditions.

Since day-to-day variations in the response of the pituitary gland to equivalent doses of LH-RH have been described in the same subjects (Lafuenti *et al.*, 1973; Schwarzstein *et al.*, 1975), it seemed of importance to establish if the deficient responses were a fortuitous or a constant characteristic of each subject. As shown in Fig.3, repeated daily injections of equivalent doses of LH-RH produced similar patterns of responses, thus demonstrating that the deficient responses were in fact a peculiarity of the endocrine habit of these subjects.

In two of these subjects a pituitary LH discharge test was performed infusing 50 mcg of LH-RH over a period of 5 h. As can be seen in Fig.4 compared to values obtained in a normal subject, this test confirmed a clearly deficient release of LH.

Since exogenous administration of synthetic decapeptide should act directly on the pituitary gland, it has been stated that the injection of LH-RH should allow

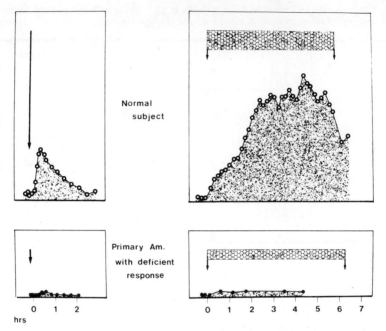

Fig.4 Comparison between responses to rapid i.v. injection of LH-RH (single arrow) and to prolonged LH-RH infusion (400 mcg × 6 h) obtained in a normal subject and in a patient affected by P.A. and with deficient response to LH-RH test.

precise localization of the primary defect at either the hypothalamic or pituitary level, the latter being characterized by a lack of response to LH-RH (Czygan et al., 1974). Considering our results, should this hypothesis be correct, our subjects with 'deficient' responses to LH-RH rapid i.v. test should have been considered as primarily affected at the pituitary gonadotrope. However, in the light of the role that LH-RH may have in inducing not only the release, but also the synthesis of LH (Schally et al., 1973), doubts arise as to whether this defect is an expression of a primary pituitary lesion or a consequence of higher centers hypofunction.

To verify this hypothesis patients with deficient responses to a first provocative LH-RH test have been treated with the synthetic decapeptide infused for 8 h daily for 3 days. Control tests were performed on successive days with equivalent rapid i.v. LH-RH tests. Figure 5 shows the results obtained using this treatment in 10 patients so far studied. As may be seen, an increment in the LH response has been obtained in all cases but one (patient No.10), with doses ranging from 400 to 25 mcg infusion per day. 'Normalization' of the responses has been obtained in all 5 cases treated with 400 mcg of LH-RH per infusion, and in 2 cases treated with 100 and 50 mcg (patients 8 and 7). These findings extend and confirm our previous report on the possibility of enhancing the pituitary response by administering LH-RH (Serra et al., 1975).

If the basal tests demonstrated that the deficient responses were a steady endocrine character in our patients, control tests performed on successive days have shown that the enhancement of the pituitary response to normal levels remains stable for at least 9-10 days (Fig.6) before returning to the pretreatment

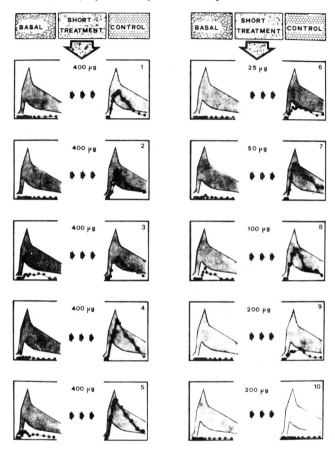

Fig.5 The LH responses to the rapid i.v. injection of LH-RH in 10 patients with P.A. and deficient responses, stimulated before (left) and after (right) 3 days of slow infusion with LH-RH (at doses of 400, 200, 100, 50 and 25 mcg). Dashed areas indicate the range of responses found in normal subjects after single injection of 25 mcg of LH-RH.

levels, as found 2 months later (Fig.7).

Although the FSH response after 25 mcg of LH-RH even in normal subjects appear to be very poor, in some of these subjects with 'deficient' LH responses, it appeared to be significant. No changes however were observed before and after treatment (Fig.8).

Total estrogens, measured before, during and after the LH-RH infusions, failed to show significant changes in their excretion (Fig.9).

The large percentage of cases in which the responses found were 'normal' in spite of the presence of P.A., the failure to induce ovulation in some of these patients using high doses of LH-RH, and finally the permanence of the clinical symptoms even after the 'normalization' obtained after the 3 days treatment with LH-RH infusions, have aroused several doubts as to whether a 'normalized' response could be considered proof of an acquired normal pituitary function. In this regard it seemed of interest to prolong the treatment checking if, to the

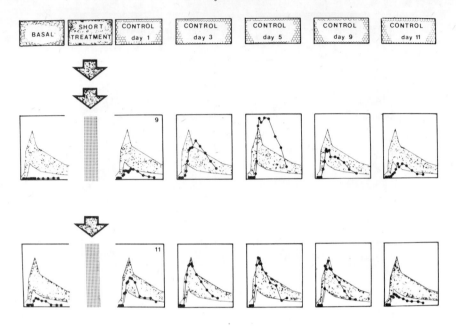

Fig. 6 Basal LH responses before (left) the slow infusion with LH-RH × 3 days in 2 patients with P.A. Control tests have been performed on successive days. Dashed areas indicate the range of responses found in normal subjects after a single injection of 25 mcg of LH-RH.

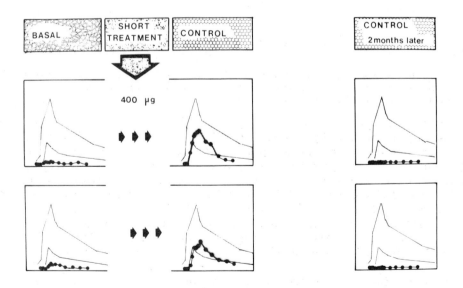

Fig. 7 LH responses before (left) and 12 h after treatment with LH-RH infusion × 3 days in 2 patients with P.A. On the right the LH responses found two months later.

Modification of Gonadotropin Release

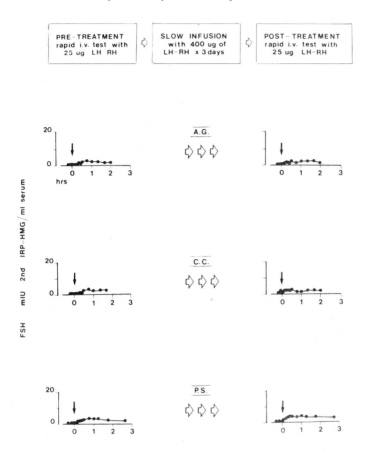

Fig. 8 The FSH responses to the rapid i.v. injection of LH-RH in 3 patients with hypogonadotropic hypogonadism and P.A., stimulated before (left) and after (right) 3 days of slow and prolonged infusion with LH-RH (Serra *et al.*, 1975).

'normalization' of the rapid i.v. test, corresponded a 'normalization' of the pituitary discharge test obtained by long-term LH-RH infusion. Two patients with P.A. and 'deficient' responses to a first provocative test have been therefore treated with varying and increasing doses of LH-RH, from 500 mcg to the dose of 5 mg per day, received during the whole treatment a total of 22.5 mg of synthetic decapeptide. Figure 10 shows the modification of the rapid i.v. tests, performed before the daily infusions of LH-RH, obtained on days 1-4-7-9 and 13. As may be seen, starting from deficient values, 'normalization' of the responses progressively occurs during the treatment. When considering the LH values measured during the 1st LH-RH infusion, it must be noted that the values of LH remain very low in spite of the treatment with the synthetic decapeptide and that this deficiency in the release remains throughout the period of the study although higher levels of LH could have been measured on days 9 and 15. The discrepancy in day 9 between the rapid i.v. test and the discharge test seems of importance: while the first appears to be within the 'normal' range, the second shows clear signs of relative pituitary

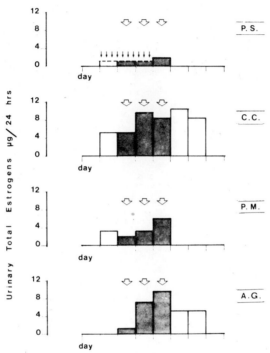

Fig. 9 Urinary steroid excretion before, during and after slow and prolonged infusion of 400 mcg of LH-RH × 3 days in 4 patients with hypogonadotropic hypogonadism and P.A.. Dashed areas indicate the days of infusion. Small and black arrows designate all values below 2 mcg/24 h (by Serra *et al.*, 1975).

exhaustion, confirmed on a successive discharge test performed on day 15.

Total estrogens measured in urine and the 'ferning' of the cervical mucus remained unchanged during the treatment.

Figure 11 shows similar results obtained in a second patient. Again, a significant enhancement to the acute stimulus, as shown by the serial tests performed on days 1-4-7-9 and 13, can be noted. A significant increase is present also on the discharge test monitored on days 1-4-9 and 15. Again, however, a clear exhaustion of LH release during the LH-RH infusion is evident.

This exhaustion of the pituitary gland on the release of LH is clearly in contrast with our findings in normal subjects (Fig.4), and may be considered as a sign of pituitary hypofunction. Considering the possibility of having an abnormal response to a long-term LH-RH infusion even in presence of a 'normal' response to the acute stimulus, the 'infusion discharge test' appears to be more appropriate in demonstrating a lesion of the diencephalic-pituitary system, and, *vice versa*, a 'normal' response to an acute stimulus may not be assumed as proof of normal

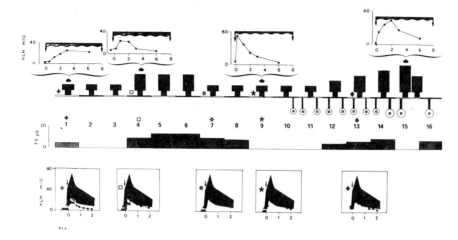

Fig.10 Daily treatment × 8 h with varying doses of LH-RH in a patient with hypogonadotropic hypogonadism and P.A.. Infusions of LH-RH were performed with doses of 500 mcg i.v. on days 1, 2, 3: 1500 mcg i.v. on days 4,5,6; 500 mcg i.v. on days 7,8,9; 500 mcg i.v. plus 500 i.m. on day 10; 500 mcg i.v. plus 1000 mcg i.m. on day 11; 1000 mcg i.v. plus 1000 mcg i.m. on day 12; 1500 mcg i.v. plus 1000 mcg on day 13; 2000 mcg i.v. plus 1500 mcg i.m. on day 14; 4000 mcg i.v. plus 1000 mcg i.m. on day 15; 1000 mcg i.m. on day 16. Rapid i.v. tests with LH-RH are shown below, and pituitary discharge tests on the upper part.

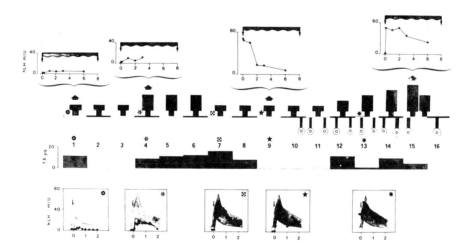

Fig.11 See legend Fig.10.

pituitary function.

As the 'deficient' response exhibited a major frequency in patients affected by P.A., the 'exaggerated' responses have been found exclusively in subjects affected by S.A. and oligomenorrhea.

Again, as demonstrated in the case of 'deficient' responses, daily administration of equivalent doses of LH-RH elicited LH releases qualitatively similar (Fig.12).

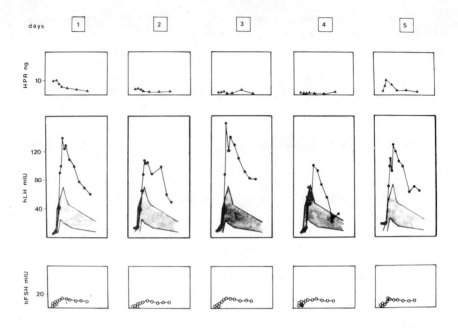

Fig.12 "Exaggerated" LH responses (closed circles) after daily administration of 25 mcg of LH-RH in a patient affected by S.A.. Open circles indicate FSH levels and triangles indicate HPRL levels.

Thus also proving that the 'exaggerated' release may be considered a relatively constant characteristic of the endogenous hormonal environment. Although, in our study, using 25 mcg of LH-RH, the 'exaggerated' responses have been found only for LH, the possibility of an FSH hyper-response cannot be excluded.

Considering the P.A. and the S.A. as relatively similar clinical syndromes, in the latter group of patients a deficiency in endogenous LH-RH secretion has been suspected and the increased release interpreted as expression of an enhanced pituitary responsiveness. These patients, therefore, in analogy with the treatment used for subjects with P.A. have also been treated with LH-RH infusion for 8 h for 3 days.

Figure 13 shows the patterns of release found in 5 patients both before and 12 h after treatment with LH-RH. As can be seen in all cases 'normalization' of the responses was obtained. Moreover, in two of these patients restoration of ovulatory cycles was observed. These data seem to validate the hypothesis of a reduced secretion of endogenous LH-RH as primary cause of the 'exaggerated' response.

On the basis of the effects that ovarian steroids have on pituitary release (Jaffe and Keye, 1974; Keye and Jaffe, 1974) the possibility exists that long-term infusion may produce an increment of the LH plasma levels, able to stimulate steroidogenesis and in turn reduce pituitary responsiveness. Figure 14 shows the excretion of urinary estrogens in two subjects during the period of study. It may be seen that in one case the excretion of these steroids exhibited only a slight increase, unlikely to be of significance. Further investigation however is in progress

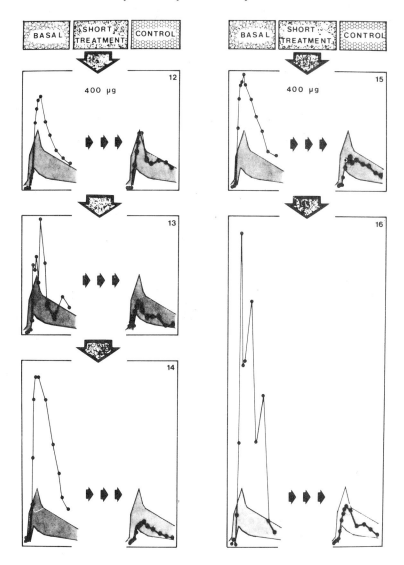

Fig.13 The LH responses to the rapid i.v. injection of LH-RH in 5 patients with S.A. and exaggerated responses, stimulated before (left) and after (right) 3 days of short LH-RH treatment (400 mcg infused daily × 3 days × 6 h).

measuring estrogen levels in plasma.

Although several alternative speculations may be proposed to understand the 'exaggerated' release and its 'normalization' after LH-RH treatment, the hypothesis based on a change in pituitary responsiveness produced by exogenous or endogenous LH-RH is well in agreement with previous reports from this laboratory on the increase or decrease of pituitary responsiveness following different LH-RH acute stimulations (Serra and Lafuenti, 1974). Further support of this hypothesis derives from the work of Roth and his group (Gavin *et al.*, 1974)

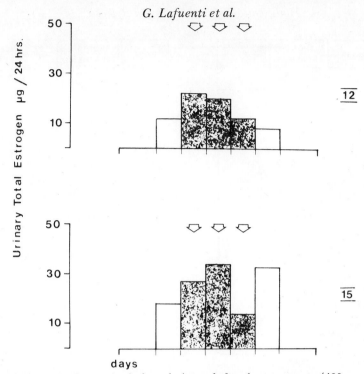

Fig.14 Urinary steroid excretion before, during and after short treatment (400 mcg of LH-RH infused × 3 days × 6 h) in 2 patients (No.12 and No.15 of Fig.13) with hypogonadotropic hypogonadism, S.A. and pretreatment exaggerated responses. Dashed areas indicate the days of infusion.

demonstrating *in vitro* a reduction of hormonal receptor concentrations following chronic exposure to large amount of insulin, thus suggesting a general reverse relationship between circulating hormonal levels and the number of receptors. Following this hypothesis it could be assumed that, as shown in Fig.15A, to 'normal' endogenous LH-RH levels would correspond a particular receptor concentration so that an exogenous stimulus would produce what we consider a 'normal' response. In the presence of a reduced LH-RH secretion (Fig.15B) the receptor concentration would increase, thus maintaining a 'normal' or slightly reduced plasma level of LH. An exogenous stimulus, however, acting upon an increased number of receptors, would produce in this case an 'exaggerated' response. Exposure to elevated LH-RH levels, as during the infusion treatment, could in turn reduce the number of receptors, leading to a 'normal' response to an acute stimulus.

In normal subjects, as shown in Fig.16, 12 h after the 3 days treatment with LH-RH, the pituitary response to an acute stimulus with synthetic LH-RH appears to be within the normal range. This could imply that, in the presence of a 'normal' diencephalic-pituitary axis, 12 h after the treatment, the pituitary responsiveness was grossly unchanged, even if a second treatment was performed.

Interesting results have been obtained in patients with an 'exaggerated' response submitted to two successive LH-RH treatments. As can be seen in Fig.17, once the LH release was 'normalized' a second equivalent treatment did not fur-

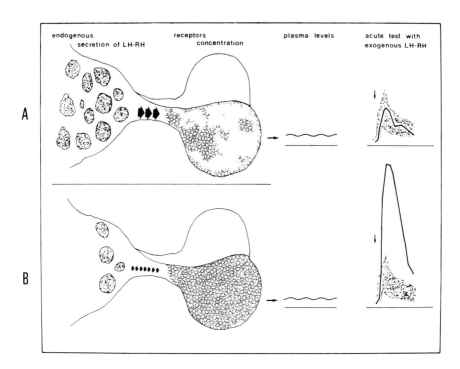

Fig.15 See the text.

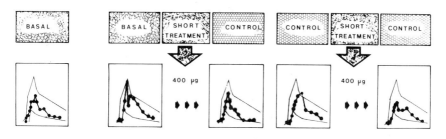

Fig.16 LH responses in a normal subject with exaggerated responses before and after two successive treatments with 400 mcg of LH-RH infused daily × 3 days × 6 h.

ther modify the pituitary responsiveness that remained within the range of normality. This may suggest that when the pituitary gland is exposed to a particular concentration of exogenous LH-RH it acquires a responsiveness no more related to the initial production of endogenous LH-RH, but mainly dependent on the new levels produced.

CONCLUSIONS

The acute stimulation with LH-RH has produced 3 different response patterns characterized by a 'normal' or 'deficient' or 'exaggerated' release of LH in

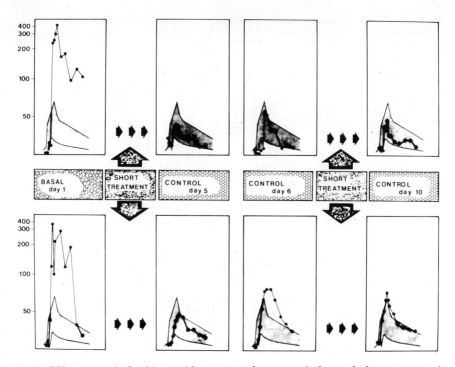

Fig.17 LH responses in 2 subjects with exaggerated responses before and after two successive treatments with 400 mcg of LH-RH infused daily × 3 days × 6 h.

response to a bolus of synthetic LH-RH. On the basis of our knowledge, however, these responses are not able to differentiate between hypothalamic or pituitary primary failure. Since 'deficient' and 'exaggerated' responses appear to be susceptible to 'normalization' after treatment with LH-RH it seems possible to hypothesize a hypothalamic or higher centers defect in most of the cases. Failure to obtain regression of the clinical syndromes in cases with P.A. even after prolonged treatment with LH-RH leads to the hypothesis that chronic deprivation of LH-RH may have produced an irreversible lesion in the pituitary gland. On the other hand, restoration of ovulatory cycles in two of our patients with S.A. suggests that a minor defect may have been involved in these cases.

In some subjects, the finding of a 'normal' response to an acute stimulus and of an 'exhausted' release during a prolonged infusion with the synthetic decapeptide, seems to demonstrate that 'normal' response to the acute stimulus may not be assumed as proof of normal pituitary function. In other patients, however, the 'normal' response may be expression of an inhibited, although maintained, hypothalamic-pituitary gonadotrophin function. This can be assumed in those cases with galactorrhea in which treatment with CB154 either acting upon prolactin or ovarian cells, restores cyclicity of the gonadotropin pattern, producing ovulation as demonstrated by the consequent pregnancy (Fig.18).

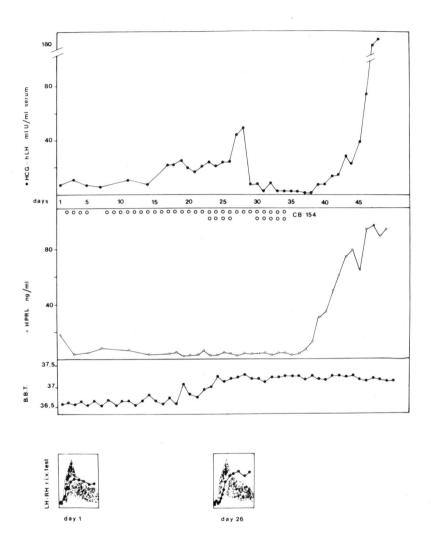

Fig.18 Induced ovulation and successive pregnancy in a patient affected by amenorrhea and galactorrhea with 'normal' LH response to 25 mcg of LH-RH before (left) and during (right) CB154 treatment.

REFERENCES

Czygan, P.G., Breckwoldt, M., Lehmann, F., Langefeld, R. and Bettendorf, G. (1974). *Acta endocr. (Copenh.)* **75**, 428-434.
Gavin, J.R., Roth, J., Neville, D.M., de Meytes, P. and Buell, D.N. (1974). *Proc. natn. Acad. Sci. U.S.A.* **71**, 84-88.
Jaffe, R.B. and Keye, W.R, Jr. (1974). *J. clin. Endocr. Metab.* **39**, 850.
Keye, W.R., Jr. and Jaffe, R.B. (1974). *J. clin. Endocr. Metab.* **38**, 805.

Lafuenti, G., Serra, G.B. and Palla, G.P. (1973). *Folia Endocrin. (Rome)* **26**, 180.
Schally, A.V., Arimura, A., Debeljuk, L., Redding, T.W. and Reeves, J.J. (1973). *In* "Hypothalamic-hypophysiotropic Hormones" (E. Rosenberg, ed) pp.3-11. Excerpta Medica, Amsterdam.
Schwarzstein, L., de Laborde, N.P., Aparicio, N.J., Turner, D., Mirkin, A., Rodriguez Lhullier, F. and Rosner, J.M. (1975). *J. clin. Endocr. Metab.* **40**, 313.
Serra, G.B. and Lafuenti, G. (1974). *Folia Endocrin. (Rome)* **27**, 158-167.
Serra, G.B., Muscatello, P., Menini, E., Lafuenti, G. and Caniglia, R. (1975). *Obstet. Gynec.* **45**, 523.

DISORDERS OF THE HYPOTHALAMIC-PITUITARY-OVARIAN AXIS IN ANOREXIA NERVOSA

A. Borghi, G. Forti, S. Fusi, C. Vigiani, E. Calabresi and G. Cattani

INTRODUCTION

"True" or "primary" anorexia nervosa is a peculiar form of psychosomatic disease characterized by an active refusal of food which gradually leads as an aim and as a consequence to a very significant loss of body weight.

This disease is typical of girls and young women, the age of onset lying between 13 and 20 years, and amenorrhea is a constant feature: normal menstrual function in anorexia nervosa is exceptional.

The whole clinical picture is potentially reversible even in the most serious and long-lasting cases: this fact suggests that amenorrhea is mainly due to a *functional* disorder of the hypothalamo-pituitary-ovarian axis.

Nevertheless to date there is no agreement on whether this disorder is due to psychological factors or is a consequence of starvation. On the basis of our studies we think that the first interpretation is true for some patients, the second interpretation is true for other patients, and both interpretations are true in other cases.

AMENORRHEA: TIME OF ONSET

Amenorrhea is primary in a few cases, when the disease began before puberty, but in most cases amenorrhea is secondary, and often occurs suddenly after a normal menarche and after a period of normal menstrual function.

This fact seems to indicate that the hypothalamic-pituitary-ovarian axis of these patients is perfectly normal before the onset of the disease; a potentially normal function of the neuro-endocrine system is also confirmed by the fact that

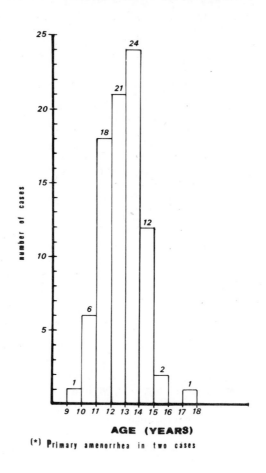

Fig. 1

menstruation and ovulation restart when the physical and psychological conditions of these patients improve.

The onset of amenorrhea can precede by many months, or be contemporary with, or follow the weight loss and the active refusal of food.

In our study amenorrhea preceded the onset of the other symptoms in 17.1% of cases, was contemporary in 48.7% and followed in 34.1%: these percentages are in agreement with the data in the literature (Borghi et al., 1975).

The early onset of amenorrhea in a large number of cases proves that anorexia nervosa is not a simple matter of undernourishment. Indeed, it seems correct to consider amenorrhea as a consequence of starvation when it occurs after a far advanced loss of weight (as happens in the so-called "hunger disease"). But in the other cases amenorrhea seems due to psychological factors, even if later it can be maintained by starvation; this interpretation is supported by the fact that menses

PLASMA LEVELS OF ESTROGENS IN ANOREXIA NERVOSA

CASE	AGE (YEARS)	AMENORRHEA (YEARS)	ESTRONE ng%ml	ESTRADIOL-17β ng%ml
F. S.	15	½	1.90	1.70
V. T.	18	2	2.00	4.10
B. R.	19	5	1.50	1.37
L. C.	19	2	2.20	2.90
P. E.	20	3	1.50	1.30
B. L.	22	3	1.50	0.98
NORMAL WOMEN (EARLY FOLLICULAR PHASE)			3.0 ± 2.6	6.35 ± 5.30
ANOREXIA NERVOSA			1.7 ± 0.3	2.05 ± 1.20

Fig. 2

resume only when the body weight of these patients has returned to normal.

HORMONAL STUDIES

To date only a few hormonal studies, performed with R.I.A. methods, are available on hypothalamic-pituitary-ovarian function in anorexia nervosa.

A) *Estrogens*

Urinary estrogens are usually very low and constant: urinary levels of estrone, estriol and estradiol-17β are similar to those found in the post-menopausal period, in agreement with atrophy of vaginal epithelium and of endometrium, typically found in anorexia nervosa.

Very low levels of plasma estrogens have been found by some authors (Franchimont, 1975; Mecklenburg *et al.*, 1974; Travaglini *et al.*, 1974); the recovery in general condition is followed by normalization of the plasma estrogens (Beumont *et al.*, 1974).

In 6 cases of anorexia nervosa with a 6 months to 5 years period of amenorrhea, we found low mean values of estrone and estradiol-17β, but not significantly different from the mean values found in 5 normal women in the early follicular phase.

**PLASMA LEVELS OF FSH AND LH
IN ANOREXIA NERVOSA**

CASE	AGE (YEARS)	AMENORRHEA (YEARS)	WEIGHT LOSS	NUMBER of ASSAYS	FSH ng/ml (mean)	LH ng/ml (mean)
C. A.	18	2	−36	2	1,09	0,78
B. R.	19	5	−21	7	4,54	0,84
B. P.	19	3/4	−42	5	0,84	1,42
C. P.	20	3/4	−18	2	1,92	0,85
B. L.	22	3	−41	2	1,45	0,22
S.M.G.	34	7	−21	2	1,55	1,04
NORMAL WOMEN (EARLY FOLLICULAR PHASE)					2,40 ± 0,32	0,99 ± 0,20
ANOREXIA NERVOSA					2,22 ± 1,94	0,94 ± 0,20

Fig.3

Nevertheless these data are not sufficient to confirm that estrogens are produced by the ovary: in fact they could also be of adrenal origin.

Moreover we think that single determinations of these hormones could not give reliable results. In fact we measured estrone and estradiol-17β for seven consecutive days in a patient; as shown in Fig.3, estradiol-17β levels were nearly constant, but those of estrone showed large variations from day to day.

This particular phenomenon is probably related both to episodic secretion of these hormones and to modifications of their metabolic clearance rate.

B) *Gonadotrophins*

Urinary gonadotrophin levels are usually very low, with the exception of a few cases.

Studies on circulating levels of gonadotrophins are rare: some authors found very low and constant levels of LH, while FSH levels were often less reduced (Beumont et al., 1974; Lundberg et al., 1972; Travaglini et al., 1974; Warren and Vande Wiele, 1973; Warren et al., 1975). The recovery in general condition is followed by normalization of the FSH and LH values (Beumont et al., 1974; Frankel and Jenkins, 1975).

In our laboratory we measured LH and FSH levels by radioimmunoassay for two or more consecutive days in 6 patients: the mean values were low, but not significantly different from the mean values found in 8 normal women in the early follicular phase.

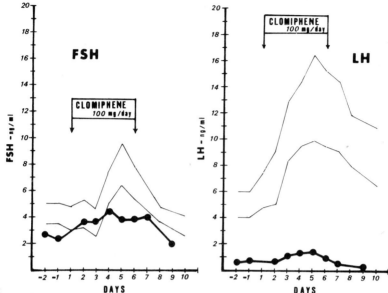

Fig. 4

DYNAMIC TESTS

There is general agreement on the central origin of amenorrhea in anorexia nervosa. Recent studies have demonstrated that the functional disorder is mainly at the hypothalamic level.

A) *Response to Clomiphene*

Clomiphene administration induces neither a rise in plasma gonadotrophin levels nor ovulation in the stationary phase of the disease; this treatment, on the contrary, becomes efficient as soon as there is an improvement in the psychological condition and a considerable weight gain of the patients (Besser, 1975; Beumont et al., 1973; Marshall and Russell-Fraser, 1971; Noacco, 1972; Travaglini et al., 1974).

These data show that hypothalamic activity, impaired in the acute phase of the disease, returns toward normality with the improvement of the general condition of these patients; a positive response to clomiphene treatment is considered a good prognostic element by some authors (Besser, 1975; Franchimont et al., 1974).

B) *Responsiveness to LH-RH*

The lack of action of clomiphene could be due to primary hypothalamic impairment, but also to impaired pituitary responsiveness: the results obtained with LH-RH stimulation show that the pituitary response can be very variable: usually normal (Beumont et al., 1973; Mecklenburg et al., 1974; Wiegelmann and Solbach, 1972), sometimes it can be increased (Besser, 1975, Franchimont et al., 1975), and sometimes it can be very poor (Besser, 1975; Yoshimoto et al., 1975;

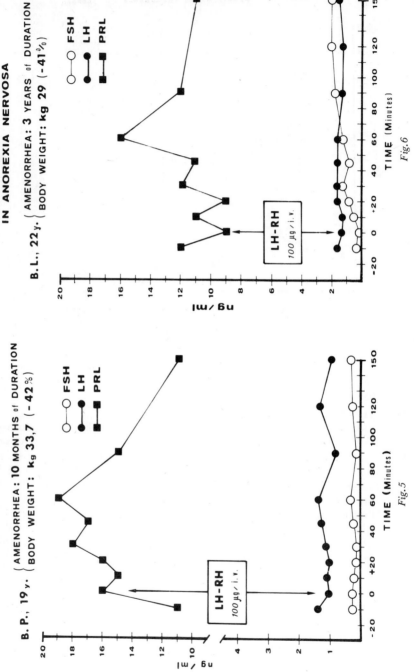

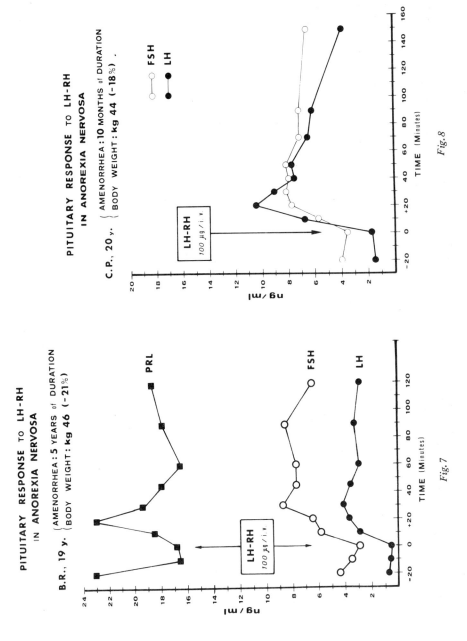

Fig. 7

Fig. 8

PNEUMOPELVOGRAPHIC PATTERNS
IN ANOREXIA NERVOSA

	UTERUS	OVARIES	Number of Cases
I	Normal Size	Enlarged	1
II	Normal Size	Normal Size	1
III	Small	Normal Size	3
IV	VERY SMALL	VERY SMALL	22 (♦)

(♦) One patient showed a dermoid cyst at the right ovary

Fig. 9

Frankel and Jenkins, 1975; Warren et al., 1975).

Yoshimoto et al. (1975) correlated this variability of response with the duration of the disease and particularly with the duration of amenorrhea: in our opinion this correlation should be made also and especially with the degree of weight loss.

No LH and FSH release can be seen after LH-RH stimulation in severely underweight patients.

A significant rise of FSH and LH is shown in other patients in rather good general condition, even with long-standing amenorrhea.

In the last patient a peculiar feature reported by other authors is also evident: a significant rise of FSH with a poor response of LH to LH-RH stimulation.

This type of response to LH-RH is quite similar to that reported in girls before puberty: therefore the pituitary responsiveness to LH-RH in anorexia nervosa would revert to the typical prepubertal "functional immaturity" (Franchimont et al., 1974).

These data agree with those of Boyar et al. (1974), who observed a 24-hour secretory pattern of FSH and LH in anorexia nervosa quite similar to that of prepuberal or puberal girls. After recovery from disease this gonadotrophin secretory pattern returns to normal, showing that this hypothalamic-pituitary disorder is reversible, and therefore functional.

Yoshimoto et al. (1975) reported that the pituitary response gradually increases after prolonged stimulation with LH-RH (from 3 to 6 days), and sometimes became supra-normal: for this reason we think that impaired pituitary responsiveness in the stationary phase of anorexia nervosa is probably due to the absence of a normal hypothalamic function, to the low estrogen levels (quite similar to the early follicular phase), and especially to the effects of starvation.

Fig.10

In fact, Warren et al. (1975) reported that after LH-RH stimulation, even in high doses, gonadotrophin release is practically absent when the loss of weight is 25% or more, but recovers with the improvement in general condition. Travaglini et al. (1974) reported that menstrual bleeding restarted in four of seven patients (after refeeding), 15-20 days after a single administration of LH-RH.

All these data supported the view of a predominant impairment of the hypothalamic-pituitary axis in anorexia nervosa. From other data we may also suspect impaired responsiveness of target organs and tissues to the hormones. Very often in fact, in the stationary phase of the disease, the ovaries of these patients do not respond to gonadotrophin administration, while the endometrium and the vaginal epithelium do not respond to natural estrogens. Therefore some authors consider that the central nervous system could have a direct inhibiting action on the target tissues of estrogens and gonadotrophins (Netter et al., 1957).

PNEUMOPELVOGRAPHIC PATTERNS

This interpretation is supported by the observation that the utero-ovarian hypoplasia found in these patients is much more marked than that found in secondary amenorrhea merely due to an acquired gonadotrophin defect.

Only few data are available in the literature on this matter, and all concern autopsy cases: a marked hypoplasia of uterus and ovaries was always present, and on microscopic examination a complete lack of follicular maturation was found.

As these data concern only severely underweight subjects, we could suppose

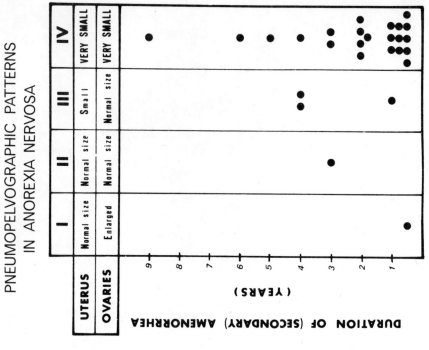

Fig. 12

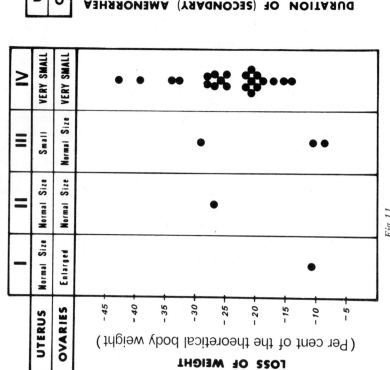

Fig. 11

PLASMA LEVELS OF TESTOSTERONE IN ANOREXIA NERVOSA

Case	Age (years)	Weight Loss	TESTOSTERONE ng/100ml	urinary 17-KS mg/24h
F. S.	15	-27%	<10	5,4
V. T.	18	-39%	<10	3,8
L. C.	19	-16%	12	4,8
C. A.	18	-36%	13	3,3
T. G.	19	-17%	13	8,4
B. L.	22	-41%	17	2,9
P. E.	20	-33%	17	5,2
C. P.	20	-18%	25	7,9
B. R.	17	-28%	28	5,7
S. M. G.	34	-21%	30	9,6
T. L.	27	-43%	44	4,8
B. E.	18	-27%	80	12,2

Fig. 14

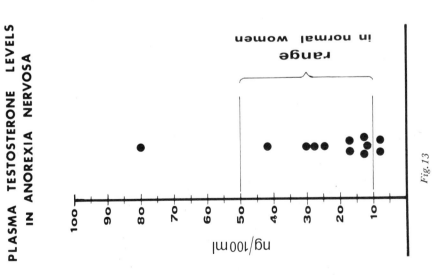

Fig. 13

that the hypoplasia of the internal genitalia occurs only in the terminal stage of the disease, as in the so-called "hunger disease".

To clarify this aspect, we performed pneumopelvography in 27 random patients. Results are shown in Fig.11.

The marked hypoplasia of the uterus and of the ovaries found in 85% of patients did not correlate either with the degree of weight loss or with the duration of amenorrhea.

In Fig.12 it is also evident that this marked hypoplasia of uterus and ovaries is found also in patients with amenorrhea of only a few months.

As we have no data on the pneumopelvographic pattern of these patients before the onset of the disease, we cannot draw any conclusion: however, as all of our patients had previously normal menstrual function, it seems logical to think that the utero-ovarian regression has been very rapid, and probably due not only to endocrine factors.

This observation seems to be supported by the fact that, in other forms of hypothalamic amenorrhea without anorexia, even of long duration, the pneumopelvographic aspect of the internal genitalia was nearly always normal.

PROLACTIN LEVELS

Plasma prolactin levels have been recently measured by radioimmunoassay in patients with anorexia nervosa, and have been found normal (Beumont et al., 1974; Mecklenburg et al., 1974), or elevated (Travaglini et al., 1974).

Prolactin levels measured by radioimmunoassay in our laboratory in 5 cases were normal; repeated measurements showed large fluctuations from day to day in one case, and no significant variations in another.

The normal levels of prolactin could contribute to an explanation of the lack of mammary regression in most cases of anorexia nervosa: breast hypotrophia, when present, is due only to the reduction of fat tissue, which is never as reduced in the breasts as in the rest of the body. The mammary gland is normal as shown by mammographic observations, showing a particular "ivory-like" aspect, quite similar to that of puberal breasts.

PLASMA TESTOSTERONE LEVELS

Some years ago Graef and Winckler (1971) reported an increased urinary testosterone-glucuronide excretion in 6 cases of anorexia nervosa. Crisp and Casey (1965) found plasma testosterone levels at the upper limits of the normal range in 5 cases, while Travaglini et al. (1974) found normal levels in 16 cases. As the ovarian function of these patients is very low or absent, most probably testosterone production comes from peripheral conversion of precursors secreted by the adrenals.

In our laboratory, as shown in Fig.13, we found normal plasma testosterone levels in 9 cases, low levels in 2, and only one high value.

Plasma testosterone levels do not correlate with either the 17-ketosteroid excretion, or with the duration of the disease, or the clinical aspect of these patients, particularly the presence or absence of the hirsutism of the so-called "lanugo type" so often found in these patients.

These data are of interest as some authors ascribed the above-mentioned "lanugo" to an increased androgen production. But this form of hirsutism has not the features of the forms due to an increased androgen production.

DISCUSSION

In conclusion, the function of the hypothalamic-pituitary-ovarian axis seems to be severely impaired in the stationary phase of anorexia nervosa.

There is no doubt that psychological factors can induce a direct inhibitory effect on the hypothalamic structures regulating gonadotrophin release in most cases; nevertheless it is also true that after the onset of the disease the effects of starvation are important and sometimes become prevalent at the pituitary level.

In fact, the impairment of the pituitary responsiveness is gradually restored when body weight nears the normal value.

However, as in the stationary phase of the disease, the cyclic administration of gonadotrophins at physiological doses is nearly always ineffective, we can also deduce the existence of an impaired ovarian responsiveness, partly due to a central inhibition and partly to the starvation effects.

However, we want to underline, once again, that *all* these modifications are potentially reversible. In our opinion this fact proves that the hypothalamic-pituitary-ovarian modifications found during the disease are always functional, even if due to a different pathogenesis.

REFERENCES

Besser, G.M. (1975). This volume, p.261.
Beumont, P.J.V., Carr, P.J. and Gelder, M.G. (1973). *Psychol. Med.* 3, 495.
Beumont, P.V.J., Friesen, H.G., Gelder, M.G. and Kolakowska, T. (1974). *Psychol. Med.* 4, 219.
Borghi, A., Salvestroni, R. and Fusi, S. (1975). "L'anoressia nervosa". Società Editrice Universo, Rome.
Boyar, R.M., Katz, J., Finkelstein, J.W., Kapen, S., Weiner, H., Weitzmann, E.D. and Hellman, L. (1974). *New Engl. J. Med.* 291, 861.
Crisp, A.H. and Casey, J.H. (1965). (Unpublished data, quoted by Crisp, A.H. (1970). *World Rec. Nutr. Diet.* 12, 452.)
Franchimont, P., Valcke, J.C., Schellens, A.M.C., Demoulin, A. and Legros, J.J. (1973). *Ann. Endocrin.* 34, 491.
Franchimont, P., Valcke, J.C. and Lambotte, R. (1974). "Female gonadal dysfunction. Clinics in Endocrinology and Metabolism" Vol.3, No.3, p.533. Saunders, London.
Franchimont, P. (1975). "Simposio su Problemi di Endocrinologica Clinica Ipotalamo-ipofisaria", Turin, June 17.
Frankel, R.J. and Jenkins, J.S. (1975). *Acta endocr. (Copenh.)* 78, 209.
Frisch, R.E. and Revelle, R. (1971). *Arch. Dis. Child.* 46, 695.
Graef, V. and Winkler, G. (1971). *Klin. Wschr.* 49, 881.
Johanson, A. (1974). *New Engl. J. Med.* 271, 904.
Lundberg, P.Q., Walinder, J., Werner, L. and Wide, L. (1972). *Europ. J. clin. Invest.* 2, 150.
Marshall, J.C. and Russell Fraser, T. (1971). *Brit. Med. J.* IV, 590.
Mecklenburg, R.S., Loriaux, D.L., Thompson, R.H., Andersen, A.E. and Lipsett, M.B. (1974). *Medicine (Baltimore)* 53, 147.
Netter, A., Lambert-Netter, A., Lumbroso, P., Mantel, O. and Faure (1957). *Ann. Endocrin.* 18, 1014.
Noacco, C. (1972). "L'amenorrea dell'anoressia psicogena primaria: risposta gonadotropinica al trattamento con clomifene", Tesi di Specializzazione della Scuola di Endocrinologia dell'Università di Firenze.
Travaglini, P., Ferrari, C., Ambrosi, S., Severgnini, A., Beck-Peccoz, P., Paracchi, A. and Faglia, G. (1974). "Alcuni aspetti della funzione ipotalamo-ipofisaria nell'anoressia nervosa". Giornate Endocrinologiche Senesi, Siena, December 5-7.

Warren, M.P. and Vande Wiele, R.L. (1973). *Am. J. Obstet. Gynec.* **117**, 435.
Warren, M.P., Jewelewicz, R., Dyrenfurth, I., Ans, R., Khalaf, S. and Vande Wiele, R.L. (1975). *J. clin. Endocr. Metab.* **40**, 601.
Wiegelmann, W. and Solbach, H.G. (1972). *Horm. Metab. Res.* **4**, 404.
Yoshimoto, Y., Moridera, K. and Imura, H. (1975). *New Eng. J. Med.* **292**, 242.

PROLACTIN AND OVARIAN FUNCTION

R. Rolland[1]*, J.M. Hammond[1], L.A. Schellekens[2], R.M. Lequin[3] and F.H. de Jong[4]

[1] *Department of Medicine, Division of Endocrinology*
The Milton S. Hersey Medical Center
The Pennsylvania State University, Hershey, Pa. 17033, USA

[2] *Department of Obstetrics and Gynecology*
De Wever Hospital, Heerlen, The Netherlands

[3] *Department of Obstetrics and Gynecology*
University Hospital, Nijmegen, The Netherlands

[4] *Department of Biochemistry*
(Division of Chemical Endocrinology), Medical Faculty
Erasmus University, Rotterdam, The Netherlands

INTRODUCTION

Only since 1971 has the existence of prolactin as a separate hormone in humans been generally accepted (Friesen *et al.*, 1970; Lewis *et al.*, 1971; Hwang *et al.*, 1972). This led to the immediate development of specific radioimmunoassays capable of measuring circulating levels of this hormone in plasma (Hwang *et al.*, 1971; Sinha *et al.*, 1973). It has been shown that elevated levels of prolactin are commonly associated with disorders of ovarian function although the pathophysiological basis for this is still not clear (Rolland *et al.*, 1974; Del Pozo *et al.*, 1974). From a theoretical point of view, prolactin could interfere with gonadal function at at least three different levels. These include the hypothalamic, the hypophyseal, and the ovarian level as illustrated in Fig.1.

The Hypothalamic Level: Pituitary prolactin secretion appears to be under control of a Prolactin Inhibiting Factor from the hypothalamus (Sulman, 1970). Evidence suggests that dopaminergic pathways are involved in the hypothalamic

---

* Fellow from The Netherlands Organization for the Advancement of Pure Research (Z.W.O.).

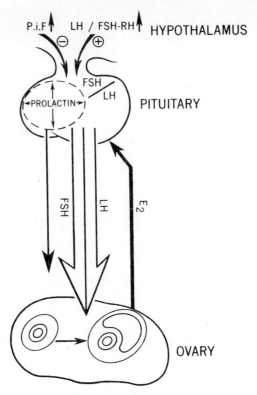

Fig.1 Possible mechanisms by which prolactin may interfere with gonadal function. (For further details, see text.)

regulation of both PIF and gonadotropin-releasing hormone (GRH) (Kambergi et al., 1970, 1971a,b). Although not proved, it has been postulated that the release of PIF parallels the release of GRH and *vice versa* (Ganong, 1972; Varga et al., 1973). Thus, an increase in pituitary released gonadotropins would be consistent with low peripheral levels of prolactin; hyperprolactinemia, on the other hand, would be associated with hypogonadotropism.

The Hypophyseal Level: Pituitary tumors are known to lead to hypopituitarism due to compression of remaining tissue (Batzdorf and Stern, 1973; Nieman et al., 1967). Growth hormone and gonadotropin secretion usually decrease first followed at a later stage by TSH and ACTH. If the tumor is a prolactin producing adenoma as indicated in Fig.1, hyperprolactinemia in association with hypogonadotropism could be expected.

The Ovarian Level: Prolactin could also have a direct influence on the ovary, interfering with steroidogenesis, with follicular maturation or with both. Hyperprolactinema in this case might be associated with normal or high gonadotropin levels. There is also abundant evidence of a facilitory role of gonadal steroids on LH secretion (Tsai and Yen, 1971; Yen et al., 1972). If prolactin blocked this positive feedback loop, LH values might fall despite reduced ovarian function.

Although hypothalamic or pituitary factors may be of importance in patients with hyperprolactinemia and gonadal dysfunction, the emphasis of the present report will be on evidence from our clinical and *in vitro* studies which indicate a direct effect of prolactin on the ovary. In many other species this hormone is known to act on the ovary in addition to stimulating the mammary gland (Denamur *et al.*, 1973; Nicoll and Bern, 1972).

CLINICAL STUDIES

Prolactin and Menstrual Disturbances

The syndromes of galactorrhea and oligo- or amenorrhea have been recognized for many years (Chiari *et al.*, 1958; Frommel, 1958; Argonz and Del Castillo, 1953; Forbes *et al.*, 1954). Recent studies indicate that the common denominator in these syndromes is elevated circulating levels of prolactin (Zárate *et al.*, 1973; Rolland *et al.*, 1974). An etiological role of prolactin in gonadal dysfunction is also suggested by the fact that lowering of plasma prolactin levels pharmacologically restores gonadal dysfunction (Varga *et al.*, 1973; Zárate *et al.*, 1973). For this purpose, bromergocryptine (2-Br-a-ergocryptine, CB154, Sandoz, Basel, Switzerland) which is a cyclic tripeptide, seems to be the drug of choice (Rolland *et al.*, 1974). This agent, a dopamine agonist, blocks prolactin secretion by the pituitary gland and probably also interferes with *de novo* synthesis of the hormone (Pasteels *et al.*, 1971; Del Pozo *et al.*, 1973, 1974).

Our own studies began in September 1972. Eleven patients suffering from galactorrhea and menstrual disturbances were treated with bromergocryptine at a dose of 2.5 mg twice daily. Treatment was considered successful when milk secretion disappeared and the hormonal and ovulatory functions of the ovaries normalized. In Table I some of the pretreatment findings are summarized. Plasma prolactin levels were clearly above 30 ng/ml plasma (30 ng/ml plasma = upper normal limit in cyclic women). In addition to galactorrhea, all patients had absence of clinical estrogen effect, as assessed by "spinnbarkeit", fern-pattern of the cervical mucus and vaginal smears.

Table II summarizes the results of treatment. Within 1-3 weeks the galactorrhea disappeared in all subjects and gynecological examination revealed signs of normal estrogen production. Normal ovarian function was conclusively demonstrated by pregnancy in all the patients who wanted to conceive.

The history and the results of treatment in three of these cases are of particular interest with regard to the postulated influence of prolactin *per se* on ovarian function. Their data are summarized in Table III. Ovarian biopsies in two subjects showed only primordial follicles despite (high) normal gonadotropin secretion. Two patients demonstrated unusual resistance to exogenous gonadotropin treatment: the first did not respond to exogenous Human Menopausal Gonadotropin (HMG) and Human Chorionic Gonadotropin (HCG) therapy; the second ovulated only after the fifth treatment cycle when high doses of gonadotropins were given [HMG (Humegon®, Organon), 12 days 300 i.u./day followed by HCG (Pregnyl®, Organon, Oss, The Netherlands), 3 days 5,000 i.u./day].

Although low levels of gonadotropins have frequently been reported in patients with galactorrhea and menstrual dysfunction, normal or even elevated values similar to those reported here have also been encountered (Thorner *et al.*, 1974; Rankin *et al.*, 1969). These findings and the (partial) unresponsiveness to exogenous gonadotropin administration demonstrated by two of our patients are

Table I. A summary of pretreatment findings in patients with galactorrhea and menstrual disturbances.

No. of Pat.	Age*	Duration of Galactorrhea*	Duration of Amenorrhea*	Pretreatment Prolactin Levels (ng/ml)	Gonadotropin Excretion (24 h urine)**	Estrogen Excretion (24 h urine)
11	Mean: 26 Range: 20-32	3 1/2 1 1/12-10	3 7/12-7	32-260 (Range***)	High Normal: 2 Normal: 4 Below Normal: 5	Normal: 1 Below Normal: 7

* Years. ** Mouse uterus test. *** Measured in 6 patients.

Table II. A summary of the results of bromergocryptine treatment in patients with galactorrhea and menstrual disturbances.

No. of Pat.	Disappearance of Galactorrhea Within (Weeks)	Normo-Estrogenic Within (Weeks)	Ovulation Noticed Within (Days)	Pregnancy
11	Mean: 2 Range: 1-3	2 1/2 1/2-3	< 30: 4 30-50: 6 > 50: 1	8/8*

* Pregnancy wanted in 8 women.

Table III. A summary of pretreatment findings and results of bromergocryptine treatment in three patients with galactorrhea and menstrual disturbances.

Patient	Urinary Excret. Gonadotr.	Ovarian Biopsy	Previous Therapy	Result	Results of Bromergocryptine Treatment		
					Ovulation Within (Days)	Pregnancy	Disappearance of Galactorrhea
Ke.C.	Normal	None	HMG/HCG* (3X)	None	42	+	+
G.K.	High Normal	Primordial Follicles	Clomiphene HMG/HCG* (5X)	One Pregnancy	23	+	+
Kw.C.	High Normal	Primordial Follicles	Clomiphene	None	45	+	+

* Humegon®/Pregnyl®

consistent with primary ovarian failure. The immediate restoration of ovarian function to normal in those patients when treated with bromergocryptine strongly suggests a direct effect by prolactin itself on the ovary.

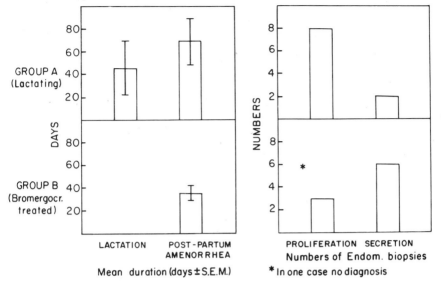

Fig. 2 Left panel: the mean duration (± SEM) of lactation in 10 lactating women and the mean duration (± SEM) of post-partum amenorrhea in 10 lactating and 10 non-lactating (bromergocryptine-treated) women. Right panel: the stage of endometrial development at the moment of the first vaginal bleeding.

Prolactin and the Post-Partum Period

To investigate further the influence of prolactin at the ovarian level, it was important to study a group of women with hyperprolactinemia in whom there was little or no suspicion of hypothalamic-pituitary dysfunction. For this reason, women during the physiological period of hyperprolactinemia and amenorrhea post-partum were selected. Plasma levels of prolactin, FSH, LH, 17β-estradiol (E_2) and progesterone were monitored during this period in an attempt to establish which component of the hypothalamic-hypophyseal-ovarian axis was the most refractory. Bromergocryptine, which also is a potent inhibitor of puerperal lactation (Rolland and Schellekens, 1973) was administered to some women. It was hoped that administration of this drug would substantiate the hypothesis that the presence of prolactin is responsible for the delayed restoration of cyclic ovarian function post-partum. A quicker restoration of ovarian function could be expected when this drug was given.

Twenty healthy volunteers who had given proof of adequate lactation for more than 2 weeks during a previous puerperium were studied. Ten were normal lactating women and in 10 lactation was suppressed by bromergocryptine 2.5 mg twice daily from immediately after delivery until the occurrence of the first vaginal bleeding. All women then underwent an endometrial biopsy. Those women with no bleeding after 3 months were biopsied at that time and the study was closed.

Blood samples were drawn frequently from the 36th week of pregnancy throughout the study for later determination of the above-mentioned hormones. (For further details, see Rolland et al., 1975a,b).

The duration of lactation and post-partum amenorrhea and the findings of

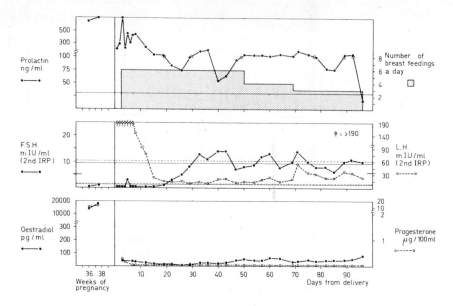

Fig.3 The relationship between breast-feeding, peripheral plasma levels of prolactin, FSH, LH, 17β-estradiol and progesterone during the puerperium (Subject Va). Horizontal lines indicate the range in normal cycling women (from Rolland et al., 1975a,b with permission).

the endometrial biopsies are summarized in Fig.2. The mean duration of lactation was 46 days with a wide range from 9 to more than 96 days. Two women were still lactating when the study was closed after 3 months. The mean duration of amenorrhea in lactating women was 68 days (range: 40 > 96); 3 women did not menstruate at all. For those cases, as well as for the women who were still lactating, the day the study was closed has been used for the calculations. In bromergocryptine-treated women, the duration of amenorrhea was reduced to only 35 days (range: 37-48). Secretory activity was observed in 6 cases as opposed to only 2 in the lactating women.

Lactating women

The hormonal changes in one of the women who breast-fed and remained amenorrheic throughout the observation period are shown in Fig.3. Prolactin remained clearly above the upper limit of 30 ng/ml plasma although there is a decrease during the first 2 weeks. Initially high LH levels are due to cross-reaction with HCG in the assay system. From day 14 to 70, LH is consistently in the low-normal range; thereafter, clearly within the normal range. After FSH has returned to normal between day 10 and 20, it remains consistently high-normal or distinctly elevated throughout the observation period. In spite of this, 17β-estradiol remains very low indicating the absence of follicular development.

The findings in another subject who showed post-ovulatory signs in the endometrial biopsy, although no menses occurred, are summarized in Fig.4. The stage of hyperprolactinemia is characterized by high FSH levels, low LH levels, and little or no 17β-estradiol. Thereafter, 17β-estradiol increases followed by ovulatory LH, FSH, and 17β-estradiol peaks on day 78. As demonstrated by the

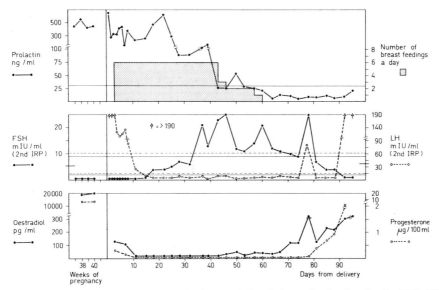

Fig. 4 The relationship between breast-feeding, peripheral plasma levels of prolactin, FSH, LH, 17β-estradiol and progesterone during the puerperium (Subject Xa). Horizontal lines indicate the range in normal cycling women (from Rolland et al., 1975a, with permission).

secondary rise in LH (HCG), this woman became pregnant following this ovulation. In Fig. 5, the hormonal data are shown of a women who menstruated on day 58. Again, high FSH levels, low LH levels, and almost no 17β-estradiol are found during the stage of hyperprolactinemia which exists until approximately day 20. Thereafter, ovarian function was restored with FSH and LH concentrations within the cyclic range. Interestingly, the rise in 17β-estradiol preceded that of LH.

As the period of lactation and the period of hyperprolactinemia were extremely variable in all 10 lactating women, their hormonal data differed markedly. However, one of the most consistent findings was the low level of 17β-estradiol during the hyperprolactinemic stage, even though FSH was found to be supranormal within a few weeks post-partum. This is shown in Table IV which summarizes the most noteworthy changes in the hormonal status of each individual.

Non-Lactating Women

The hormonal levels in 9 of the 10 women who used bromergocrytpine are summarized in Fig. 6. One subject has been excluded as she also received drugs other than bromergocryptine known to interfere with central regulation of prolactin. The effect of bromergocryptine on each of the hormones can be summarized as follows:

Prolactin was effectively suppressed by bromergocryptine with values within the normal cyclic range 24-48 h after birth. As expected, initiation of lactation was also completely inhibited.

FSH was either undetectable or extremely low during late pregnancy and immediately post-partum but returns to normal between day 6 and 12 in all subjects. The further increase to elevated levels, as noticed in lactating women, occurs

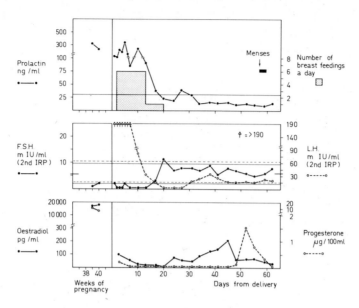

Fig. 5 The relationship between breast-feeding, peripheral plasma levels of prolactin, FSH, LH, 17β-estradiol and progesterone during the puerperium (Subject IIA). Horizontal lines indicate the range in normal cycling women (from Rolland *et al.*, 1975a, with permission).

Table IV. A summary of the most noteworthy changes in the hormonal status of ten women during physiological lactation.

Patient	Duration of lactation (days)	Prolactin >30 ng/ml until day:	Ended HCG excretion at day: (approx.)	FSH>2 mIU/ml at day: (approx.)	FSH≥cyc. range (days)	LH≤cyc. range (days)	Oestradiol increase (day)	Progesterone increase (day)	Day of vaginal bleeding
I	30	41	10	15	20–44	12–30	30	44	52
II	20	20	13	17	20–38	13–31	24	48	57
III	9	15	9	12	15–29	12–22	19	36	42
IV	46	49	11	18	32–74	15–60	58	78	87
V	>96	96	15	18	29–85	15–60	Not clear	None	>96
VI	35	35	11	18	25–57	14–66	39	57	65
VII	>93	93	8	7	15–93	8–89	None	None	>93
VII	26	31	6	7	13–60	7–28	40	49	60
IX	42	Not clear	11	14	14–42	11–28	21	None	40
X	60	53	11	18	37–81	11–74	67	78	>94

(From Rolland *et al.*, 1975a, with permission.)

only sporadically. Between day 23 and 29, however, some women showed FSH values consistent with an ovulatory FSH surge. After day 29, FSH levels decrease to the range normally seen during the luteal phase.

LH: HCG is fully cleared within 12-16 days as in the lactating group. Thereafter, LH is found within the normal cyclic range until approximately day 23. At that time, 4 women showed values suggesting an ovulatory LH surge. Others demonstrated definite increases compared to the previous values, although the

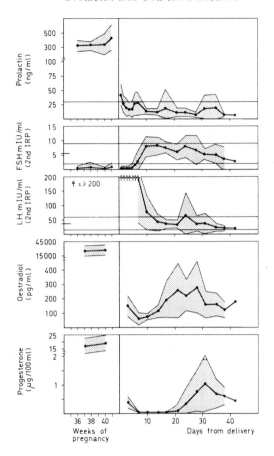

Fig. 6 The mean and SD of plasma prolactin-, LH-, FSH-, 17β-estradiol- and progesterone-concentrations during late pregnancy and the puerperium during bromergocryptine treatment in 9 healthy women. Horizontal lines indicate the normal range of cycling women. (From Rolland et al., 1975b, with permission.)

increments were not sufficient to be regarded as ovulatory peaks. As for FSH, a decrease in LH is noticed in all subjects thereafter.

17β-Estradiol excretion during the first 7 days post-partum is similar to that of the lactating group. At this time, levels consistent with the early proliferative stage of the cycle are found. In contrast to the lactating group where there was a further decline to nearly undetectable levels, 17β-estradiol increases in all cases to concentrations indicative of the pre-ovulatory surge. Subsequently, a decrease is noticed to luteal phase levels.

Progesterone excretion during the first 7 days post-partum reaches nearly undetectable levels similar to those of lactating women. Between day 19 and 29 an increase in plasma progesterone above 0.3 μg/dl was noticed in all cases indicating that ovulation had taken place.

The data of one of these 9 subjects are summarized in Fig. 7 and are included to emphasize that during the puerperium in the absence of hyperprolactinemia,

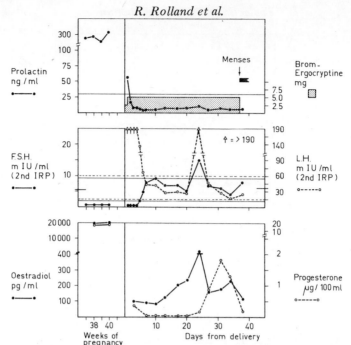

Fig. 7 The relationship between peripheral plasma levels of prolactin, FSH, LH, 17β-estradiol and progesterone during late pregnancy and the puerperium during bromergocryptine treatment (Subject Xb). Horizontal lines indicate the normal range in cycling women (from Rolland et al., 1975b, with permission).

the ovaries respond immediately to FSH. In this case, ovulatory LH, FSH and 17β-estradiol peaks were noticed at day 24 which is in agreement with the endometrial biopsy taken at day 37 showing endometrium with pseudo-decimal reaction.

The temporal sequence of hormonal changes during the puerperium, as described here, is complex. Therefore, a schematic diagram of these changes has been constructed (Fig.8) which shows that:

(1) The stage of hyperprolactinemia is characterized by high-normal or even elevated FSH levels with low-normal LH levels. During this stage, 17β-estradiol and progesterone remain low. This is a consistent finding in all 10 lactating women.

(2) After weaning, prolactin falls to the normal cyclic range, at which time 17β-estradiol increases, as does LH. In 6 women, the estradiol rise definitely preceded the rise in LH; in 2, it occurred concomitantly; and the last 2 women were still in the stage of lactation when the study was closed.

(3) The increase in 17β-estradiol and LH is followed by restoration of normal ovulatory function, although the first luteal phase tends to be rather short. Two features of this sequence are central to the interpretation of this state as one of ovarian rather than hypothalamic-hypophyseal refractoriness: (a) FSH levels are supranormal during most of this period, indicating ovarian resistance to this hormone; and (b) 17β-estradiol returns to normal before LH, indicating that hypo-estrogenemia may be the cause rather than the effect of the low LH levels.

(4) In contrast, when pituitary prolactin release is blocked immediately after birth with bromergocryptine, the entire axis returns to ovulatory function within 25 days. 17β-Estradiol starts to rise between days 8 and 10 while FSH is

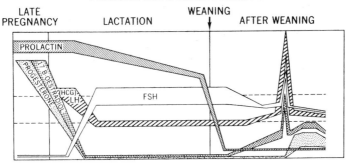
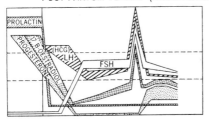

Fig.8 Upper panel: a schematic diagram of the changes in the hormonal status of women during physiological lactation, weaning and after weaning. Lower panel: a similar diagram for bromergocryptine-treated women in the post-partum period.

returning to normal. At this point, HCG is still present. Normal ovarian function is restored thereafter as the first ovulation occurs within approximately 16 days after FSH has reached normal levels. This is within the anticipated range of the proliferative stage of the menstrual cycle.

These observations suggest that the presence or absence of hyperprolactinemia plays a major role both in the syndrome of galactorrhea and menstrual disturbances and in the restoration of normal cyclic ovarian function post-partum.

Studies performed by others have shown that the presence of 17β-estradiol synergistically enhances the amount of LH released by the pituitary (Tsai and Yen, 1971; Yen et al., 1972; Reiter et al., 1974). As the data presented so far indicated a discrepancy between FSH and LH levels during *physiological* hyperprolactinemia, the following hypothesis seems to give the most plausible explanation for the observed hormonal events (Fig.9). During the stage of hyperprolactinemia, FSH is unable to stimulate the early stages in follicular development which allow steroidogenesis to take place. These steps normally initiate at low LH concentrations. The result is a lack of positive feedback of 17β-estradiol on pituitary LH release. When pituitary prolactin release is blocked by bromergocryptine, FSH immediately initiates follicular development; this is followed by normal steroid production which, in turn, allows further cyclic LH release to take place and ovarian function is restored within a short time.

Another mechanism which has been postulated by others is the possibility of a direct inhibitory action of prolactin itself ("short negative feedback loop") on LH secretion (Boyar et al., 1974; Tyson et al., 1975). However, this does not explain the fact that 17β-estradiol increased previous to that of LH. This mechanism would also be in disagreement with the concept that both LH and FSH

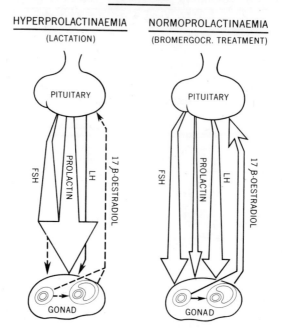

Fig. 9 Hypothetical role of prolactin in regulating ovarian function. (For further details, see text.)

secretion are regulated by one single hypothalamic gonadotropin-releasing hormone (G-RH) (Schally, 1975). Exogenous administration of synthetic G-RH in patients with hyperprolactinemia has shown a normal pituitary response with increased release of both LH and FSH (Corbey and Lequin, 1975; Thorner et al., 1974).

In an attempt to study the effect of prolactin *per se* on the ovary, an *in vitro* model has been developed to complement ongoing clinical studies.

IN VITRO STUDIES

In view of the evidence presented so far, it was pertinent to investigate prolactin binding to ovarian steroid producing cells. Prolactin, like many other protein hormones, probably initiates its biological action on the cell surface (Turkington, 1970). The presence of binding sites or receptors on these cells would be compatible with an action of prolactin on them.

Although *in vitro* studies with human material would be preferred, porcine material has been used to date because of the ease with which it can be obtained. However, as in human, the post-partum period in this species is characterized by a strong lactationally-induced anestrus with high FSH levels, low LH levels, and signs of hypoestrogenism (Crighton and Lamming, 1969).

Porcine ovaries were obtained at a local slaughterhouse and granulosa cells aspirated from those without corpora lutea or corpora albicantia. Initially, the

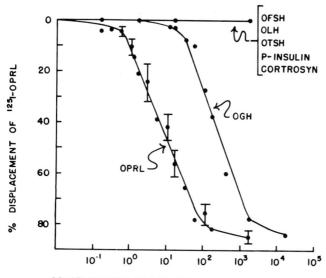

Fig.10 Specificity of binding of ^{125}I-OPRL to porcine granulosa cell receptors. Vertical bars in the displacement curve indicate the mean ± SEM from 3 different experiments. (From Rolland and Hammond, 1975a, with permission.)

cells were pooled from follicles of all sizes. For most studies, however, cells from ovaries containing either small (< 3 mm), medium (3-7 mm), or large follicles (> 7 mm) were collected and analyzed separately.

Corpora lutea of the reproductive cycle and of pregnancy, corpora hemorrhagica, and corpora albicantia were dissected, decapsulated, homogenized and fractionated by means of differential centrifugation. In general, the resuspended 15,000 X g pellets were used for binding studies.

Ovine prolactin (OPRL) was iodinated by a lactoperoxidase method (Thorell and Johansson, 1971), purified by ion-exchange chromatography, and validated in a mammary gland system (Shiu et al., 1973; Shiu and Friesen, 1974). Whole granulosa cells or subcellular particles were incubated with ^{125}I-OPRL in the absence or presence of unlabeled OPRL or other hormones. For further details about the methodology of these studies, see Rolland and Hammond (1975a, b).

Porcine granulosa cells indeed demonstrated a specific receptor for prolactin as illustrated in Fig.10. Of the other hormones tested, only ovine growth hormone (GH) competed with prolactin for the binding site. This can be explained by the contamination of the GH preparation with prolactin.

The relationship between cell number and specific binding of ^{125}I-OPRL for cells from various size follicles is shown in Fig.11. The binding of ^{125}I-OPRL to the receptor is linear with the number of cells per tube; however, the binding for each concentration of cells decreases with increasing follicular size. To gain a more quantitative understanding of this change in prolactin binding with follicular maturation, displacement curves were generated with cells from various size follicles, as shown in Fig.12. These data were transformed into Scatchard

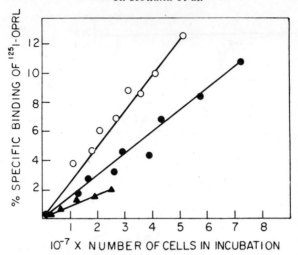

Fig.11 Effect of the number of granulosa cells harvested either from small follicles (o), medium follicles (•), or large follicles (▲) on specific binding of ^{125}I-OPRL. (From Rolland and Hammond, 1975a, with permission.)

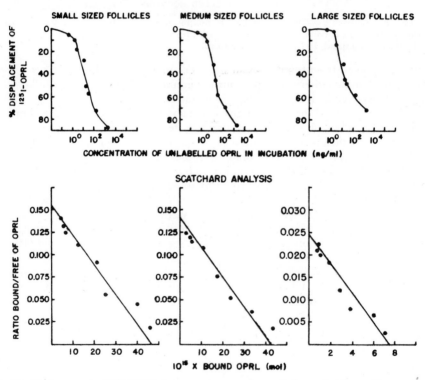

Fig.12 Displacement of ^{125}I-OPRL from porcine granulosa cells harvested from either small, medium or large follicles as a function of unlabeled OPRL. A Scatchard plot has been obtained from each curve after subtracting the non-specific binding of ^{125}I-OPRL from each point of the curve. (From Rolland and Hammond, 1975a, with permission.)

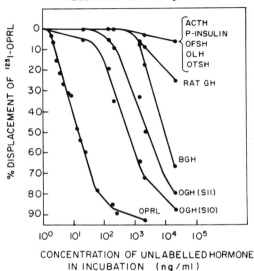

Fig.13

plots with the ratio of bound over free prolactin plotted against the amount of bound hormone. From the slope of these plots, dissociation constants (K_d's) could be calculated for the various cell types. The parallel lines of these plots indicate that these are quite similar. The calculated values bear this out as K_d's of 7.55×10^{-10} M, 7.37×10^{-10} M and 7.66×10^{-10} M were found for cells from small, medium and large follicles, respectively. However, when the binding sites per cell are calculated from the intercept of the curve on the abscissa, there appears to be a nearly 50% decrease in binding sites per cell from 555 on cells from small follicles to 356 on cells from medium sized follicles to only 301 in cells from large follicles.

The specific binding of iodinated prolactin to 15,000 × g fractions from corpora hemorrhagica or corpora albicantia was too small to allow accurate displacement experiments to be performed. However, well-established corpora lutea of the reproductive cycle and of pregnancy showed substantial binding. A displacement curve utilizing cell fragments from corpus luteum of pregnancy is shown in Fig.13. More than 90% displacement of initially bound ^{125}I-OPRL is obtained by excess cold hormone. The competition of other hormones with ^{125}I-OPRL for the binding sites is due to contamination of these preparations with prolactin.

Calculated dissociation constants of the two corpus luteum fractions were quite similar in repeated experiments and were found around 6×10^{-10} M, which is in good agreement with the K_d's of the granulosa cell receptors. However, the corpus luteum of pregnancy showed an approximate 6-fold increase in binding capacity per equal amount of binding material versus corpus luteum of the reproductive cycle.

Summarizing the data from the *in vitro* studies so far (Fig.14), it appears that the granulosa cells in small follicles clearly bind prolactin, that this binding

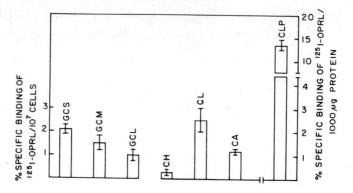

Fig.14 Specific binding (mean ± SEM of 3 or more experiments) of ^{125}I-OPRL to porcine ovarian steroid producing tissue during the reproductive cycle and early pregnancy. (G.C.S, G.C.M., and G.C.L. = granulosa cells from small, medium or large follicles; C.H. = corpus hemorrhagicum; C.L. = corpus luteum of the reproductive cycle; C.A. = corpus albicans; and C.L.P. = corpus luteum of pregnancy.)

capacity decreases as the follicles develop toward ovulation to reach a minimum just after the ovulation has taken place in the formed corpus hemorrhagicum. As the corpus luteum develops, the binding capacity again increases. If fertilization does not occur, the corpus luteum regresses into a corpus albicans which binds less prolactin. If pregnancy occurs, the specific binding increases substantially. These changes take place without alteration of the affinity of prolactin to the receptor but are due to changes in the number of receptors per cell.

FINAL COMMENT

The clinical data presented here indicate that prolactin plays an important role in the regulation of female reproduction. Although plasma prolactin levels, as measured during a normal menstrual cycle, do not show a specific cyclic pattern (McNeilly and Chard, 1974), studies with human granuolsa cells *in vitro* have shown prolactin to be essential for maintenance of steroidogenesis by these cells (McNatty et al., 1974). However, as the plasma concentration of prolactin increases, as in the inappropriate lactation syndromes or during post-partum lactation, an inhibition of ovarian steroid production seems to take place. This is most likely due to a direct effect by prolactin itself on the ovary. Thus, this hormone seems to demonstrate a biphasic effect dependent on the circulating amount present.

The existence of specific receptors for prolactin on the surface of ovarian steroid producing cells, and more importantly, the apparent changes in prolactin binding during the reproductive cycle and pregnancy, supports this view and may provide a basis for a better understanding of the regulatory role of prolactin in ovarian function.

REFERENCES

Argonz, J. and Del Castillo, E.B. (1953). *J. clin. Endocr. Metab.* 13, 79-87.
Batzdorf, U. and Stern, E. (1973). *In* "Diagnosis and Treatment of Pituitary Tumours" (P.O. Kohler and G.T. Ross, eds) pp.17-25. Excerpta Medica, Amsterdam.

Boyar, R.M., Kapen, S., Finkelstein, J.W., Perlow, M., Sassin, J.F., Fukushima, D.K., Weitzman, E.D. and Hellman, L. (1974). *J. clin. Invest.* **53**, 1588-1598.
Chiari, J., Braun, C. and Spaeth, J. (1958). In "Obstetrics and Gynecologic Milestones" (H.H. Speert, ed) Chap. 46. MacMillan, New York.
Corbey, R. and Lequin, R.M. (1975). Personal communication.
Crighton, D.B. and Lamming, G.E. (1969). *J. Endocrinol.* **43**, 507-519.
Del Pozo, E., Friesen, H.G. and Burmeister, P. (1973). *Schwiezerische Medizinische Wochenschrift* **103**, 847-853.
Del Pozo, E., Varga, L., Wyss, H., Tolis, G., Friesen, H.G., Wenner, R., Veiter, L. and Vetwiller, A. (1974). *J. clin. Endocr. Metab.* **39**, 18-26.
Denamur, R., Martinet, J. and Short, R.V. (1973). *J. Reprod. Fertil.* **32**, 207-220.
Forbes, A.P., Henneman, P.H., Griswold, G.C. and Albright, F. (1954). *J. clin. Endocr.* **14**, 265-271.
Friesen, H.G., Guyda, H. and Hardy, J. (1970). *J. clin. Endocr. Metab.* **31**, 611-624.
Frommel, J.T. (1958). In "Obstetrics and Gynecologic Milestones" (H.H. Speert, ed) Chap. 46. MacMillan, New York.
Ganong, W.F. (1972). In "Topics in Neuroendocrinology" (J.A. Kappers and J.P. Schade, eds). Progress in Brain Research, **Vol.38**, p.46. Elsevier Pub. Co., Amsterdam.
Hwang, P., Guyda, H. and Friesen, H.G. (1971). *Proc. natn. Acad. Sci. U.S.A.* **68**, 1902-1906.
Hwang, P., Guyda, H. and Friesen, H.G. (1972). *J. biol. Chem.* **247**, 1955-1958.
Kamberi, I.A., Mical, R.S. and Porter, J.C. (1970). *Endocrinology* **87**, 1-12.
Kamberi, I.A., Mical, R.S. and Porter, J.C. (1971a). *Endocrinology* **88**, 1003-1011.
Kamberi, I.A., Mical, R.S. and Porter, J.C. (1971b). *Endicrinology* **88**, 1012-1020.
Lewis, V.J., Singh, R.N.P. and Seavey, B.K. (1971). *Biochem. Biophys. Res. Commun.* **44**, 1169-1176.
McNatty, K.P., Sawers, R.S. and McNeilly, A.S. (1974). *Nature* **250**, 653-655.
McNeilly, A.S. and Chard, T. (1974). *Clin. Endocr.* **3**, 105-112.
Nicoll, C.A. and Bern, H.A. (1972). In "Lactogenic Hormones" (G.E.W. Wolstenholme and J. Knight, eds) pp.299-317. Churchill, London.
Nieman, E.A., Landon, J. and Wynn, V. (1967). *Quart. J. Med.* **36**, 357-392.
Pasteels, J.L., Danguy, A., Frérotte, M. and Ectors, F. (1971). *Ann. Endocrin.* **32**, 188-192.
Rankin, J.S., Goldfarb, A.F. and Rakoff, A.E. (1969). *Obstet. Gynec.* **33**, 1-10.
Reiter, E.O., Kulin, H.E. and Hamwood, S.M. (1974). *Ped. Res.* **8**, 740-745.
Rolland, R. and Schellekens, L.A. (1973). *J. Obstet. Gynaec. Brit. Commw.* **80**, 23-29.
Rolland, R., Schellekens, L.A. and Lequin, R.M. (1974). *Clin. Endocr.* **3**, 155-165.
Rolland, R., Lequin, R.M., Schellekens, L.A. and De Jong, F.H. (1975a). *Clin. Endocr.* **4**, 15-25.
Rolland, R., De Jong, F.H., Schellekens, L.A. and Lequin, R.M. (1975b). *Clin. Endocr.* **4**, 27-38.
Rolland, R. and Hammond, J.M. (1975a). *Endocr. Res. Commun. (in press)*.
Rolland, R. and Hammond, J.M. (1975b). Manuscript in preparation.
Schally, A.V. (1975). *Obstet. and Gynec. Survey* **30**, 122-123.
Shiu, R.P.C., Kelly, P.A. and Friesen, H.G. (1973). *Science* **180**, 968-970.
Shiu, R.P.C. and Friesen, H.G. (1974). *Biochem. J.* **140**, 301-311.
Sinha, Y.N., Selby, F.W., Lewis, V.J. and VanderLaan, W.P. (1973). *J. clin. Endocr. Metab.* **36**, 509-516.
Sulman, F.C. (1970). In "Hypothalamic Control of Lactation" (F.G. Sulman, ed). Monographs on Endocrinology, **Vol.3**, pp.7-13. Springer Verlag, New York, Heidelberg, Berlin.
Thorell, J.I. and Johansson, B.G. (1971). *Biochim. biophys. Acta* **251**, 363-369.
Thorner, M.O., McNeilly, A.S., Hagan, C. and Besser, G.M. (1974). *Brit. Med. J.* **2**, 419-422.
Tsai, C.C. and Yen, S.S.C. (1971). *J. clin. Endocr. Metab.* **33**, 917-923.
Turkington, R.W. (1970). *Biochem. Biophys. Res. Commun.* **41**, 1362-1367.
Tyson, J.E., Khojandi, M., Huth, J., Smith, B. and Thomas, P. (1975). *Am. J. Obstet. Gynec.* **121**, 375-379.
Varga, L., Wenner, R. and Del Pozo, E. (1973). *Am. J. Obstet. Gynec.* **117**, 75-79.
Yen, S.S.C., Vanden Berg, G., Rebar, R. and Ehara, Y. (1972). *J. clin. Endocr. Metab.* **35**, 931-934.
Zárate, A., Canales, E.S., Jacobs, L.S., Manerio, J.P., Soria, J. and Daughaday, W.H. (1973). *Fertil. Steril.* **24**, 340-344.

PROLACTIN AND OVULATION

P.G. Crosignani, A. D'Alberton, G.C. Lombroso, A. Attanasio, E. Reschini
and G. Giustina

Department of Obstetrics and Gynecology
and
Department of Internal Medicine
University of Milan, Milan, Italy

In some species prolactin plays an essential role for the maintenance of the corpus luteum (Denamur *et al.*, 1973), while its importance in regulating the human menstrual cycle is still a matter for discussion.

The levels of prolacin in males and in females are essentially the same until puberty, when the increasing estrogen secretion in females leads to a slight increase of prolactin values, but there is a wide overlap between the serum prolactin ranges of adult man and woman, and the advent of menopause does not change prolactin levels (Friesen *et al.*, 1973).

The menstrual cycle, the puerperium and the anovulation associated with hyperprolactinemia are situations in which plasma prolactin and ovulation appear to be theoretically-linked parameters.

CONCENTRATION OF PLASMA PROLACTIN DURING THE MENSTRUAL CYCLE

Estrogens are potent stimulators of prolactin secretion in animals (Meites, 1969) while man is less sensitive to this effect of estrogens. Thus, it has been reported that high dosages of estrogens caused an increase of prolactin levels in postmenopausal women (L'Hermite, 1973) and the effect of estrogens both on the basal prolactin and on its response to TRH is also evident in primary amenorrhea[*] (Fig.1).

---

[*] The radioimmunoassay of FSH, LH, and prolactin was carried out using the materials supplied by the National Pituitary Agency, National Institutes of Health, U.S.A. Results for LH and FSH

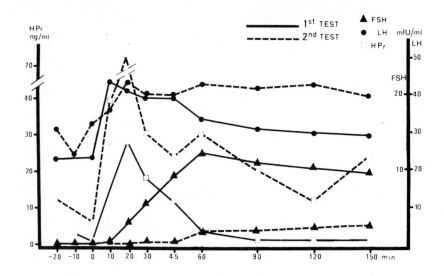

Fig.1 Gonadotropin and prolactin response to LRH-TRH (200 μg i.v.) in a patient with primary amenorrhea before and during treatment with ethinyl-estradiol (0.1 mg/day).

Studies of the urinary excretion of prolactin by bioassay during the cycle showed higher hormone levels in the luteal than in the follicular phase (Coppedge and Segaloff, 1951), while Hwang et al. (1971) using a radioimmunoassay method failed to observe a similar pattern in human plasma. Vekemans et al. (1972) reported in a group of 18 women with ovulatory cycles a modest peak of the hormone at midcycle and observed that prolactin secretion was significantly increased in the luteal phase. But other authors were unable to find significant prolactin changes either around midcycle (Hwang et al., 1971; Jaffe et al., 1973; McNeilly and Chard, 1974) or in the luteal phase (Jaffe et al., 1973). Prolactin response to TRH in a given subject may change during the cycle (Fig.2) suggesting a modulation of the response exerted by plasma estrogen fluctuations.

Hyperprolactinemia induced by sulpiride lowers or abolishes the LH ovulatory peak and decreases the plasma progesterone levels in the luteal phase of the cycle (Delvoye et al., 1974), while elevated plasma prolactin levels decrease pulsatility patterns of LH secretion suggesting an antigonadotropic effect of prolactin (Boyar et al., 1974). On the other hand granulosa cells cultured *in vitro* show a decreased progesterone production in presence of high levels of prolactin in the medium (McNatty et al., 1974) strongly suggesting an effect of the hormone upon the ovary. Jewelewicz et al. (1974) in a recent study reported that the administration of 20 mg b.i.d. of TRH significantly decreased the progesterone profile in 3 out of 5 subjects studied. On the contrary, the high PRL plasma levels induced by TRH (20 mg t.i.d.) in 6 women for 8 consecutive days beginning between days 8 and 9 of the menstrual cycle failed to influence the normal profile of gonadotropins and ovarian steroids (Zárate et al., 1974).

are expressed in milli International Units (mIU) of the second International Reference Preparation of Human Menopausal Gonadotropin (2nd IRP-HMG) supplied by Medical Research Council, Mill Hill, London. Prolactin values are expressed in ng of the Friesen No.1 preparation given by the National Pituitary Agency.

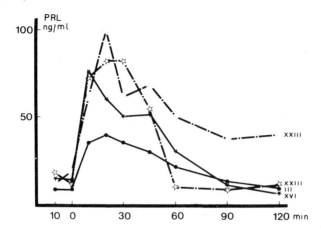

Fig.2 Variability of prolactin response to TRH (200 μg i.v.) during the cycle in a normally menstruating woman.

Similarly Del Pozo et al. (1975) studying the effect of bromocriptin (CB-154) on 7 normally menstruating women did not observe alteration of the hormonal profiles during the menstrual cycle, leading to the conclusion that prolactin has no action in the regulation of the human menstrual cycle. These data agree with the fact that after ablation of the pituitary a woman may conceive after an induced ovulation but does not lactate (Jørgersen et al., 1973).

LACTOTROPIC AND GONADOTROPIC ACTIVITY IN THE PUERPERIUM

The puerperium is characterized by spontaneous anovulation and high plasma prolactin levels or frequent bursts of prolactin secretion. FSH is consistently low for the first 10 days (Canales et al., 1974; Jeppsson et al., 1974); then (around day 15) it returns to the midfollicular phase level. On the contrary, LH concentration at day 15 is still below the normal range for the midfollicular phase. Moreover puerperal women show a characteristic refractoriness to LRH stimulation (Fig.3) during the first two weeks after delivery. Thereafter there is an enhanced FSH and a diminished LH release in response to LRH (Jeppsson et al., 1974; Canales et al., 1974). A similar reversal of FSH and LH response to LRH has been observed in prepubertal children (Roth et al., 1972). Bromocriptin (2.5 mg b.i.d.) administered in order to suppress lactation lowers prolactin levels but does not change the response of lactotropes to TRH and does not incude a response of gonadotropes to synthetic LRH (Fig.3).

Said and Wide (1973) did not find any difference in basal FSH and LH levels between lactating and non-lactating women. The hormonal consequence of lactation appears evident, however, in comparing women who breast-feed their babies from those who do not. Menstruation returns on the average within 8 weeks postpartum in non-lactating women and within 12 weeks in lactating women (Vorherr, 1974). Bonte and Van Balen (1969) showed that within 9 months after delivery 74% of non-lactating Egyptian women became pregnant whereas the pregnancy rate was only 8% among lactating mothers during the same time span.

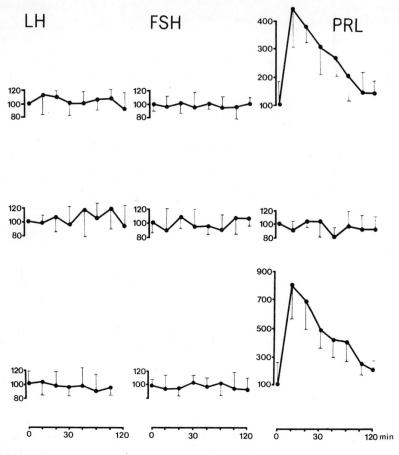

Fig.3 LH (left), FSH (middle) and PRL (right) response (% of the basal level ± SE) to LRH and TRH (200 μg i.v.) at day 5 of the puerperium. *Top:* LRH-TRH combined test (3 women). *Middle:* LRH test (3 women). *Bottom:* LRH-TRH combined test during bromocriptin treatment (5 women).

These results are similar to those obtained by Berman *et al.* (1972) in a recent study of puerperal Alaskan Eskimos, the average duration of postpartum amenorrhea being of 10 months in nursing mothers, but of only 2 months in non-nursing women. In the subjects who had been breast-feeding for 9 months or more the pregnancy rate was 10% compared to 75% in non-nursing mothers. In women in whom menstruation reappeared while they were still breast-feeding, 29% of the subjects reported irregular bleeding, suggesting that a high proportion of anovulatory cycles was occurring.

Menstruations during the late puerperium or the postpuerperal period are usually preceded by ovulation. Regardless of whether the mother is nursing or not, usually no ovulation takes place before the fourth week postpartum.

The lack of pituitary response to LRH in early puerperium, followed by the reversal of FSH/LH release, may indicate an imperfect balance of hypothalamic-

Table I. Therapeutic effect of bromocriptin (CB-154) in 8 amenorrhea-galactorrhea patients.

Subject	Age	Parity	Years of A-G	Prolactin (ng/ml) Before CB-154	Prolactin (ng/ml) During CB-154	Days Between Beginning of CB-154 & Menses
FA	28	0	3	100	10	47
BE	25	0	5	110	10	30
CC	31	1	3	105	10	30
BG	24	1	8	50	20	55
MM	27	0	7	150	20	*
BA	19	0	3	1000	20	**
AC	23	0	4	35	5	31
GN	31	1	5	10	5	41

* This patient became pregnant after two months of CB-154 therapy.
** This patient failed to menstruate, but reverted from clomiphene-negative to clomiphene-positive after two months of CB-154 therapy.

pituitary-ovarian function. As the refractoriness is more prolonged in nursing women (Fourner et al., 1974; Canales et al., 1974) it seems reasonable to assume that the periodic bursts of prolactin secretion which follow suckling may counteract and delay the resumption of pituitary gonadotropic activity.

The ovarian responsiveness to exogenous gonadotropins in puerperium is on the other hand still a matter of discussion. The administration of 2250 IU of menopausal goandotropins, as a total dose from the 6th to the 11th day after delivery, did not raise the ovarian output of estrogens and pregnanediol (Zárate et al., 1972). On the contrary, in a recent study (Nakano et al., 1975) the administration of a total dose of 750 IU from postpartum days 4 through 8 induced an ovarian response in 4 out of 5 subjects. Therefore both delayed gonadotrope recovery and some ovarian insensitivity may contribute to amenorrhea and to the anovulatory cycles in puerperal women.

ANOVULATION IN HYPERPROLACTINEMIC PATIENTS

The incidence of galactorrhea is higher in anovulatory women than in ovulatory ones (Jones and Gentile, 1975). Galactorrhea is not otherwise always associated with elevated plasma prolactin levels. Most of the patients with chromophobe adenoma and exceptionally elevated PRL values quite often do not have inappropriate lactation. It is well known that normal PRL concentrations may be found in women with galactorrhea (Fournier et al., 1974).

Amenorrhea associated with galactorrhea represents a major disturbance of hypothalamic-pituitary function and is characterized by: 1) usually elevated PRL secretion (Tolis et al., 1974); and 2) refractoriness of lactotropes to TRH stimulation (Del Pozo et al., 1974); and 3) positive response of gonadotropes to LRH (Crosignani et al., 1974).

Most of the cases are due to a suprapituitary lesion. This is in agreement with the restoration of ovarian function in patients with the amenorrhea-galactorrhea syndrome after long-term therapy with L-DOPA (Zárate et al., 1973). More favourable results have been obtained recently with the use of bromocriptin, a drug which lowers PRL secretion (Del Pozo et al., 1974).

Table I shows the treatment schedule and the outcome of 8 amenorrhea-

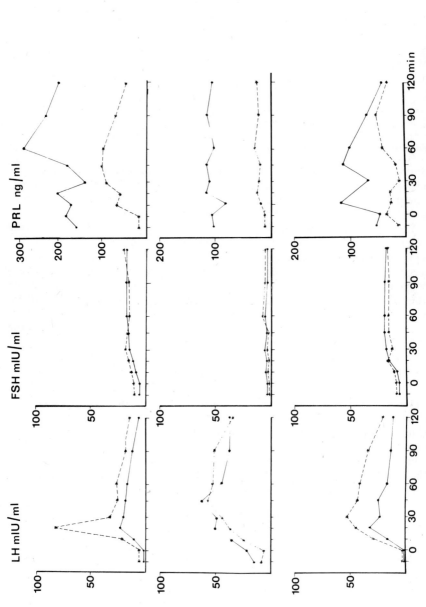

Fig. 4 Changes in the response to LRH and TRH before (solid line) and during (dashed line) bromocriptin treatment registered in 3 hyperprolactinemic women with amenorrhea-galactorrhea.

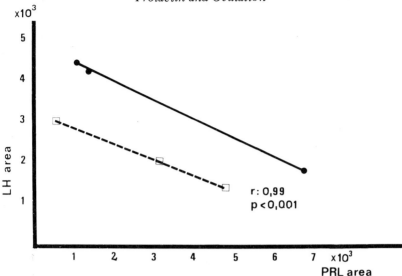

Fig. 5 Relationship between LH and PRL pituitary "reserve" (see text) before (dotted) and during (solid line) bromocriptin treatment.

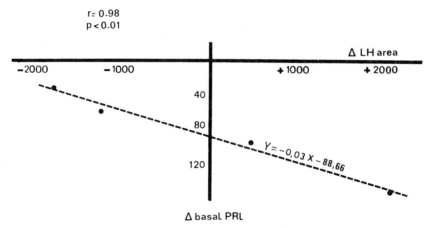

Fig. 6 Relationship between the decrease of plasma prolactin concentration during CB-154 treatment (Δ Basal PRL), and the change of pituitary LH "reserve" (Δ LH area).

galactorrhea patients. It is worthy of note that one patient became pregnant, 6 menstruated and one (probably affected by pituitary microadenoma) shifted from negative to positive clomiphene test. Three of the women have still a menstrual cycle after discontinuation of the therapy as reported by others (Tyson *et al.*, 1975).

Figure 4 shows the changes induced by CH-154 treatment in 3 amenorrhea-galactorrhea women with high PRL levels. The improvement of LH response is notable in all women while prolactin response improved in two. The possible link between these parameters has been investigated by measuring the pituitary "reserve" for LH and PRL before and after bromocriptin treatment. As a measure of

Table II. Ratios of LH and FSH pituitary reserve (see text) in 4 A-G patients with high prolactin before and during bromocriptin treatment.

	$\frac{LH}{FSH}$ PITUITARY RESERVE	
	BEFORE	DURING
Mean	5.4	42.2
Range	1.3 – 13.6	6.4 – 119

$p < 0.03$ (Mann Whitney "U" test)

pituitary reserve we have used the surface area circumscribed by the response curve (induced by LRH and TRH, 200 μg i.v.) and by the line corresponding to the basal LH and PRL levels.

Figure 5 shows the inverse relationship existing at least in some of these patients between LH and PRL pituitary reserve both before and after therapy, while the improvement in the LH response appears correlated to the extent of the decrease in plasma prolactin induced by bromocriptin (Fig.6).

The cessation of galactorrhea and return of the menses suggest that elevated plasma prolactin levels may somehow interfere with the hypothalamic-pituitary gonadal axis and prevent the normal discharge of FSH and LH (Fournier et al., 1974). In this context it is interesting to observe the reversal of pituitary LH and FSH reserve during CB-154 therapy (Table II). It should be noted that in another study FSH secretion was significantly decreased also by L-DOPA (Tyson et al., 1975). Whatever the level may be at which bromocriptin is active, the result is the starting of the cyclic center. According to McNeilly and Chard (1974), the withdrawal of the inhibitory effect of prolactin on ovarian steroidogenesis may *per se* explain the changes induced by the treatment. More difficult to explain is the favorable effect exerted by bromocriptin in normo-prolactinemic patients and in women suffering from secondary amenorrhea without galactorrhea, although the possible non-coincidence of radioimmunoassayable and bioassayable prolactin, as pointed out by Besser (in this book), may have a bearing. In both the latter categories, the change in LH and FSH release appears quite variable and unpredictable, while prolactin response to TRH is always lacking during CB-154 treatment. At any rate, we cannot exclude the possibility that an effect different from the decrease of PRL secretion may be exerted by bromocriptin and there is some evidence that the drug may change the levels of catecholamines in the hypothalamus (Hockfelt, 1972).

ACKNOWLEDGEMENTS

This study was supported by a Grant from the Population Council.

LRH preparation was generously supplied by Istituto Serono (Rome), Farbwerke Hoechst AG (Frankfurt), and Ayerst Laboratories International (Montreal).

TRH preparation was manufactured and kindly donated by Istituto Serono (Rome) and Farbwerke Hoechst AG (Frankfurt).

REFERENCES

Berman, M.L., Hanson, K., Hellman, I.L. (1972). *Am. J. Obstet. Gynec.* 114, 524.
Bonte, M. and van Balen, H. (1969). *J. Biosoc. Sci.* 1, 97-100.
Boyar, R.M., Kapen, S., Finkelstein, J.W. *et al.* (1974). *J. clin. Invest.* 53, 1588.
Canales, E.S., Zárate, A., Garrido, J. *et al.* (1974). *J. clin. Endocr. Metab.* 38, 1140.
Coppedge, R.L. and Segaloff, A. (1951). *J. clin. Endocr. Metab.* 11, 465.
Crosignani, P.G., Reschini, E., D'Alberton, A. *et al.* (1974). *Am. J. Obstet. Gynec.* 120, 376-384.
Delvoye, P., Taubert, H.-D., Jüugensen, O. *et al.* (1974). *Acta endocr. (Copenh.)* **Suppl. 184**, 110.
Del Pozo, E., Varga, L., Wyss, H. *et al.* (1974). *J. clin. Endocr. Metab.* 39, 18.
Del Pozo, E., Goldstein, M., Friesen, H. *et al.* (1975). *J, clin. Endocr. Metab.* 40, 315.
Denamur, R., Martinet, J. and Short, R.V. (1973). *J. Reprod. Fertil.* 32, 207.
Fournier, P.J.R., Desjardins, P.D. and Friesen, H.G. (1974). *Am. J. Obstet. Gynec.* 118, 337-343.
Friesen, H.G., Fournier, P. and Desjardins, P. (1973). *In* "Clinical Obstetrics and Gynecology" (H.J. Osofsky and G. Schaefer, eds) **Vol.XVI**, pp.25-45. Harper & Row, Hagerstown.
Hokfelt, T. and Fuxe, K.(1972). *Neuroendocrinology* 9, 100.
Hwang, P., Guyda, H. and Friesen, H. (1971). *Prof. natn. Acad. Sci.* 68, 1902.
Jaffe, R.B., Yuen, B.H., Keye, W.R. *et al.* (1973). *Am. J. Obstet. Gynec.* 15, 757-773.
Jeppsson, S., Rannevik, G. and Kullander, S. (1974). *Am. J. Obstet. Gynec.* 120, 1029-1034.
Jewelewicz, R., Dyrenfurth, I., Warren, M. *et al.* (1974). *J. clin. Endocr. Metab.* 39, 387-390.
Jones, J.R. and Gentile, G.P. (1975). *Obstet. Gynec.* 45, 13-14.
Jørgensen, P.E., Sele, V., Buus, O. *et al.* (1973). *Acta endocr. (Copenh.)* 73, 117.
L'Hermite, M. (1973). *Clinics in Endocr. Metab.* 2, 423-449.
McNatty, K.P., Sawers, R.S. and McNeilly, A.S. (1974). *Nature* 250, 653.
McNeilly, A.S. and Chard, T. (1974). *Clin. Endocr.* 3, 105.
Meites, J. (1969). *General and Comparative Endocrinology* **Suppl. 2**, 245-252.
Nakano, R., Mori, A., Kayashima, F. *et al.* (1975). *Am. J. Obstet. Gynec.* 121, 187-192.
Roth, J.C., Kelch, R.P., Kaplan, S.L. *et al.* (1972). *J. clin. Endocr. Metab.* 36, 817.
Said, S. and Wide, L. (1973). *Acta Obstet. Gynec. Scand.* 52, 361.
Tolis, G., Somma, M., Van Campenhout, J. *et al.* (1974). *Am. J. Obstet. Gynec.* 118, 91-101.
Tyson, J.E., Khojandi, M., Huth, J. *et al.* (1975). *Am. J. Obstet. Gynec.* 121, 375-379.
Vekemans, M., Delvoye, P., L'Hermite, M. *et al.* (1972). *C.R. Acad. Sc. Paris* D 275, 2247-2250.
Vorherr, H. (1974). "The Breast". Academic Press, New York and London.
Zárate, A., Canales, E., Soria, J. *et al.* (1972). *Am. J. Obstet. Gynec.* 112, 1130.
Zárate, A., Canales, E., Jacobs, L.S. *et al.* (1973). *Fertil. Steril.* 24, 340-344.
Zárate, A., Schally, A.V., Soria, J. *et al.* (1974). *Obstet. Gynec.* 43, 487-489.

RECENT ADVANCES IN THE NEUROENDOCRINE REGULATION OF GONADOTROPIN SECRETION

S.S.C. Yen, B.L. Lasley, C.F. Wang and A. Lein

Department of Reproductive Medicine, School of Medicine
University of California, San Diego, La Jolla, California 92093, USA

INTRODUCTION

During the past few years tremendous progress has been achieved in the understanding of the neural control of endocrine homeostasis. The regulation of the cyclic gonadotropin release by the pituitary represents a unique example of rhythmic neuroendocrine integrations and productive research is being conducted in primates including human subjects. The use of human disease models has provided invaluable information concerning the dominating role of CNS in the control of gonadotropin output. The feedback signal from the ovary and its modulatory role on the functional changes of the pituitary gonadotrophs has been characterized. This exciting new knowledge should provide clues in the assessment of hypothalamic-pituitary-ovarian system in normal and abnormal conditions and in the attainment of new means of fertility control. On the other hand, the existence of wide gaps in our knowledge of neuroendocrine regulation of production function is recognized.

HYPOTHALAMIC CONTROL OF GONADOTROPIN RELEASE

The neurovascular chemotransmitter hypothesis first proposed by Green and Harris (1947) has now become well-established as the first-order control mechanism for secretion from the adenohypophysis. With the purification, structural elucidation, and synthesis of the LRF decapeptide (Matsuo et al., 1970; Burgus et al., 1971), the concept of neurosecretory hypothalamic control acquired reality and the stimulatory nature of hypothalamic control of LH and FSH release was confirmed in all species studied, except rhesus monkeys (see Review, Yen, 1975a).

The prevailing view regarding the CNS regulation of gonadotropin secretion is that diverse neural inputs converge upon the medial basal hypothalamus (MBH) and operate through neurotransmitters which lead to the release of LRF by the neurosecretory neurons and subsequent pituitary release of LH and FSH (Wurtman, 1971; Halasz, 1972).

1. The Brain as an Endocrine Organ

There is growing recognition that the brain can be viewed as a new master gland through which the integration of the neural and endocrine systems prevail. The hormones of the target gland have profound effects on the brain, some permanent, some transient. The gonadal steroids control sexual differentiation, initiate puberty, regulate sexual behavior as well as modulate gonadotropin secretion. The recent demonstration of specific receptors (Davies et al., 1975) or binding sites for gonadal steroids suggests that distinctive areas of the brain participate in specific neuroendocrine events (Stump et al., 1975; McEwan et al., 1974). Further, the brain, particularly the hypothalamus possesses the capacity in steroid conversions especially from androgen to estrogens (Naftolin et al., 1975).

The brain not only acts as the hormonal target but also is capable of *de novo* synthesis and secretion of numerous peptide hormones. Studies of the recently identified hypothalamic peptides that regulate secretion of anterior pituitary tropic hormones, TRF, LRF, and somatostatin have suggested that these peptides are synthesized by the so-called releasing factor neurosecretory neurons located in or near the MBH (Vale and Rivier, 1975). However, the recent localization of these peptides in many regions of the CNS and the nerve terminals has stimulated the postulation of the existence of a system of peptidergic neurons in the brain, analogous to those of the catecholaminergic systems, with axon terminations both on the pituitary portal system and on the neurons. This hypothesis (Martin, Renaud and Brazeau, 1975) emphasizes the significance of these peptide hormones in the regulation of biological rhythms within the brain.

The brain has billions of neurons and nerves that communicate to one another by means of neurotransmitters. Many neurotransmitters have been identified in the CNS; the catecholamine system (dopamine and norepinephrine); the indoleamines (serotonin and its derivative melatonin); acetylcholine, histamine and gamma-aminobutyric acid (GABA). Among the neurotransmitters, catecholamines are best known in their role of neuroendocrine regulation. It is established that dopamine and norepinephrine systems in the CNS are intimately involved in the modulation of mood, behavior as well as hypothalamic functions (Axelrod, 1975). Thus, some endocrine disorders ascribed to hypothalamic dysfunctions may not represent a primary defect in the hypothalamic neurosecretory cells but are attributable to disturbances of catecholamine systems.

2. The Dominating Role of CNS-Hypothalamic Complex in the Regulation of Gonadotropin Release

Until recently, most of the information concerning the hypothalamic control of the adenohypophysis in humans was extrapolated from the results of animal experimentation principally in rodents and lately in infra-human primates. These early works provided the basis for the more recent studies of the neuroendocrine regulation of reproductive cycles in the human. The results of these investigations disclosed the pulsatile pattern of release of gonadotropins and the peculiar relationship of sleep to pituitary function; the fundamental role of the hypothalamus was also delineated. Of particular importance is the availability of

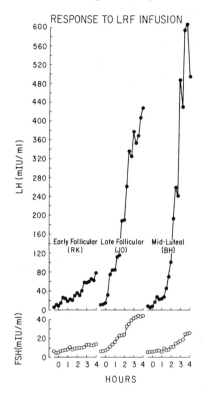

Fig.1 The presence of spontaneous fluctuation of LH during the rapid gonadotropin release induced by constant LRF infusion (0.2 µg/min × 4 h) in the early and late follicular and during the mid-luteal phase of the cycle.

synthetic hypothalamic hormones (or factors), which have made possible the design and execution of quantitative studies on hormonal factors influencing the response of the pituitary to the hypothalamus. Although limited by the kinds of experimental manipulations which may be performed in human subjects, much information concerning the operating characteristics of the hypothalamic-hypophyseal system has been generated.

Gonadotropin secretion by the pituitary is composed of two components; a relatively constant basal secretion and, superimposed on it, an episodic or pulsatile release (Midgley and Jaffe, 1971; Yen et al., 1972a). In the absence of feedback, the hypothalamus-pituitary system forms a functional unit, releasing large amounts of gonadotropin in periodic pulses characterized by a high frequency (about every 1-2 h) and large amplitude involving both FSH and LH. Thus, the elevated gonadotropin levels in ovariectomized women and in hypogonadal subjects, appear to be maintained by an increase in the magnitude of pulsatile pituitary discharge of LH and FSH (Yen et al., 1972a; Yen et al., 1972b; Santen and Bardin, 1973). This accelerated release (and synthesis) of FSH and LH, is in a large measure, determined by the hypophysiotropic effect of an increased LRF secretion; an elevated LRF release, in the absence of gonadal feedback, has been found in rats (Piacsek and Meites, 1966; Ben-Jonathan et al., 1973), rhesus monkey (Carmel et al., 1975)

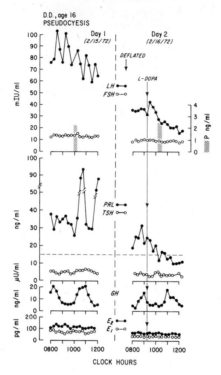

Fig.2 Basal pituitary and ovarian hormone concentrations and their pulsatile patterns in a 16-year-old patient with pseudocyesis (38 week duration) studied before (day 1) and after (day 2) the clinical resolution (deflation). Note: L-dopa suppression was also tested on day 2.

as well as in humans (Seyler and Reichlin, 1973).

The presence of pulsatile fluctuations of LH concentration, in spite of constant infusion of LRF (Fig.1), probably represents an expression of pituitary response to pulses of endogenous LRF, although a local pituitary mechanism has not been excluded. Similar observations have been reported in the rhesus monkey (Ferin *et al.*, 1973) and the rat (Dowd *et al.*, 1975). Several lines of evidence are now available to support a hypothalamic mechanism for pulsatile gonadotropin release: 1) a pulsatile pattern of LRF concentration in the portal blood of the rhesus monkey has been observed (Carmel *et al.*, 1975); 2) pulsatile secretion of LH and FSH can be abolished by the administration of antiserum to LRF (McCormack and Knobil, 1975); and 3) pulsatile LH release was observed only with pulsed delivery of LRF but not by constant infusion during superinfusion of isolated rat pituitaries (Osland *et al.*, 1975). Thus, the pulsatile release of gonadotropins appears to be directed primarily by episodic LRF secretion.

In view of the possible involvement of catecholaminergic neuronal input to the hypothalamic LRF secreting mechanisms, it is interesting to find changes in the LH pulses associated with certain emotional and psychological states. A marked increase in the amplitude of LH and PRL pulses is observed in patients with pseudocyesis (Fig.2) whereas, a decrease is found in patients with anorexia nervosa (Boyar *et al.*, 1974a) and certain hypothalamic dysfunctions (Fig.3).

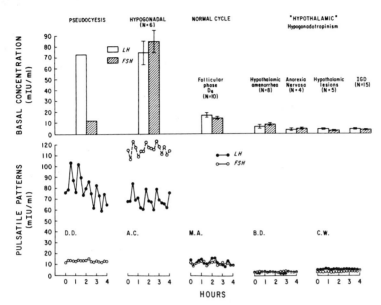

Fig.3 The basal concentrations of LH and FSH and their pulsatile pattern during normal follicular phase of the cycle as contrasted to the exaggerated pattern in subjects without ovarian feedback (hypogonadal) and in pseudocyesis and to the absence of pulsatile fluctuations in hypothalamic hypogonadotropinism (responsive to LRF).

Evidence for the role of hypothalamic catecholamines in the regulation of gonadotropin release by the pituitary in humans is necessarily limited. Attempts to increase hypothalamic and/or pituitary dopamine by the administration of L-dopa (Leblanc and Yen, 1975a) and dopamine infusion (Leblanc et al., 1975b) have yielded no demonstrable changes in basal gonadotropin secretion. Measurement of catecholamine levels and of enzymatic activities such as tyrosine hydroxylase (for the conversion of tyrosine to dopa) and dopamine β-hydroxylase (for the conversion of dopamine to norepinephrine) during acute gonadotropin release, either spontaneous or estrogen and/or progesterone induced, has not been reported. On the other hand, plasma monoamine oxides (MAO) activity (using C^{14} tyromine as a substrate in the assay) has been found to be significantly greater in the postovulatory than in the preovulatory phase of the cycle. The MAO activity in plasma of amenorrheic and postmenopausal women is significantly greater than in postovulatory subjects. These elevated levels of MAO can be reduced by estrogen treatment, but estrogen plus progestin treatment significantly increased the MAO activity over that found during estrogen treatment alone (Klaiber et al., 1971). These data imply that estrogens tend to inhibit MAO activity in human plasma which is a finding consistent with the changes in hypothalamic MAO activity in rats under similar hormonal environments (Kobayashi et al., 1964). Since plasma MAO activity is believed to be inversely related to the level of central adrenergic functioning, they reasoned that estrogen, which appears to inhibit MAO activity, may also enhance central adrenergic processes. Progesterone may be assumed to effect a reversal of this process. It should be noted that the real clinical significance of these findings cannot be assessed since measurements of enzymatic activities in circulation may have no relationship to the cellular events of catecholamine metabolism.

The mechanism responsible for the onset of puberty has been, until recently, an unsolved puzzle. The enigma has resided in the apparent development of secondary sex characteristics prior to the detection of an increase in plasma levels of sex steroids. However, the recent disclosure of a sleep-associated increase in pulsatile pituitary LH-release at the time of puberty, which regresses after sexual maturation (Boyar et al., 1972; Parker et al., 1975), has advanced our understanding of the basis for sexual maturation. A progressive increase in the pulsed-release of LH during sleep occurs from puberty stage I through stage V, and a temporally associated nocturnal testosterone secretion is also found (Judd et al., 1974; Boyar et al., 1974b; Parker et al., 1975). Since this augmented LH secretion during sleep is independent of gonadal activity, as is found in gonadal dysgenesis (Boyar et al., 1973), a concept of a "CNS-program" for the pubertal activation of the hypothalamic-pituitary system has been proposed. To what extent the catecholamine and LRF neuronal activities are involved in this pubertal CNS program, remains to be determined. Nonetheless, these human models, such as pseudocyesis, anorexia nervosa, certain hypothalamic dysfunctions and puberty are outstanding examples for the dominating role of CNS in the regulation of gonadotropin output.

OVARIAN MODULATION OF THE HYPOTHALAMIC-PITUITARY SYSTEM IN THE RELEASE OF GONADOTROPIN

1. Cyclic Gonadotropin Release: "CNS Clock" vs "Ovarian Clock"

Under a variety of experimental conditions, ovarian estradiol has been shown to exert a stimulatory feedback action on gonadotropin release in the rat (Krey and Everett, 1973). The most interesting feature of these observations in the rat is that estrogen-induced surges of LH were observed only in the afternoon of proestrus, suggesting the involvement of a "CNS clock". Since the LH surge can be blocked by the administration of antibodies to 17β-estradiol (Ferin et al., 1969), it was proposed that in the rat, ovarian estradiol determines the day of the surge while the CNS determines its hour of onset (Bogdanove, 1972).

Recent studies of gonadotropin secretion revealed a remarkable similarity between the menstrual cycles in humans and in the rhesus monkey. They appear to be different from the rat in that the evidence supports an "ovarian clock" rather than a "CNS clock" in the regulation of cyclic gonadotropin surge. The background of the endocrine relationships of the menstrual cycle has been extensively reviewed both for the human (Vande Wiele et al., 1970; Ross et al., 1970; Yen et al., 1974a; Yen et al., 1975) and for the rhesus monkey (Knobil et al., 1974).

2. Approaches in the Functional Analysis of Gonadotrophs as Target Cells

There is good experimental evidence to indicate that both the pituitary and the hypothalamus (Orias et al., 1974) are involved in the feedback action of estrogen. Further, estrogen receptors have been found in both the hypothalamus and the pituitary (Davies et al., 1975). Although much information has been obtained concerning the pituitary gonadotropin responses to LRF under varied steroid environments (Yen et al., 1974b; Jaffe and Keye, 1974), assessments of the quantitative and qualitative relationships between individual controllers (LRF and

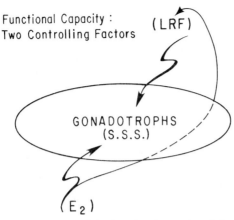

Fig. 4 A diagrammatic illustration of the functional capacity of the gonadotrophs, as determined by the two controllers, LRF and E_2.

gonadal steroids) and the status of gonadotrophs has been a recent approach. It should be recognized that these functional dynamics are extremely complex and characterization of gonadotrophs, as target cells, in the human is hampered by the lack of meaningful *in vitro* information. These studies are based on the premise that the functional capacity of the gonadotrophs is determined by the relative inputs of hypothalamic LRF and ovarian E_2 (Fig.4). The gonadotropin response of the pituitary to pulse injections of LRF (5 times) at two-hour intervals with large (150 μg), small (10 μg), incremental (10-300 μg) and decremental (300-10 μg) doses was assessed in eugonadal males with a relatively stable hypothalamic-pituitary-gonadal system (Lasley et al., 1975a). This experimental design was employed (a) in an effort to simulate a pulsatile hypothalamic LRF input assumed to be immediately responsible for the pulsatile pituitary gonadotropin output and (b) to determine whether the pituitary detects small, incremental and decremental changes of LRF input at two-hour intervals. At constant doses, the rates at which serum LH increase and decrease appear remarkably similar to the rates characteristic of spontaneous pulses in normal male subjects (Naftolin et al., 1972). It is apparent that the pituitary can detect relatively small increments or decrements of LRF as reflected by corresponding changes in circulating LH concentrations. At no time was there a suggestion of a refractory period, an augmentation or a reduction of sequential responses when tested at two-hour intervals. These data indicate that graded doses of LRF induce a graded variation in the quantity of LH released and suggest, as indicated earlier, that variations in the amount of LRF delivered represent a significant factor in the control of the pituitary gonadotropin output.

The technique of administering "pulses of LRF" and "LRF infusion" were employed in further studying the feedback action of estrogen and progesterone on pituitary response in women. It is assumed that the magnitude of initial, or 1st phase response is dependent upon the previous history or "set" of gonadotrophs, which is determined by the relative inputs of LRF and E_2. The rapidity and quantity (Δ) of this initial release will probably reflect the size of the acute releasable pool and may be a reasonable measure of "sensitivity". The second phase of release also appears to depend on the previous "set" of gonadotrophs.

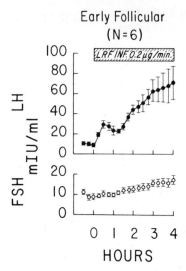

Fig.5 The two components of LH release elicited by constant LRF infusion in subjects during the early follicular phase of the cycle (Mean ± SE).

It reflects the size of the 2nd pool (pituitary reserve) but, in addition, is influenced by the rate of new synthesis which may be augmented by the time-related inputs of both E_2 and LRF at an appropriate ratio. These premises are instrumental in the functional characterization of the gonadotrophs, as target cells, relative to the changes of sensitivity/capacity and the variables of controlling inputs; LRF, estrogen and progesterone.

3. Ovarian Modulations of the Pituitary Capacity
    a. The concept of two pools of pituitary gonadotropin

The biphasic pattern of serum LH response to continuous LRF stimulation supports the concept of the presence of two components of releasable LH in the anterior pituitary, one immediately releasable and the other requiring continued stimulus input (Fig.5). Similar data have been reported for man (Bremner and Paulson, 1974) and for pubertal (but not pre-pubertal) children (Reiter et al., 1975). It is likely that the initial rise in serum LH during the first 30 minutes results from release of the first pool and with continued LRF stimulation, the second or reserve pool is tapped with a secondary rise in serum LH concentration. In the case of pulsed LRF administration, the first pulse would appear to induce LH release from the first pool, and the increased release from subsequent pulses of LRF may be a result of activations of the reserve pool of LH.

b. Pituitary sensitivity versus reserve

Although this two-pool concept remains arbitrary, it may nevertheless be useful in the analysis of the functional meaning of the pituitary sensitivity and reserve. When the LRF stimulation is small and brief, such as a 10 μg pulse, the initial increment at peak serum concentration (Δ_i) may be a reasonable measure of pituitary sensitivity and probably reflects the size of the 1st pool or acutely releasable gonadotropin in the pituitary. On the other hand, estimation of pituitary reserve may require a longer duration of stimulation, and the quantitative release as a function of time is likely an approximation of the size of the 2nd or storage

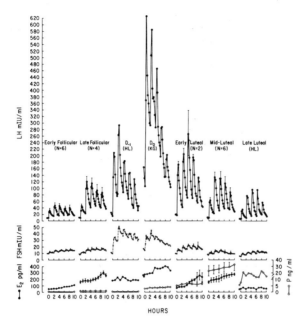

Fig. 6 Basal concentrations and the release of LH and FSH in response to pulses of LRF (10 μg at 2 h intervals × 5) during various phases of the ovulatory cycles. Basal levels of estradiol (E_2) and progesterone (P) and their changes during the gonadotropin rise are also depicted (mean ± SE). D_1 = the day before midcycle surge. D_0 = the day of midcycle surge.

pool which may include a component of, yet unmeasurable amount, newly synthesized gonadotropins. When a large single dose of LRF is used (100-150 μg), the components of sensitivity and reserve cannot be separately appreciated since a large dose provides a longer action than a small dose and, thereby, induces a release not only from the acutely releasable pool but also from the 2nd pool (VandenBerg et al., 1974; Wang et al., 1975a).

c. *Changes in pituitary sensitivity and reserve as related to E_2 and P*

Major changes in the pituitary sensitivity and reserve occur during the menstrual cycle (Figs 6, 7 and 8). In general, these changes are in synchrony with the cyclicity of ovarian steroid levels. During the early follicular phase, both pituitary sensitivity and reserve seem to be at a minimum, but with increasing levels of E_2 a preferential increase in pituitary reserve than in sensitivity ($P < 0.005$) in response to LRF pulses was eliciated. The finding that pituitary reserve (but not sensitivity) in hypogonadal women is enhanced by estrogen treatment (Lasley et al., 1975b), and negated by the administration of anti-estrogen (clomiphene) during the late follicular phase (Wang and Yen, 1975b) suggests that increasing reserve in the late follicular phase of the cycle is causally related to the rising levels of E_2. This preferential augmented pituitary reserve requires that the rate of synthesis exceeds the rate of release, thereby, increasing the pituitary gonadotropin store. This build-up in pituitary store, in all probability, constitutes the prerequisite for the "development" of the midcycle surge.

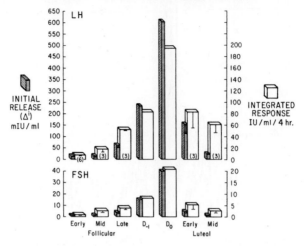

Fig. 7 Analyses of the changes in pituitary sensitivity (Δ_1) and reserve (integrated response) elicited by pulses of LRF during different phases of the menstrual cycle (mean ± SE).

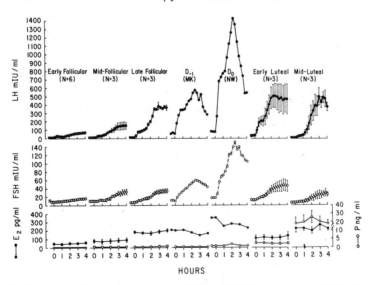

Fig. 8 Basal concentrations and the release of LH and FSH in response to LRF infusion (0.2 µg/min) during various phases of the ovulatory cycles. Basal levels of estradiol (E_2) and progesterone (P) and their changes during the gonadotropin rise are also depicted (mean ± SE). D_1 = the day before midcycle surge. D_0 = the day of midcycle surge.

Pituitary sensitivity and reserve continued to be high during the early luteal phase but reduced progressively thereafter. The presence of progesterone as well

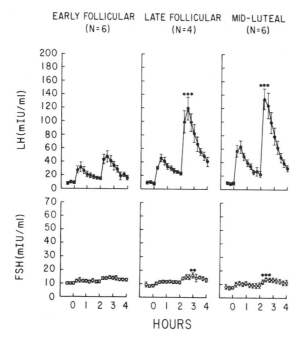

Fig. 9 The quantitative and qualitative comparison between gonadotropin responses to the first and second pulses of LRF during the low (early follicular) and the high (late follicular and mid-luteal) estrogen phases of the cycle (mean ± SE). ** = P < 0.01. *** = P < 0.005.

as high circulating E_2 levels during the mid-luteal phase was associated with almost identical functional capacity (sensitivity and reserve) of the gonadotrophs, as that found in the presence of high E_2 levels alone (Fig.6). This finding provides good evidence that progesterone at a relatively high level does not impair pituitary sensitivity or reserve. On the contrary, we have recently demonstrated that progesterone at a relatively low circulating level (2-5 ng/ml) enhances both sensitivity and reserve in the estrogen-augmented pituitary (Lasley et al., 1975b). Thus, progesterone is probably responsible, in part, for maintaining pituitary sensitivity and reserve at relatively high levels during the early luteal phase. The pituitary sensitivity and reserve during the late luteal phase are reduced (Fig.6), but remain higher than those observed during the early follicular phase at which time they are at the lowest level found during the entire cycle. It should be noted that although the FSH responses are less obvious, they are remarkably parallel to the pattern of LH responses.

d. *Self-priming effect of LRF*

In the late follicular and in the early to mid-luteal phases of the menstrual cycle particularly, the second and subsequent gonadotropin responses to pulsed LRF always exceed the response to the first pulse (Fig.6). One is tempted to postulate that the first pulse not only induces release from the first pool but also "activates" the reserve pool so that subsequent pulses of LRF may induce release of "stored" or reserve gonadotropin. The pattern of gonadotropin release during LRF infusion is also in accord with this concept (Fig.8). This "self-priming" effect

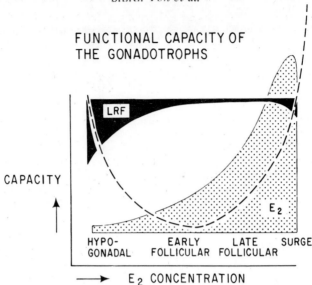

Fig.10 The relationship between the functional capacity of the gonadotrophs and E_2 as a function of concentration and time. The postulated endogenous LRF input is also depicted.

of LRF occurs only during those phases of the menstrual cycle when the serum E_2 concentration is high (Fig.9) and, in fact, can be produced in hypogonadal women through estrogen treatment (Lasley et al., 1975b). Although the physiological significance of the self-priming effect of LRF is not clear, it seems likely that it may serve to activate the reserve pool and render its gonadotropins more readily releasable; this is then revealed as an increase in sensitivity. The finding of a more rapid increase in sensitivity than in reserve from the late follicular to the midcycle surge is consistent with this postulate and is displayed by a dramatic and stepwise increase in the pituitary sensitivity on the day before and during the midcycle surge (Figs 6, 7 and 8). This confirmation of changes in pituitary function may play an important role in the development of midcycle surge. A remarkably similar finding of self-priming by LRF in the rat has been reported (Aiyer et al., 1974). It is concluded that the functional capacity of the gonadotrophs exhibits a remarkable cyclic change and that the adenohypophysis represents a critical feedback site in the development of pre-ovulatory gonadotropin surge.

4. *The Contribution of Endogenous LRF*

These dramatic changes in pituitary sensitivity and reserve provide a basis for rationalizing the hypothalamic component in the operating characteristics of the pituitary. During the late follicular and mid-luteal phases of the cycle, when sensitivity and reserve are found to be high and the self-priming effects of LRF found to be at a maximum, the relatively low basal gonadotropin secretion normally found requires that endogenous LRF release be very low. The modest increase in basal LH secretion observed just prior to the onset of the midcycle surge (Midgley and Jaffe, 1968; Ross et al., 1970; Yen et al., 1970) may reflect the beginnings of incremental LRF secretion; increased amounts of LRF have been found at the time of the midcycle surge in the portal blood of rhesus monkeys (Carmel et al., 1975) and in the peripheral blood of humans (Malacara et al., 1972; Arimura et al., 1974)

The increased pituitary sensitivity and reserve, the development of the self-priming effect of LRF, and the increments in LRF release may represent the essential combination required to induce the stepwise release of LH at the time of the midcycle surge.

A progressive decrease in sensitivity and reserve characterizes pituitary function from the mid-luteal to late luteal phases and into the early follicular phase of an ensuing cycle. This is probably largely the result of the progressive decline in concentration of ovarian steroids on which sensitivity and reserve appear to be dependent. Thus, a continued low level of endogenous LRF may be speculated. However, the elevation in basal LH and FSH release in the face of reduced pituitary sensitivity and reserve, observed during the early follicular phase requires that an increased LRF release be postulated for this period of the cycle. Additionally, a decrease in the speculated ovarian inhibitory factor with preferential action on FSH secretion should be considered in order to account for the higher FSH than LH seen at this time.

THE PARADOX OF THE NEGATIVE AND POSITIVE FEEDBACK CONTROL SYSTEMS ON GONADOTROPIN RELEASE

When the role of estradiol as a feedback modulator of LH release is studied through the administration of this hormone to otherwise normal women, a somewhat puzzling observation is made. Whereas estradiol inhibits the pituitary response to LRF in hypogonadal women (Yen et al., 1974b), in normal women the response of the pituitary to LRF is enhanced (as discussed above). Thus, ovarian steroids are said to exert both a "negative" and "positive" feedback on gonadotropin secretion, and for many years numerous attempts were made to sort out and explain these effects. The positive feedback effect of estradiol on gonadotropin release is viewed as being paradoxical since, under most circumstances, estrogen inhibits gonadotropin secretion. From the results of a series of recent investigations in which estrogen was employed and circulating gonadotropin concentrations measured, it is proposed that the negative and positive feedback action of estradiol is not discontinuous but represents an overlapping event (Yen et al., 1974a). When changes in pituitary capacity (sensitivity and reserve) are being analyzed, the feedback event is represented by a continuum in the form of a U-shaped curve (Yen et al., 1975b). Figure 10 describes this U-shaped functional capacity of the gonadotrophs as influenced by the relative inputs of LRF and E_2. Moderate concentrations of estradiol, such as are found in the early follicular phase of the menstrual cycle, inhibit gonadotropin output but a change in estradiol concentration *in either direction* tends to diminish that inhibition and permit increased activity of the hypothalamic-hypophyseal complex. The normal, unopposed hypothalamic-pituitary system (as in the hypogonadal and ovariectomized woman) produces gonadotropin at a relatively high rate under the influence of increased hypophysiotropic effect of LRF; small amounts of estradiol suppress the gonadotropin output (via inhibition of LRF release) but when the estradiol concentration is further elevated, the ability to suppress the gonadotropin release is no longer operative. Instead, an augmentation of the pituitary capacity occurs principally through the increase in sensitivity and reserve within the population of the gonadotrophs.

ACKNOWLEDGEMENTS

This work was supported by Rockefeller Foundation Grant RF. 75029. We are grateful for the excellent assistance provided by M. Sigel, P. Malcom, J. Aurand, T. Vargo, N. Czekala, B. Hopper and G. Laughlin.

Our special thanks to Dr. R. Guillemin for the generous gift of synthetic LRF and for the continuous collaborations, to J. Hoff for his special assistance.

REFERENCES

Aiyer, M.S., Shiappa, S.A. and Fink, G. (1974). *J. Endocrinol.* **62**, 573.
Arimura, A., Kastin, A.J. and Schally, A.V. (1974). *J. clin. Endocr. Metab.* **38**, 510.
Axelrod, J. (1975). *Rec. Progr. Horm. Res.* **31**, 1.
Ben-Jonathan, N., Mical, R.S. and Porter, J.C. *Endocrinology* **93**, 497.
Bogdanove, E.M. (1972). In "Reproductive Biology" (Balin and Glasser, eds) p.5. Excerpta Medica, Amsterdam.
Boyar, R.M., Finkelstein, J., Roffwarg, H., Kapen, S., Weitzman, E.G. and Hellman, L. (1972). *New Engl. J. Med.* **287**, 582.
Boyar, R.M., Finkelstein, J.W., Roffwarg, H., Kapen, S., Weitzman, E.G. and Hellman, L. (1973). *J. clin. Endocr. Metab.* **37**, 521.
Boyar, R.M., Katz, J., Finkelstein, J.W., Kapen, S., Weiner, H., Weitzman, E.G. and Hellman, L. (1974a). *New Engl. J. Med.* **291**, 861.
Boyar, R.M., Finkelstein, J.W., Roffwarg, H., Kapen, S., Weitzman, E.D. and Hellman, L. (1974b). *J. clin. Invest.* **54**, 609.
Bremner, W.J. and Paulson, C.A. (1974). *J. clin. Endocr. Metab.* **39**, 811.
Burgus, R., Butcher, M., Amoss, M., Ling, N., Monahan, M., Rivier, J., Fellows, R., Blackwell, R., Vale, W. and Guillemin, R. (1972). *Proc. natn. Acad. Sci. U.S.A.* **69**, 278.
Carmel, P.C., Araki, S. and Ferin, M. (1975). *Endocrinology* **96**, 107A.
Davies, I.J., Naftolin, F., Ryan, K.J. and Siu, J. (1975). *J. clin. Endocr. Metab.* **40**, 909.
Dowd, A.J., Barofsky, A.-L., Chanhuri, N., Lloyd, D.W. and Weisz, J. (1975). *Endocrinology* **96**, 243.
Ferin, M., Tempone, A., Zimmering, P.E. and Vande Wiele, R.L. (1969). *Endocrinology* **85**, 1070.
Ferin, M., Khalaf, S., Warren, M., Dyrenfurth, I., Jewelwiez, R., White, W. and Vande Wiele, R.L. (1973). *Endocrinology* **93**, A-221.
Green, J.D. and Harris, G. (1947). *J. Endocrinol.* **5**, 136.
Halaz, B. (1972). *Progr. in Brain Res.* **38**, 97.
Jaffe, R.B. and Keys, W.R., Jr. (1974). *J. clin. Endocr. Metab.* **39**, 850.
Judd, H.L., Parker, D.C., Siler, T.M. and Yen, S.S.C. (1974). *J. clin. Endocr. Metab.* **38**, 710.
Klaiber, E.L., Kobayashi, Y., Broverman, D.M. and Hall, F. (1971). *J. clin. Endocr. Metab.* **33**, 630.
Knobil, E. (1974). *Rec. Progr. Horm. Res.* **30**, 1.
Kobayashi, T., Kobayashi, T., Kato, J. and Menaguchi, H. (1964). *Endocrinol. Jap.* **11**, 283.
Krey, L.C. and Everett, J.W. (1973). *Endocrinology* **93**, 377.
Lasley, B.L., Wang, C.F. and Yen, S.S.C. (1975a). *J. clin. Endocr. Metab., submitted.*
Lasley, B.L., Wang, C.F. and Yen, S.S.C. (1975b). *J. clin. Endocr. Metab. (in press).*
Leblanc, H. and Yen, S.S.C. (1975a). *J. clin. Endocr. Metab., submitted.*
Leblanc, H., Abu-Fadil, S. and Yen, S.S.C. (1975b). *J. clin. Endocr. Metab., submitted.*
Malacara, J.M., Seyler, L.E., Jr. and Reichlin, S. (1972). *J. clin. Endocr. Metab.* **34**, 271.
Martin, J.B., Renaud, L.P. and Brazeau, P. (1975). *Lancet* **II**, 393.
Matsuo, H., Baba, Y., Nair, R.M.G., Arimura, A. and Schally, A.V. (1971). *Biochem. Biophys. Res. Commun.* **43**, 1334.
McCormack, J.T. and Knobil, E. (1975). *Endocrinology* **96**, 108A.
McEwan, B.S., Denef, C.J., Gerlach, J.L. and Plapinger, L. (1974). "Chemical Studies of the Brain as a Steroid Hormone Target Tissue". The Neurosciences 3rd Study Program (F.O. Schmitt and F.G. Worden, eds) p.599. MIT Press, Cambridge, Massachusetts.
Midgley, A.R., Jr. and Jaffe, R.B. (1968). *J. clin. Endocr. Metab.* **28**, 1699.

Midgley, A.R., Jr. and Jaffe, R.B. (1971). *J. clin. Endocr. Metab.* 33, 962.
Naftolin, F., Yen, S.S.C. and Tsai, C.C. (1972). *Nature New Biol.* 236, 92.
Naftolin, F., Ryan, K.J., Davies, I.J., Reddy, V.V., Flores, F., Petro, Z. and Kuhn, M. (1975). *Rec. Progr. Horm. Res.* 31, 295-319.
Osland, R.B., Gallo, R.V. and Williams, J.A. (1975). *Endocrinology* 96, 1210.
Parker, D.C., Judd, H.L., Rossman, L.G. and Yen, S.S.C. (1975). *J. clin. Endocr. Metab.* 40, 1099.
Piacsek, B.E. and Meites, J. (1966). *Endocrinology* 79, 432.
Reiter, E.O., Duckett, G.E. and Root, A.W. (1975). *Pediat. Res.* 9, 294 (A-223).
Ross, G.T., Cargille, C.M., Lipsett, M.B., Rayford, P.L., Marshall, J.R., Strott, C.A. and Rodbard, D. (1970). *Rec. Progr. Horm. Res.* 26, 1.
Santen, R.J. and Bardin, C.W. (1973). *J. clin. Invest.* 52, 2617.
Seyler, L.E., Jr. and Reichlin, S. (1973). *J. clin. Endocr. Metab.* 37, 197.
Stump, W.E., Sar, M. and Keeper, D.A. (1975). *Advances in the Biosciences* 15, 77.
Vale, W. and Rivier, C. (1975). *In* "Neuroendocrine Relationships" (L. Fisher, ed) *(in press).* Raven Press, N.Y.
VandenBerg, G., DeVane, G. and Yen, S.S.C. (1974). *J. clin. Invest.* 53, 1750.
Vande Wiele, R.L., Bogumil, J., Dyrenfurth, I., Ferin, M., Jewelewicz, R., Warren, M., Rixkallah, T. and Mikhail, G. (1970). *Rec. Progr. Horm. Res.* 26, 63.
Wang, C.F., Lasley, B.L. and Yen, S.S.C. (1975a). *J. clin. Endocr. Metab., submitted.*
Wang, C.F. and Yen, S.S.C. (1975b). *J. clin. Invest.* 55, 201.
Wurtman, R.J. (1971). *Neurosc. Res. Program Bull.* 9, 171.
Yen, S.S.C., Vela, P., Rankin, J. and Littell, A.S. (1970). *J. Am. Med. Assn.* 211, 1513.
Yen, S.S.C., Tsai, C.C., Naftolin, F., VandenBerg, G. and Ajabor, L. (1972a). *J. clin. Endocr. Metab.* 34, 671.
Yen, S.S.C., Tsai, C.C., VandenBerg, G. and Rebar, R. (1972b). *J. clin. Endocr. Metab.* 35, 897.
Yen, S.S.C., VandenBerg, G., Tsai, C.C. and Siler, T.M. (1974a). *In* "Biorhythms and Human Reproduction" (M. Ferin *et al.*, eds) Chapter 14, p.219. John Wiley & Sons, Inc., New York.
Yen, S.S.C., VandenBerg, G. and Siler, T.M. (1974b). *J. clin. Endocr. Metab.* 39, 170.
Yen, S.S.C. (1975a). *In* "Annual Review of Medicine: Selected Topics in the Clinical Sciences" (W.P. Creger *et al.*, eds) Vol.26, p.403. Annual Reviews Inc., Palo Alto, California.
Yen, S.S.C., Lasley, B.L., Wang, C.F., Leblanc, H. and Siler, T.M. (1975b). *Rec. Progr. Horm. Res.* 31, 312.

PITUITARY-OVARIAN RELATIONSHIPS IN DISORDERS OF MENSTRUATION

D.T. Baird

*M.R.C. Unit of Reproductive Biology, 39 Chalmers Street
Edinburgh EH3 9ER, Scotland*

SUMMARY

The control of the secretion of gonadotrophins by ovarian steroids is discussed. Cyclical ovarian activity is dependent on the ability of oestrogen from the mature Graafian follicle to provoke an ovulatory discharge of LH (positive feedback). This positive feedback effect of oestrogen is restricted to the female and in primates cannot be demonstrated before puberty. If the positive feedback control in the hypothalamus fails to mature post-pubertally, anovular cycles with associated dysfunctional uterine bleeding may result.

Failure of ovulation and disturbances of menstrual pattern also occurs in women with polycystic ovarian disease. Almost all the oestrogen in this condition is derived by extraglandular conversion of androstenedione, which is secreted in increased amounts by ovaries and/or adrenals. Although both negative and positive feedback effects of oestrogen can be demonstrated, the basal concentration of LH is elevated and relatively constant with an absence of the preovulatory LH surge. The raised level of LH further stimulates the secretion of androstenedione from the ovaries. It is suggested that the constant extraglandular production of oestrogen from androstenedione inhibits the secretion of FSH so that normal follicular development cannot be sustained.

INTRODUCTION

In response to cyclical stimulation by ovarian hormones, bleeding occurs at regular intervals from the endometrium of the uterus in the mature woman. The occurrence of menstrual cycles of normal duration and frequency therefore are dependent on the functional integrity of the feedback mechanisms operating between the hypothalamus, anterior pituitary and the ovary (H.P.O. axis). Disorders of the H.P.O. axis usually present clinically as disorders of menstruation, e.g. amenorrhoea, menorrhagia etc. (Baird and Fraser, 1973). In this chapter I shall summarise the feedback effects of ovarian steroids which are important in maintaining ovulatory menstrual cycles before considering some examples of how a failure in one or more of these mechanisms can lead to a clinical disorder of menstruation.

NORMAL PITUITARY-OVARIAN RELATIONSHIPS

Although the ovary secretes a variety of steroid hormones, only oestradiol-17β and progesterone are known to have established roles in controlling the secretion of pituitary gonadotrophins.

Oestradiol-17β is the main oestrogen secreted by both the Graafian follicle and the corpus luteum in women (Baird and Fraser, 1974). In both phases of the ovarian cycle it is the principal steroid inhibiting the release of gonadotrophins (LH and FSH) from the anterior pituitary (*negative* feedback). The release of FSH is particularly sensitive to negative feedback of oestrogen and a low dose (10 µg per day) will suppress the secretion of FSH without affecting LH (Vaitukaitis *et al.*, 1971). In most species, e.g. the sheep, cow, rat, rabbit, etc. the Graafian follicle is the principal source of oestrogen at all stages of the ovarian cycle (Baird *et al.*, 1975a). Throughout the luteal phase follicular development continues because the release of FSH and LH is controlled by oestradiol - the follicular hormone. In the human, however, and probably in other primates (Knobil, 1974) in the luteal phase over 90 per cent of the oestradiol is secreted by the corpus luteum (Baird and Fraser, 1974). In women therefore in the luteal phase the steroids controlling the release of gonadotrophins originate from the corpus luteum and not the Graafian follicles as in other species. Thus control of FSH and LH is effectively removed from the developing follicles which become atretic (Baird *et al.*, 1975a).

Oestradiol in the female will also provoke a discharge or surge of LH (*positive* feedback). This positive feedback effect of oestrogen is restricted to the female and in primates cannot be demonstrated before the onset of puberty (Dierschke *et al.*, 1974; Reiter *et al.*, 1974). A failure of positive feedback will result in a failure of ovulation and accounts for the occurrence of anovulatory cycles in the few months after the menarche in rhesus monkeys (Dierschke *et al.*, 1974) and in adolescent girls (Doring, 1969).

Progesterone *per se* has little effect on the release of gonadotrophins but inhibits the ability of oestrogen to provoke positive discharge (Dierschke *et al.*, 1973). It may enhance or reinforce the negative feedback effect of oestradiol resulting in suppression of FSH and LH during the luteal phase.

Ovarian androgens (mainly androstenedione) may influence the release of gonadotrophins either directly or by extraglandular aromatization to oestrogens. In sheep neutralization of the biological effect of androstenedione by active immunization, results in an elevation in the secretion of LH and an increase in the rate of ovulation (Martenz *et al.*, 1975). The amount of oestrone produced from androstenedione by extraglandular aromatization (about 40 µg per day) in normal women is small compared to the quantity of oestradiol and oestrone secreted by the ovary (Baird et al., 1969).

In addition to feedback at the hypothalamus, oestradiol has effects on the anterior pituitary and the ovary (Everett, 1961). Oestrogens alter the sensitivity of the anterior pituitary to LH/RH (Keye and Jaffe, 1974); and of the follicle to gonadotrophins (Goldenberg et al., 1972). The modulation of the sensitivity of the hypothalamus and/or pituitary by steroid hormones accounts for the variation in response to LH/RH and oestrogen provocation at different stages of the menstrual cycle (Nillius and Wide, 1972).

DYNAMIC TESTS OF H.P.O. AXIS

The negative and positive feedback effects of oestrogen can be tested by means of an "oestrogen provocation test" (Tsai and Yen, 1971; Marshall et al., 1973; Baird et al., 1975b). We have administered the synthetic oestrogen, ethinyl oestradiol rather than oestradiol benzoate (Nillius and Wide, 1971) because it is then possible to observe any change in ovarian function by measuring the concentration of endogenous oestradiol in plasma. Ethinyl oestradiol (200 μg per day) is given by mouth on Days 5, 6 and 7 and samples of blood are collected every day for 10 days and analysed for LH, FSH, oestradiol, testosterone and androstenedione.

The pituitary responsiveness is tested with synthetic gonadotrophin releasing hormone (LH-RH) 50 μg injected intravenously (e.g. Keye and Jaffe, 1974). Both tests are performed during a period of minimal ovarian activity (excretion of total oestrogens less than 12 μg per 24 h) and the results compared to those found in normal women during the early follicular phase of the cycle. In normal women there is a release of LH and FSH within 5 min of the injection with a peak value occurring at 20-30 min.

In some patients the dynamic relationships of the H.P.O. axis have been investigated by administration of the anti-oestrogen, clomiphene (50-100 mgm per 5 days). By interfering with the negative feedback of endogenous oestrogen, clomiphene causes an elevation in the concentration of FSH and LH (Greenbladt et al., 1968). If follicular development and ovulation occurs in response to clomiphene, it can be assumed that both negative and positive feedback mechanisms are intact.

DISORDERS OF THE H.P.O. AXIS

A. Failure of Positive Feedback

The essential role that follicular oestradiol plays in stimulating the surge of LH necessary for ovulation has already been mentioned. Maturation of the positive feedback mechanism occurs relatively late in puberty so that the first few menstrual cycles are often anovulatory (Doring, 1969). The painless menstrual bleeding characteristic of anovulatory cycles, is often irregular and heavy so that the menstrual interval in the first two years after menarche shows wide variation (Treloar et al., 1967). It has been known for many years that dysfunctional uterine bleeding i.e. frequent and/or excessive uterine bleeding not due to recognizable local pelvic or systemic diseases, is commoner at the extremes of reproductive life, i.e. adolescence and in the pre-menopausal period. Serial measurements of

PITUITARY-OVARIAN RELATIONSHIPS

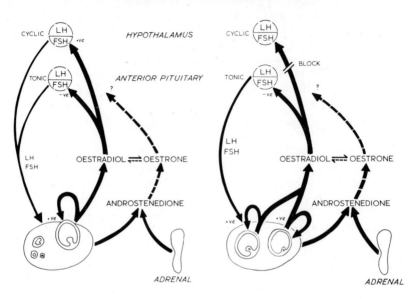

Fig.1 Pituitary-ovarian relationships in dysfunctional uterine bleeding. The release of LH and FSH from the anterior pituitary is controlled by two hypothalamic centres - "tonic" and "cyclic". In the follicular phase of the normal cycle oestradiol from the developing Graafian follicle inhibits and stimulates release of gonadotrophins by feedback at the tonic and cyclic centres respectively. In women with dysfunctional uterine bleeding several follicles may be functionally active. Negative feedback control of gonadotrophins is normal but there is a partial or complete failure of positive feedback. In both situations the amount of oestrogen produced from androstenedione by extraglandular aromatization is negligible (as indicated by the dashed lines) compared to the ovarian secretion of oestradiol (cf. Fig.2).

urinary oestrogen and pregnanediol confirmed the anovulatory nature of this condition (Brown and Matthew, 1962). The concentrations of FSH and LH in plasma are in the normal range found during the follicular phase of the cycle but there is a partial or complete lack of the mid-cycle LH peaks characteristic of ovulatory cycles (Fraser *et al.*, 1973). Administration of clomiphene in patients with adolescent anovulatory dysfunctional bleeding stimulates further follicular development although ovulation fails to occur due to the absence of an LH surge. Direct testing by means of an oestrogen provocation test confirms the virtual absence of positive feedback although negative feedback is intact (Van Look *et al.*, 1975). These patients have a normal response to LH-RH indicating that the failure to release LH in response to oestrogen is not due to pituitary failure.

The functional defect in adolescent girls with anovulatory dysfunctional uterine bleeding can be considered to be a failure of the maturation of the hypothalamic cyclic centre (Fig.1). Ovarian function as indicated by the secretion of oestradiol is apparently normal although multiple follicular development may point to a defect in the mechanisms whereby normally all but one follicle becomes atretic (Fraser et al., 1973; Fraser and Baird, 1974). In the absence of ovulation, the endometrium hypertrophies under the influence of unopposed oestrogen and cystic glandular hyperplasia may develop. About 25 percent of girls who present in adolescence with dysfunctional uterine bleeding and cystic glandular hyperplasia of the endometrium, subsequently develop signs of polycystic ovarian disease with associated anovulatory infertility (Fraser and Baird, 1972). Persistent anovulation may predispose to the development of sclerocystic changes in the ovaries.

Anovulatory dysfunctional uterine bleeding and associated cystic glandular hyperplasia of the endometrium is particularly common in the few years prior to the menopause (Fraser and Baird, 1972). A failure of positive feedback has been demonstrated in these patients similar to that described above in those presenting in adolescence. However the defect appears to be partial with ovulatory cycles occurring sporadically after long periods of anovulation (Baird and Fraser, 1974). Serial measurements of gonadotrophins and ovarian steroids over several months have shown that anovulatory cycles are frequently followed by periods of temporary ovarian failure during which the concentrations of FSH and LH rise towards the menopausal range (P.F.A. Van Look et al., 1975, unpublished results). The aetiology of these periods of ovarian inactivity is unknown although they may indicate that the supply of developing follicles responsive to gonadotrophins is restricted some time before the menopause when complete ovarian failure occurs.

Failure of positive feedback occurs also in about 30 percent of patients presenting with seconday amenorrhoea associated with functional suppression of the hypothalamus (Marshall et al., 1973). The associated failure of tonic release of gonadotrophins results in secondary ovarian failure and amenorrhoea. If gonadotrophin release and follicular development is stimulated by the administration of clomiphene, ovulation does not occur due to the inability of oestrogen to provoke an LH surge (Baird, 1974).

In summary the clinical symptoms of failure of positive feedback mechanism will depend on the associated disorders. If it occurs as an isolated defect then anovulatory infertility and menstrual irregularity will occur. In association with failure of tonic gonadotrophin release, secondary amenorrhoea may be partially or totally unresponsive to treatment with clomiphene.

B. Disturbance of Negative Feedback

The cyclical changes in tonic gonadotrophin secretion occur in response to changes in negative feedback of ovarian steroids. For example the elevation in the concentration of FSH, which is thought to be responsible for initiation of development of the Graafian follicle, occurs at the end of each cycle in response to the decline in secretion of oestradiol (and progesterone) from the corpus luteum as it regresses (Ross et al., 1970). If oestrogens (or androgens) are secreted or produced from outside the ovary or from structures in the ovary other than the Graafian follicle or the corpus luteum, the normal feedback mechanisms may be disturbed.

The adrenal normally secretes very little oestrogen and the quantity of androgens (androstenedione, testosterone and dehydroepiandrosterone) is insufficient to suppress gonadotrophins directly. In pathological situations e.g. congenital adrenal hyperplasia, Cushing's syndrome or adrenal carcinoma the adrenal secretion of androgens and less frequently oestrogens, may be sufficiently high to interfere with H.P.O. relationships. The adrenal production of androstenedione and

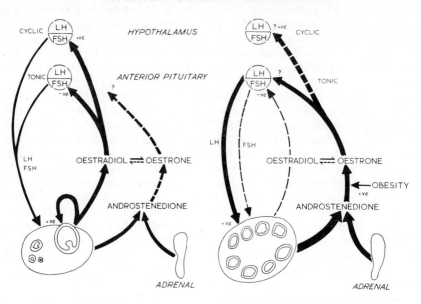

PITUITARY-OVARIAN RELATIONSHIPS

NORMAL FOLLICULAR   POLYCYSTIC OVARY

Fig. 2 Pituitary-ovarian relationships in polycystic ovarian disease. In the follicular phase of the normal cycle the developing Graafian follicle secretes oestradiol which has positive and negative feedback effects at the cyclic and tonic hypothalamic centres respectively. The ovary and adrenal both secrete androstenedione but the amount of oestrogen produced by extraglandular aromatization is small in comparison to the secretion of ovarian oestradiol. In women with polycystic ovarian disease the ovaries (and adrenals) secrete relatively large amounts of androstenedione but very little oestradiol. The large amount of oestrogen (mainly oestrone) produced by extraglandular aromatization of androstenedione, results in constant inhibition of the tonic centre (and FSH secretion) and possibly low grade stimulation of the cyclic centre. The resulting raised basal level of LH stimulates further secretion of androstenedione from the ovary.

testosterone in untreated girls with congenital adrenal hyperplasia, produces virilism with associated amenorrhoea. The high secretion of androstenedione provides a source for extraglandular production of oestrone. When treatment with cortisol is started, the excess androgen production is reduced and cyclical ovarian function usually occurs (Wilkins *et al.*, 1950).

In patients with polycystic ovarian disease (P.C.O.) the ovaries secrete excessive quantities of androstenedione and testosterone at the expense of oestrogens (Mahesh and Greenbladt, 1964). This syndrome as defined originally by Stein and Leventhal in 1935, is characterized by obesity, hirsutism, menstrual irregularity and enlarged polycystic ovaries. Sclerocystic ovaries and oligomenorrhoea is probably the end result of a variety of conditions including Cushing's syndrome (Ianacone *et al.*, 1959) and chronic failure of ovulation (Fraser and Baird, 1972).

In spite of excessive androgen production most women with polycystic ovarian disease, show no evidence of oestrogen deficiency. Indeed the endometrium may be stimulated to such a degree that irregular dysfunctional uterine bleeding occurs (Gemzell *et al.*, 1956). The amount of oestrogen excreted in the urine (10-

25 µg per 24 h) and the total production rate (Jeffcoate et al., 1968) are similar to those found in normal women during the mid-proliferative phase of the cycle (Brown and Matthew, 1962).

A clue to this anomalous situation i.e. minimal ovarian secretion of oestrogen in the presence of a normal production rate is found by studying the concentration of oestrogen in peripheral plasma (Baird, 1973). In contrast to normal women, the concentration of oestrone in plasma from women with polycystic ovarian disease, exceeds that of oestradiol. This finding is similar to that in postmenopausal women in whom the bulk of oestrogen is not secreted as oestradiol but arises indirectly by extraglandular peripheral conversion of androstenedione.

Isotope dilution studies have confirmed that virtually all the oestrogen in women with polycystic ovarian disease is derived from extraglandular conversion of androstenedione (MacDonald et al., 1969). Although the conversion from androstenedione to oestrone is normal (1.3 percent), the elevated production rate of androstenedione provides an abundant source of precursor to be converted to oestrogen. This relatively constant extraglandular production of oestrogen, may have significant effects on control of gonadotrophin secretion.

On the basis of bioassays in urine, it was suggested nearly 20 years ago that the production of LH was raised in polycystic ovarian disease (McCarthur et al., 1958). More recent studies using radioimmunoassay have confirmed that the basal concentration of LH in plasma is about double that found during the normal follicular phase but that there is an absence of the normal mid-cycle peaks (Yen et al., 1970). The concentration of FSH is within the low normal range. Dynamic tests of pituitary-ovarian function in women with polycystic ovarian disease have shown normal negative and positive feedback to oestrogen (Yen et al., 1970) and an exaggerated response to LH–RH (Baird et al., 1975b). These patients are known to be particularly sensitive to clomiphene suggesting that endogenous oestrogen is responsible for inhibiting the secretion of FSH. FSH is more sensitive to negative feedback effect of oestrogen than LH so that the constant low level of oestrogen may be sufficient to inhibit FSH secretion to a level below that necessary to support adequate follicular growth. The concentration of LH found during administration of oestrogen in small amounts is the net result of both negative and positive feedback. There is virtually no information about the effect of oestrone and/or androstenedione on modulating the effect of feedback control of gonadotrophins in women, but the continuous administration of the synthetic oestrogens, results in a sustained rise in the basal secretion of LH (Swerdloff and Odell, 1969).

It is suggested therefore that the basic pathology in polycystic ovarian disease is an excess production of extraglandular oestrogen from androstenedione (Fig.2). This may be due to excess secretion of androstenedione from the ovary and/or adrenal. The relatively constant oestrogen level results in an elevation in basal concentration of LH which further stimulates the ovarian secretion of androstenedione so that the condition is aggravated. HCG is known to stimulate the secretion of androstenedione from the ovaries (Baird et al., 1974) and both the theca interna and stroma show histological evidence of hyperplasia in polycystic ovaries (Morris and Scully, 1958). Obesity by increasing extraglandular aromatization (MacDonald et al., 1969) and stress by stimulating adrenal secretion of androstenedione might be involved in the pathogenesis of the condition. Any therapy which reduces the production of androstenedione e.g. wedge resection of the ovaries or suppression of the adrenal with dexamethasone, will interrupt the vicious circle of events and lead to restoration of cyclical ovarian function.

In a cyclical system such as the menstrual cycle, interruption of any one link in the chain will inevitably lead to a disruption of subsequent events. In one of the examples mentioned in this chapter (anovulatory dysfunctional uterine bleeding) changes in pituitary gonadotrophins and ovarian steroids appear to be secondary to a primary defect in the hypothalamus. It is suggested in polycystic ovarian disease that alterations in hypothalamic-pituitary function may be produced by abnormalities in the secretion of steroids secreted by the ovaries and/or adrenals. Further research is needed to elucidate the underlying endocrine pathology of other disorders of menstruation e.g. short luteal phase.

REFERENCES

Baird, D.T. (1973). *Proceedings of the IV International Congress of Endocrinology*, Washington. 18-24 June 1972, pp.851-856. Excerpta Medica, Amsterdam.
Baird, D.T. (1974). *Brit. J. Hosp. Med.* 12, 49-56.
Baird, D.T. and Fraser, I.S. (1973). *Clinics Endocr. Metab.* 2, 469-488.
Baird, D.T. and Fraser, I.S. (1974). *J. clin. Endocr. Metab.* 38, 779-787.
Baird, D.T., Horton, R., Longcope, C. and Tait, J.F. (1969). *Rec. Progr. Horm. Res.* 25, 611-656.
Baird, D.T., Burger, P.E., Heavon-Jones, G.D. and Scaramuzzi, R.J. (1974). *J. Endocrinol.* 63, 201-202.
Baird, D.T., Baker, T.G., McNatty, K.P. and Neal, P. (1975a). *J. Reprod. Fertil. (in press).*
Baird, D.T., Corker, C.S., Fraser, I.S., Hunter, W.M., Michie, E.A. and Van Look, P.F.A. (1975b). *J. Endocrinol.* 64, 53-54P.
Brown, J.B. and Matthew, G.D. (1962). *Rec. Progr. Horm. Res.* 18, 337-373.
Dierschke, D.J., Yamaji, T., Karsch, F.J., Weick, R.F., Weiss, G. and Knobil, E. (1973). *Endocrinology* 92, 1496-1501.
Dierschke, D.J., Weiss, G. and Knobil, E. (1974). *Endocrinology* 94, 198-206.
Doring, G.K. (1969). *J. Reprod. Fertil.* **Suppl. 6**, 77-81.
Everett, J.W. (1961). *In* "Sex and Internal Secretion" (W.C. Young, ed) **Vol.I**, pp.497-555. Ballière, Tindall & Cox, Ltd.
Fraser, I.S. and Baird, D.T. (1972). *J. Obstet. Gynec. Brit. Commonw.* 79, 1009-1015.
Fraser, I.S. and Baird, D.T. (1974). *J. clin. Endocr. Metab.* 39, 564-570.
Fraser, I.S., Michie, E.A., Baird, D.T. and Wide, L. (1973). *J. clin. Endocr. Metab.* 37, 407-412.
Gemzell, C., Tillinger, K.G. and Westman, A. (1956). *Acta Obstet. Gynec. Scand.* 35, 42-46.
Goldenberg, R.L., Vaitukaitis, J.L. and Ross, G.T. (1972). *Endocrinology* 90, 1492-1498.
Greenbladt, R.B., Zárate, A. and Mahesh, V.B. (1968). *In* "Clinical Endocrinology II" (E.B. Astwood and C.E. Cassidy, eds) pp.630-642. Grune & Stratton, New York.
Ianacone, A., Gabrilove, J.L., Sohval, A.R. and Soffer, L. (1959). *New Eng. J. Med.* 261, 775-780.
Jeffcoate, S.L., Brooks, R.V., London, D.R., Smith, P.M., Spathis, G.S. and Prunty, F.T.G. (1968). *J. Endocrinol.* 42, 213-228.
Keye, W.R., Jr. and Jaffe, R.B. (1974). *J. clin. Endocr. Metab.* 38, 805-810.
Knobil, E. (1974). *Rec. Progr. Horm. Res.* 30 *(in press).*
MacDonald, P.C., Grodin, J.M. and Siiteri, P.K. (1969). *In* "Progress in Endocrinology" (C. Gual, ed) **Vol.184**, pp.770-776. Excerpta Medica, Amsterdam.
McArthur, J.W., Ingersoll, F.M. and Worcester, J. (1958). *J. clin. Endocr. Metab.* 18, 1202-1215.
Mahesh, V.B. and Greenbladt, R.B. (1964). *Rec. Progr. Horm. Res.* 20, 341-379.
Marshall, J.C., Morris, R., Court, J. and Butt, W.R. (1973). *J. Endocrinol.* 59, xxiv (**Abs.**).
Martenz, N.D., Scaramuzzi, R.J., Van Look, P. and Baird, D.T. (1975). *J. Endocrinol.* 64, 46-47P.
Morris, J.M. and Scully, R.E. (1958). *In* "Endocrine Pathology of the Ovary" p.46. C.V. Mosby Co.
Nillius, S.J. and Wide, L. (1971). *J. Obstet. Gynaec. Brit. Commonw.* 78, 822-827.

Nillius, S.J. and Wide, L. (1972). *J. Obstet. Gynaec. Brit. Commonw.* **79**, 865-873.
Reiter, E.O., Kulin, H.E. and Hammond, S.M. (1974). *Paed. Res.* **8**, 740-745.
Ross, G.T., Cargille, C.M., Lipsett, M.B., Rayford, P.L., Marshall, J.R., Shott, C.A. and Rodbard, D. (1970). *Rec. Progr. Horm. Res.* **126**, 1-62.
Stein, I.F. and Leventhal, M.L. (1935). *Am. J. Obstet. Gynec.* **29**, 181-191.
Swerdloff, R.S. and Odell, W.D. (1969). *J. clin. Endocr. Metab.* **29**, 157-163.
Treloar, A.E., Boynton, R.E., Behn, B.G. and Brown, B.W. (1967). *Inter. J. Fert.* **12**, 77-126.
Tsai, C.C. and Yen, S.S.C. (1971). *J. clin. Endocr. Metab.* **33**, 917-923.
Vaitukaitis, J.L., Bermudez, J.A., Cargille, C.M., Lipsett, M.B. and Ross, G.T. (1971). *J. clin. Endocr. Metab.* **32**, 503-508.
Van Look, P.F.A., Fraser, I.S., Hunter, W.M., Michie, E.A. and Baird, D.T. (1975). *J. Endocrinol.* **64**, 54-55P.
Wilkins, L., Lewis, R.A., Klein, R. and Rosemberg, E. (1950). *Bull. Johns Hopk. Hosp.* **86**, 249-252.
Yen, S.S.C., Vela, P. and Rankin, J. (1970). *J. clin. Endocr. Metab.* **30**, 435-442.

HORMONAL REGULATION IN NORMAL AND ABNORMAL MENSTRUAL CYCLES

S.G. Korenman and B.M. Sherman

Department of Medicine, UCLA San Fernando Valley Medical Program
The V.A. Hospital, Sepulveda, California, USA
and
University of Iowa

INTRODUCTION

The menstrual cycle of the human as well as reproductive cycle in all other mammals exhibit regulatory processes quite different from the equilibrium or fluctuating equilibrium processes characteristic of other endocrine systems. The serum Ca^{++} and glucose are retained within a narrow range by hormonal mechanisms. The serum free cortisol is maintained at a fluctuating equilibrium concentration. By contrast, the menstrual or estrus cycle encompasses two climactic events — ovulation and corpus luteum regression which have no parallels elsewhere and whose regulation has many unexplained features with extensive ramifications into other non-equilibrium states such as cell division, differentiation and perhaps neoplasm formation.

As knowledge moved ahead in studies involving various perturbations and their effects on the secretion of reproductive hormones, on the intimate fluctuations of hormone concentrations diurnally, and on the mechanisms of hormonal control at the follicle or corpus luteum, questions relating to the characteristics of various clinical types of menstrual abnormalities remained unanswered. New technology made it possible for us to carry out daily measurements of the four principal reproductive hormones in large numbers of patients over long periods of time. What follows is a description primarily of our own work which has led to a better understanding of the normal, abnormal, and premenopausal menstrual cycle and to more extensive understanding of the processes underlying its regulation. Cyclic features underlying some pathological cycles, so-called experiments in nature, have contributed to our understanding of hormonal regulation.

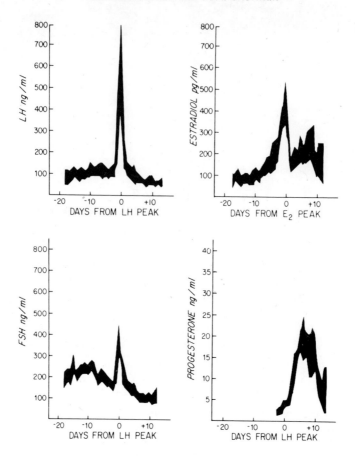

Fig.1 Hormonal features of the normal menstrual cycle in young women. Values are means ± 2 standard errors for 10 cycles of women, ages 18-30.

METHODS

FSH and LH were measured by radioimmunoassay employing reagents obtained from the National Pituitary Agency (Odell *et al.*, 1968a; Odell *et al.*, 1968b). The development of the use of activated charcoal to separate bound from free estrogen (Korenman, 1968) and radioligand procedures employing the uterine estrogen receptor (Korenman *et al.*, 1969; Tulchinsky and Korenman, 1970) permitted the measurement of the serum estrone (E_1) and estradiol (E_2). More recently, a very simple radioimmunoassay system for E_2 has been employed (Korenman *et al.*, 1974). Progesterone (P) has been radioimmunoassayed using a high specific antiserum supplied by Gordon Niswender. It was the advanced procedures, carried out without the need for chromatography which permitted the large number of determinations (> 15,000) made.

All of the studies were carried out on ambulatory volunteers who reported every morning for blood drawing. Many were paid. Some refused payment in the

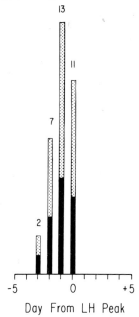

Fig. 2 Incidence of E_2 peak in relation to the LH peak. The solid bars represent cycles studied in our laboratory and the shaded areas, data from published cycles.

interest of scientific advancement. None underwent any significant changes in their life styles. The unstinting cooperation of the women of Iowa is gratefully acknowledged.

THE NORMAL MENSTRUAL CYCLE IN YOUNG WOMEN

Features of the normal menstrual cycle in young women were the subject of a number of studies. Subjects were selected with a history of regular menses. They were characterized by a biphasic basal body temperature curve, a midcycle LH peak, a luteal phase greater than 11 days and were found to have an early follicular phase increase of FSH (Fig.1). The hormone measurements in these cycles are arrayed with the midcycle LH peak as day 0 except for the estradiol which is centered about the E_2 peak. The shaded areas constitute the mean ± standard errors of the mean. In all studies where cycles are compared to the normals, the mean and range for the group are presented. The extreme values shown are representative for the day and do not characterize a particular cycle.

In the normal cycles, the ovulatory LH and FSH peaks (Odell *et al.*, 1968a; Odell *et al.*, 1968b) were seen uniformly. They were either preceded by or concomitant with a substantial rise in E_2 which characterized the second half of the follicular phase (Korenman *et al.*, 1969; Korenman *et al.*, 1970; Tulchinsky and Korenman, 1970; Korenman and Sherman, 1973). The E_2 peak was demonstrated to be on or during the 3 days prior to the LH peak raising the question of a causal relation between E_2 secretion and LH (Fig.2). This concept was further supported by studies employing sequential birth control pills (Swerdloff and Odell, 1969) and by estrogen administration during the early follicular phase (Yen

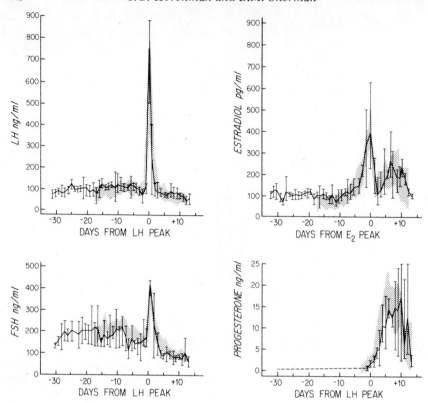

Fig.3 The long follicular phase. Values are means and *range* for 4 cycles in young women with infrequent menses.

and Tsai, 1972). Several important points were raised. The concept of the hypothalamic-pituitary axis as the master gland controlling a passive ovary became untenable in the light of the fact that ovarian secretion exhibited a positive feedback on the hypothalamic-pituitary axis. This made good teleological sense because it was important for the midcycle bursts of LH secretion to take place when the ovary was prepared to ovulate a mature follicle. Other interesting features of the late follicular phase are rapidly fluctuating rising E_2 concentrations in the absence of comparable gonadotropin changes and a striking fall of E_2 secretion clearly prior to ovulation (if ovulation must follow the LH peak) in many cases (Korenman and Sherman, 1973). These findings indicate a substantial degree of ovarian self-regulation. Other mechanisms at play, however, may include a progressive induction of LH receptors as a result of FSH (Zeleznik *et al.*, 1974).

In addition to the anticipated enormous output of progesterone, the luteal phase was characterized by very substantial estradiol secretion. The importance of luteal phase E_2 secretion in species with long reproductive cycles is not established but it is known that one action of estrogens is to induce the uterine progesterone receptor (Milgrom *et al.*, 1972; Illingworth *et al.*, 1975; Janne *et al.*, 1975).

The normal menstrual cycle terminates with corpus luteum atrophy and cessation of its steroid hormone secretion. The basis for the limited and fixed life

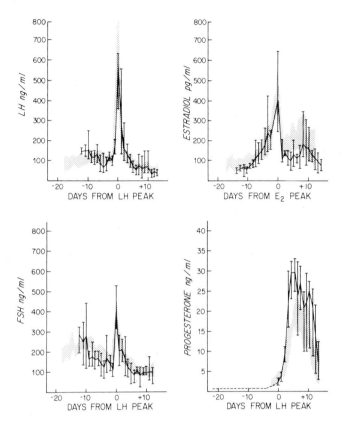

Fig.4 Normal cycles in 40-year-old women.

span of the corpus luteum in the absence of a pregnancy is not understood in humans.

The concentration of estrone fluctuates during the normal menstrual cycle with a small peak in parallel with the estradiol peak (Tulchinsky and Korenman, 1970). This estrogen is derived mainly from the peripheral metabolism of androstenedione and from the oxidation of E_2. Its concentration is subject to substantial diurnal variation. Since the serum concentration of estrone changes little after menopause when effective estrogen concentration is remarkably reduced, its importance as a hormone is questionable.

THE NORMAL MENSTRUAL CYCLE THROUGHOUT LIFE

Today a woman in the United States can expect to have about 40 years of reproductive cycles. Menarche has been occurring at a younger age (Zacharias and Wurtman, 1969), while cessation of menses is occurring later and later in life. In his classic study of menstrual interval throughout life, Treloar (1967) noted increased cycle variability in the initial and terminal years of reproductive cycles. He also noted a progressive decline in mean cycle duration during the middle men-

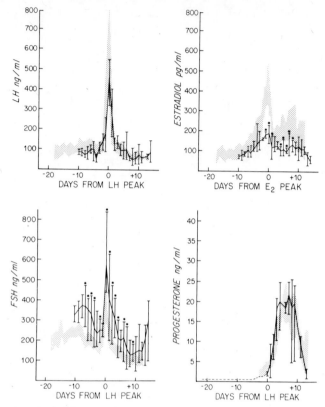

Fig.5 Normal cycles in 50-year-old women.

strual years. Studies of cycles of various lengths during reproductive years by daily measurement have allowed us to characterize these differences (Sherman and Korenman, 1975).

In young women of normal weight with long cycle intervals (Fig.3), it was found that the entire difference was due to prolongation of the follicular phase. Each cycle was terminated by an episode of follicular maturation and corpus luteum formation which was entirely normal with respect to serum FSH, LH, E_2 and progesterone levels. These cycles are termed long follicular phase cycles. It may be concluded from these observations that the concept that anovulation is a typical feature of the long menstrual cycles of young girls can no longer be held.

As women get older, their typical menstrual cycles change significantly in character. In the group of five women age 40-41 shown (Fig.4), we see a shortened follicular phase and significantly reduced luteal phase E_2 concentrations in the face of entirely normal progesterone levels. The mean follicular and luteal phase durations in women of various ages as defined as the interval from the first day of menses to the LH peak and from the LH peak for the first day of menses are indicated in Table I. The striking reduction of follicular phase with age is apparent. In the group of 40-year-old women, we observed in addition a reduction in estrogen secretory response in the luteal phase. Short cycles from 6 women ages 46-51

Hormonal Regulation in Normal and Abnormal Menstrual Cycles 365

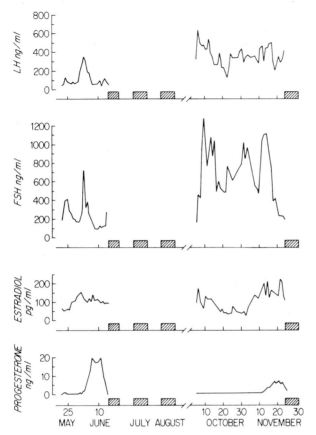

Fig.6 The menopausal transition. Hormone concentrations arrayed by calendar date, in a 50-year-old woman with recent initiation of prolonged menstrual cycles.

Table I.

Age	Follicular Phase	Luteal Phase	Total Length
18-30	16.9 ± 3.7	12.9 ± 1.8	30.0 ± 3.6
40-41	10.4 ± 2.9	15.0 ± 0.9	25.4 ± 2.3
46-51	8.2 ± 2.8	15.9 ± 1.3	24.2 ± 2.9

(Fig.5) were strikingly different from those of the younger women. LH and progesterone concentrations were no different from those previously seen. However E_2 secretion was markedly diminished during both the late follicular phase and the mid-luteal phase. Why the maturing follicle and corpus luteum of the older woman had defective aromatization is not known. The most striking finding was the persistently elevated FSH found throughout the cycle despite a normal LH. This dissociation between FSH and LH concentration was also noted during the

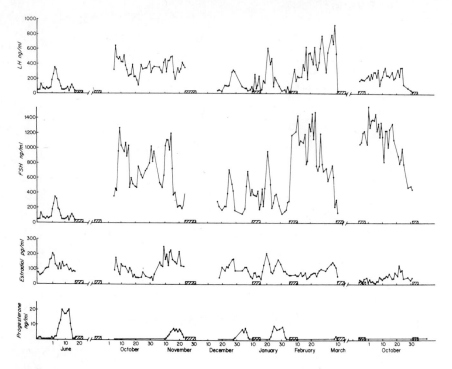

Fig. 7 The menopausal transition. Hormonal features of 10 cycles in a woman undergoing menopause.

menopausal transition (see below).

There are two principal physiological interpretations of these data. Reflecting on the decreased estrogen concentrations, it could be argued that FSH secretion is more sensitive to the negative feedback control of estrogens at the pituitary level than LH. Certainly a pituitary rather than a hypothalamic control point makes sense because all the evidence points to a single GnRH which has a much greater effect on LH than FSH secretion. The second, and to us more attractive hypothesis, is that there is an ovarian *inhibin* similar to the postulated testicular inhibin which specifically exerts negative feedback control over FSH secretion. Late in reproductive life, inhibin secretion would be diminished because of the reduced number of follicles present. There are several lines of evidence to support such a suggestion. On theoretical teleological grounds, it is important for there to be a follicular signal that the appropriate number of follicles for the species are undergoing maturation. Otherwise, problems of super-ovulation would be much more common. Analogy with the male is also supportive. There are no hormones unique to one sex. The evidence for inhibin, particularly to explain the high levels of FSH and normal levels of LH in oligospermic males is growing.

Studies of the menopause and castration have indicated a proportionate elevation of both gonadotropins as well as a proportionate response to estrogen administration (Franchimont *et al.*, 1972; Ostergard *et al.*, 1970; Sherman and

Hormonal Regulation in Normal and Abnormal Menstrual Cycles

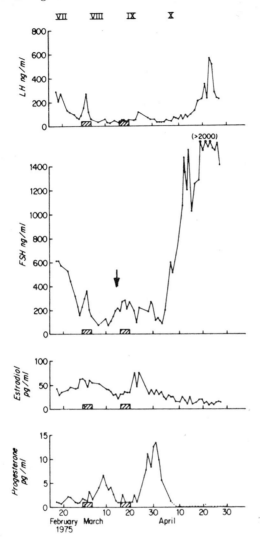

Fig. 8 The menopausal transition. Hormonal features of 10 cycles in a woman undergoing menopause.

Korenman, 1975; Wallach et al., 1970; Wise et al., 1973), suggesting that the hormonal dissociation noted in the older women must have an additional regulatory control. Studies of the menopausal transition were also informative with regard to control mechanisms.

Two women were willing to have blood drawn daily over a number of cycles. Figure 6 illustrates cycles with hormonal concentrations arrayed by calendar date in a 49-year-old woman (Sherman et al., submitted). Both FSH and LH values are quite high and estradiol levels hover in the range of 50 pg/ml. In the first two cycles which were long, a modest rise in E_2 was associated with a rapid roughly proportionate fall in both gonadotropins. Vaginal bleeding followed a fall in estro-

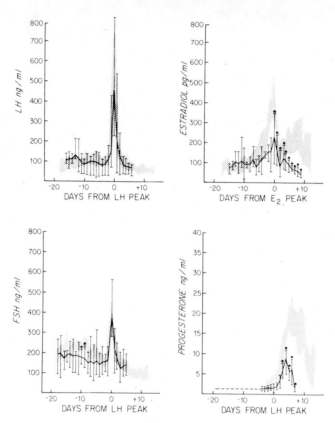

Fig. 9 The short luteal phase. Values were obtained during 7 cycles in 4 subjects.

gen levels and was not preceded by progesterone concentrations greater than 1 ng/ml. These cycles and three of those studied in an additional premenopausal woman constitute the only truly anovulatory cycles seen in our studies. The next cycle studied in this woman several months after establishing anovulation was a short cycle with low normal LH values, a very high FSH, low estrogen values and ample evidence of corpus luteum formation as attested to by progesterone levels.

In Figs 7 and 8, ten cycles in a woman who eventually underwent menopause are shown. Similar features were noted.

The short cycles terminated after a period of progesterone secretion. Three long cycles were anovulatory. In the presumed ovulatory cycles, the FSH-LH dissociation was again apparent. A new feature was the demonstration of what we call luteal phase inadequacy in three cycles. This condition, to be described in detail below, features a luteal phase of normal length with diminished progesterone secretion. This woman underwent menopause after the last cycle shown. Hot flushes began in January 1975 despite the fact that she had substantial FSH elevation for the prior year and long periods of LH elevation as well. In fact, the symptoms began and persisted in the absence of any change in pattern of hormone secretion.

These studies of the longitudinal pattern of hormonal secretion indicate

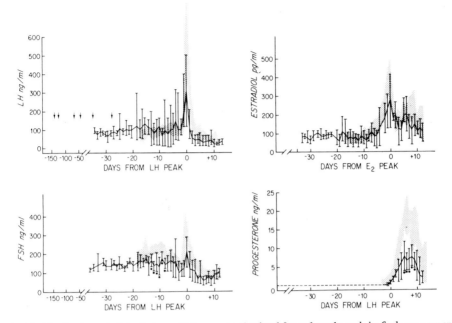

Fig.10 The inadequate luteal phase. Values were obtained from 1 cycle each in 6 obese women, 3 of whom were hirsute.

Table II.

NAME	CLINICAL PRESENTATION
Long Follicular Phase	Long cycles in young women of normal weight.
Short Luteal Phase	Short or normal length cycles in women of normal weight, may be associated with infertility.
Inadequate Luteal Phase	Oligomenorrhic women with obesity in the presence or absence of hirsutism — also during the menopausal transition.
Short Follicular Phase	Progressively short ovulatory cycles with age.
Anovulatory	Many long cycles during the menopausal transition.

the importance of proper stratification. They also serve to test proposed computer models of the menstrual cycle.

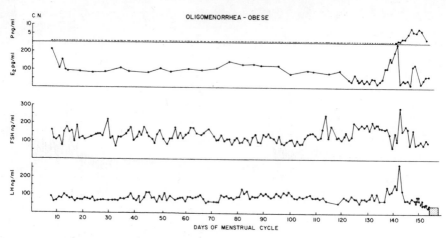

Fig.11 A menstrual cycle in obesity.

ABNORMAL MENSTRUAL CYCLES

Systematic analyses were undertaken in groups of subjects with abnormal cycles defined, in advance, on clinical grounds alone. We were impressed by the uniformity of the pattern found in each group and developed a classification of menstrual cycles based on our findings (Table II). The classification is not intended to be exclusive. It does not include periods characterized by frequent spotting, for example, nor does it identify features of regular cycles in the obese.

The long follicular phase cycles previously described (Fig.3) were completely normal in every other way. Although this finding may not characterize all long cycles near menarche, it certainly serves to explain the common occurrence of pregnancy in sexually active teenagers with oligomenorrhea.

A short luteal phase defined as an interval from the LH peak to the onset of menses of less than 10 days was found in 11 cycles in four women (Sherman and Korenman, 1974a). Each was normal weight and was not hirsute. Each had regular cycles. Two had evidence of infertility by history. The characteristic feature of their cycles are presented in Fig.9. Luteal phase progesterone and E_2 levels were very low. There was also a low concentration of E_2 during the late follicular phase and a significant reduction in the early follicular phase concentration of FSH. These data strongly support the suggestion of Ross (1970) that inadequacy of the corpus luteum had as its basis, defective follicular maturation which could have been due to deficient FSH secretion. Short luteal phases were found to be a repetitive feature of the menstrual cycles of three of the four subjects, suggesting a systematic defect responsible for infertility. Despite the low luteal phase progesterone levels, significant BBT elevation occurred in most cycles for a few days. In women with regular cycles in whom short luteal phases are suspected to be responsible for infertility, the diagnosis may be made by finding an extremely low progesterone level 5 or 6 days before the expected onset of menses.

A new type of abnormal cycle, the inadequate luteal phase was found when six obese women, three of whom were hirsute, were studied (Sherman and Korenman, 1974b). Five had primary infertility and one developed oligomenorrhea and hirsutism after three pregnancies. The characteristics of their menstrual cycles in comparison with the normal group is shown in Fig.10. Note the long latency in five of the six cycles shown prior to initiation of follicular maturation. When these women had menses, they were always preceded by evidence of follicular maturation with normal estrogen and LH fluctuations. As in the short luteal phases, there were deficient early follicular FSH levels. In each case, however, a luteal phase of normal length was associated with low normal estrogen and grossly defective progesterone secretion. In only one case did any daily progesterone level reach 10 mg/ml.

The disparty between estrogen and progesterone secretion by the corpus luteum suggests defective ovarian regulation. However, recent evidence that FSH exerts regulatory control over the number of LH receptors during follicular maturation indicates how difficult it is to assign regulatory primacy in this ecosystem (Zeleznik *et al.*, 1974). In any case, the defect cannot simply be due to FSH deficiency because luteal phase inadequate perimenopausally is related to elevated FSH values. These results give no clue as to a reason for infrequent follicular maturation although in one case, Fig.11, an inexplicable reduction in the *serum* E_2 which had persisted for months in the range of 100 pg/ml resulted in a prompt elevation of FSH with subsequent follicular maturation. It seems likely that these cycles are ovulatory. Infertility then is due to infrequency of ovulation, probably compounded by a poor setting for implantation resulting from inadequate progesterone secretion. It may well be that HCG administration improves the fertility results of cycles induced by clomiphene in such women by improving luteal phase progesterone secretion.

CONCLUSION

Intensive study of the human menstrual cycle over the past ten years has provided considerable understanding of the interrelationships among the component glands, hormones and ovarian cell types to result in this uniquely complex cyclical behavior. These simple descriptive experiments add new elements of complexity to the theory of reproductive cycle regulation. The element of ovarian age must be added (perhaps relating to the number of ancillary follicles available). The elevations of FSH in later years suggests a role for inhibin. Repetitive short and inadequate luteal phases as well as long follicular phase cycles must be incorporated. Factors responsible for pulsatile secretion of the major reproductive hormones independent of fluctuating stimuli must be clarified and causes for the limited life span of the normal and abnormal non-pregnant corpus luteum identified. The reproductive cycle can be viewed as an ecosystem of interdependent components receiving information from and responding to each other which is influenced by external elements such as psychological states, nutrition, stress, and acute and chronic illness.

The study of external influences on closely related body processes is an important goal for the seventies.

REFERENCES

Franchimont, P., Legros, J.J. and Meurice, J. (1972). *Horm. Metab. Res.* 4, 288.
Illingworth, D.V., Wood, G.P., Flickinger, G.L. and Mikhail, G. (1975). *J. clin. Endocr.* 40, 1001-1008.
Janne, O., Kontula, K., Luukkainen, T. and Vinko, R. (1975). *J. Steroid Biochem.* 6, 501-509.
Korenman, S.G. (1968). *J. clin. Endocr.* 28, 127.
Korenman, S.G., Perrin, L.E. and McCallum, T. (1969). *J. clin. Endocr.* 29, 879.
Korenman, S.G., Perrin, L.E., Rao, B. and Tulchinsky, D. (1970). *Res. Steroids* 4, 287.
Korenman, S.G., Stevens, R.H., Carpenter, L.A., Robb, M., Niswender, G.D. and Sherman, B.M. (1974). *J. clin. Endocr.* 38, 718-720.
Korenman, S.G. and Sherman, B.M. (1973). *J. clin. Endocr.* 36, 1205-1209.
Milgrom, E., Atger, M., Perrot, M. and Baulieu, E.E. (1972). *Endocrinology* 90, 1071.
Odell, W.D., Parlow, A.F., Cargille, C. and Ross, G.T. (1968a). *J. clin. Invest.* 47, 2551.
Odell, W.D., Ross, G.T. and Rayford, P.L. (1968b). *J. clin. Invest.* 47, 655.
Ostergard, D.R., Parlow, A.F. and Townsend, D.E. (1970). *J. clin. Endocr. Metab.* 31, 43.
Ross, G.T., Cargille, C.M., Lipsett, M.G., Rayford, P.L., Marshall, J.R., Strott, C.A. and Rodbard, D. (1970). *Rec. Progr. Horm. Res.* 26, 1.
Sherman, B.M. and Korenman, S.G. (1974a). *J. clin. Endocr.* 38, 89-93.
Sherman, B.M. and Korenman, S.G. (1974b). *J. clin. Endocr.* 39, 145-149.
Sherman, B.M. and Korenman, S.G. (1975). *J. clin. Invest.* 55, 699-706.
Sherman, B.M. and Korenman, S.G. (1975) *J. clin. Endocr.*, submitted.
Swerdloff, R.S. and Odell, W.D. (1969). *J. clin. Endocr.* 29, 157.
Treloar, A.E., Boynton, R.E., Benn, B.G. and Brown, B.W. (1967). *Int. J. Fertil.* 12, 77-126.
Tulchinsky, D. and Korenman, S.G. (1970). *J. clin. Endocr.* 31, 76.
Wallach, E.E., Root, A.W. and Garcia, C.R. (1970). *J. clin. Endocr. Metab.* 31, 376.
Wise, A.J., Gross, M.A. and Schalch, D.S. (1973). *J. Lab. Clin. Med.* 81, 28.
Yen, S.S.C. and Tsai, C.C. (1971). *J. clin. Invest.* 50, 1149.
Yen, S.S.C. and Tsai, C.C. (1972). *J. Clin. Endocr.* 34, 298.
Zacharias, L. and Wurtman, R.J. (1969). *NEJM* 280, 868-876.
Zeleznik, A.S., Midgley, A.R. Jr. and Reichert, L.E. Jr. (1974). *Endocrinology* 95, 818-825.

FUNCTIONAL ABERRATIONS OF THE HYPOTHALAMIC-PITUITARY SYSTEM IN POLYCYSTIC OVARY SYNDROME: A CONSIDERATION OF THE PATHOGENESIS

S.S.C. Yen, C. Chaney and H.L. Judd

Department of Reproductive Medicine, School of Medicine University of California, San Diego, La Jolla, California 92093, USA

INTRODUCTION

Despite extensive investigations made during the past forty years, the pathophysiology of polycystic ovary syndrome (Stein and Leventhal, 1935), remain elusive. An erratic or inappropriately elevated LH secretion with a relatively constant and low FSH release has been found (McArthur et al., 1958; Keetel et al., 1957; Yen et al., 1970). It is not known whether this pattern of gonadotropin secretion is due to a primary defect in the hypothalamus-pituitary system, or as a result of inappropriate feedback occasioned by the abnormal steroidogenesis of the polycystic ovary (Warren and Salhanick, 1961; Mahesh and Greenblatt, 1962). Recent advances in the understanding of neuroendocrine regulation of the adenohypophysis in humans have afforded a renewed effort in the delineation of the functional aberrations in this syndrome. In addition, the rapidly expanding knowledge in the mechanism of human puberty has permitted a formulation in the genesis and maintenance of the polycystic ovary syndrome (PCO).

THE MAINTENANCE OF CHRONIC ANOVULATION

The central issue in PCO syndrome is chronic anovulation. This is associated with elevated levels of circulating testosterone (T), androstenedione (A) (Bardin et al., 1968; Horton and Neisler, 1968; Kirschner and Jacob, 1973), dehydroepiandrosterone (DHEA) and dehydroepiandrosterone sulphate (DS) and estrone (E_1) but not estradiol (E_2)(De Vane et al., 1975) (Figs 1 and 2).

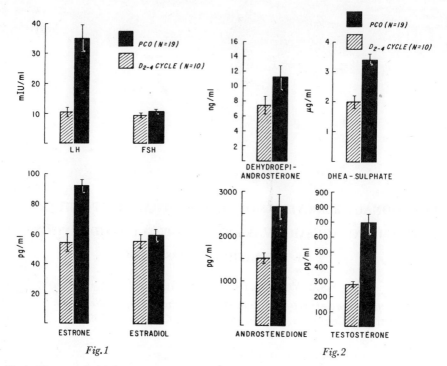

Fig.1 The mean (±S.E.) concentrations of LH, FSH, E_1 and E_2 in 19 patients with PCO and 10 normal subjects between day 2 to 4 of their menstrual cycles.

Fig.2 The mean (± S.E.) concentrations of T, A, DHEA, and DS in 19 patients with PCO and 10 normal subjects between day 2 to 4 of their menstrual cycles.

The adrenal contribution to the excessive androgen production, in addition to the ovarian source, has recently been amplified; the hirsute woman who had an exaggerated diurnal swing of A, also had an exaggerated response to ACTH (Givens et al., 1975). Similar observations have been made for DHEA and DS (Barnett and Yen, in preparation). These findings suggest an increased responsiveness of adrenal androgen to endogenous ACTH and support the hypothesis that adrenal hyperfunction plays a significant role in the pathophysiological event of PCO syndrome.

1. Inappropriate Gonadotropin Secretion

Inappropriately elevated LH release and low FSH secretion were present in most patients with PCO syndrome (Yen et al., 1970). The random LH levels exhibited an even distribution without evidence of clumping in either the high or low range (Fig.1). This finding suggests that a separation of PCO patients into high and low LH groups cannot be meaningfully accomplished as attempted by others. This high circulating LH level appears to be maintained by and temporally related to the exaggerated pulsatile LH discharge (Fig.3), either in the form of enhanced amplitude or increased frequency (oscillations). This preferential augmented LH release exhibits marked day-to-day variation in a random fashion, with alternating periods of release resulting in LH levels comparable to normal cycling women to concentrations similar to those observed in postmenopausal women.

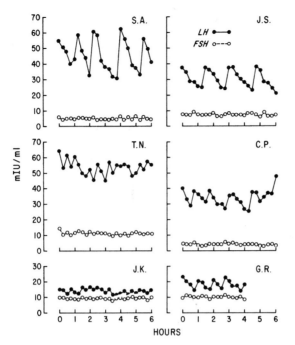

Fig.3 The variability and exaggerated patterns of pulsatile LH release (but not FSH) in 6 patients with PCO.

Frequently, daily LH excursions are so great they resemble, in magnitude, the spontaneous midcycle surge. The concomitant surge of LH and FSH sometimes preceded by a rapid increase in endogenous circulating E_2 levels (Fig.4) was reminiscent of the temporal event of the positive feedback action of E_2 and the acute gonadotropin release at midcycle. In this patient, ovulation did not occur, probably due to a lack of matured follicles (Goldzieher and Green, 1962). However, a rapid ovulatory advancement could be appreciated where few follicles near maturity are present and account for the occasional occurrence of ovulation in these patients. The exaggerated estradiol secretion, in normal women than in patients with PCO, in response to comparable clomiphene induced gonadotropin rises (Fig.5), suggests that events leading to follicular maturation and ovulation are not simple functions of gonadotropin stimulation, and that subsequent increases in steroidogenesis, especially estrogen and androgen may respectively exert an intraovarian influence in the enhancement or inhibition of gonadotropin-ovarian interaction. At any rate, it is demonstrated that the positive feedback action of estradiol, either of endogenous or exogenous source, is operationally intact in PCO patients (Figs 5 and 6).

That the high LH levels were not the result of a defect in the normal negative feedback of estrogen on gonadotropin release in PCO patients was shown by two experiments. Firstly, E_2 infusions resulted in an acute reduction of baseline levels and attenuation of pulses of LH (Fig.7). This response is similar to the ones observed in normal cycling women and in hypogonadal subjects (Tsai and

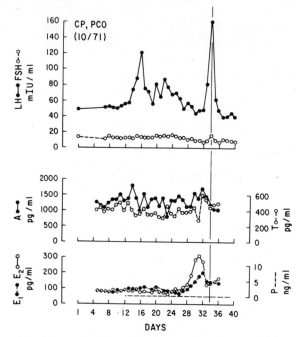

Fig.4 Long-term daily fluctuation of gonadotropins, androstenedione (A), testosterone (T), estradiol (E_2), estrone (E_1) and progesterone levels in a PCO patient studied for 34 days.

Yen, 1971; Yen et al., 1972b). Secondly, the administration of the anti-estrogen, clomiphene, resulted in similar qualitative and quantitative rises of LH and FSH in both PCO patients and normal controls (Fig.5), presumably, through competitions of estradiol binding at the hypothalamus and pituitary sites.

In response to maximal LRF stimulation (150 μg), the pituitary sensitivity and the integrated release for both LH and FSH are several-fold greater in patients with PCO as compared to responses during the low and high estrogen phases of the normal cycle (Fig.8a). The degree of pituitary responses to LRF appears to be dependent on the levels of basal LH concentration; the higher the basal LH, the greater the LRF mediated LH release. The increased pituitary sensitivity to LRF can also be elicited by a smaller dose of LRF (Fig.8b). The first bolus of LRF (10 μg) induced a response of LH release, which was four times greater than those seen in normal women during the early follicular phase (Yen et al., 1975a).

The progressive decrease in pituitary responses to pulses of LRF stimulation may reflect a partial depletion of pituitary gonadotropin store occasioned by the enhanced release as a consequence of an increased pituitary sensitivity. Hence, the pituitary reserve when viewed as the availability of releasable LH pool may actually be increased in these patients. These interpretations are derived from observations made in identical experiments during the early and mid follicular phase of the cycle where the sensitivity is much lower and LH increments in response to successive pulses of LRF are found to be stable (Yen et al., 1975a). Nevertheless, the demonstration of a heightened pituitary sensitivity to LRF offers sufficient explanation for the occurrence of the exaggerated pulsatile LH release in this syndrome without implicating an associated increase in endogenous LRF secretion.

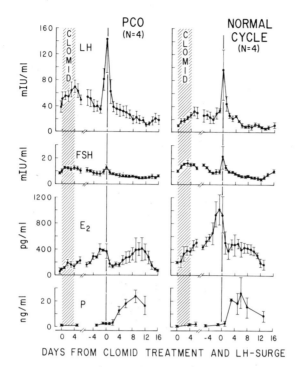

Fig.5 The effects of clomiphene (100 mg/d × 5 days) on the release of gonadotropins and associated changes in ovarian steroids in 4 PCO patients as contrasted to 4 normal women receiving same treatments during the early follicular phase of the cycle (mean ± S.E.).

The increased pituitary sensitivity to LRF is likely related to the chronically inappropriate estrogen levels found in these patients (DeVane et al., 1975). A positive correlation between E_1 and E_2 levels and the basal LH concentration has been demonstrated and a significant correlation between E_1 and LH increments to LRF was also found (DeVane et al., 1975; Rebar et al., 1975). The recent demonstration that the pituitary sensitivity to LRF is amplified during the late follicular phase and can be augmented by estrogen administration in normal and hypogonadal women (Yen et al., 1975a), offers direct support for an estrogen induced, high pituitary sensitivity in PCO patients resulting in an altered feedback "set point". Siiteri and MacDonald (1973) have demonstrated that more than half of estrone production is derived from peripheral conversion of excessive amounts of androstenedione found in this syndrome, and the quantitative conversion is related to body weight. They have speculated that this extraglandular source of estrogen may be etiologically important in the maintenance of chronic anovulation. Our data provide evidence for such causal relationships.

The disparity between LH and FSH levels found in this syndrome may be explained by 1) preferential inhibition of FSH than LH by elevated estrogen; 2) relative insensitivity of FSH to LRF (Yen et al., 1975b); 3) the secretion of an "inhibin-like" substance by the multi-follicular cyst which preferentially inhibits FSH secretion and finally 4) the presence of a dysfunction of a separate FSH-releasing factor cannot be excluded.

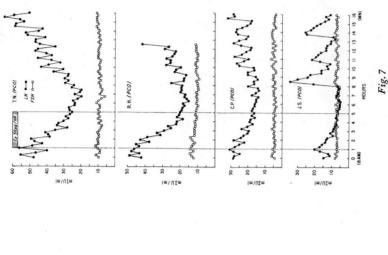

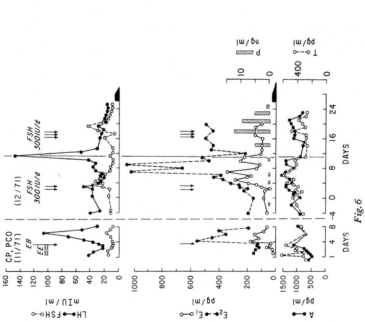

Fig. 6 The gonadotropin, E_1, E_2, and P responses to exogenous estrogen administration (EE 50 μg/d × 3 days followed by 1 mg i.m. injection of EB, Nov, 1971) and to exogenous FSH administration (Dec., 1971) in 1 PCO patient.

Fig. 7 The negative feedback effect of 17β-estradiol infusion (50 μg/h × 4 h) on the pulsatile release and the decline of basal concentrations of LH but no FSH changes were noted.

The Hypothalamic-Pituitary System in Polycystic Ovary Syndrome

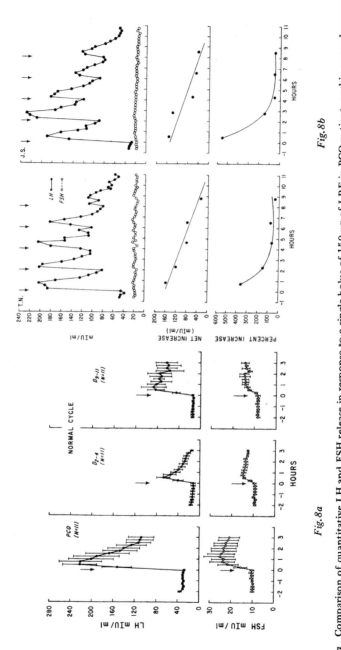

Fig. 8a Comparison of quantitative LH and FSH release in response to a single bolus of 150 μg of LRF in PCO patients and in normal women during the early and late follicular portions of their cycle (mean ± S.E.).

Fig. 8b Changes in serum concentrations, the net increase, and percent increments of LH and FSH in response to pulses of LRF (10 μg) at 2 h intervals.

PROPOSED MECHANISM FOR PERSISTENT ANOVULATION
IN POLYCYSTIC OVARY SYNDROME

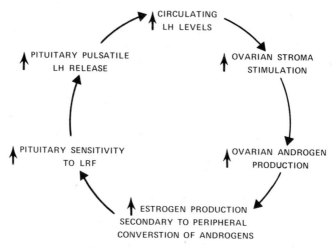

Fig. 9 A proposed mechanism for the persistent anovulation in PCO syndrome.

2. *Pathophysiology of Chronic Anovulation*

We have presented a series of experiments to gain quantitative and qualitative information concerning the hypothalamic-pituitary function in PCO patients. Our findings suggest that the abnormal gonadotropin secretion seen in these patients is not due to an inherent defect of the hypothalamic-pituitary system, but represents a functional derangement consequent to the chronic inappropriate estrogen feedback. Based on these observations, it would appear that a vicious cycle is present which perpetuates chronic anovulation; high levels of LH stimulate the ovary to secrete increasing amounts of androgen, particularly androstenedione and its subsequent conversion to estrogen, which in turn, augments the pituitary sensitivity and an increased LRF mediated LH release. Consequently, exaggerated pulsatile LH release resulted which led to an inappropriately elevated circulating LH. This event is illustrated in Fig. 9. According to current understanding of the time and dose-related feedback action of estrogen, interruption of this altered "set-point" of the hypothalamic-pituitary system should and does occur in a predictable fashion by an increase in circulating ovarian signal, estradiol levels, promoted either by exogenous administration or by endogenous secretion via the addition of FSH stimulation and clomiphene treatment. The lack of follicular maturation in these ovaries may be a consequence of a relative deficiency of FSH or it is likely due to an inhibition of follicular maturation by the increased intraovarian androgen concentration. This hypothesis has been advanced for some time, but good evidence has become available only recently. Louvet *et al.* (1975) showed that in hypophysectomized rats, the ovarian weight response to DES treatment is inhibited by small doses of HCG or LH but not by FSH. This effect is negated by the treatment with anti-androgen or testosterone antiserum. The hypothesis of ovarian enzymatic deficiency for the conversion of androgen to

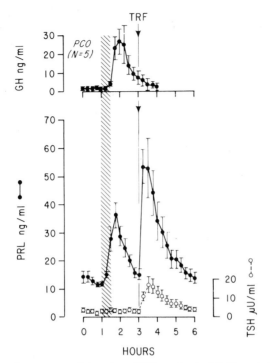

Fig.10 Responses of serum growth hormone (GH), prolactin (PRL), and thyrotropin (TSH) levels (mean ± S.E.) in 5 PCO patients receiving sequential arginine infusion 0.5 mg/kg × 30 min (hatched area) and TRF (500 μg) administrations.

estrogen as a primary etiological importance in this syndrome is no longer tenable, since normal ovulatory function can be restored by the availability of appropriate gonadotropin stimulation and by wedge resection. Hypothalamic-pituitary function, other than gonadotropin regulation, appears to be within normal limits (Fig.10) but the integrity of ACTH-adrenal axis, not evaluated in this study, remains a critical issue to be investigated.

THE GENESIS OF PCO SYNDROME

1. *Clinical and Biochemical Characteristics*

Chronic anovulation, androgen excess and the manifestation of anabolism (obesity) are salient features of the PCO syndrome. Although patients with this syndrome are heterogeneous (Goldzieher and Green, 1962), certain common denominators can be found. Analyses of the clinical characteristics in 98 consecutive patients seen by us have revealed 1) the mean age at menarche was 12.3 years (Fig.11), as compared to 12.9 years in the normal population (Frisch and Revelle, 1970); 2) menstrual irregularity suggesting the presence of anovulation is reported in a vast majority of cases (Fig.12); 3) the clinical definable excessive hair growth, as corroborated by both patients and their mothers, is noticed either before or around the time of menarche in most instances (Fig.13) and 4) heaviness prior to

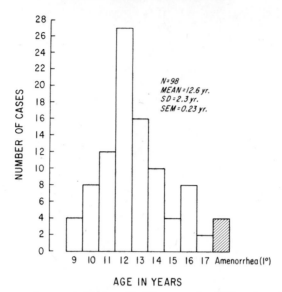

Fig.11 The mean age of menarche analysed in 98 consecutive PCO patients (documented by laparoscopy).

menarche is a constant feature (90 of 98 cases).

All these clinical characteristics are pointing to the occurrence of biochemical aberrations chronologically before puberty and prior to the final maturation of the hypothalamic-pituitary-ovarian system. The onset of normal puberty is heralded by an event of adrenarche (Wilkins, 1957) with an increase in adrenal size (Nathanson et al., 1941) and a rise in adrenal steroid production; in girls, a rapid increase in circulating DHEA was observed between ages 11-12 and reached adult levels by age 12. This was accompanied by a slow and progressive increase in DS, reaching adult levels by age 15 (Hopper and Yen, 1975). Since both DHEA (Kirschner et al., 1973) and DS (Loriaux and Noall, 1969) can be converted to Δ^4 androstenedione and to biological active testosterone, and DS can produce estrogenic and androgenic activities in man (Drucker et al., 1972), the androgenic manifestations during early puberty in girls may be causally related to the increased adrenal activity, which includes cortisol secretion. Thus, our working hypothesis in the pathogenesis of PCO syndrome is based on an exaggerated adrenarche.

2. *PCO A Functional Disorder at Puberty*

The mechanism for the postulated enhanced adrenal androgen secretion during the early phase of puberty remains obscure. The recent disclosure of the sleep-associated increase in pituitary LH secretion which occurs selectively at the time of puberty has drastically revised our understanding of the pubertal event (Boyar et al., 1973; Judd et al., 1974; Parker et al., 1975). Since this augmented LH release during sleep is independent of gonadal activity, as was found in gonadal dysgenesis, a renewed concept of the "CNS-program" for the activation of the hypothalamic-pituitary system as a critical step in the initiation of puberty has been proposed. Based on the finding that in patients with congenital adrenal hyperplasia

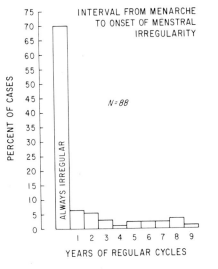
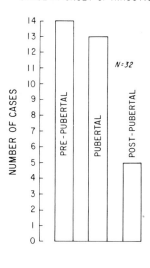

Fig.12   *Fig.13*

Fig.12 The high incidence of menstrual irregularities analysed in 88 patients with PCO.

Fig.13 Excessive hair growth frequently occurring before or around the time of menarche.

an excessive adrenal androgen was associated with earlier initiation of this "pubertal program" (Boyar et al., 1973), these investigators have considered the possibilities that adrenal androgen may be the initiating factor for the CNS program. The recent concept of a critical body weight, required for the onset of menarche, stems from the observation that the mean weights at menarche are not significantly different between girls with early and late pubertal onset (Frisch and Revelle, 1970). This phenomenon may well reflect the effect of increasing levels of adrenal anabolic hormones in concert with other metabolic hormones of the anterior pituitary in the regulation of the initial growth spurt. Consideration must be given to the possibility that the CNS (Naftolin et al., 1975), as well as peripheral conversion of androgen to estrogen (Siiteri and MacDonald, 1973), may be a critical step in the enhancement of CNS maturation. The increased body surface area, especially adipose tissue, should and does facilitate this conversion. Thus, adrenarche, attainment of critical body weight and consequently the availability of the optimal pool size for the conversion of androgen to estrogen may all be integrated with the sleep induced gonadotropin output to account for the multifactorial event of human puberty. Overproduction of adrenal androgen in this scheme may result in an elevation of estrogen, especially within the CNS-pituitary system, consequently, inappropriate feedback and anovulation.

Based on the foregoing findings and discussion, an adrenal hyperfunction at the time of adrenarche may constitute the initiating factor for the development of exaggerated anabolism (obesity), androgen stimulation (hirsutism) and inappropriate estrogen feedback due to conversion of androgen to estrogen (anovulatory cycle) at the time of puberty. This hypothesis is illustrated in Fig.14 and represents

PROPOSED PATHOGENESIS OF PCO

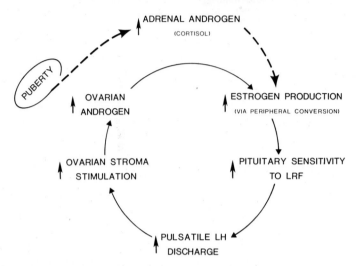

Fig.14 The role of exaggerated adrenarche in the initiation of chronic anovulation and the development of PCO syndrome.

our direction in the future delineation of the pathogenesis of the PCO syndrome.

ACKNOWLEDGEMENT

The work was supported by The Rockefeller Foundation Grant RF-75029.

REFERENCES

Bardin, C.W., Hembree, W.C. and Lipsett, M.B. (1968). *J. clin. Endocr. Metab.* **28**, 1300.
Boyar, R.M., Finkelstein, J.W., David, R., Rottwarg, H., Kapen, S., Weitzman, E.D. and Hellman, L. (1973). *New Engl. J. Med.* **289**, 282.
DeVane, G.W., Czekala, N.C., Judd, H.L. and Yen, S.S.C. (1975). *Am. J. Obstet. Gynec.* **121**, 496.
Drucker, W.D., Blumberg, J.M., Gandy, H.M., Davis, R.R. and Verde, A.L. (1972). *J. clin. Endocr. Metab.* **35**, 48.
Frisch, R.E. and Revelle, R. (1970). *Science* **169**, 397.
Givens, J.R., Anderson, R.N., Ragland, J.B., Wiser, W.L. and Umstot, E.S. (1975). *J. clin. Endocr. Metab.* **40**, 988.
Goldzieher, J.W. and Green, J.A. (1962). *J. clin. Endocr. Metab.* **22**, 325.
Hopper, B.R. and Yen, S.S.C. (1975). *J. clin. Endocr. Metab.* **40**, 458.
Horton, B. and Neisler, J. (1968). *J. clin. Endocr. Metab.* **28**, 479.
Judd, H.L., Parker, D.C., Siler, T.M. and Yen, S.S.C. (1974). *J. clin. Endocr. Metab.* **38**, 710.
Keetel, W.C., Bradbury, J.T. and Stoddard, F.J. (1957). *Am. J. Obstet. Gynec.* **73**, 954.
Kirshner, A. and Jacobs, J. (1971). *J. clin. Endocr. Metab.* **33**, 199.
Kirschner, M.A., Sinhamahapatra, S., Zucker, I.R., Loriaux, L. and Nieschlag, E. (1973). *J. clin. Endocr. Metab.* **37**, 183.
Loriaux, D.L. and Noall, M.W. (1969). *Steroids* **13**, 43.

Louvet, J-P., Harman, S.M., Schreiber, J.R. and Ross, E.T. (1975). *Endocrinology* **97**, 366.
Mahesh, V. and Greenblatt, R.B. (1962). *J. clin. Endocr. Metab.* **22**, 441.
McArthur, J.W., Ingersoll, F.M. and Worcester, J. (1958). *J. clin. Endocr. Metab.* **18**, 1202.
Nathanson, I., Towne, L.E. and Aub, J.C. (1941). *Endocrinology* **28**, 851.
Naftolin, F., Ryan, K.J., Davies, I.J., Reddy, V.V., Flores, F., Petro, Z. and Kuhn, M. (1975). *Rec. Progr. Horm. Res.* **31** *(in press)*.
Parker, D.C., Judd, H.L., Rossman, L.G. and Yen, S.S.C. (1974). *J. clin. Endocr. Metab.* **40**, 1099.
Rebar, R., Judd, H.L., Yen, S.S.C., Rakoff, J., VandenBerg, G. and Naftolin, F. (1975). *J. clin. Invest.*, submitted.
Siiteri, P.K. and MacDonald, P.C. (1973). *In* "Handbook of Physiology-Endocrinology" (R.O. Greep and E. Astwood, eds) **Vol.II**, Part I, p.615. American Physiological Society, Washington, D.C.
Stein, I.F. and Leventhal, M.L. (1935). *Am. J. Obstet. Gynec.* **29**, 181.
Tsai, C.C. and Yen, S.S.C. (1971). *J. clin. Endocr. Metab.* **32**, 766.
Warren, J.C. and Salbanick, H.A. (1961). *J. clin. Endocr. Metab.* **21**, 1218.
Wilkins, L. (1957). "The diagnosis and treatment of endocrine disorders in childhood and adolescence", 2nd Ed., Springfield, Ill. Charles C. Thomas.
Yen, S.S.C., Vela, P. and Rankin, J. (1970). *J. clin. Endocr. Metab.* **30**, 435.
Yen, S.S.C., Tsai, C.C., Naftolin, F., VandenBerg, G. and Ajabor, L. (1972a). *J. clin. Endocr. Metab.* **34**, 671.
Yen, S.S.C., Tsai, C.C., VandenBerg, G. and Rebar, R. (1972b). *J. clin. Endocr. Metab.* **35**, 897.
Yen, S.S.C., VandenBerg, G., Rebar, R. and Ehara, Y. (1972c). *J. clin. Endocr. Metab.* **35**, 931.
Yen, S.S.C., Lasley, B.L., Wang, C.F., Leblanc, H. and Siler, T.M. (1975a). *Rec. Progr. Horm. Res.* **31**, 321.
Yen, S.S.C.,(1975b). *In* "Ann. Rev. Med.: Selected Topics in the Clinical Sciences" (W.P. Creger *et al.*, eds) **Vol.26**, p.403. Annual Reviews Inc., Palo Alto, California.

CORTICOID TREATMENT OF THE SHORT LUTEAL PHASE:

A CASE REPORT

E. Friedrich, E. Keller and A.E. Schindler

*Abteilung Gynäkologie und Geburstshilfe der RWTH
Aachen und Universitäts-Frauenklinik, Tübingen, W. Germany*

INTRODUCTION

The short luteal phase was defined as a period of 8 days or less from the LH peak* to the onset of menstrual bleeding with progesterone levels below 5 ng/ml (Ross et al., 1970; Strott et al., 1970). In previous investigations evidence was obtained that short luteal phase cycles showed subtle changes not only of progesterone concentrations, but also in the sequence of hormonal events preceding the LH peak. FSH levels were found to be below normal or in the low normal range (Ross et al., 1970; Sherman and Korenman, 1974). Midcycle FSH/LH ratios of these cycles were significantly lower than normal controls. Moreover, Sherman and Korenman (1974) reported decreased estradiol peak values in 4 women with a luteal phase of less than 10 days. Abnormal patterns of 17-OH-progesterone were observed by Ross et al. (1970). The possible role of androgens in the pathogenesis of the short luteal phase however has not been discussed in the literature. We therefore wish to report a case of inadequate luteal phase that did not seem to fit into any of the categories of patients previously described and was successfully treated with 16-methylprednisolone.

CASE REPORT

A 22-year-old woman of mediterranean origin presented with a regular 21-23 day cycle with a short luteal phase. Menarche began at 17 and menstruation had been irregular for 1 year and subsequently regular with 28 day intervals. At the age of 21, intervals shortened to 21-23 days and menstrual bleeding increased in quantity and duration (8 days). With the exception of a 3 months' period at the

---

* The following abbreviations are used: Follicle stimulating hormone: FSH; Luteinizing hormone: LH; Androstenedione: Δ^4; Dehydroepiandrosterone: DHEA.

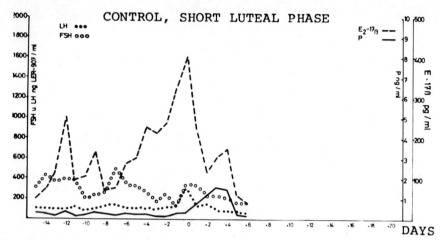

Fig.1 Correlations of FSH, LH, estradiol-17β (E_2-17β) and progesterone (P) plasma concentration patterns centered on the LH peak (day 0) throughout a pretreatment cycle.

age of 20 she had not used hormonal contraception and she had never been pregnant.

Body weight and blood pressure were normal. Galactorrhea could not be demonstrated and a minor mustache was the only sign of hirsutism. Pelvic examination was unremarkable and clitoromegaly could not be demonstrated.

Venous blood was obtained daily throughout a 22 day cycle without medication (control). After this control cycle the patient received 100 mg clomid daily for the first 5 days of the following cycle. This cycle lasted for 18 days and during the next cycle no medication was given. Subsequently the patient received 16-methylprednisolone, 6 mg daily for 3 complete cycles. Venous blood was again drawn daily throughout the clomid cycle and during the third corticoid treatment cycle and plasma samples were kept at $-20°C$ until assayed.

METHODS

Plasma FSH and LH were determined by the double antibody technique. LER 960 (HLH) and LER 1366 (HFSH) were used for iodination, LER 907 as standard reference preparation and anti-HLH (batch 1) as well as anti-HFSH (batch 3) as antisera. Specific radioimmunoassays for estradiol-17β, progesterone, Δ^4, DHEA and testosterone were developed in our laboratory. In brief, antisera against 6-keto-17β-estradiol, 6 CMO:BSA, 4-pregnen-11α-ol-3,20-dione hemisuccinate: BSA, 4-androsten-6β-ol-3,17-dione hemisuccinate: BSA, 5androsten-3β-ol-17-one hemisuccinate: BSA and 4-androsten-17β-ol-3-one 3-CMO: BSA were raised in rabbits. Because of the high specificity of the estradiol and progesterone antisera, dried plasma extracts were assayed without further purification. For the Δ^4, DHEA and testosterone assay plasma extracts were submitted to column chromatography using Sephadex LH-20. The dried plasma extracts and chromatographic fractions were incubated with 100 μl of the corresponding antiserum and 100 μl of the corresponding tritiated steroid solved in 0.1 per cent gelatin-PBS, pH = 7.0.

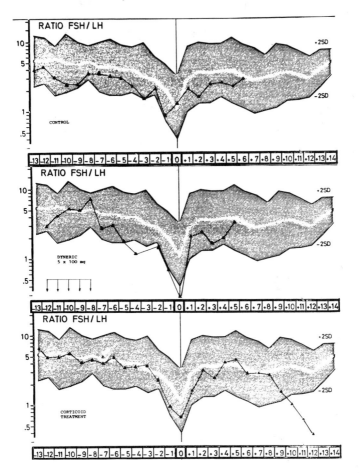

Fig.2 Shaded area: Normal range (mean ± 2SD) of FSH/LH ratios on semilog scale obtained from 24 apparently ovulatory cycles. ▲—▲—▲— FSH/LH ratios during pretreatment cycle (upper), clomid treatment (middle), corticosteroid treatment (lower).

After charcoal separation, the bound fraction was counted by LSC and the raw data were further processed by means of a Hewlett Packard Calculator (HP 9810A). Plasma cortisol was determined as described by Vecsei et al. (1972).

RESULTS

Pretreatment Cycle (Fig.1)

FSH and LH of the short luteal phase cycle were found to be within the normal range (mean ± 2SD) when compared with values obtained from 24 apparently ovulatory cycles. LH concentrations however were consistently above the mean of normal controls. As a consequence the FSH/LH ratios were in the low normal range (Fig.2). In contrast to normal cycles the lowest value of the FSH/LH ratio was observed one day before the LH peak. Estradiol levels indicated follicular

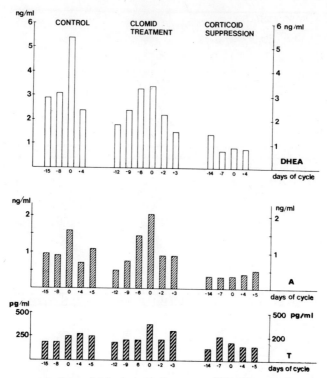

Fig.3 Plasma concentrations of dehydroepiandrosterone (DHEA), androstenedione (A) and testosterone (T) during a control cycle (left), clomid treatment (middle) and corticosteroid treatment (right).

maturation resulting in an estradiol peak value of 400 pg/ml. The FSH/LH peak (day 0) was followed by a luteal phase of 6 days with progesterone values of 1.5 ng/ml. Coincident with the estradiol, FSH and LH peak elevated DHEA and Δ^4 concentrations were noted, but with testosterone no significant changes occurred throughout this cycle (Fig.3).

Clomid Treatment (Fig.4)

An attempt to induce ovulation with clomid 100 mg daily for 5 days resulted in a pronounced effect on FSH, LH and estradiol but the luteal phase was again inadequate. The FSH/LH ratios were slightly below normal prior to the day of the LH peak (day 0, Fig.2). As demonstrated in Fig.3, DHEA and Δ^4 increased progressively to midcycle peak values (day 0) with Δ^4 concentrations elevated to 2.05 ng/ml. The effect on testosterone was less significant.

Corticosteroid Treatment (Fig.5)

Treatment with 16-methylprednisolone for three cycles resulted in an ovulatory pattern and pregnancy was confirmed after the third treatment cycle. LH values were slightly lower than during the untreated cycle. Since FSH remained unaltered the FSH/LH ratios were nearly identical with the mean values of normal cycles (Fig.2). Δ^4 concentrations of 0.35-0.50 ng/ml indicated a considerable suppression by 16-methylprednisolone. The same effect was observed for DHEA,

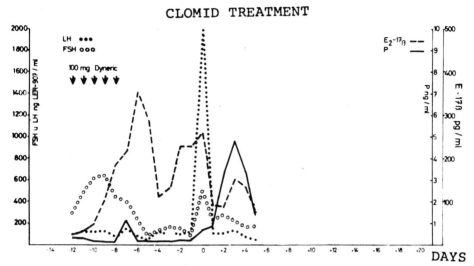

Fig. 4 Correlation of FSH, LH, estradiol-17β (E$_2$-17β) and progesterone (P) plasma concentrations centered on the LH peak (day 0) during a clomid (Dynerid®) treated cycle.

whereas the suppression of testosterone was less obvious. Plasma cortisol was within the normal range before and during clomid treatment and moderately suppressed during corticosteroid medication.

DISCUSSION

Studies performed by Abraham (1974) with peripheral plasma indicating a clear cut Δ^4 peak value on the day of the LH peak have not been fully confirmed by others. Judd and Yen (1974) as well as Ribeiro et al. (1974) did not observe a consistent pattern of Δ^4 during the menstrual cycle, but mean values seemed to indicate a trend towards elevated values at the time around ovulation.

On the basis of composite graphs of several cycles, testosterone levels followed a similar pattern with the higher values during the middle of menstrual cycles (Judd and Yen, 1974; Abraham, 1974; Goebelsmann, 1974). Individual patterns however showed wide deviations from the mean. According to Abraham (1974) DHEA has no consistent cyclic pattern but wide day-to-day variations.

In view of these data it is difficult to decide whether the patterns of Δ^4 and DHEA of the short luteal phase cycle reported herein are within normal limits. A progressive increase of Δ^4 and DHEA concentrations from the beginning of the cycle to the day of the LH peak however is obvious. Since corticosteroids have been shown not to interfere with LH secretion in normal ovulatory cycles (Kim et al., 1974), it is unlikely that the different patterns of the FSH/LH ratios before and during corticoid suppression represented a direct effect of 16-methylprednisolone on hypothalamic-pituitary function. In this context it is noteworthy to mention that the estradiol pattern of the untreated cycle was comparable to normal ovulatory cycles until day 0. It appears therefore that maturation of the follicle was not impaired. Androgens, by local mechanisms, however, may interfere

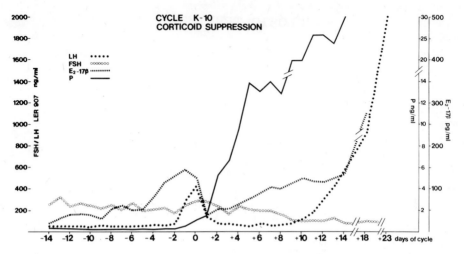

Fig.5 As in Fig.4, during the third treatment cycle with 16-methylprednisolone, 6 mg daily.

with ovulation (Payne et al., 1956; Hoffmann and Meger, 1965). This would explain why clomid treatment which produced a high LH peak as well as elevated Δ^4 and DHEA concentrations was unsuccessful in producing a normal luteal phase. It is therefore suggested that corticosteroid-induced suppression of elevated androgens at the time around midcycle is the treatment of choice in selected cases of short luteal phase. This view is further supported by catheterization studies performed by Kirschner and Jacobs (1971) who demonstrated a direct suppressive effect of dexamethasone on ovarian androgen secretion.

ACKNOWLEDGEMENT

The reagents for radioimmunoassay of FSH and LH were kindly supplied by NIAMD, Bethesda, USA.

REFERENCES

Abraham, G.E. (1974). *J. clin. Endocr. Metab.* 39, 340-346.
Goebelsmann, U., Arce, J., Thorneycroft, I.H. and Mishell, D.R. (1974). *Am. J. Obstet. Gynec.* 119, 445-452.
Hoffman, F. and Meger, C. (1965). *Geburtsh. Frauenheilk.* 25, 1132-1137.
Judd, H.L. and Yen, S.S.C. (1973). *J. clin. Endocr. Metab.* 36, 475-481.
Kim, M.H., Hosseinian, A.H. and Dupon, C. (1974). *J. clin. Endocr. Metab.* 39, 706-712.
Kirschner, M.A. and Jacobs, B. (1971). *J. clin. Endocr. Metab.* 33, 199-209.
Payne, R.W., Hellbaun, A.A. and Owens, J.N. (1956). *Endocrinology* 59, 306.
Ribeiro, W.O., Mishell, D.R. and Thorneycroft, I.H. (1974). *Am. J. Obstet. Gynec.* 119, 1026-1032.
Ross, G.T., Cargille, C.M., Lipsett, M.B., Rayford, P.L., Marshall, J.R., Strott, C.A. and Rodbard, D. (1970). *In* "Recent Progress in Hormone Research" (E.B. Astwood, ed) 26, pp.1-48. Academic Press, New York and London.
Sherman, B.M. and Korenman, S.G. (1974). *J. clin. Endocr. Metab.* 38, 89-93.
Strott, C.A., Cargille, C.M., Ross, G.T. and Lipsett, M.B. (1970). *J. clin. Endocr. Metab.* 30, 246-251.
Vecsei, P., Penke, B., Katzy, R. and Baek, L. (1972). *Experientia* 28, 1104-1105.

USING ANTIBODIES TO STUDY THE ROLE OF LH-RH IN OVULATION

S.L. Jeffcoate

Department of Chemical Pathology, St. Thomas's Hospital, London, England

INTRODUCTION

The isolation, sequencing and subsequent synthesis of the hypothalamic decapeptide, luteinising hormone-releasing hormone (LH-RH) has led to many investigations both clinical and experimental that have increased the understanding of the factors controlling the pituitary secretion of gonadotrophins. There is still much that is not known, however, particularly regarding the precise role of LH-RH in the control of the pre-ovulatory surge of LH. In our studies we have used antibodies raised against LH-RH as specific binding reagents to study this question both *in vivo* and *in vitro*.

PREPARATION AND PROPERTIES OF ANTIBODIES AGAINST LH-RH

Our experience in the preparation and assessment of anti-LH-RH sera has been recently reviewed (Jeffcoate et al., 1975) but can be summarized again. Either unconjugated LH-RH or conjugates of the decapeptide with protein carriers can be used as immunogens. Antibodies have been successfully raised in rabbits, rats, mice and chickens. The *titre* is assessed by binding of ^{125}I-labelled LH-RH and reaches a peak 8-12 weeks after primary intradermal immunisation; booster injections can be used to increase the titre if necessary. The specificity of the antisera is assessed by comparison of the cross-reaction in the *in vitro* radioimmunoassay of analogues and fragments of LH-RH compared with that of the synthetic decapeptide. All our antisera show similar results being highly specific for the C-terminus of the molecule but show some cross-reaction with peptides differing from LH-RH in the N-terminal positions. *Affinities* of the antibodies measured by Scatchard plots of radioimmunoassay data range from 1.1×10^9 to 2×10^{11} l/mole.

What causes the LH surge?

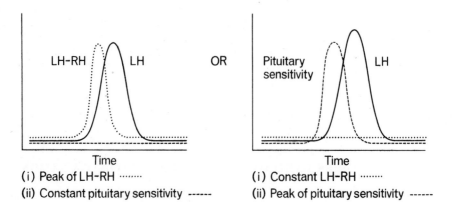

Fig.1

Antibodies to LH-RH can be used to study hypothalamic LH-RH and its role in reproduction in a number of different ways. These include: immunofluorescence mapping of LH-RH containing cells in the hypothalamus; *in vivo* inhibition of LH-RH either by active immunisation or after passive transfer of antibodies; *in vitro* inhibition of LH-RH stimulation of pituitary tissue; development of radioimmunoassays.

INHIBITION OF LH-RH *IN VIVO* BY PASSIVE TRANSFER OF ANTISERA

These experiments were carried out by H.M. Fraser in Dundee (Fraser, 1975; Fraser and Gunn, 1973) and the results are similar to those of Koch *et al.* (1973). If anti-LH-RH serum is injected into the normal female rat during the 'critical period' on the afternoon of proestrus then both the preovulatory surge of LH and subsequent ovulation are abolished. This indicates that LH-RH is essential for these events but does not necessarily imply that there is an LH-RH surge preceding them.

INHIBITION OF LH-RH IN ACTIVELY IMMUNISED ANIMALS

The hormone results of active immunisation against LH-RH have been studied in male rats (Fraser *et al.*, 1974) and rabbits (Fraser, 1975) and ovariectomised rats (Fraser *et al.*, 1975). In male animals there is involution of the testes associated with absent spermatogenesis, low plasma testosterone levels and low levels of both LH and FSH both in serum and in the pituitary gland. The changes in LH and FSH levels in serum of the ovariectomised rats (Fraser *et al.*, 1975) were studied sequentially. As the antibody titre rose from 3 to 7 to 12 weeks after immunisation there was a progressive and coincidental fall in LH and FSH. It was concluded that these results supported the proposal of Schally and his colleagues (Schally *et al.*, 1971) that the decapeptide, LH-RH, controls both the synthesis and release of both LH and FSH.

INHIBITION OF LH-RH ACTION *IN VITRO*

The absence of an immunologically distinct FSH-releasing hormone in extracts of rat median eminence has also been shown in experiments in which LH-RH stimulation of LH and FSH release by rat pituitary tissue was inhibited by antisera *in vitro* (M. Shahmanesh and S.L. Jeffcoate, unpublished results). Although FSH release is inhibited to a lesser extent than LH release as was also shown by Shin and Kraicer (1974) the results obtained by inhibition of the synthetic decapeptide are the same and there is no evidence for further separate FSH-releasing material in the median eminence extracts.

LH-RH RADIOIMMUNOASSAY

Anti-LH-RH sera can also be used to develop radioimmunoassays for LH-RH. These can be readily established in buffer solution but their application to body fluids or extracts is associated with a number of problems more fully discussed elsewhere (Jeffcoate et al., 1975).

LH-RH LEVELS DURING FEMALE REPRODUCTIVE CYCLES

Attempts have been made in several species to establish the relationship between LH-RH secretion and the pre-ovulatory LH surge by measurement of LH-RH by radioimmunoassay. So far there is no conclusive evidence for a peak of LH-RH secretion at this time.

The only claim for a peak of LH-RH in a reproductive cycle (Arimura et al., 1974) is unconvincing. In this study, in 5 women, single samples were obtained at mid-cycle and the concentrations appeared to be higher than those at other times. Our own data (Jeffcoate, 1975) do not confirm this. There appeared to be "pulses" of secretion but no evidence for increased levels at mid-cycle. In the rat, too, no correlation has been found between LH levels and levels of immunoreactive LH-RH (Meyer et al., 1974; Eskay et al., 1975) in either peripheral or portal blood. These results should not be taken as conclusive evidence that such a correlation does not exist, only that it has not yet been demonstrated.

Conflicting results have been reported in the sheep. Kerdelhue et al. (1973) in an early study reported occasional high peaks but Nett et al. (1974) found no evidence for any peaks related to LH release (or unrelated to LH release). In our own studies (Crighton et al., 1973; Foster et al., 1974) we have found evidence for episodic secretion of LH-RH-like immunoreactivity throughout the oestrous cycle. At oestrus, peaks occurred at intervals of 90-120 min; these were rapidly cleared. No correlation with LH was observed. Although we subsequently showed (Jeffcoate and Holland, 1974) that this immunoreactive material was not the LH-RH decapeptide, we feel that the probability is that it is a metabolite(s) formed in the pituitary (Jeffcoate et al., 1975).

IS THERE A PRE-OVULATORY LH-RH SURGE?

Conceptually, there are two possible mechanisms that could lead to a pre-ovulatory LH surge (Fig.1) (in addition, a combination of the two is possible). These are: a surge of LH-RH with a constant pituitary sensitivity to LH-RH stimulation or a constant secretion of LH-RH with a preovulatory increase in pituitary sensitivity. There is now considerable evidence for the latter in the rat (Cooper *et al.*, 1973; Aiyer *et al.*, 1974; Gordon and Reichlin, 1974) as well as in human and non-human primates. This increase in pituitary sensitivity can be induced by exogenous oestrogen in the rat (Aiyer *et al.*, 1974), rhesus monkey (Karsch *et al.*, 1973) and human (Jaffe and Keye, 1974). The clear implication is that the physiological LH surge is caused, at least in part, by a greatly increased sensitivity of the pituitary to LH-RH and that this is a result of the pre-ovulatory peak of oestradiol.

That LH-RH is necessary in the pre-ovulatory period, at least in the rat, is shown by Fraser's passive immunisation experiments mentioned earlier. Whether there is an increase of LH-RH secretion in addition, remains to be seen.

REFERENCES

Aiyer, M.S., Fink, G. and Greig, F. (1974). *J. Endocrinol.* **60**, 47.
Arimura, A., Kastin, A.J., Schally, A.V., Saito, M., Kumasaka, T., Yaoi, Y., Nishi, N. and Ohkura, K. (1974). *J. clin. Endocr. Metab.* **38**, 510.
Cooper, K.J., Fawcett, C.P. and McCann, S.M. (1973). *J. Endocrinol.* **57**, 187.
Crighton, D.B., Foster, J.P., Holland, D.T. and Jeffcoate, S.L. (1973). *J. Endocrinol.* **59**, 373.
Eskay, R.L., Oliver, C., Ben-Jonathan, N. and Porter, J.C. (1975). *In* "Hypothalamic Hormones" (M. Motta, P.G. Crosignani and L. Martini, eds) pp.125-137. Academic Press, London.
Foster, J.P., Holland, D.T., Jeffcoate, S.L. and Crighton, D.B. (1974). *J. Endocrinol.* **61**, lxiii.
Fraser, H.M. (1975). *In* "Physiological Effects of Immunity Against Reproductive Hormones" (R.G. Edwards and M. Johnson, eds). Cambridge University Press *(in press)*.
Fraser, H.M. and Gunn, A. (1973). *Nature* **77**, 160.
Fraser, H.M., Gunn, A., Jeffcoate, S.L. and Holland, D.T. (1974). *J. Endocrinol.* **63**, 399.
Fraser, H.M., Jeffcoate, S.L., Holland, D.T. and Gunn, A. (1975). *J. Endocrinol.* **64**, 191.
Gordon, J.H. and Reichlin, S. (1974). *Endocrinology* **94**, 974.
Jaffe, R.B. and Keye, W.R. (1974). *J. clin. Endocr. Metab.* **39**, 850.
Jeffcoate, S.L. and Holland, D.T. (1974). *J. Endocrinol.* **62**, 333.
Jeffcoate, S.L., Holland, D.T. and Fraser, H.M. (1975). *In* "Physiological Effects of Immunity Against Reproductive Hormones" (R.G. Edwards and M. Johnson, eds). Camrbidge University Press *(in press)*.
Jeffcoate, S.L., Holland, D.T., White, N., Fraser, H.M., Gunn, A., Crighton, D.B., Foster, J.P., Griffiths, E.C., Hooper, K.C. and Sharp, P.J. (1975). *In* "Hypothalamic Hormones" (M. Motta, P.G. Crosignani and L. Martini, eds) pp.279-297. Academic Press, London.
Jeffcoate, S.L. (1975). *In: Proceedings of European Colliquium on Hypothalamic Hormones* (D. Gupta and W. Voelter, eds). Verlag Chemie Weinheim *(in press)*.
Karsch, F.J., Weick, R.F., Butler, W.R., Dierscke, D.J., Krey, L.C., Weiss, G., Hotchkiss, J., Yamaji, T. and Knobil, E. (1973). *Endocrinology* **92**, 1740.
Kerdelhue, B., Jutisz, M., Gillessen, D. and Studer, R.O. (1973). *Biochim. biophys. Acta* **297**, 540.
Koch, Y., Chobsieng, P., Zor, U., Fridkin, M. and Lindner, H.R. (1973). *Biochim. biophys. Res. Commun.* **55**, 623.
Meyer, M.H., Masken, J.F., Nett, T.M. and Nisurender, G.D. (1974). *Neuroendocrinology* **15**, 32.
Nett, T.M., Akbar, A.M. and Niswender, G.D. (1974). *Endocrinology* **94**, 713.
Schally, A.V., Arimura, A., Kastin, A.J., Matuso, H., Baba, Y., Redding, T.W., Nair, R.M.G., Debeljuk, L. and White, W.F. (1971). *Science* **173**, 1036.
Shin, S.H. and Kraicer, J. (1974). *Life Sci.* **14**, 281.

THE BIOCHEMISTRY OF OVARIAN CANCER

K. Griffiths and W.J. Henderson

Tenovus Institute for Cancer Research, Welsh National School of Medicine Heath, Cardiff CF4 4XX, Wales

Carcinoma of the ovary now results in more than 10,000 deaths each year in the United States of America (Burbank, 1971) and in Britain, has the highest incidence of all gynaecological cancers (Campbell, 1975). The particularly malignant nature of this type of tumour is reflected in the overall 5-year survival figures which have been reported as low as 20.6% (Stone et al., 1963). These rather depressing mortality rates are related to the difficulties in diagnosing the condition in its early stages before the tumour disseminates and it has been stated that approximately 50% have distant metastases when diagnosed. Furthermore, data on ovarian cancer in England and Wales indicates that the mortality rate is increasing steadily each year at a level of 15% each decade (Table I) (Campbell, 1975). Little is known concerning the aetiology of the disease and apart from detailed investigations of tumours removed from patients who presented with some associated endocrine disturbance, our knowledge and understanding of ovarian cancer is relatively limited.

The lack of specific information on ovarian cancer may well relate to the difficulties pathologists and gynaecologists have had in establishing a reasonably standardised form of tumour classification. Reviews by Morris and Scully (1958), Teilum (1971) and Novak and Woodruff (1974) discuss the problems of ovarian tumour classification. It would seem, however, that some form of international agreement has now been reached on this problem and the World Health Organisation (WHO) (1973) together with the Pathology Committee of the International Federation of Gynaecologists and Obstetricians (1971) have produced a standardised form of classification which will more readily allow the comparison of data on ovarian cancer between different research and medical centres. In particular, such a development will facilitate the accumulation of information on the individual histological types of ovarian cancers, which may have different forms of behaviour, biochemistry and aetiology, rather than on a group of tumours, col-

Table I. Ovarian cancer records - England and Wales.

Deaths from Cancer of the Ovary Annual death rate/1,000,000 women				Deaths from Cancer of the Ovary (by age at death) — 1971	
1931-1935	65	1961	127	Age	No. of deaths
1936-1940	80	1962	126	<15	2
1941-1945	92	1963	125	15-24	19
1946-1950	102	1964	129	25-34	31
1951-1955	113	1965	130	35-44	190
1956-1960	124	1966	133	45-54	686
1961-1965	128	1967	132	55-64	1064
1966-1970	138	1968	136	65-74	1062
		1969	140	75-84	525
		1970	144	>85	109
		1971	146	All ages	3688

lectively referred to as ovarian neoplasms.

A major proportion of the ovarian tumours fall into four of the categories used in the WHO classification: (i) common epithelial tumours, for example, the serous and mucinous cystadenomata, cystadenocarcinomata, adenofibromas and endometrioid tumours; (ii) the sex cord-stromal tumours or mesenchyme type tumours, which includes the oestrogen secreting granulosa-theca cell and the virilising Sertoli-Leydig cell tumours. These virilising tumours which contain the testicular elements, Leydig cells and Sertoli cells in various stages of maturity, are generally referred to as arrhenoblastomas. A more rare type of neoplasm, the lipoid cell tumour (iii) is separately classified from the 'functional tumours' of group (ii) despite their general association with virilising symptoms. This group of tumours includes the adrenal-rest cell tumour, the hilus cell tumour, the masculinovoblastoma and luteoma. Those tumours referred to as germ cell tumours (iv), the dysgerminoma, choriocarcinoma, teratoma and embryonal carcinomas, although comparatively rare, contribute significantly to the ovarian cancer mortality statistics in younger people (Lingeman, 1974).

Many types of 'functional tumours' synthesising and secreting steroid hormones have been studied in considerable detail, especially during the 1960s. It was during this particular period that most impressive advances were made in our knowledge and understanding of steroid hormone biosynthesis by normal ovarian tissue. Despite the complex nature of the ovary with its various cell types and its cyclical changes of structure and secretory activity, experiments *in vitro* were found of value in defining the biosynthetic role of individual cells. In particular, the work of Savard and his colleagues (Savard et al., 1965; Savard, 1968) and also of Smith (1961) clearly confirmed using such techniques, the principal secretory function not only of the follicle and corpus luteum, but also of the ovarian stroma. There was a major interest in the various biosynthetic pathways (Fig.1) by which steroid hormones were produced by the various cell types and experiments generally involved incubating precursor radioactive steroids with sliced, minced or homogenised ovarian tissue and determining the products formed after extraction, separation, identification and quantitation. The precise mechanisms however that

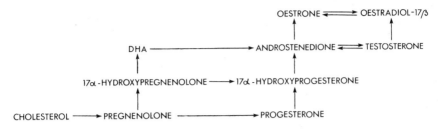

Fig.1 Pathways of steroid biosynthesis.

control the course of steroid biosynthesis in the various types of cells still remain to be determined.

Studies *in vitro* of ovarian tumours were undertaken in similar manner to determine the steroid biosynthetic capacity of such neoplastic tissue and generally the data obtained related well to the associated clinical symptoms or endocrine disturbance of the patient. Such information together with urinary and plasma steroid analyses, provided certain useful biochemical parameters in the search for a rational basis for tumour classification. The steroid producing potential of histologically well-defined tumour cells was probably of value in the discussions concerned with determining whether classification should be based on histology or on the resultant endocrine effect (Teilum, 1971).

Granulosa-theca cell tumours constitute approximately 10% of primary ovarian carcinomas (Novak, 1952). The clinical symptoms associated with these neoplasms, precocious puberty, menstrual disorders or postmenopausal bleeding, are attributable to the synthesis and secretion of oestrogen by the tumours. Increased urinary oestrogen excretion has been shown in patients with granulosa-theca cell tumours (Dorfman, 1960; Bruk *et al.*, 1960; Besch *et al.*, 1966). Marsh and his colleagues (Marsh *et al.*, 1962) similarly observed in a 26-month old girl with a luteinised granulosa cell tumour, urinary oestrogen excretion values higher than the normal range for a mature woman and it was not unreasonable therefore that their studies *in vitro* indicated a high capacity of the tumour tissue to convert testosterone to oestrone and oestradiol-17β. More recently Jenner *et al.* (1972) reported in a 9 year old girl with a granulosa cell tumour, a preoperative plasma oestradiol-17β concentration of 413 pg/ml. This is substantially higher than the concentration at any stage of the menstrual cycle and a girl at a similar stage of prepubertal development would have a plasma oestradiol-17β concentration of approximately 20 pg/ml. Postoperatively, the plasma level decreased to 22 pg/ml by the 7th day and to 12 pg/ml after two months. Investigations *in vitro* on the biosynthetic capacity of granulosa cell tumours have been described with a variety of tumour types, luteinised and non-luteinised and also using varying, rather arbitrary incubation conditions. Although Griffiths *et al.* (1964) demonstrated the conversion of androstenedione to oestrogens by luteinized granulosa-theca cell tumour tissue, progesterone was not found to be a precursor of either C_{19}- or C_{18}-steroids. These experiments, together with certain other subsequent investigations *in vitro* with granulosa cell tumour tissue (Kase, 1964; Griffiths *et al.*, 1966; MacAulay and Weliky, 1968) tended to suggest that the 'alternative' biosynthetic pathway for C_{19}- and C_{18}-steroid production, which does not include progesterone but involves dehydroepiandrosterone (DHA) (Fig.1), may be principally concerned

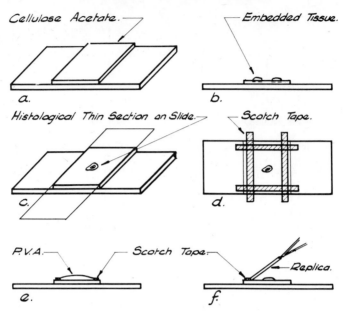

Fig. 2 Extraction-replication technique. Tissue is embedded onto acetone softened cellulose acetate mounted on a microscope slide. Once the cellulose acetate hardens, the tissue is outlined with tape to form a 'well' and polyvinyl alcohol (PVA) solution layered onto the specimen. On hardening this forms a replica of the tissue surface and this can be shadowed with platinum and carbon. Particles in the tissue are extracted onto the PVA film. PVA is removed in hot water and the replica mounted on microscope grids.

in oestrogen synthesis. Other experiments (Besch *et al.*, 1966) with a luteinised granulosa cell tumour suggested that progesterone was concerned as an essential intermediate. Whether luteinisation of these cells affects the biosynthesis of steroids remains to be seen and despite considerable study, it would seem that there is no acceptable biochemical interpretation of the process of luteinisation.

Although together, these reports make up an interesting study of granulosa cell tumour tissue, when carefully examined, they do clearly emphasise the difficulties inherent in such *in vitro* experimentation and the care required when interpreting results. Although patterns of steroid biosynthesis tend to emerge, the results are not strictly comparable. Incubation times vary, tissue preparations used by the various laboratories are different and obviously the use of co-factors in incubation media will almost certainly affect the results. Certain research groups for example (Ryan, 1963; Savard *et al.*, 1965) have reported that in the human corpus luteum, oestrogens are synthesized by the pathway involving progesterone. This would appear not to be so in similar tissue of other animals (Huang and Pearlman, 1962; Mahajan and Samuels, 1963; Bjersing and Cartensen, 1964; Savard *et al.*, 1965) and in our laboratories (Fahmy *et al.*, 1967, 1968; Griffiths *et al.*, 1971) we have been unable to confirm that progesterone was a precursor of C_{19}- and C_{18}-steroid synthesis in the human corpus luteum.

Although the theca cell tumour or thecoma is also an oestrogen-producing tumour and occurs with an incidence similar to that of the granulosa cell tumour (Novak *et al.*, 1971) there have been few biochemical studies, either with the tissue

The Biochemistry of Ovarian Cancer

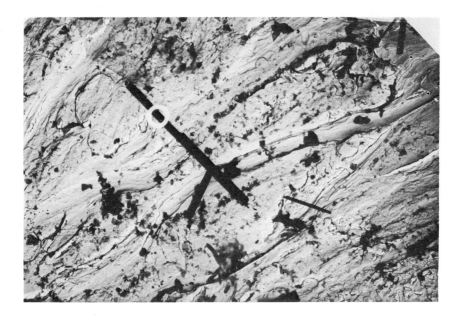

Fig.3 Asbestos fibre recovered from a mesothelioma by the extraction replication technique. The elemental analysis of the area of the fibre outlined is shown in Fig.7.

or the patients from whom the tumours were removed. In contrast our knowledge of the virilising ovarian tumours referred to as arrhenoblastomas, or androblastomas is more complete. These tumours are composed of the testicular elements, the Leydig and Sertoli cells in different stages of maturity and are distinct from the hilus cell tumour which is a tumour of pure Leydig cells, also associated with the clinical symptoms of virilisation. By 1960, 240 cases of arrhenoblastoma had been reported in the literature (Pedowitz and O'Brien, 1960), the greatest incidence occurring in the third decade and although characteristically virilising, certain arrhenoblastomas produce no recognisable endocrine effects, in keeping with the principle that morphology should constitute the major criterion for diagnosis (Morris and Scully, 1958).

Furthermore, Pedowitz and O'Brien (1960) reported that urinary 17-oxosteroid excretion, routinely measured in patients with virilising arrhenoblastomas was only elevated in 13 of the 35 cases studied. The urinary 17-oxosteroid excretion is however a poor reflection of the plasma androgen concentration, and subsequent investigations of Mahesh et al. (1970) and Greenblatt et al. (1972) clearly demonstrated in women with arrhenoblastomas high plasma testosterone and androstenedione levels which markedly decreased postoperatively. Similarly, Laatikainen et al. (1972) showed that despite apparently normal urinary 17-oxosteroid excretion in a patient with an arrhenoblastoma, detailed GLC-mass spectrometric analysis of certain plasma steroids, indicated abnormal concentrations. In particular, the levels of the monosulphates of androsterone, epiandrosterone and DHA were between 2-10 times those normally found in females.

The metabolism of various radioactive steroids by preparations of tissue from arrhenoblastomas has also indicated what can probably be considered extensive

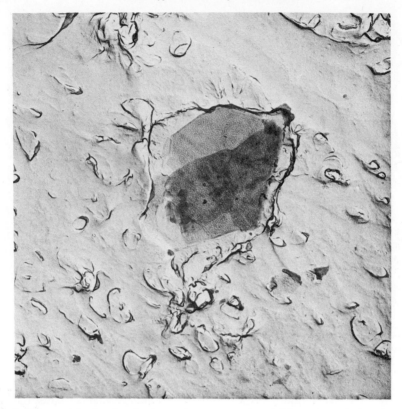

Fig. 4 Talc, particle, showing the decorative pattern, extracted from an ovarian carcinoma. Elemental analysis is given in Fig.8.

formation of androgens by the tissue *in vitro* (Wiest *et al.*, 1959; Savard *et al.*, 1961; Kase and Conrad, 1964; Sato *et al.*, 1969).

It has been recognised for many years that the hilus of the ovary contains nests of cells which are morphologically similar if not identical to the Leydig cells of the testes (Berger, 1922; Sternberg, 1949). Hyperplasia, or neoplasia of such cells, forming a hilus cell tumour is usually associated with the clinical manifestations of virilisation (Berger, 1942). These tumours are usually small, less than 5.0 cm diameter, encapsulated and are composed of closely packed polyhedral cells with eosinophilic, finely granular cytoplasm and distinctive round nuclei (Morris and Scully, 1958). The rather limited number of 60 tumours which were reported in the literature by 1972, have been reviewed (Boivin and Richart, 1965; Dunnihoo *et al.*, 1966; Anderson, 1972).

The clinical symptoms of virilisation associated with the presence of a hilus cell tumour therefore suggest that these tumours are synthesizing and secreting excessive amounts of testosterone or, alternatively, a precursor which is peripherally converted to testosterone. The first reported biochemical study with this tumour tissue (Corral-Gallardo *et al.*, 1966) indicated that a minced preparation, in the presence of various co-factors, converted incubated radioactive progesterone to both androstenedione (5.6%) and testosterone (0.3%). Interesting, however,

The Biochemistry of Ovarian Cancer

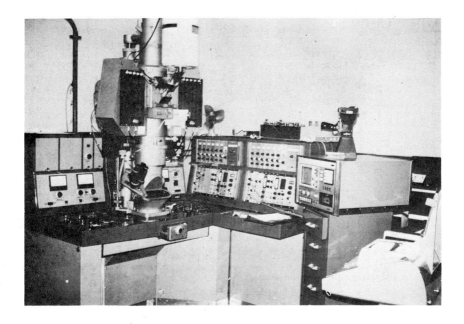

Fig.5a EMMA-4 (AEI, Manchester, England) as installed at the Tenovus Institute showing the associated Tracor computer unit.

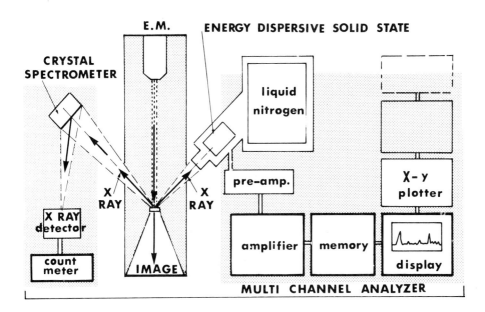

Fig.5b Diagrammatic representation of EMMA-4.

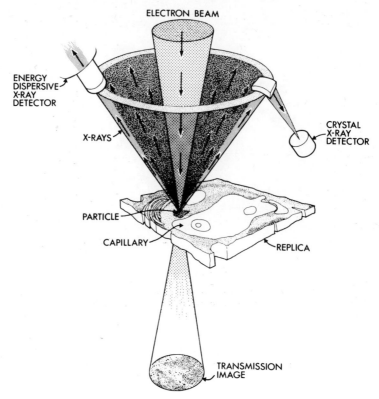

Fig. 6 Diagrammatical representation of the focussed electron beam in EMMA.

was the formation of 11-deoxycorticosterone (DOC) from progesterone and of 11β-hydroxyandrostenedione from incubated androstenedione. The 11β- and 21-hydroxylases are essentially adrenal enzymes, although the former enzyme system had previously been shown to be present in testicular neoplasms (Savard et al., 1960; Dominguez, 1961; Besch et al., 1964) and 21-hydroxylation has been described in association with a masculinovoblastoma (lipoid cell tumour) (Brysson et al., 1962). The lack of corticosterone formation after incubating tissue with progesterone would suggest that the 11β-hydroxylase was specific for C_{19}-steroids.

A similar biochemical investigation *in vitro* (Fahmy et al., 1968) with tissue from a hilus cell tumour provided adequate evidence that the necessary enzymic capacity for androgen biosynthesis was present in the tumour tissue. After incubation of radioactive pregnenolone with chopped tissue in the absence of co-factors, 78.5% of the radioactivity was isolated in androstenedione and 1.2% in testosterone. Furthermore, DHA sulphate, normally present in high concentrations in plasma, was actively metabolised to androstenedione by the tissue although other studies, both *in vivo* (Loriaux et al., 1970) and *in vitro* (Sandberg et al., 1966) indicated that this conjugated plasma steroid was not a precursor of steroid hormone synthesis in arrhenoblastomas. In considering steroid conjugates, however, it is noteworthy that Echt and Hada (1968) described the presence of urinary testosterone sulphate in concentrations up to 10,000 times the normal

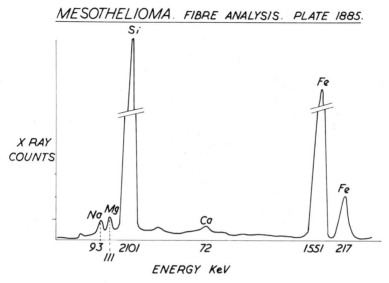

Fig. 7 Elemental analysis of the asbestos fibre outlined in Fig. 3. The sodium associated with iron, magnesium and silicon confirms the identification as crocidolite asbestos.

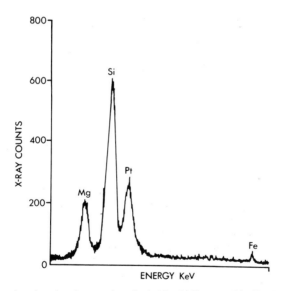

Fig. 8 Spectrum showing the elemental analysis of talc illustrated in Fig. 4. The 3:1 silicon: magnesium ratio with low iron content is characteristic of talc. The platinum present was added as part of the shadowing procedure.

level in a woman with a metastatic hilus cell tumour. They suggested that the sulphate was produced by the tumour.

Some evidence for oestrogen synthesis by hilus cell tumour tissue has also been obtained (Fahmy *et al.*, 1968; Jeffcoate and Prunty, 1968) and it is interesting

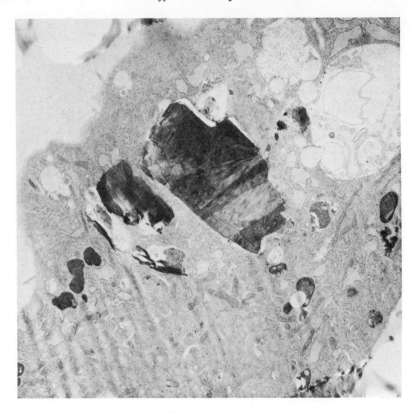

Fig. 9a Talc particle embedded in a cultured lung fibroblast.

that Novak and Mattingly (1960) directed attention to the association between endometrial hyperplasia and virilisation in a number of patients with these tumours. Elevated urinary oestrogen excretion has been reported (Hawkins and Lawrence, 1965) in a patient with hilus cell hyperplasia of the ovary although Brown and his colleagues (Bhargava et al., 1969) indicated that in a study of various steroid secreting ovarian tumours, vaginal and endometrial cytology did not relate to the oestrogen excretion data.

It is clear that there have not yet been sufficient comparative studies to establish that particular ovarian tumours of histologically defined cell types have characteristic steroid biosynthetic patterns. Results from studies *in vitro* have however correlated well with the observed clinical disturbance of the patient, although these biochemical investigations have at present provided data of only limited value to our understanding of tumour histogenesis.

The term 'lipoid cell tumours' is general and non-committal and is used to include those usually virilising tumours with cells resembling the zona fasciculata of the adrenal cortex, Leydig cells of the testis and luteinised stromal cells of the ovary. Pathologists have long debated whether the different cell types can be differentiated, especially in the presence of neoplasia (Teilum, 1971) and Taylor and Norris (1967) consider with similar clinical and gross pathological findings, histological subdivision is of doubtful value. They believe that on the basis of the

Fig. 9b Particles of talc embedded in cells from normal ovarian stroma cultured as explants. Thin sectioning for electron microscopy with particles present results in tissue disintegration.

available evidence, the most likely origin of the lipoid cell tumour is from ovarian stromal cells within the medulla. Hughesdon (1966) also considers that the presence of intraovarian adrenal-rest cells from which a tumour might originate has yet to be conclusively demonstrated. Certainly, despite the striking morphological resemblances of certain lipid cell tumours to adrenal tissue and the presence of certain signs and symptoms of Cushing's syndrome in many of these patients with such tumours (Scully, 1963) biochemical investigations have only shown the synthesis of androgens by the adrenal-like ovarian cells (Sandberg *et al.*, 1962) and not adrenocorticosteroids (Rosner *et al.*, 1964). Other investigations (Morrow *et al.*, 1968) have suggested that lipoid cell tumours respond to ACTH as well as HCG, although this might reflect decreased specificity of membrane receptor sites for protein hormones. There has also been a report of an arrhenoblastoma responding to ACTH (Gallagher *et al.*, 1962).

Although the generally high excretion of urinary 17-oxosteroids in patients with adrenal-like lipoid cell tumours (Pedowitz and Pomerance, 1962; Prunty, 1967) has been shown to be related to an increased formation of 11-oxygenated 17-oxosteroids (Dorfman, 1960; Morrow *et al.*, 1968), the observations that the hilus cell tumour and also a gynandroblastoma, an arrhenoblastoma synthesizing

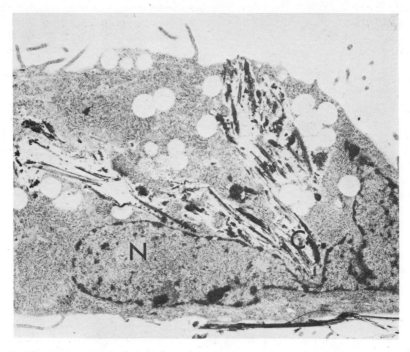

Fig.10 Fibroblast with ingested asbestos chrysotile fibres. The fibres have penetrated the cell membrane but the nuclear membrane remains intact (× 8,500).

androgen and oestrogen (Nocke, 1965) possess the 11β-hydroxylating enzyme system would appear to render invalid, the concept that 11β-oxygenation is a parameter of value in distinguishing between adrenal and ovarian tissue.

It is interesting that of the many biochemical investigations *in vitro* that have been completed, only one (O'Malley *et al.*, 1967) has produced a steroid biosynthetic pattern which appears unique. This interesting study of a luteoma of pregnancy, which caused virilisation of mother and child, suggested that the predominant pathway for testosterone synthesis involved pregnenolone—17α-hydroxypregnenolone—17α-hydroxypregnenolone sulphate—DHA sulphate—androstenediol sulphate—androstenediol—testosterone.

Until recently, only those ovarian tumours of mesenchymal origin were considered responsible for endocrine changes in the patient. Over the past few years, however, evidence has now accumulated to indicate that endocrine effects can also occur in the presence of ovarian tumours not classicaly considered 'functional' and menstrual disturbances, postmenopausal bleeding, virilism, changes in endometrial and vaginal cytology have been reported in association with epithelial tumours, Brenner tumours, Krukenberg tumours and various others (Wren and Frampton, 1963; Fox, 1965; Fathalla, 1968). Hormone synthesis has been ascribed to an ovarian stromal reaction resulting from the stimulus from the invasive epithelial elements of either a primary tumour of metastatic tissue from possibly a primary gastrointestinal tumour. That the ovaries develop large secondary tumours, termed Krukenberg tumours (Israel *et al.*, 1965; Scott *et al.*, 1967; Connor *et al.*, 1968) sometimes before the primary tumour is evident suggests a favourable environment for growth. This localised stromal reaction results in the formation in close relation to the tumour of theca-lutein cells, or luteinised stromal cells, which have been considered responsible for the synthesis of oestrogen (Hughesdon, 1958; Pfleiderer et al., 1968). The presence of such cells may well have been responsible for the steroid metabolising activity reported in homogenates of certain 'non-functional' ovarian cancers (Plotz et al., 1967). Endometrial hyperplasia has often been reported to occur in the postmenopausal woman when these tumours are present (Wren and Frampton, 1963; Fox, 1965; Eddie, 1967) and elevated urinary oestrogens have also been found in these women (Bhargava et al., 1969; Procopé, 1969; Edwards et al., 1971). Unfortunately, there are few biochemical studies on the relationship between gonadotrophin secretion and action and ovarian cancer. It is also difficult to determine whether the luteinised-stromal cell synthesises and secretes oestrogen or whether an increased production of androstenedione, resulting from the stromal reaction or hyperplasia is peripherally converted to oestrogen as indicated by Dr. Edman during this conference. It has been suggested previously (Griffiths et al., 1971) that any localised oestrogen formation by the luteinised stromal cells may influence the growth of primary or secondary tumour tissue. This could be particularly important in relation to the histogenesis of ovarian cancer from endometriotic lesions (Scully et al., 1966; Stevenson, 1970; Aure et al., 1971).

It is patently obvious however that despite the considerable interest during the past two or three decades of both endocrinologist and pathologist, there is still little real information concerning the aetiology of ovarian cancer. Although the 'functional' tumour with its associated endocrine effects offers a fascinating study to the biologist, these generally benign and comparatively rare tumours make up however only a small proportion of all ovarian cancers. Reports from various cancer registries and from the cancer agencies (Doll et al., 1970) indicate that in many countries, the serous cystadenocarcinoma is the most frequent type of ovarian neoplasm comprising something of the order of 50% of the total ovarian cancers (Table II). Because of the differences in terminology and tumour classification between various countries, there are few good epidemiological studies of ovarian cancer and little knowledge of possible factors that might relate to the aetiology of the disease. There would now seem little doubt however that various forms of cancer, previously considered as spontaneous, may be caused or influenced by environmental factors (Lingeman, 1974). With the exception of Japan, the highest incidence rates of ovarian cancer are found in industrialised developed countries particularly in the west. It was of interest therefore that Graham and Graham (1967) directed attention firstly to the reported association between asbestosis and ovarian cancer (Keal, 1960) and secondly, to the histological similarity between certain ovarian tumours and mesotheliomas, peritoneal and pleural tumours considered to be induced by crocidolite, an amphibole type of asbestos (Wagner et al., 1960; Elwood and Cochrane, 1964). They suggested that asbestos may be the environmental factor present in developed countries which has a marked influence on ovarian metabolism.

An extraction-replication technique (Fig.2), developed in our laboratories to investigate foreign particles in biological tissues (Henderson and Griffiths, 1972) and used to identify crocidolite (Fig.3) in mesotheliomas (Henderson et al., 1969) and various forms of particulate material in stomach tumours from Japanese males (Henderson et al., 1975) failed to demonstrate the presence of asbestos in

Fig.11 A particle of talc in contact with the surface of the cell. The "fuzzy coat" appears to form around the end of the particle. Microfilaments can also be seen running parallel to the cell surface (× 12,800).

Table II. Malignant cancers of the ovary (Connecticut Tumour Registry: 1960-1964).

Types of Tumour	No.	%
Serous carcinomas	422	46
Various unspecified adenocarcinomas	174	19
Mucinous carcinomas	87	10
Malignant granulosa cell tumours	38	4
Malignant teratomas and dysgerminomas	21	2
Other tumours	173	19
Total:	915	

(See Lingeman, 1974)

ovarian tumour tissue. The technique did however clearly provide evidence of talc contamination of both neoplastic and normal ovarian tissue (Henderson *et al.*, 1971). Talc, a magnesium silicate with an elemental composition similar to asbestos, was recognised by the distinct decoration pattern induced by the evaporation of platinum *in vacuo* on the crystal surface (Fig.4). Although there are reports that commercial talcs are contaminated with particles of asbestos (Lingeman, 1974), an intensive study in our laboratories of various forms of natural talc from many

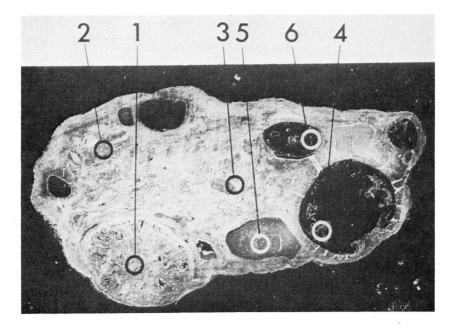

Fig.12 Histological section of ovary. After ashing the tissue on the slide, the areas outlined were transferred for particle analysis onto numbered electron microscopy 3 mm grids using polyvinyl alcohol films (Henderson and Griffiths, 1972).

countries has failed to provide any evidence which would confirm such reports.

Although particle shape has for many years been the sole means of identifying foreign particulate material, more sophisticated techniques are now being developed and in our laboratories, analytical electron microscopy can now be used to provide complete identification. The electron microscope microanalysis (EMMA-4, AEI, Manchester; Fig.5) (Chandler, 1971) is used to identify individual particles or fibres extracted from biological tissues by the extraction replication procedure. The technique makes use of the fact that electrons passing through the specimen on the grid to form an electron image, also generate from the irradiated specimen area, X-rays which have energies characteristic of the elements of which the specimen is composed (Fig.6). The illuminating electron beam can then be focussed on to crystalline particles visible on the transmission image. The particle then becomes the only part of the specimen emitting X-rays. Silicon atoms of talc then emit X-ray photons with a well-defined energy which differs from that of magnesium. These X-rays are collected by both a twin crystal spectrometer system and also a solid state Si(Li) energy dispersive detector linked to a 1000 channel analyser. The spectrometer provides an accurate quantitative analysis whereas the energy dispersive detector offers a rapid qualitative and semiquantitative procedure (see Figs 7 and 8 for spectra of crocidolite and talc). Such an approach therefore provided unequivocal proof that talc is present in ovarian tissue.

The mode of entry of talc into the ovary and the biological effects of particles such as magnesium silicates within the cell are obviously of particular interest. Talc is certainly widely used in the pharmaceutical and cosmetic industry and

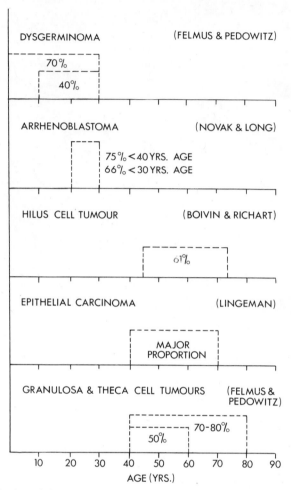

Fig.13 An indication of the ages at which certain tumours appear to occur with greater incidence rates. (E.R. Novak and J.H. Long (1965). *Am. J. Obstet. Gynec.* 92, 1082; L.B. Felmus and P. Pedowitz (1967). *Obstet. Gynec.* 29, 344.)

is used in association with certain contraceptives. Furthermore, although it has been suggested that asbestos fibres but not the plate-like crystal of talc might be expected to penetrate cell membranes, Figs 9 and 10 show talc particles embedded within cultured fibroblasts and ovarian cells. Figure 11 illustrates the contact between cellular material and the talc particle, the "fuzzy coat" appearance being characteristic of pinocytosis and some recent collaborative studies were concerned with the effect of ingested particles on cell ultrastructure and collagen synthesis (Richards *et al.*, 1974; Henderson *et al.*, 1975). Moreover, Fig.12 and Table III indicate that the amount of talc present in ovarian tissue could be of the order of 2×10^5 particles/mm^3.

It is obvious that talc should not be present within the cells of the ovary and probably other organs of the female genital tract but its presence in no way incri-

Table III. Ashed ovarian section. Analysis for areas 1-6 are shown and the number of talc particles found in 10 randomly selected squares of one grid. The size range of the particles found and recorded is also given. There are 565 squares/grid; on average therefore 8 × 56.5 = 452 particles/grid - approximately 0.7 mm^2 of the tissue section, indicating (with 5 μm thick section) 452 × 200 × 3.3 talc particles/mm^3, i.e. 271,200 particles.

Area	1	2	3	4	5	6	Mean
No. talc particles/10 square	2	8	8	18	6	4	8
Size range (μ)	0.5–1.0	0.2–2.0	0.3–0.8	0.2–2.5	0.4–7.5	0.2–1.5	

minates the particle in the process which results in the development of ovarian neoplasia. However, in this field of cancer research in which little is known about the aetiology of the disease, new concepts are necessary. Speculation that granulomatous changes in the ovary induced by talc particles may result in disturbances leading to stromal reaction, theca-lutein cell formation, an imbalance of steroid biochemistry, possible co-carcinogenic effects of steroid and protein hormones should at least be considered. Certainly records indicate that many ovarian cancers, in particular the epithelial tumours occur (Fig. 13, Table I) when ovarian function is decreasing and plasma gonadotrophin concentration is high. It is interesting that although, overall, ovarian tumours occur predominantly in the 50's, careful analysis would suggest that certain different histological types of ovarian neoplasm may be associated with different aetiological factors. Little is known about the hormone dependence of the various histological types of ovarian tumours and carefully established detailed studies are required in this aspect of cancer research.

ACKNOWLEDGEMENT

The authors gratefully acknowledge the generous financial support of the Tenovus Organisation.

REFERENCES

Anderson, H.C. (1972). *J. clin. Pathol.* 25, 106-110.
Aure, J.Chr., Høeg, K. and Kolstad, P. (1971). *Acta Obstet. Gynec. Scand.* 50, 63.
Berger, L. (1922). *C.r.Hebd.Seance. Acad. Sci., Paris* 175, 498.
Berger, L. (1942). *Revue Can. Biol.* 1, 539-566.
Besel, P.K., Barry, R.D., Miller, R.R. and Watson, D.J. (1964). *J. clin. Endocr.* 24, 1339.
Besel, P.K., Watson, D.J., Vorys, N., Hamuri, G.J., Barry, R.D. and Barnett, E.B. (1966). *Am. J. Obstet. Gynec.* 96, 466-477.
Bhargava, V.L., Buscher, N.A., Brown, J.B. and Townsend, L. (1969). *Aust. N.Z. J. Obstet. Gynaec.* 9, 108.
Boivin, T. and Richart, R.M. (1965). *Cancer, N.Y.* 18, 231.
Bruk, I., Dancaster, C.P. and Jackson, W.P.U. (1960). *Br. med. J.* 2, 26.
Brysson, M.J., Dominquez, O.V., Kaiser, I.H., Samuels, L.T. and Sweat, M.L. (1962). *J. clin. Endocr. Metab.* 22, 773.
Burbank, F. (1971). *Natl. Cancer Inst. Monogr.* 33.
Campbell, H. (1975). *In* "Gynaecological Malignancy: Clinical and Experimental Studies" (M.G. Brush and R.W. Taylor, eds) p.111. Bailliere Tindall, London.
Chandler, J.A. (1971). *American Laboratory* 3, 50.
Connor, T.B., Ganis, F.M., Levin, H.S. Migeon, C.J. and Martin, L.G. (1968). *J. clin. Endocr. Metab.* 28, 198.

Corral-Gallardo, J., Acevedo, H.A., Perez de Salazar, J.L., Loria, M. and Goldzieher, J.W. (1966). *Acta endocr. (Copenh.)* 52, 425.
Doll, R., Muir, C. and Waterhouse, J. (1970). "International Union Against Cancer" Vol.2. Springer-Verlag, Berlin.
Dominquez, O.V. (1961). *J. clin. Endocr.* 21, 663.
Dorfman, R.I. (1960). *In* "Biological Activities of Steroids in Relation to Cancer" (G. Pincus and E.P. Volmer, eds) p.445. Academic Press, London.
Dunnihoo, D.R., Grieme, D.L. and Woolf, R.B. (1966). *J. Obstet. Gynaec. Br. Commonw.* 27, 703.
Echt, C.R. and Hada, H.E. (1968). *Am. J. Obstet. Gynec.* 100, 1055-1061.
Eddie, D.A.S. (1967). *J. Obstet. Gynaec. Br. Commonw.* 74, 283.
Edwards, R.L., Nicholson, H.O., Zoidis, T., Butt, W.R. and Taylor, C.W. (1971). *J. Obstet. Gynaec. Br. Commonw.* 78, 467.
Elwood, P.C. and Cochrane, A.L. (1964). *Brit. J. Indust. Med.* 21, 304.
Fahmy, D., Griffiths, K. and Turnbull, A.C. (1967). *Eur. J. Steroids* 2, 483.
Fahmy, D., Griffiths, K. and Turnbull, A.C. (1968). *Biochem. J.* 107, 725.
Fahmy, D., Griffiths, K., Turnbull, A.C. and Symington, T. (1968). *J. Endocrinol.* 41, 61.
Fathalla, M.F. (1968). *J. Obstet. Gynaec. Br. Commonw.* 75, 78.
Fox, M. (1965). *Cancer* 18, 1041.
Gallagher, T.F., Spencer, H., Bradlow, H.L., Allen, L. and Hellman, L. (1962). *J. clin. Endocr. Metab.* 22, 970.
Graham, J. and Graham, R. (1967). *Environ. Res.* 1, 115.
Greenblatt, R.B., Mahesh, V.B. and Gambrell, R.D. Jr. (1972). *Obstet. Gynec.* 39, 567-576.
Griffiths, K., Grant, J.K. and Symington, T. (1964). *J. Endocrinol.* 30, 247.
Griffiths, K., Grant, J.K., Browning, M.C.K., Cunningham, D. and Barr, G. (1966). *J. Endocrinol.* 35, 299.
Griffiths, K., Fahmy, D., Henderson, W.J., Turnbull, A.C. and Joslin, C.A.F. (1971). *In* "Control of Gonadal Steroid Secretion" (D.T. Baird and J.A. Strong, eds) pp.282-291. Edinburgh Univ. Press.
Hawkins, D.F. and Lawrence, C.M. (1965). *J. Obstet. Gynaec. Br. Commonw.* 72, 285-291.
Henderson, W.J., Harse, J. and Griffiths, K. (1969). *J. Cancer* 5, 621-624.
Henderson, W.J., Joslin, C.A.F., Turnbull, A.C. and Griffiths, K. (1971). *J. Obstet. Gynec. Br. Commonw.* 78, 266.
Henderson, W.J. and Griffiths, K. (1972). *In* "Shadow Casting and Replication: Principles and Techniques of Electron Microscopy" Vol.2, 5, 151 (M.A. Hayat, ed). Nostrand Reinhold, New York.
Henderson, W.J., Evans, D.M.D., Davies, J. and Griffiths, K. (1975). *Environmental Research* 9, 240.
Henderson, W.J., Blundell, G., Richards, R., Hext, P.M., Volcani, B.E. and Griffiths, K. (1975). *Environmental Research* 9, 173.
Hughesdon, P.E. (1958). *J. Obstet. Gynaec. Br. Emp.* 65, 702.
Hughesdon, P.E. (1966). *Obstet. Gynec. Survey* 21, 245.
International Federation of Gynecologists and Obstetricians (1971). *Acta Obstet. Gynec. Scand.* 50, 1-7.
Israel, S.L., Helsel, E.V. and Hausman, D.H. (1965). *Am. J. Obstet. Gynec.* 93, 1094.
Jeffcoate, S.L. and Prunty, F.T.G. (1968). *Am. J. Obstet. Gynec.* 101, 684.
Jenner, M.R., Kelch, R.P., Kaplan, S.L. and Grumbach, M.M. (1972). *J. clin. Endocr. Metab.* 34, 521-530.
Kase, N. (1964). *Am. J. Obstet. Gynec.* 90, 1262.
Kase, N. and Conrad, S.H. (1964). *Am. J. Obstet. Gynec.* 90, 1251.
Keal, E.E. (1960). *Lancet* 2, 1211.
Lingeman, C. (1974). *J. National Cancer Instit.* 53, 1603-1618.
Loriaux, D.L., Malinok, L.R. and Noall, M.W. (1970). *J. clin. Endocr. Metab.* 31, 702-704.
MacAulay, M.A. and Weliky, I. (1968). *J. clin. Endocr. Metab.* 28, 819.
Mahajan, D.K. and Samuels, L.T. (1963). *Fed. Proc. Fedr. Am. Soc. Exp. Biol.* 22, 2214.
Mahesh, V.B., McDonough, P.G. and Deles, C.A. (1970). *Am. J. Obstet. Gynec.* 107, 183-187.
Marsh, J.M., Savard, K., Baggett, B., Judson, J., Van Wyk, J.J. and Talbert, L.M. (1962). *J. clin. Endocr. Metab.* 22, 1196.
Morris, J.M. and Scully, R.E. (1958). "Endocrine Pathology of the Ovary". C.V. Mosby Co., St. Louis.

Morrow, L.B., Thompson, R.J. and Millinger, R.C. (1968). *J. clin. Endocr. Metab.* **28**, 1756.
Nocke, W. (1965). *In* "Proceedings Second International Congress of Endocrinology", p.1304. Excerpta Medica, Amsterdam.
Novak, E. (1952). "Gynecology and Obstetric Pathology", p.416. W.B. Saunders Co.
Novak, E.R., Kutch Meshgi, J., Mupas, R.S. and Woodruff, J.D. (1971). *Obstet. Gynec.* **38**, 701-713.
Novak, E.R. and Mattingly, R.F. (1960). *Obstet. Gynaec. N.Y.* **15**, 425.
Novak, E.R. and Woodruff, J.D. (1974). "Novak's Gynecologic and Obstetric Pathology with Clinical and Endocrine Relations", 7th edn, pp.367-485. Saunders, Philadelphia, Penn.
O'Malley, B.W., Lipsett, M.B. and Jackson, M.A. (1967). *J. clin. Endocr. Metab.* **27**, 311.
Pedowitz, P. and O'Brien, F.B. (1960). *Obstet. Gynec.* **16**, 62-77.
Pedowitz, P. and Pomerance, W. (1962). *Obstet. Gynec.* **19**, 183.
Pfeiderer, A. and Teufel, G. (1968). *Am. J. Obstet. Gynec.* **102**, 997.
Plotz, E.J., Wiener, M., Stein, A.A. and Hahn, B.D. (1967). *Am. J. Obstet. Gynec.* **97**, 1050.
Procopé, B.-J. (1969). *A.E.* **60**, Suppl. 135.
Prunty, F.T.G. (1967). *J. Endocrinol.* **38**, 85.
Richards, R.J., Hext, P.M., Blundell, G., Henderson, W.J. and Volcani, B.E. (1974). *Br. J. exp. Path.* **55**, 275-281.
Rosner, J.M., Conte, N.F., Horitz, S. and Forsham, P.M. (1964). *Am. J. Med.* **37**, 638.
Ryan, K.J. (1963). *Acta endocr. (Copenh.)* **44**, 81.
Sandberg, A.A., Slaunwhite, W.R., Jackson, J.E. and Frawley, R.F. (1962). *J. clin. Endocr. Metab.* **22**, 929.
Sandberg, E.C., Jenkins, R.C. and Trifon, H.M. (1966). *Steroids* **8**, 249.
Sato, T., Shinada, T. and Matsumoto, S. (1969). *Am. J. Obstet. Gynec.* **104**, 1124-1130.
Savard, K. (1968). *In* "The Ovary" (H.C. Mack, ed) p.911. C.C. Thomas, Springfield.
Savard, K., Dorfman, R.I., Baggett, B., Fielding, L.L., Engel, L.L., McPherson, H.T., Lister, L.M., Johnson, D.J., Hamblen, E.C. and Engel, F.L. (1960). *J. clin. Invest.* **39**, 534.
Savard, K., Gut, M., Dorfman, R.I., Gabrilove, J.L. and Soffer, J.L. (1961). *J. clin. Endocr. Metab.* **21**, 165.
Savard, K., Marsh, J. and Rice, B.F. (1965). *Rec. Progr. Horm. Res.* **21**, 285.
Scott, J.S., Lumsden, C.E. and Levell, M.J. (1967). *Am. J. Obstet. Gynec.* **97**, 161.
Scully, R.E. (1963). *In* "The Ovary" (H.G. Evady and E.C. Smith, eds) 1st edn, p.143. Williams and Wilkins, Co., Baltimore.
Scully, R.E., Richardson, G.S. and Barlow, J.F. (1966). *Clin. Obstet. Gynec.* **9**, 384.
Smith, O.W. (1961). *Endocrinology* **69**, 869.
Sternberg, W.H. (1949). *Am. J. Path.* **25**, 493.
Stevenson, C.S. (1970). *Obstet. Gynec.* **36**, 443.
Stone, M.L., Weingold, A.B., Sall, S. and Sonnenblick, B. (1963). *Surg. Gynec. Obstet.* **116**, 351.
Taylor, H.B. and Norris, H.J. (1967). *Cancer* **20**, 1953.
Teilum, G. (1971). "Special Tumours of Ovary and Testis and Related Extragonadal Lesions. Comparative Pathology and Histological Classification", pp.135-263. Lippincott, Philadelphia, Penn.
Wagner, J.C., Sleggs, K.A. and Marchand, P. (1960). *Brit. J. Indust. Med.* **17**, 260.
Wiest, W.G., Zander, J. and Holmstrom, E.G. (1959). *J. clin. Endocr. Metab.* **19**, 297.
World Health Organisation (1973). "International Histological Classification of Tumours, No.9" (S.E. Serov and R.E. Scully, eds). Geneva.
Wren, B.G. and Frampton, J. (1963). *Br. Med. J.* **2**, 842.

HORMONAL STUDIES IN A VIRILIZING "LUTEOMA" OF PREGNANCY

("Nodular theca-luteinic hyperplasia of pregnancy")

G. Forti, R. Buratti, M. Colafranceschi, G. Fiorelli, M. Pazzagli, G. Lombardi
M. Marchionni and M. Serio

*Endocrinology Unit, Department of Obstetrics and Gynecology
and Department of Pathology, University of Florence, Italy*

SUMMARY

A rare case of virilizing so-called luteoma of pregnancy is described in a primigravida 22 year old woman. The symptoms of virilization were evident from the 11th week of gestation. The mother was operated, and aborted two hours after the resection of the tumor: the male fetus showed no anomalies. Plasma levels of testosterone (T), dihydrotestosterone (DHT), androstenedione (Δ) and dehydroepiandrosterone (DHA) were measured in the mother before and after the removal of the tumor. Before surgery T and DHT levels were more than a hundred times higher than the corresponding levels found in normal women at the same week of gestation; Δ and DHA levels were about ten times higher than the corresponding values in normal subjects. After removal of the tumor all these plasma androgens fell within the normal range. T, DHT, Δ and DHA were also measured in the "tumoral" tissue: their concentrations were found to be extremely high when compared with those found in normal corpora lutea and ovaries. The hypothesis of a testosterone dependence of the 5α-reductase activity is discussed.

INTRODUCTION

The term "luteoma of pregnancy" was first used by Sternberg in 1963 to describe an uncommon specific benign lutein cell tumor of the ovary that is usually found in late pregnancy and apparently regresses after delivery.

To date only 64 documented cases were reported in the literature (Green *et al.*, 1964; Navarrete-Reyna and Kaufman, 1964; Rachman and Tellem, 1964;

Table I. Plasma androgens before and after surgery (ng/100 ml).

PLASMA ANDROGENS BEFORE AND AFTER SURGERY (ng/100 ml)					
	T	DHT	Δ	DHA	P
Before surgery	7749	1392	2200	3350	295
3 days after surgery	15	12	--	--	--
60 " " "	11	19	195	553	558
Pregnant women (17th week of gestation. Mean±SD)	53 ± 9	11 ± 8	161 ± 35	526±404	3481±714
Normal women (Mean±SD)	27 ±10	12 ± 8	90 ± 44	394±277	20 ± 13 * 434 ± 272 **
Normal men (Mean±SD)	527±226	54 ± 19	84 ± 47	308±209	21 ± 8

* follicular phase
** luteal phase

Malinak and Miller, 1965; Krause and Sternbridge, 1966; Sternberg and Barclay, 1966; Mandell et al., 1967; Norris and Taylor, 1976; O'Malley et al., 1967; Jenkins et al., 1968; Laffargue et al., 1968; Shuster and Leake, 1968; Jewelewicz et al., 1971; Thomas et al., 1972; Wolff et al., 1973): the relative rarity of this lesion is probably related to the spontaneous regression of the tumor after delivery and to the fact that the tumor is usually asymptomatic and discovered fortuitously during surgery for cesarean section or tubal ligation at delivery or in the immediate post-partum period.

Other typical features of the "luteoma of pregnancy" are the high parity of most of the patients and the higher incidence in Negro women: this fact is probably related to multiparity, to the high percentage of Negro admissions in the Hospitals in which cases were found and to the higher frequency of cesarean sections and tubal ligations in Negro women, rather than to racial predisposition (Sternberg and Barclay, 1966).

Luteomas of pregnancy are often multinodular, sometimes occur bilaterally and regress spontaneously post-partum. For these reasons the luteoma of pregnancy is usually considered a hyperplastic phenomenon rather than a true neoplasma (Malinak and Miller, 1965; Sternberg and Barclay, 1966; Krause and Sternbridge, 1966; Mandell et al., 1967; Norris and Taylor, 1967; Kerber et al., 1969; Jewelewicz et al., 1971; Thomas et al., 1972).

A very rare phenomenon associated with the luteoma of pregnancy is virilization of the mother and/or the fetus due to an increased androgen production by the tumor. Of the cases reported in the literature only 12 showed evidence of maternal virilization and only 3 were associated with signs of virilization of a female fetus (Malinak and Miller, 1965; Jenkins et al., 1968; Wolff et al., 1973). In the cases so far reported steroid determinations were performed only occasionally.

Hormonal Studies in a Virilizing "Luteoma" of Pregnancy

In this chapter we report one case of virilizing luteoma in which the increased androgen production was evident from the 11th week of gestation, with measurements of T, DHT, Δ, DHA and progesterone (P) plasma levels before and after the resection of the tumor. The concentrations of the steroids mentioned above in the "tumoral" tissue are also reported.

CASE REPORT

A 22 year old Caucasion woman, primigravida, was admitted to the Obstetric Service of our University during the 15th week of gestation because she had noted increased hair growth on her face and back with acneform lesions, flushing and itching beginning about a month before. The menstrual history of the patient prior to pregnancy was normal and she had received no medication including steroid hormones during gestation. On physical examination hirsutism and acne were evident especially on the face and on the back. No temporal balding was observed and there was no history of deepening of the voice or atrophy of breasts. External genitalia were normal; no hypertrophy of the clitoris was present; on pelvic examination a large, firmed, fixed mass was palpable behind the uterus in the posterior cul de sac. Blood pressure, ECG and routine blood chemistry studies were normal. Because of the very high levels of plasma testosterone (Table 1), in view of the risk of virilization of the fetus, the patient was operated on December 18, 1972. Laparotomy showed the right ovary deformed by a large, smooth, pink-yellowish lobulated round mass with a diameter of about 11 cm, apparently encapsulated and rather soft. The left ovary appeared normal. The mass was resected and right oophorectomy was performed. Two hours after surgery, in spite of antispastic therapy that had been started before laparotomy, the patient aborted: the fetus weighed 250 g, was of male sex and showed no anomalies. Placenta and membranes appeared normal. The patient had an uncomplicated post-operative course: the itching and flushing of the skin disppeared in a few days. Hirsutism and acne began to improve within two months and disappeared entirely within eight months.

PATHOLOGICAL FINDINGS

The tumor was a round mass, with a diameter of 11 cm, apparently encapsulated and lobulated (Fig.1). The external surface was yellowish with small pink areas; the cut surface (Fig.2) was homogeneously yellow. No areas of necrosis or hemorragic degeneration were present. On microscopic examination the tumor was composed of masses of polygonal cells uniform in size, without a particular pattern (Fig.3). With higher magnification (Fig.4) cells with oval, vesicular nuclei and a distinct nucleolus are evident; the cytoplasm is finely granular and without distinct borders. Some cells with small, pycnotic and hyperchromatic nuclei are present; cells with the so-called "ballooning" degeneration are also evident. The histopathological diagnosis was "luteoma of pregnancy".

The histogenesis of this lesion is as yet uncertain. Sternberg and Barclay (1966) suggest an origin either from the theca lutein cells of atretic follicles or from stromal lutein cells, whilst Norris and Taylor (1967) think that these lesions develop from theca-lutein cysts.

Fig.1 Macroscopic appearance of the luteoma.

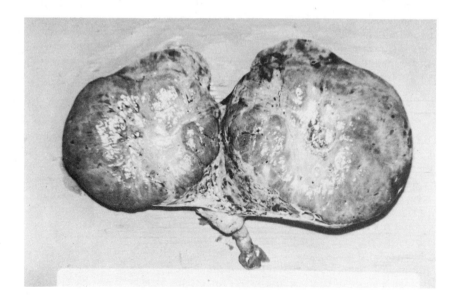

Fig.2 Cut surface of the tumor.

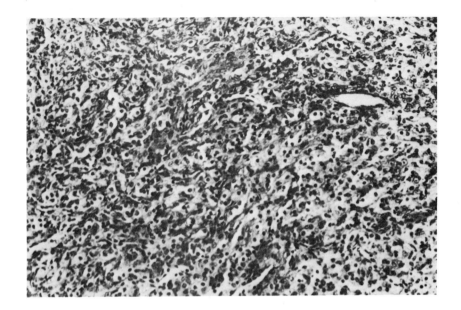

Fig.3 Photomicrograph showing sheets of uniform, polygonal cells (hematoxylin and eosin × 75).

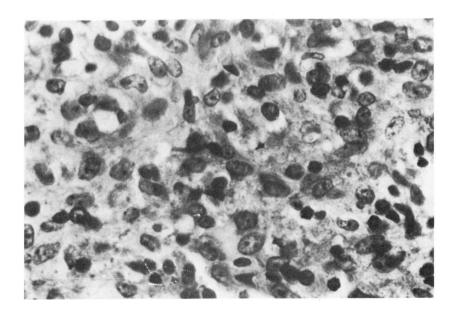

Fig.4 Photomicrograph showing cells with the so-called "ballooning" degeneration.

Table II. Steroid content of "tumoral tissue", normal corpus luteum (N=2) and normal ovarian cortex (N=4); values are expressed as ng/gr of tissue.

STEROID CONTENT OF "TUMORAL" TISSUE, NORMAL CORPUS LUTEUM (n=2) AND NORMAL OVARIAN CORTEX (n=4)
(ng/gr of tissue)

	T	DHT	Δ	DHEA	P
Luteoma	7,740	1,880	18,600	4,660	4
Corpus Luteum (n=2; mean)	60	4.5	123	22	199
Normal Ovarian Cortex (n=4; mean)	11	1.7	79	22	35

Table III. Steroid content of the luteoma in the present case and in the case of O'Malley et al. (1967); values are expressed as µg/gr of tissue.

	Steroid content (µg/gr of tissue)				
	T	DHT	Δ	DHEA	P
LUTEOMA (our case)	7.74	1.88	18.6	4.66	0.004
LUTEOMA (case of O'Malley et al.)	7.60	—	0.46	3.10	0

HORMONAL STUDIES IN THE PLASMA

Plasma levels of T, DHT, Δ, DHA and P were determined before and after surgery: results are shown in Table I. Radioimmunoassay methods previously described were used to measure T, DHT, Δ and P (Forti et al., 1974; Pazzagli et al., 1975; Fiorelli et al., 1975; Calabresi et al., 1974). Plasma DHA levels were determined with a radioimmunoassay in which a chromatographic step in the Bush A system is included.

The levels of all the androgens measured in the plasma of the patient before the resection of the tumor were extremely high: plasma T level was more than a hundred times the corresponding value of normal pregnant women at the same week of gestation; a similar increase was observed in DHT levels, while Δ and DHA were about tenfold the corresponding normal values. On the other hand P levels in the plasma were very low.

Table IV. T and DHT levels in peripheral and spermatic venous plasma of normal men (Fiorelli *et al.*, 1975); values are expressed as µg/100 ml (Mean ± SD).

T AND DHT LEVELS IN PERIPHERAL AND SPERMATIC VENOUS PLASMA OF NORMAL MEN (Mean ± SD; µg/100 ml)

N	Age range (years)	Spermatic plasma		Peripheral plasma	
		T	DHT	T	DHT
24	17-80	32.2 ± 26.3	0.42 ± 0.33	0.33 ± 0.73	0.03 ± 0.01

Correlation coefficient between T and DHT:
Spermatic venous plasma $r = +0.5098$ ($p < 0.05$)
Peripheral venous plasma $r = +0.5585$ ($p < 0.01$)

Three days after the removal of the tumor plasma T and DHT levels fell within the normal range for females. Δ and DHA levels were not measured at that time, but two months after surgery they were also within the normal range.

HORMONAL STUDIES IN THE TUMORAL TISSUE

A 6.50 g sample of the tumor was homogenised in 100 ml of distilled water; the homogenate was then extracted twice with 500 ml of diethyl ether. The ether extract was dried, taken up in 100 ml of absolute ethanol and kept at $-20°C$ in 10 ml aliquots. The steroid concentrations of the extract have been measured with the methods mentioned above and the results are shown in Table II where the steroid content of 2 normal corpora lutea and of 4 normal ovarian cortex are also shown for comparison. The small progesterone content of the luteoma is probably due to its capacity to metabolize progesterone.

DISCUSSION

This is in our knowledge the first case of maternal virilization resulting from a luteoma of pregnancy in which T, DHT, Δ, DHA and P have been measured in plasma before and after the removal of the tumor, and in the tumoral tissue. The levels of the four androgens measured in the plasma were exceptionally high, especially T and DHT; plasma T levels were measured in only two other cases of luteomas of pregnancy: one of these cases had plasma T levels in the normal range for males (Wolff *et al.*, 1973), the other one had levels very similar to our case (Jewelewicz *et al.*, 1971).

The steroid levels of the tumoral tissue of our case agree, with the exception of Δ, with the data found by O'Malley and coworkers (1967) in another virilizing luteoma in which no plasma measurements were performed (Table III).

Table V. Plasma androgens in two adrenal virilizing tumors; values are expressed as ng/100 ml.

Case	Age (years)	Diagnosis	T	DHT	Δ	DHA
1	6	Carcinoma	84	33	432	1085
2	68	Adenoma	525	57	90	244

The virilization of our patient appeared to be related to the increased production of androgens by the tumor: in fact they fell within the normal range immediately after the removal of the tumor.

We think that the interest of this case is due not only to its rarity but also to the fact that it can be considered a model, even if not physiological, for the study of testosterone metabolism.

Our studies on androgen secretion by the human ovary have shown that usually the ovary does not produce DHT; however when, in the luteal phase, T levels in the ovarian blood overcome 100 ng/100 ml a small gradient may be found between DHT levels in ovarian and peripheral blood (Serio et al., 1975, this volume, Table VI). A significant correlation was also found between T and DHT in ovarian venous plasma during the luteal phase. So it would seem that a 5α-reductase activity in the ovary could be initiated by high levels of testosterone. A 5α-reductase activity has been found in normal human ovaries in the luteal phase (Smith and Zuckermann, 1973).

The testis is another organ in which the 5α-reductase activity seems to be induced by testosterone: it is known that in the male about 80% of blood production rate (BPR) comes from peripheral conversion of T and 20% from testicular secretion (Fiorelli et al., 1975), while in the female the whole BPR of DHT comes from the peripheral conversion of T (Ito and Horton, 1971). In our laboratory Fiorelli and coworkers have demonstrated a significant correlation between T and DHT either in spermatic or in peripheral venous blood of male subjects (Fiorelli et al., 1975; see Table IV).

Our hypothesis of the testosterone dependence of the 5α-reductase activity is in agreement with the recent studies of Kuttenn and Mauvais-Jarvis (1975) who showed that the induction of this enzymatic activity in the human pubic skin depends upon the presence of intracellular active androgens.

In the luteoma we have described, even if we could not measure the DHT levels in the venous blood coming from the tumor, the DHT content of the tumoral tissue was so high that it seems correct to think that the tumor was also able to produce DHT: it is impossible to say whether DHT was directly secreted by the same tumoral cells secreting T or produced by other tumoral cells transforming T in DHT.

Our data concerning two other cases of adrenal virilizing tumors in women (Table V) with high T levels show that DHT levels were also increased: in these cases DHT could be produced by the peripheral transformation of T, but also directly secreted by the tumor as in the luteoma we have described.

We would make only two comments: 1) the so-called pregnancy luteoma is doubtless a more common lesion than is generally suspected. Since this condition is benign and regresses following delivery no surgical treatment is required: so most lesions of this nature, when encountered at laparotomy or at cesarean section, could be correctly diagnosed by frozen section and unnecessary ovariectomies avoided; 2) when sudden virilization occurs during pregnancy the possibility of this lesion must be kept present.

ACKNOWLEDGEMENT

This work was supported by a grant from the Administrative Board of the University of Florence (Item XI B).

REFERENCES

Calabresi, E., Pazzagli, M., Fiorelli, G., Forti, G. and Serio, M. (1974). *J. Nucl. Biol. Med.* 1, 30-37.
Forti, G., Pazzagli, M., Calabresi, E., Fiorelli, G. and Serio, M. (1974). *Clin. Endocr.* 3, 5-17.
Fiorelli, G., Borrelli, D., Forti, G., Gonnelli, P., Pazzagli, M. and Serio, M. (1975). *J. Steroid Biochem. (in press).*
Greene, R.R., Holzwarth, D. and Roddick, J.W. (1964). *Am. J. Obstet. Gynec.* 88, 1001-1011.
Ito, T. and Horton, R. (1971). *J. clin. Invest.* 50, 1621-1627.
Jenkins, M.E., Surana, R.B. and Russel-Cutts, C.M. (1968). *Am. J. Obstet. Gynec.* 101, 923-928.
Jewelewicz, R., Perkins, R.P., Dyrenfurth, I. and Vande Wiele, R.L. (1971). *Am. J. Obstet. Gynec.* 109, 24-28.
Kerber, I.J., Bell, J.S., Camacho, A.M. and Fish, S.A. (1969). *Southern Med. J.* 62, 1343-1350.
Krause, D.E. and Sternbridge, V.A. (1966). *Am. J. Obstet. Gynec.* 95, 192-206.
Kutten, F. and Mauvais-Jarvis, P. (1975). *Acta endocr. (Copenh.)* 79, 164-176.
Laffargue, P., Payan, H. and Rampal, M. (1968). *Presse méd.* 76, 155-158.
Malinak, L.R. and Miller, G.V. (1965). *Am. J. Obstet. Gynec.* 91, 251-259.
Mandell, G.H., Floyd, S.W., Cohn, S.L. and Goodman, P.A. (1967). *Am. J. clin. Path.* 47, 148-153.
Navarrete-Reyna, A. and Kaufman, R.H. (1964). *Am. J. Obstet. Gynec.* 90, 521-525.
Norris, H.J. and Taylor, H.B. (1967). *Am. J. clin. Path.* 47, 557-566.
O'Malley, B.W., Lipsett, M.B. and Jackson, M.A. (1967). *J. clin. Endocr.* 27, 311-319.
Pazzagli, M., Forti, G., Cappellini, A. and Serio, M. (1975). *Clin. Endocr.* 4, 513-520.
Rachman, R. and Tellem, M. (1964). *Am. J. Obstet. Gynec.* 88, 132-134.
Serio, M., Dell'Acqua, S., Calabresi, E., Fiorelli, G., Forti, G., Cattaneo, S., Lucisano, A., Lombardi, G., Pazzagli, M. and Borrelli, D. (1975). This volume, pp.471-479.
Shuster, E. and Leake, F.M. (1968). *Obstet. Gynec.* 32, 637-642.
Smith, O.W. and Zuckerman, N.G. (1973). *Steroids* 22, 379-399.
Sternberg, W.H. (1963). "The Ovary", pp.209-254. International Academy of Pathology. Williams and Wilkins, Cp., Baltimore.
Sternberg, W.H. and Barclay, D.L. (1966). *Am. J. Obstet. Gynec.* 95, 165-184.
Thomas, E., Mestman, J., Henneman, C., Anderson, G. and Hoffman, R. (1972). *Obstet. Gynec.* 39, 577-584.
Wolff, E., Glasser, M., Gordon, G.G., Jaime, O. and Southren, A.L. (1973). *Am. J. Med.* 54, 229-233.

RELEASE OF C_{21}, C_{19} AND C_{18} STEROIDS BY HUMAN OVARIES STIMULATED *IN VIVO* BY HCG

S. Dell'Acqua, G.B. Serra, A. Lucisano, A. Montemurro, B. Cinque
E. Arnò and A. Bompiani

Department of Obstetrics and Gynaecology
Catholic University S. Cuore, Rome, Italy

During the last fifteen years many data have been accumulated demonstrating that the human ovary is capable of synthesizing and releasing a variety of C_{21}, C_{19} and C_{18} steroids. This conclusion has been reached through experiments carried out mostly *in vitro*. However, before using these data in order to design the definitive map of the biosynthetic reactions present in the ovary during the various phases of the menstrual cycle, and therefore to define the terms of its function as an endocrine gland, it is necessary to prove that the reactions demonstrated *in vitro* are actually operating *in vivo*.

It is unnecessary to point out the obvious advantages which the *in vitro* experiments offer. In fact some results may be obtained only by means of these techniques (homogenates, tissue slices, tissue cultures, perfusions, superfusions, etc.), particularly for the ovary, if we consider the subunits which compose this organ. However, regarding physiological events, the state of tissue organization obviously limits the value of the experiments *in vitro*.

In the ovary *in situ* substrates and metabolic energy arrive at the cells in the best way for the function of the organ. Furthermore, they must be involved at the same time in the regulatory mechanisms, together with the concentration and the state of oxidation of the cofactors and with the intermediate and final steroids synthesized, that accumulate in different quantity in the cells. Moreover, as regards the ovary, the transport of intermediate steroids from one subunit to another may constitute an additional regulatory mechanism.

All these systems need therefore, in order to produce physiological conditions as a whole, intact cellular structure, intact organ structure and intact blood circulation, both for the uptake of substrates and energetic materials, and for the release of metabolic products.

By using *in vivo* techniques, many of these difficulties may be overcome. Obviously the plasma levels of the steroids released from the ovary can be measured only by means of *in vivo* techniques. It should be kept in mind, however, that the *in vivo* experimental approach presents new and serious problems. The most important concerns general anesthesia, which can influence ovarian steroid metabolism. Moreover, surgical manipulation of the organ could alter its function, and particularly its blood flow (the cannulation of the veins and especially of the ovarian arteries may present technical difficulties). Finally, in the human species, the time available for the experiment must be limited as much as possible.

In order to verify *in vivo* in the human some of the data obtained *in vitro*, we devised two experimental designs, in which the ovary, still *in situ*, could approximate as far as possible the physiological conditions.

We will refer briefly to the first group of experiments, already published (Dell'Acqua *et al.*, 1974), only as an introduction to the second group of observations, that we want to present here. On the basis of previous experiments (De Paoli and Eik-Nes, 1973, in dogs; Domanski and Dobrowolski, 1966, in sheep; Hilliard *et al.*, 1964, in rabbits; Doouss and Deshpande, 1968 and Oertel *et al.*, 1968, in humans) we have performed a perfusion technique, in which one artery and one vein of the ovary were cut and cannulated ("Type III *in vivo* perfusion system", according to Ahren *et al.*, 1971). This study was carried out on four fertile patients, candidates for radical surgery for stage one carcinoma of the uterine cervix, at mid-luteal phase of the cycle (Dell'Acqua *et al.*, 1974). Through the cannulated ovarian artery was perfused homologous oxygenated heparinized blood, in which 45 μCi of tritiated cholesterol was mixed. From the cannulated vein three 15 minute samples of blood were collected. In two cases, 15 minutes after the beginning of the perfusion, a dose of 991 I.U. of highly purified human urinary LH (Donini *et al.*, 1968), obtained through the courtesy of Dr. Donini, Serono, Rome, was rapidly injected into the ovarian artery. At the end of the experiment, the ovarian tissue and the perfusates were analysed for radioactive steroids.

In all cases, from the perfused ovary pregnenolone and progesterone were isolated: part of the radioactive cholesterol administered is therefore rapidly metabolized into progesterone, via pregnenolone. In the two experiments in which LH was injected, the amount of progesterone isolated from the ovaries was five time greater than in the other two cases and, in addition, radioactive progesterone was found in good yield in the third sample of the perfusate. This evidence demonstrates that in the human the circulating cholesterol can be an important and prompt substrate for the synthesis of progesterone; moreover, these experiments prove that LH plays an important role as an acute stimulus for the enzymes involved in the ovarian steroidosynthesis, confirming many observations in animals (Hilliard *et al.*, 1964; Hilliard and Sawyer, 1966; Armstrong, 1967; Bartosik and Romanoff, 1969).

In order to improve our knowledge of the action of gonadotrophins on ovarian steroidogenesis, at this point it seemed to us a more physiological approach, at least as to stimulation, to utilize the cubital vein, and therefore the peripheral circulation, and to measure at short intervals the progesterone level in plasma from one of the ovarian veins, excluding the use of radioactive precursors. As the biological effects obtained *in vitro* between LH and HCG are similar and considering the stimulatory action of HCG on the steroidogenesis of the corpus luteum during

Release of C_{21}, C_{19} and C_{18} Steroids by Human Ovaries

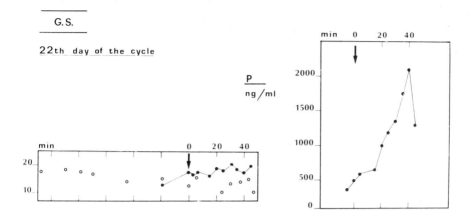

Fig.1 Progesterone (P) level in the venous blood from the ovary containing the corpus luteum (right) in the venous blood from the contralateral ovary (left) and in the peripheral venous blood (open circles, left). Arrow indicates the zero-time, i.v. injection of 30,000 I.U. of HCG.

the first weeks of gestation, in this new series of experiments purified HCG has been used. In order to damage as little as possible the vein isolated from the infundibulo-pelvic ligament, the technique of cannulation of the vessel has been substituted by insertion into the vessel of a microneedle, connected to a small tube. The needle and the catheter were maintained patent by rapidly washing the system with diluted heparin between the samples collected. The flow was kept at 1 ml/min. In all cases investigated we have tried to drain the venous blood from both ovaries, but for technical reasons it was possible to achieve this only in some patients.

After collection of two or three basal samples of ovarian venous blood, 30,000 I.U. of purified HCG, generously given by Dr. Donini, Serono, Rome, were rapidly injected into the cubital vein. Some of the antisera used (for 17β-oestradiol, antiserum to 17β-oestradiol-6-carboxymethyl-oxime-BSA; for testosterone, antiserum to testosterone-3-oxime-BSA) have been obtained through the courtesy of Dr. Sommerville, Chelsea Hospital for Women, London; some have been produced in our laboratory (for progesterone, antiserum to progesterone-11α-hemisuccinate-BSA; for 17α-hydroxy-progesterone, antiserum to 17α-hydroxy-progesterone-3-carboxymethyl-oxime-BSA; for 20α-dihydro-progesterone, antiserum to 20α-dihydro-progesterone-3-carboxymethyl-oxime-BSA).

In all cases studied the day of the menstrual cycle has been calculated from the first day of the previous menstrual period. The phase of the cycle has been confirmed, besides by direct vision at laparotomy, by endometrial biopsy. The patients studied have been submitted at laparotomy to minor surgery of the uterus.

In Fig.1 is shown the profile of the blood levels of progesterone in a patient at the 22nd day of the menstrual cycle. It is evident that while from the ovary containing the corpus luteum, the injection of HCG produces a prompt increase of the release of progesterone, reaching its maximum at 40 min from the stimulus, in the venous effluent of the contradlateral ovary and in the peripheral blood the values obtained are similar and are not modified by HCG injection.

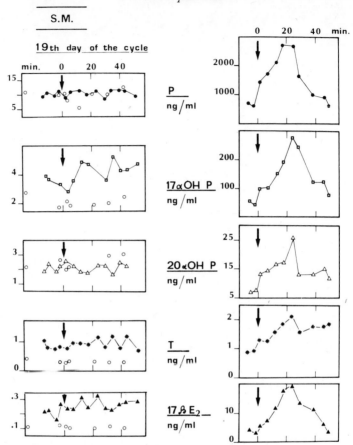

Fig. 2 Progesterone (P), 17a-hydroxy-progesterone (17a-OH-P), 20a-dihydro-progesterone (20a-OH-P), testosterone (T) and 17β-oestradiol (17β-E_2) levels in the venous blood from the ovary containing the corpus luteum (right), in the venous blood from the contralateral ovary (left) and in the peripheral venous blood (open circles, left). Arrow indicates the zero-time i.v. injection of 30,000 I.U. of HCG.

The basal values, in all cases investigated, are in agreement with those reported by others (i.e. Mikhail, 1970; Lloyd et al., 1971).

To demonstrate the ability of HCG to stimulate the release of progesterone in this system *in vivo*, it seemed interesting to examine the ovarian blood levels of other C_{21}, C_{19} and C_{18} steroids, in an attempt to better elucidate the steroidogenic potential of the ovary in this phase of the menstrual cycle. In a patient stimulated with HCG at the 19th day of the menstrual cycle, we have measured progesterone, 17a-hydroxy-progesterone, 20a-dihydro-progesterone, 17β-oestradiol and testosterone in the venous blood coming from the ovary containing the corpus luteum, in the venous blood from the contralateral ovary and in peripheral blood (Fig. 2).

The profile of the progesterone level in the venous blood from the ovary containing the corpus luteum is similar to that of the previous case. A corresponding peak, at about 20 min from the HCG stimulation, is evident also for all the

Release of C_{21}, C_{19} and C_{18} Steroids by Human Ovaries 431

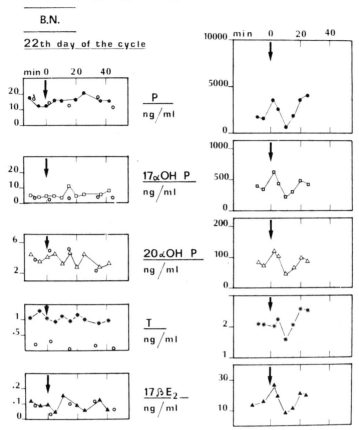

Fig. 3 Progesterone (P), 17a-hydroxy-progesterone (17a-OH-P), 20a-dihydro-progesterone (20a-OH-P), testosterone (T) and 17β-oestradiol (17β-E_2) levels in the venous blood from the ovary containing the corpus luteum (right), in the venous blood from the contralateral ovary (left) and in the peripheral venous blood (open circles, left). Arrow indicates the zero-time i.v. injection of 30,000 I.U. of HCG.

other steroids considered.

The steroid levels in the venous blood from the contralateral ovary and in the peripheral blood are very close and again they do not show significant variations after HCG stimulation. In a patient stimulated with HCG at the 22nd day of the menstrual cycle the hormonal profile appears incomplete, owing to the fact that we had to interrupt the collection of blood from the ovary containing the corpus luteum after a short time from the stimulus (Fig. 3).

The steroid levels in the venous blood from the contralateral ovary and in the peripheral blood are very similar to that of the previous case and still there is no evidence of any significant variation after HCG stimulation.

In confirmation of the necessity of prolonging the blood collection as long as possible we show a case (Fig. 4) at the 22nd day of the menstrual cycle in which the elevation above the basal values after the injection of HCG does not decrease, even after 40 min have passed.

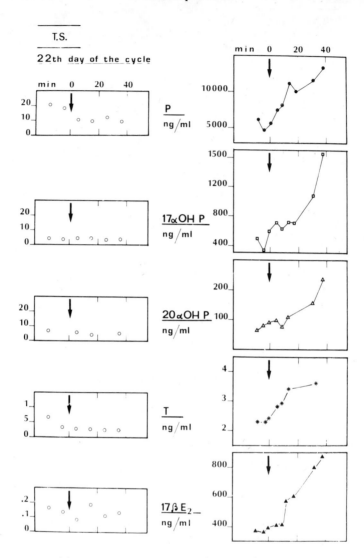

Fig.4 Progesterone (P), 17a-hydroxy-progesterone (17a-OH-P), 20a-dihydro-progesterone (20a-OH-P), testosterone (T) and 17β-oestradiol (17β-E_2) levels in the venous blood from the ovary containing the corpus luteum (right) and in the peripheral venous blood (open circles, left). Arrow indicates the zero-time i.v. injection of 30,000 I.U. of HCG.

In accordance with the previous case, also in a patient at the 23rd day of the menstrual cycle (Fig.5), it is evident, after HCG stimulation, that there is an increase of the levels of progesterone, 17a-hydroxy-progesterone and 17β-oestradiol in the venous blood from the ovary containing the corpus luteum, going on over 40 min. It is of interest, considering that the day of the menstrual cycle is more advanced as compared with the other cases, to note the discrepancy between the blood levels of testosterone and 20a-dihydro-progesterone, which decrease earlier

Release of C_{21}, C_{19} and C_{18} Steroids by Human Ovaries 433

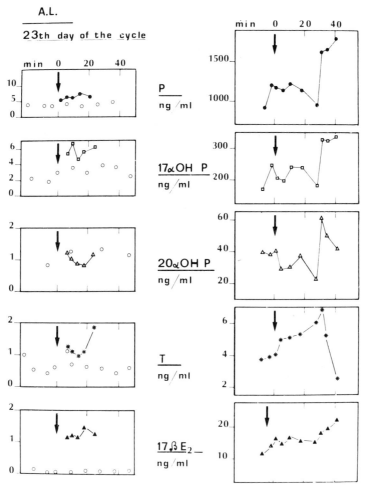

Fig. 5 Progesterone (P), 17a-hydroxy-progesterone (17a-OH-P), 20a-dihydro-progesterone (20a-OH-P), testosterone (T) and 17β-oestradiol (17β-E_2) levels in the venous blood from the ovary containing the corpus luteum (right), in the venous blood from the contralateral ovary (left) and in the peripheral venous blood (open circles, left). Arrow indicates the zero-time i.v. injection of 30,000 I.U. of HCG.

than those of the other three steroids. The hormonal levels in the venous blood from the contralateral ovary, measured for only 20 min do not allow any conclusion about the HCG stimulus.

From the examination of these five cases in the luteal phase of the menstrual cycle, some conclusions may be drawn. In confirmation of the *in vivo* experiments, it is clear that there is difficulty in obtaining homogeneous results. Yet, some concepts can be drawn from the results obtained so far:

1. According to the data obtained from others (i.e. Mikhail, 1970; Lloyd et al., 1971) the basal values of the steroids analysed seem to show that in this phase of the menstrual cycle the ovary containing the corpus luteum plays a predominant role in steroidogenesis compared with the contralateral one, at

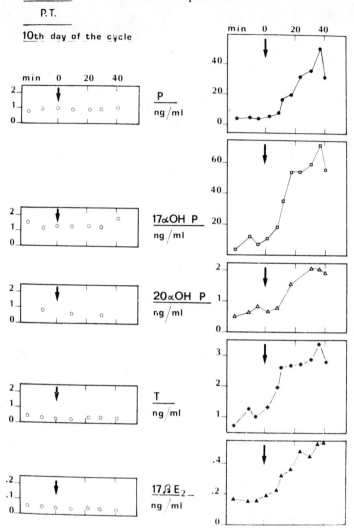

Fig.6 Progesterone (P), 17α-hydroxy-progesterone (17α-OH-P), 20α-dihydro-progesterone (20α-OH-P), testosterone (T) and 17β-oestradiol (17β-E$_2$) levels in the venous blood from the ovary containing the growing follicle (right) and in the peripheral venous blood (open circles, left). Arrow indicates the zero-time i.v. injection of 30,000 I.U. of HCG.

least for progesterone, 17α-hydroxy-progesterone, 20α-dihydro-progesterone and 17β-oestradiol.

2. The observation that only the ovary containing the corpus luteum responds to the HCG stimulation strengthens the importance of the previous consideration.

3. The HCG seems to act as an acute stimulus to the release of the steroids analysed.

4. Finally, even if this experimental design does not give us any information about the mechanism of action of HCG on the biosynthesis and the release of the

Release of C_{21}, C_{19} and C_{18} Steroids by Human Ovaries

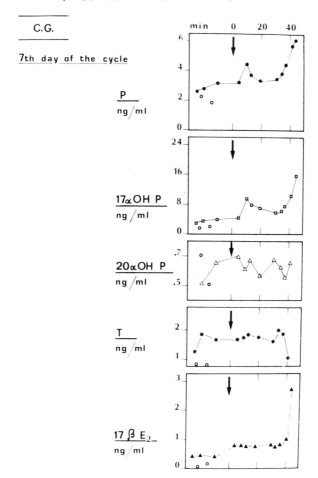

Fig. 7 Progesterone (P), 17a-hydroxy-progesterone (17a-OH-P), 20a-dihydro-progesterone (20a-OH-P), testosterone (T) and 17β-oestradiol (17β-E_2) levels in the venous blood from one ovary and in the peripheral venous blood (open circles). Arrow indicates the zero-time i.v. injection of 30,000 I.U. of HCG.

steroids in the ovary, it is evident that HCG in this phase of the menstrual cycle, increases in a uniform way the release of the various C_{21}, C_{19} and C_{18} steroids investigated.

We extended our observations in the follicular phase of the menstrual cycle. Only two cases have been studied so far. In a patient at day 10 of the menstrual cycle we measured the levels of the same steroids previously considered in the venous blood coming from the ovary containing the growing follicle and in the peripheral blood (Fig.6). Also in this case it is evident in the venous blood from the ovary containing the growing follicle that there is an increase, which reaches its maximum at about 40 min after HCG stimulation, in accordance with the data obtained in the luteal phase. It can be noted that, unlike the luteal phase, in this phase of the cycle 17a-hydroxy-progesterone is higher than progesterone.

In a patient at day 7 of the menstrual cycle we could not establish if the ovary cannulated was the one containing the growing follicle.

The technical impossibility of cannulating both ovaries and the lack of an ovarian biopsy make the evaluation of this case difficult (Fig.7). When it is possible to cannulate both ovaries in this phase of the menstrual cycle, it will be interesting to verify if also in the follicular phase either late or early the two ovaries respond in a different way to HCG stimulation. In this way it will be easier to evaluate also in this case the lack of increase of the plasma levels of testosterone and $20a$-dihydro-progesterone after HCG stimulation.

Although our research has used techniques which are slightly traumatic and pharmacological doses of HCG, yet we think we have at our disposal an experimental model, which reproduces dynamic conditions, suitable for the study of some aspects of ovarian steroidogenesis in the human.

REFERENCES

Ahrén, K., Janson, P.O. and Selstam, G. (1971). *In* "Perfusion Techniques" (E. Diczfalusy, ed) p.285. Karolinska Institutet, Stockholm.

Armstrong, D.T. (1967). *Nature (Lond.)* 213, 633.

Bartosik, D.B. and Romanoff, E.B. (1969). *In* "The Gonads" (K.W. KcKerns, ed) p.211. North-Holland Publishing Company, Amsterdam.

Dell'Acqua, S., Mancuso, S., Catelli, M.G. and Bompiani, A. (1974). *Acta endocr. (Copenh.)* 75, 368.

De Paoli, J.C. and Eik-Nes, K.B. (1973). *Biochim. biophys. Acta* 78, 457.

Domanski, E. and Dobrowolski, W. (1966). *Endokr. pol.* 17, 187.

Donini, P., Puzzuolli, D., D'Alessio, I., Bergisi, G. and Donini, S. (1969). *In* "Gonadotropins" (E. Rosemberg, ed) p.37. Geron X, Palo Alto, California.

Doouss, T.W. and Deshpande, N. (1968). *Brit. J. Surg.* 55, 673.

Hilliard, J. and Sawyer, C.H. (1966). International Congress Series 170, p.195. Excerpta Medica, Amsterdam.

Hilliard, J., Hayward, J.N. and Sawyer, G.H. (1964). *Endocrinology* 75, 957.

Lloyd, C.W., Lobotsky, J., Baird, D.T., McCracken, J.A., Weisz, J., Pupkin, M., Zanartu, J. and Puga, J. (1971). *J. clin. Endocr. Metab.* 32, 155.

Mikhail, G. (1970). *Gynaec. Invest.* 1, 5.

Oertel, G.W., Treiber, L., Wenzel, D., Knapstein, P., Wendlberger, F. and Menzel, P. (1968). *Experientia (Basel)* 24, 607.

ANDROGENIC HORMONES IN HIRSUTISM

G. Giusti, E. Roncoli, G. Forti, S. Cattaneo, G. Fiorelli
M. Pazzagli and M. Serio

Endocrinology Unit, University of Florence, Italy

Hirsutism is a syndrome with a complex pathogenesis which must be reconsidered in the light of recent experimental data.

Three main aspects will be discussed briefly:
a) Androgen secretion by the adrenal glands and the ovaries.
b) Plasma binding and metabolism of androgens.
c) Metabolism of androgens in the skin.

a) Glandular hypersecretion of androgenic hormones has been clearly demonstrated in several cases of hirsutism such as virilizing tumors (Kirschner and Barding, 1972), polycystic ovary syndrome (P.C.O.S.) (Stahl *et al.*, 1973) and in several cases of idiopathic hirsutism (Kirschner and Jacobs, 1971) (see also the chapter by Dr. Kirschner in this volume).

In our laboratory some cases of P.C.O.S. have been studied measuring androgenic hormones in ovarian tissue and in ovarian venous blood. Samples of ovarian tissue were taken during surgical intervention for ovarian wedge resection. A part of the ovarian tissue was taken for histological examination and the remainder was homogenized and extracted with diethyl ether. Androstenedione (Δ), dihydrotestosterone (DHT) and testosterone (T) were measured by radioimmunoassay after paper chromatography according to the methods previously described (Forti *et al.*, 1974; Pazzagli *et al.*, 1975; Fiorelli *et al.*, in press). Ovarian cortex from 8 normal ovaries during the luteal phase were taken as controls.

The concentrations of Δ ($P < 0.05$), DHT ($P < 0.01$) and T ($P < 0.1$) in polycystic ovarian tissue are higher than in normal ovarian cortex (see Fig.1). These data confirm our results on ovarian venous blood (Serio *et al.*, this volume) and demonstrate the ovarian production of androgens in P.C.O.S., which is also confirmed by the reduction of the testosterone production rate after ovarian wedge resection (Southren *et al.*, 1968). On the other hand, in the P.C.O.S.,

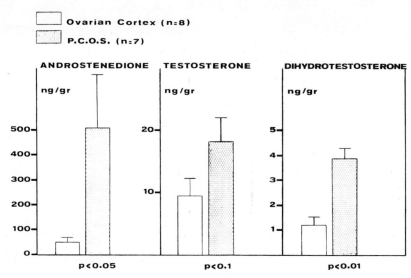

Fig.1 Concentrations of androgenic hormones in normal ovarian cortex and in polycystic ovary (see text) (mean ± S.D.).

secretion of 17β-estradiol from the ovary is very low (Serio et al., this volume). Therefore P.C.O.S. may be considered a hyperandrogenic state in which a disorder of ovarian steroidogenesis (Axelrod and Goldzicher, 1962; Mahesh and Greenblatt, 1964) may be recognized as one of the main pathogenic factors (see also chapters by Yen and by Baird in this volume). In contrast, the causes of the hyperandrogenic syndrome in idiopathic hirsutism is not yet known.

b) Several studies indicate that androgen metabolism is related to active tissue processes and to plasma binding. Testosterone metabolic clearance rate in patients with idiopathic hirsutism is greater than that of normal women and the fact that the testosterone clearance rate could be normalized by dividing MCR by the logarithm of the testosterone production rate (Bardin and Lipsett, 1967) suggests that androgens increase their own metabolism. Androgen-induced increase in clearance requires chronic rather than acute testosterone administration (Southren et al., 1968).

The origin of this increased metabolic clearance of testosterone in virilized women is important in order to understand the pathogenesis of idiopathic hirsutism. In normal women hepatic clearance is essentially equal to MCR and very little testosterone is available for extraction in extrahepatic tissues. In contrast, in virilized women 32% of plasma testosterone is metabolized in extrahepatic tissues (Kirschner and Bardin, 1972). Therefore, in idiopathic hirsutism, increased testosterone metabolism over that of normal women occurs in tissues other than the liver and as testosterone production rate increases in women, metabolic clearance rate increases at extrahepatic sites.

These studies suggest that the increased testosterone clearance rate observed in hirsute and androgen treated women is due to changes in testosterone extraction and metabolism at extrahepatic tissue level in addition to a decrease in the testosterone binding capacity in plasma (see also the chapter by Dr. Anderson in this volume).

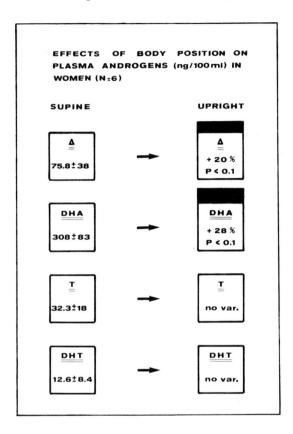

Fig. 2 The values are expressed in ng/100 ml. The increase after standing (black) is calculated in % of basal values.

c) It has been recently demonstrated that the utilization of androgens by the skin of women affected by idiopathic hirsutism is greater than in normal controls (Thomas and Oake, 1974). Moreover the skin of women affected by idiopathic hirsutism produces more 5α-dihydrotestosterone (active hydrogen at the level of target tissues) than in normal controls (Kutten and Mauvais-Jarvis, 1975). The 5α-reductase activity in the skin as well as in other target tissues depends on androgen concentration. These experimental observations could explain the increase of extrahepatic clearance of testosterone mentioned above.

We can therefore hypothesize that idiopathic hirsutism may be due either to androgen hyperproduction of glandular origin or to increased metabolism of androgen in the skin or to both these factors together.

This experimental hypothesis is difficult to prove in relation to androgen dependence of testosterone metabolism. Consequently absence of androgen hypersecretion by the glands must be demonstrated in order to prove the existence of idiopathic hirsutism as due only to a disorder of androgen metabolism in the skin.

Determination of androgens in plasma is useless for this purpose. These hormone levels change through the menstrual cycle (Abraham, 1974; Kim *et al.*, 1974)

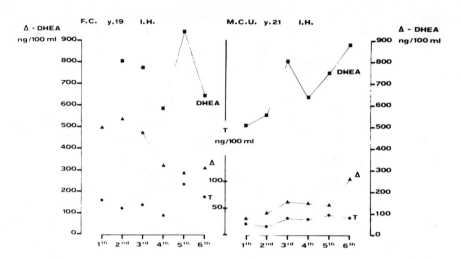

Fig.3 Plasma concentrations of androstenedione (Δ), dehydroepiandrosterone (DHEA), testosterone (T), estrone (E_1) and 17β-estradiol (E_2) at 8 p.m. of 5 consecutive days in a case of P.C.O.S.

and show important day-to-day variations in hirsute women (Ismail et al., 1974). Many factors can affect androgen plasma levels in hirsutism such as body position, episodic secretion of steroids (sometimes due to physiological factors), plasma protein binding and lastly changes in metabolic clearance rate. We have studied the effects of body position in 6 normal women by measuring Δ, DHEA, DHT and T when the subjects were supine and after 3 hours of standing. We observed a clear increase of Δ and DHEA concentrations after standing, while DHT and T remained unchanged (see Fig.2). This phenomenon is probaly related to different binding of Δ and DHEA to plasma proteins, since the 17-ketosteroids do not bind to the specific sex hormone binding globulin, but only to albumin. Consequently the values of MCR of Δ and DHEA are strictly related to hepatic blood flow (Baird et al., 1969). Therefore the variations of hepatic blood flow due to body position may change the metabolic clearance rate and consequently plasma concentrations of Δ and DHEA, while 17β-OH androgens which bind strongly with plasma proteins remain unchanged. Because of these experiments, we did not measure androgenic hormone levels in out-patients, but only in hospitalized subjects lying in bed overnight.

However, in spite of our precautions in sampling, we observed a large day-to-day variation in subjects affected by idiopathic hirsutism as well as in P.C.O.S. (Figs 3,4). The variations were greater in Δ and DHEA than in T concentrations. The variations of testosterone concentrations were generally larger in the subjects with high values of plasma testosterone than in subjects with normal values (Fig. 3).

However the increased testosterone clearance by extrahepatic tissues explains more than any other factor the paradoxical occurrence of normal plasma testosterone levels in virilized women with increased testosterone production rate (Kirschner and Bardin, 1972.).

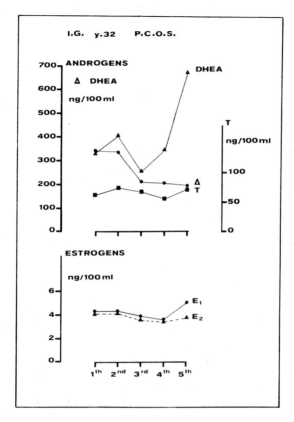

Fig.4 Plasma concentrations of androstenedione (Δ), dehydroepiandrosterone (DHEA) and testosterone (T) at 8 p.m. of 6 consecutive days in 2 cases of idiopathic hirsutism with high testosterone values (on the left) and normal testosterone values (on the right).

Measuring plasma testosterone on a fixed day of the menstrual cycle in 66 women affected by idiopathic hirsutism we found only 25 subjects with testosterone values higher than 50 ng/100 ml (Fig.5). These data are in agreement with previous observations of other authors (Kirschner and Bardin, 1972; Andre and James, 1974).

Furthermore our results confirm that plasma testosterone cannot be considered as a measure of glandular secretion of androgenic hormones in hirsutism.

On the other hand, it is not easy to measure the production rate of testosterone routinely and the evaluation of androgen secretion by direct sampling of glandular venous blood is an even more complicated procedure. Because of these experimental difficulties the relative importance of glandular secretion and skin metabolism of androgens in the pathogenesis of idiopathic hirsutism must still be investigated.

ACKNOWLEDGEMENT

This work was supported by a grant from the Administrative Board of the University of Florence (Item XI B).

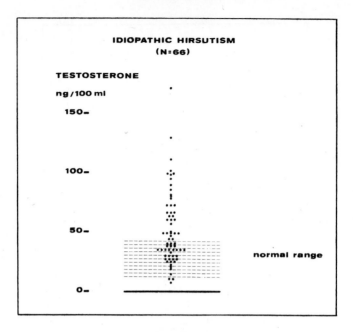

Fig.5 Plasma testosterone in 66 cases of idiopathic hirsutism.

REFERENCES

Abraham, C.E. (1974). *J. clin. Endocr. Metab.* 39, 340-346.
Andre, C.M. and James, V.H.T. (1974). *Steroids* 24, 295-309.
Axelrod, L.R., and Goldzicher, J.W. (1962). *J. clin. Endocr. Metab.* 22, 431-440.
Baird, D.T., Horton, R., Longcope, C. and Tait, J.F. (1969). *Rec. Progr. Horm. Res.* 25, 611-664.
Bardin, C.W. and Lipsett, M.B. (1976). *J. clin. Invest.* 46, 891-902.
Fiorelli, G., Borelli, D., Forti, G., Gonnelli, P., Pazzagli, M. and Serio, M. *J. Steroid Biochem.* *(in press).*
Forti, G., Pazzagli, M., Calabresi, E., Fiorelli, G. and Serio M. (1974). *Clin. Endocr.* 3, 5-17.
Ismail, A.A.A., Davinson, D.W., Sanka, A.R., Barbes, E.W., Irvine, W.J., Kilimnik, H. and Vanderbeeken, Y. (1974). *J. clin, Endocr. Metab.* 39, 81-95.
Kim, M.H., Hosseinian, A.H. and Dupon, C. (1974). *J. clin. Endocr. Metab.* 39, 706-712.
Kirschner, M.A. and Jacobs, J.B. (1971). *J. clin. Endocr. Metab.* 33, 199-209.
Kirschner, M.A. and Bardin, C.W. (1972). *Metabolism* 21, 667-689.
Kutten, F. and Mauvais-Jarvis, P. (1975). *Acta endocr. (Copenh.)* 79, 164-176.
Mahesh, V.B. and Greenblatt, R.B. (1964). *Rec. Progr. Horm. Res.* 20, 341-394.
Pazzagli, M., Forti, G., Cappellini, A. and Serio, M. (1975). *Clin. Endocr.* 4, 513-520.
Southren, A.L., Gordon, G.G. and Tochimoto, S. (1968). *J. clin. Endocr. Metab.* 28, 1105-1112.
Southren, A.L., Gordon, G.G., Tochimoto, S., Olivo, J., Sherman, D.H. and Piuzon, G. (1969). *J. clin. Endocr. Metab.* 29, 1356-1363.
Staha, N.L., Teeslink, C.R. and Greenblatt, R.B. (1973). *Am. J. Obstet. Gynec.* 117, 194-200.
Thomas, J.P. and Oake, R.J. (1974). *J. clin. Endor. Metab.* 38, 19-22.

OVARIAN AND ADRENAL VEIN CATHETERIZATION STUDIES IN WOMEN WITH IDIOPATHIC HIRSUTISM

M.A. Kirschner, I.R. Zucker and D.L. Jespersen

*Newark Beth Israel Medical Center, New Jersey Medical School
Newark, New Jersey 07112, USA*

In men, virilization is a normal occurrence in puberty and is closely related to increased testosterone production. Virilization in women is an unnatural phenomenon which has been associated with exogenous administration of androgens and with endogenous production of testosterone by ovarian and adrenal tumors. Although several C_{19} steroids with the 17β-ol configuration have been shown to have "androgenic activity" in various bioassay sytems (Dorfman and Shipley, 1956; Hilgar and Hummel, 1964), the above considerations have served to focus attention on testosterone as the androgen responsible for virilization in women, whatever the etiology.

The clinical spectrum of virilism ranges from increased coarse body hair, facial hair and menstrual disturbances to the more severe manifestations, including clitorimegaly, temporal balding, deepening of the voice, pectoral muscle hypertrophy and defeminization. Figure 1 depicts testosterone production rates in women with increasing manifestations of virilization, ranging from excessive body hair to the major signs of balding and/or pectoral muscle hypertrophy. Testosterone production rates increase in parallel with the severity of virilization, supporting the assumption that this steroid is largely responsible for virilization in women. In virtually all women with hirsutism and/or other signs of virilism studied to date by our laboratory, testosterone production rates have been found to be elevated.

The origin of testosterone in normal women has been examined previously by several laboratories and is schematically presented in Fig.2. Testosterone is secreted directly by the ovary (Horton *et al.*, 1966; Rivarola *et al.*, 1967; Gandy, 1968; Aakvaag, 1968; Lloyd *et al.*, 1971; Kirschner and Jacobs, 1971; Serio *et al.*, 1975), and by the adrenal (Burger *et al.*, 1964; Wieland *et al.*, 1965; Baird *et al.*, 1969; Kirschner *et al.*, 1971, 1973; Huq *et al.*, 1976). On the basis of testosterone con-

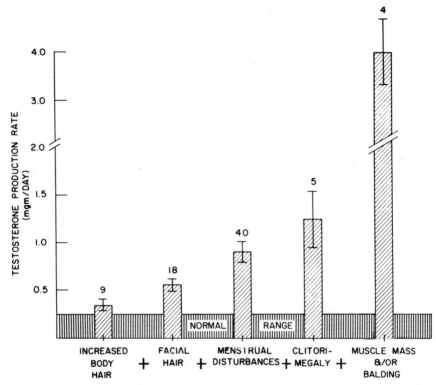

Fig.1 Testosterone production rates (± SE) in women with progressive manifestation of virilization. The number of patients examined in each category is noted above the respective bar.

centrations in ovarian and adrenal venous blood from endocrinologically normal women, we previously suggested that approximately 25% of the total testosterone production rate arises from direct adrenal secretion and 25% arises directly from ovarian secretion (Kirschner and Bardin, 1972). The remaining 50-60% arises from peripheral metabolism of prehormones, chiefly androstenedione, which is transformed at several sites including liver, fat and skin and then re-enters the circulation as testosterone (Horton and Tait, 1966; Bardin and Lipsett, 1967).

In women with unexplained hirsutism, the origin of excessive testosterone has been controversial. Bardin and Lipsett (1967) demonstrated that prehormone contribution is not increased in proportion to the elevated testosterone production rate and androstenedione metabolism accounts for only about 25% of the total testosterone produced in such women. Studies from our laboratory showed no increased conversion of other prehormones, such as dehydroepiandrosterone and androstenediol, to account for the elevated testosterone production (Kirschner *et al.*, 1973). It thus appeared that increased testosterone production noted in women with unexplained hirsutism must come from direct ovarian and/or adrenal secretion of testosterone.

Our laboratory has attempted to define the origin of excessive androgens in women with unexplained hirsutism by direct catheterization of adrenal and ovarian veins. This combined radiographic-endocrinologic approach (Jacobs, 1969) en-

ORIGIN OF TESTOSTERONE IN WOMEN

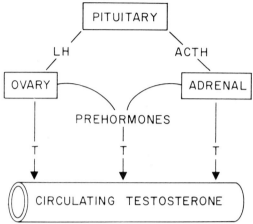

Fig. 2 Schematic diagram showing the origin of testosterone in women. Testosterone (T) is secreted directly by the ovary and adrenal. These glands also secrete prehormones (notably androstenedione) which is converted to testosterone in its peripheral metabolism.

ables us to visualize directly the organs in question and evaluate for possible tumors. It also enables us to sample directly the secretory products in the ovarian and adrenal venous effluents, permitting direct localization of the sites of hormonal secretion (Kirschner and Jacobs, 1971). To date, we have examined the origin of testosterone and other steroids in adrenal and ovarian effluents from 44 women with idiopathic hirsutism. In the following sections we shall present our data relating the magnitude of production of these steroids and our assessment as to the origins of these compounds in women with idiopathic hirsutism.

PATIENTS

Forty-four consecutive patients referred to the Newark Beth Israel Medical Center for evaluation of unexplained hirsutism are included in this study. The clinical details of these patients are presented elsewhere (Kirschner *et al.*, 1976). Testosterone production rates were found to be elevated in all patients. In all cases, the origin of each patient's virilism was unexplained, with adrenal hyperfunction, adrenal tumor and ovarian tumor having been excluded as diagnostic possibilities.

METHODS

Plasma testosterone, androstenedione, dehydroepiandrosterone, Δ^5-androstenediol and 17-hydroxyprogesterone were measured by specific radioimmunoassays, as previously described (Nieschlag and Loriaux, 1972; Kirschner, 1975). Plasma cortisol was measured by radioimmunoassay, after the method of Ruder *et al.* (1972). Total blood production rates of testosterone and androstenedione were separately determined as the product of plasma steroid concentration (mean

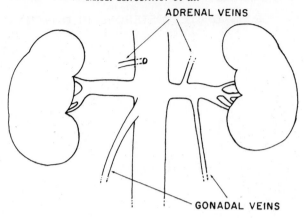

ADRENAL AND GONADAL VEIN ANATOMY

Fig.3 Schematic location of adrenal and ovarian veins.

of 3 daily samples) X metabolic clearance rate, after the method of Tait (1966), as previously described (Kirschner *et al.*, 1970).

CATHETERIZATION PROCEDURE

Venous catheterizations were performed percuatneously via the femoral vein, using a Muller guided catheter system and No.7 sized catheter, as previously reported (Jacobs, 1969; Kirschner and Jacobs, 1971). Figure 3 is a schematic representation of ovarian and adrenal venous anatomy. Although venous anatomy is far more variable than arterial locations, the origins of adrenal and ovarian veins were most frequently located as demonstrated. The right adrenal and ovarian veins are noted to enter directly into the inferior vena cava, in contrast to the left side, where these vessels more predictably terminated in the left renal vein.

Prior to the procedure, the patients were premedicated with 50 mgm hydroxyzine. Under fluoroscopic control, with image intensifications, the catheter was guided first into the left renal vein and then to the orifice of the left ovarian or adrenal vein. Minimal amounts of radiocontrast material (Renograftin-60®) were used in searching for the vessel orifices. The catheter was then threaded selectively into the effluent vein and wedged as far as possible. Five to 10 ml of lightly heparinized venous blood was slowly withdrawn over a 5-10 minute period using minimal suction. A peripheral blood sample was simultaneously drawn from the antecubital fossa. After the effluent sample was obtained, radiocontrast was injected for visualization of the ovary or adrenal, in order to rule out the possibility of tumor. A representative ovarian venogram is presented in Fig.4. Extensive collateral flow is noted across the uterine plexus to the contralateral ovarian vein. This finding was readily demonstrated on most occasions.

After removing the catheter from the effluent vessel, it was thoroughly flushed with saline before the next vessel was entered. No rigid order of vessel sampling was followed. No attempt was made to occlude other vessels during the

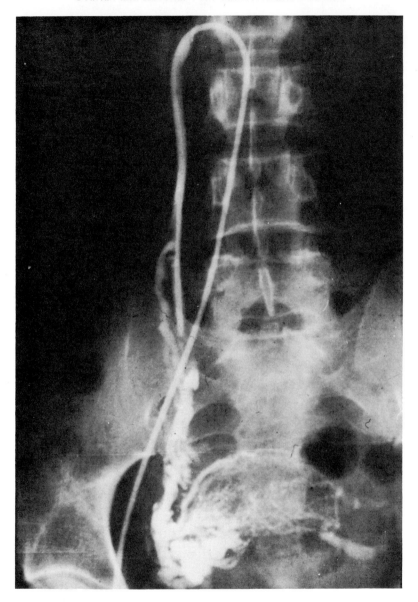

Fig.4 Representative ovarian venogram. Catheter traverses inferior vena cava and halfway down the course of the right ovarian vein. Radiocontrast material collects in the pelvis and crosses a rich uterine collateral network to the contralateral side.

sampling period. The entire procedure lasted 60-90 minutes, with minimal discomfort and morbidity to the patient.

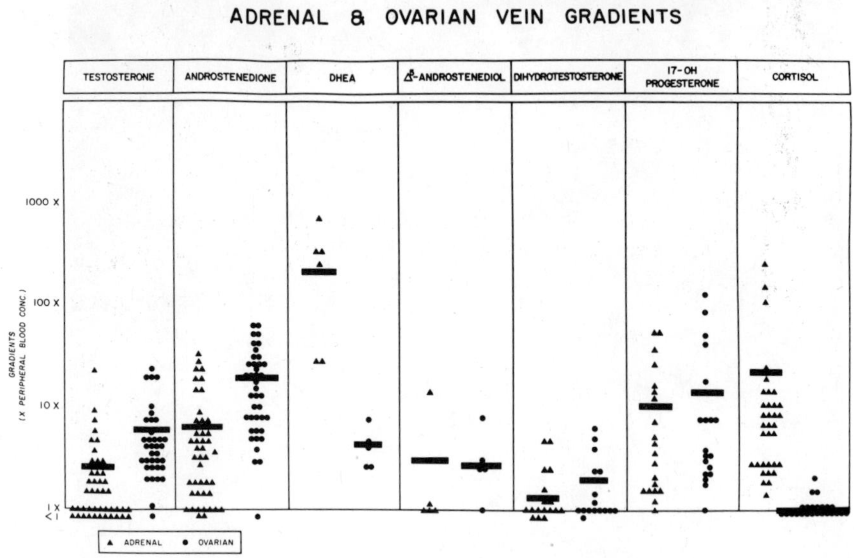

ADRENAL & OVARIAN VEIN GRADIENTS

Fig. 5 Gradients of testosterone and other steroids noted in adrenal (△) and ovarian venous effluents (○) in 44 hirsute women. Gradient = steroid concentration in effluent ÷ peripheral blood, and is plotted on a logarithmic scale. See text for details.

CATHETERIZATION DATA

In Fig. 5 are plotted the gradients of testosterone, androstenedione and cortisol found in the adrenal and ovarian venous effluents of the 44 hirsute women, as well as gradients of DHEA, androstenediol, 17-hydroxyprogesterone and dihydrotestosterone, determined in as many patients as possible. The gradient for each steroid = effluent concentration ÷ peripheral blood concentration. When samples were obtained from both adrenal veins, the value plotted represents the mean value of steroid concentration from both sides. If a considerable discrepancy between right and left-sided adrenal values was confirmed by cortisol differences, the higher value was taken. If only a single adrenal or ovarian effluent sample was obtained, this value was used in calculating the gradient.

Testosterone

The blood production rates of testosterone (P_B^T) were determined in each patient prior to catheterization. These were found to be elevated in all 44 patients, averaging 795 ± 83 SE μg/day. The mean peripheral plasma testosterone level was 97 ng/100 ml. Ovarian venous effluents gradients averaged 7.2 × peripheral blood values, with 34 of 36 samples showing significant testosterone gradients. In contrast, 22/44 adrenal ovarian effluents showed no significant gradients; the mean adrenal effluent gradient was 2.6 × peripheral blood levels.

Androstenedione (Δ^4)

Total blood production rates of androstenedione ($P_B^{\Delta^4}$) were determined in all 44 patients, and found to average 6560 ± 535 SE μg/day, approximately twice the normal value of 3300 ± 830 µg/day (Horton and Tait, 1966; Bardin and Lipsett, 1967). Ovarian Δ^4 gradients averaged 20 X the mean peripheral blood concentration of 340 ng/100 ml. Thirty-eight of 39 ovarian venous samples showed significant gradients. As in the case of testosterone, adrenal venous effluents of androstenedione were not as high, averaging 6 X peripheral blood levels. Nine of 44 adrenal venous samplings contained no significant gradient of androstenedione.

Dehydroepiandrosterone (DHEA)

Blood production rates of DHEA were measured in 11 consecutive hirsute patients. The mean DHEA production rate was 14.9 ± 2.0 SE mgm/day v. 6.04 ± 0.9 mgm/day in endocrinologically normal women of comparable age (Horton and Tait, 1967; Kirschner et al., 1973). Plasma DHEA levels averaged 899 ± 130 SE ng/100 ml and were approximately twice normal. Catheterization studies here were quite different from those noted for testosterone and androstenedione. Ovarian vein samples showed significant DHEA gradients in 4/4 women with a mean ovarian gradient of 4 X peripheral blood. These gradients, however, were not nearly as striking as the 100-fold differences noted in adrenal venous effluents of 6 hirsute women.

Δ^5-Androstenediol (Δ^5-diol)

Δ^5-androstenediol blood production rates averaged 1.27 ± 0.18 SE mgm/day in 6 hirsute women v. 0.53 ± 0.06 SE mgm/day in normal women. Plasma levels of Δ^5-diol averaged 132 ± 23 ng SE/100 ml and were not significantly different from normal women. The metabolic clearance rates of Δ^5-diol in the hirsute women were approximately twice those noted for normals (Kirschner et al., 1973). Four of 5 women had ovarian gradients of Δ^5-diol, whereas 1 of 5 women had Δ^5-diol gradients noted in adrenal venous effluents.

17-hydroxyprogesterone

Peripheral plasma levels of 17-OH progesterone were determined in 23 women and were found to average 92 ± 20 SE ng/100 ml. These values were higher than the mean of 30 ± 7 SE ng/100 ml noted in normal women in the follicular phase. Ovarian vein gradients were noted in 21/22 patients and averaged 15 X peripheral blood levels. These gradients were slightly higher than the mean adrenal vein gradient of 11X. Adrenal vein gradients were noted in 22 of 23 patients.

Dihydrotestosterone (DHT)

DHT levels in peripheral blood averaged 50.7 ± 5 SE ng/100 ml in 18 hirsute women, in contrast to a mean value of 31.4 ± 5 SE ng/100 ml noted in normal women. Ovarian venous effluents showed no gradients in 9 of the 18 women and minimal gradients in the other patients; the mean ovarian vein gradient was only 2 X peripheral blood levels. Adrenal venous effluents were similar, showing no gradients in 10 of 18 women; the mean adrenal vein gradient was 1.3 X peripheral blood levels. The highest adrenal or ovarian vein level observed was only 6 X peripheral levels. These data strongly suggest that DHT is not a secreted steroid in most women and support previous kinetic studies indicating that virtually all DHT produced in hirsute women comes from peripheral metabolism of androstenedione (Ito and Horton, 1971; Mahoudeau, 1971).

Cortisol

Ovarian and adrenal venous cortisol concentrations were determined to serve as reference values for the various C_{19} and C_{18} steroids, and to test whether

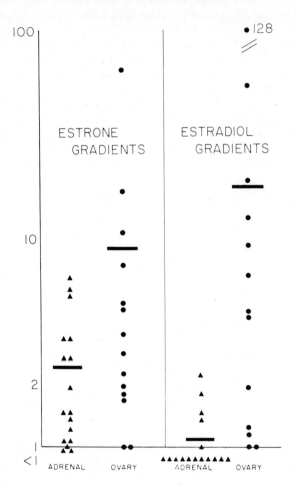

Fig. 6 Estrone and estradiol gradients noted in adrenal (△) and ovarian (○) venous effluents.

adrenal venous effluents admixed with ovarian vein samples, since cortisol is a unique secretory product of the adrenal gland. In 21/24 ovarian samples there was no cortisol gradient observed. By contrast, cortisol gradients were observed in all 44 adrenal vein samples obtained, with a mean gradient 23-fold greater than peripheral blood concentrations.

Estrone (E_1)

Estrone levels in peripheral blood averaged 7.2 ng/100 ml in 17 hirsute women. Ovarian and adrenal venous effluents of estrone and estradiol are shown in Fig. 6. Twelve of 14 hirsute women exhibited ovarian secretion of estrone, with a mean gradient of 9.5 × peripheral blood levels. Twelve of 17 adrenal venous samples showed significant E_1 gradients; the mean adrenal E_1 gradient was 2.5 × peripheral blood levels.

Estradiol (E_2)

Blood production rates of estradiol were determined in 10 hirsute women and were found to average 116 ± 22 SE μg/day. Two women in this group had production rates of 180 and 268 μg/day, suggesting that ovulation had taken place. The E_2 metabolic clearance rate averaged 1450 ± 91 SE L/day and was comparable to values previously found in normal women (Longcope, 1968; Hembree et al., 1969). Plasma E_2 levels averaged 7.7 ± 1.2 SE ng/100 ml in 17 hirsute women. Ovarian vein gradients of estradiol were noted in 11/13 hirsute women with a mean gradient 18-fold greater than peripheral blood concentrations. By marked contrast, 12/16 adrenal vein samples demonstrated no E_2 gradient. The highest gradient observed in adrenal vein samples was 2 × peripheral blood levels, supporting previous data of Baird (1969) that E_2 is not normally secreted by the adrenals.

DETERMINING THE ORIGIN OF TESTOSTERONE

Combining the above catheterization data with the total blood production rate determinations of testosterone and its chief prehormone, androstenedione, we were able to quantify the total amount of testosterone arising from ovarian v. adrenal sources, as follows:

1) The adrenal secretion rate of testosterone and androstenedione were calculated from the expression:

$$\frac{\text{adrenal testo gradient}}{\text{adrenal testo secretion rate}} = \frac{\text{adrenal cortisol gradient}}{\text{cortisol secretion rate}}$$

$$= \frac{\text{adrenal } \Delta^4 \text{ gradient}}{\text{adrenal } \Delta^4 \text{ secretion rate}}$$

These calculations assume that cortisol is secreted only by the adrenal glands, and that adrenal secretion of androgens parallels that of cortisol. Several studies support the latter assumption (Vaitukaitus et al., 1969; Rosenfeld et al., 1971; Kirschner, 1976).

2) Previous studies have shown that the transfer factor of androstenedione to testosterone in peripheral blood ($\rho_{BB}^{\Delta^4 \to T}$) is approximately 0.04 and is unchanged in hirsute women (Horton and Tait, 1966; Bardin and Lipsett, 1967). Thus, from the product of this ρ value × the total blood production rate of androstenedione, we calculated its overall contribution to testosterone. Next, we calculated the proportion of adrenal-secreted androstenedione transferred to testosterone as the product of adrenal Δ^4 secretion rate (Step No.1) × $\rho_{BB}^{\Delta \to T}$. This calculation assumes that adrenal-secreted Δ^4 follows the same metabolic fate as Δ^4 from other sources. Previous studies showed DHEA and Δ^5 steroids contribute to only 6% of the P_B^A and 8% of the P_B^T (Kirschner et al., 1973) and were thus not considered in these calculations.

3) We then determined the total ovarian contribution of testosterone and its prehormones by combining Steps 1 and 2 as follows:

$$\text{Ovarian production of testo and its prehormones} = P_B^T - \left[\text{adrenal secretion rate of} + \text{adrenal secretion of } \Delta^4 \times \rho_{BB}^{\Delta^4 \to T}\right]$$

4) As a check on the internal consistency of these calculations, we determined the ovarian Δ^4 production rate from the expression:

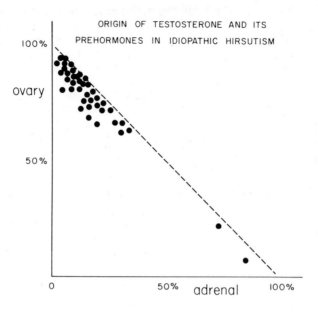

Fig. 7 Origin of testosterone and its prehormones in 44 women with unexplained hirsutism, based on catheterization data. Ovarian contribution is plotted on the ordinate; adrenal contribution is plotted on the abscissa. See text for details of calculations.

Ovarian Δ^4 secretion rate = $P_B^{\Delta^4}$ — adrenal Δ^4 secretion.

As above, these calculations assume minor prehormone contributions of DHEA to $P_B^{\Delta^4}$, (Kirschner *et al.*, 1973). Using this derived value of ovarian Δ^4 secretion rate, we calculate the ovarian testo secretion rate from the expression:

$$\frac{\text{Ovarian testo secretion rate}}{\text{Ovarian testo gradient}} = \frac{\text{Ovarian } \Delta^4 \text{ secretion rate (previously calculated)}}{\text{Ovarian } \Delta^4 \text{ gradient}}$$

In most cases the calculated values showed excellent agreement with those noted from Step No.3.

Attempts to apply the above calculations to determine the ovarian and adrenal estrone and estradiol secretion rates were unsuccessful.

ORIGIN OF TESTOSTERONE IN HIRSUTE WOMEN

Utilizing the above-described calculations, we determined the origin of testosterone and its major contributory prehormone, androstenedione, in each of the 44 hirsute women. These data are plotted in Fig. 7 and indicate that in 42 of the 44 hirsute women studied, the origin of excessive androgens was largely ovarian. Twenty of the 44 hirsute women comprising this series exhibited at least 50% decrease in peripheral venous testosterone and androstenedione concentrations after 4-5 days of exogenous glucocorticoids.

EFFECTS OF ACTH ON VENOUS EFFLUENTS

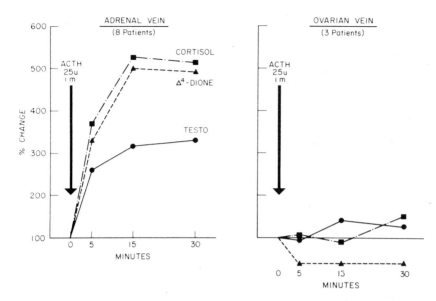

Fig.8 Effect of ACTH on adrenal and ovarian secretion of cortisol, androstenedione and testosterone. In contrast to the brisk increments noted in adrenal venous effluents, no changes in ovarian secretions were noted after ACTH.

Effects of ACTH on Adrenal and Ovarian Steroid Secretion

The acute effects of ACTH on adrenal androgen secretion over a 30 min period were examined in 8 hirsute women. Twenty-five I.U. β^{1-24} ACTH (Cortrosyn®) were given i.m. after the catheter had been wedged in the left adrenal vein. Venous effluents were sampled at 0, 5, 15 and 30 min. Figure 8 demonstrates marked 300-500% increments in adrenal vein concentrations of testosterone, androstenedione and cortisol observed over 5-15 min following ACTH. The increments, and time course of secretion suggested parallel secretion of C_{19} steroids along with cortisol (Huq et al., 1976). These studies of adrenal effluents are in contrast to earlier observations of Rivarola (1966) who found lower peripheral blood levels of testosterone in men after ACTH.

In contrast to the brisk increments of androgens and cortisol (as well as aldosterone) noted in adrenal vein blood after parenteral ACTH (Huq et al., 1976), when ovarian vein blood was sampled after ACTH in 3 patients, no significant changes in ovarian androgen concentrations were noted.

Effects of HCG on Adrenal Steroid Secretion

Figure 9 contrasts the effects of ACTH v. HCG on adrenal steroid secretions. When 2000 I.U. HCG were given intramuscularly to 7 hirsute women, adrenal venous effluents showed minor fluctuations without a discernible trend, in contrast to the brisk increments after ACTH, described above. These data make it unlikely that gonadotrophic hormones stimulate the normal adrenal gland.

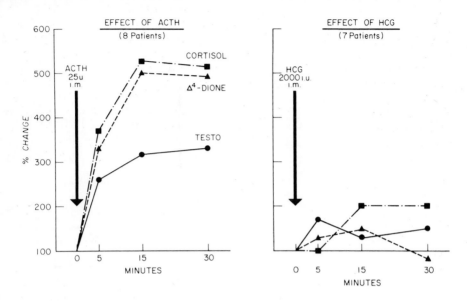

Fig. 9 Effect of ACTH v. HCG on adrenal venous secretion of cortisol, androstenedione and testosterone in hirsute women. See text for details.

COMMENTS

Previous attempts to determine the site(s) of testosterone overproduction in women with unexplained hirsutism were based largely on indirect suppression and/or stimulation studies (Mahesh and Greenblatt, 1964; Casey et al., 1966, 1967; Reforzo-Membrives et al., 1967; Lopez et al., 1967; Benjamin et al., 1970; Ettinger et al., 1971). Such studies usually led to the conclusion that the adrenal glands are the major source of androgens in many women with idiopathic hirsutism. The assignment of the adrenal glands as the site of androgen excess was based on: 1) the assumption that exogenous glucocorticoids suppress only androgens of adrenal origin. 2) the occasional finding of elevated 17-hydroxyprogesterone levels in blood, or its urinary metabolite pregnanetriol, thought to be of adrenal origin, and 3) seemingly normal ovaries without apparent abnormalities.

Our studies in 44 women with idiopathic hirsutism have attempted to localize the site of androgen production by direct sampling of adrenal and ovarian venous effluents. The raw data in Fig.5 or the derived figures of testosterone origin based on these catheterization findings (Fig.7) strongly suggest the ovaries as the source of androgen overproduction in most patients with unexplained hirsutism, including those women whose androgens are suppressed after exogenous glucocorticoids.

Our data are in contrast to those of Stahl et al. (1973), and Oake et al. (1974) whose catheterization data suggest excessive adrenal androgens. The first of these studies unfortunately suffers from lack of corroborative cortisol determinations in the adrenal venous samples, making quantification difficult. The data of Oake (1974) demonstrates higher testosterone gradients in adrenal samples v. ovarian effluents. The very high cortisol levels in adrenal effluents noted in this latter study suggests that all patients were catheterized during periods of active cortisol secretion. If one calculates adrenal testosterone secretion rates based on these testosterone and cortisol gradients, the majority of these patients would seem to have "non-adrenal" sources of their excessive testosterone.

An additional reason for implicating "adrenal" origin of excessive androgens in women with idiopathic hirsutism was the occasional finding of elevated urinary pregnanetriol excretion in such patients (Lloyd, 1968; Newark et al., 1975). This metabolite is usually excreted in great excess in patients with congenital adrenal hyperplasia, particularly with C_{21} hydroxylase deficiency. It thus seemed logical to consider this urinary metabolite or its plasma precursor, 17-hydroxyprogesterone, as an "adrenal" steroid. Strott et al. (1969) and Abraham et al. (1972), however, showed a striking parallel between plasma 17-hydroxyprogesterone levels and estrogen levels in normally menstruating women, strongly suggesting an ovarian origin for this steroid. Our studies here demonstrate minimally elevated plasma 17-hydroxyprogesterone levels in hirsute women. The catheterization data confirm that this compound is secreted by the ovary as well as the adrenals. It thus no longer seems tenable to equate 17-hydroxyprogesterone and/or pregnanetriol with adrenal secretions.

Much recent interest has been centered on dihydrotestosterone as the "true" intracellular androgen responsible for androgen activity. We can confirm and expand on the previous data of Ito and Horton (1971) indicating that DHT arises almost entirely from androstenedione, in its extragonadal metabolism, and is not a secreted steroid in most hirsute women.

REFERENCES

Aakvaag, A. and Fylling, P.A. (1968). *Acta endocr. (Copenh.)* **57**, 447.
Abraham, G.E., Odell, W.D., Swerdloff, R.S. and Hopper, K. (1972). *J. clin. Endocr.* **34**, 312.
Baird, D., Uno, A. and Melby, J.C. (1969). *J. Endocrinol.* **45**, 135.
Bardin, C.W. and Lipsett, M.B. (1976). *J. clin. Invest.* **46**, 891.
Benjamin, F., Cohen, M. and Romney, S.L. (1970). *Fertil. Steril.* **21**, 854.
Burger, H.G., Kent, J.P. and Kellie, A.E. (1964). *J. clin. Endocr.* **24**, 432.
Casey, J.H. and Nabarro, J.D.N. (1976). *J. clin. Endocr.* **27**, 431.
Casey, J.H., Burger, H.G., Kent, J.R. et al. (1966). *J. clin. Endocr.* **26**, 1370.
Dorfman, R. and Shipley, R.A. (1956). "Androgens, Biochemistry, Physiology and Clinical Significance". Wiley and Sons, New York.
Ettinger, B., Van Werden, K., Thenaers, G.C. and Forsham, P.H. (1971). *Am. J. Med.* **51**, 70.
Gandy, H.M. and Peterson, R.E. (1968). *J. clin. Endocr.* **28**, 949.
Hembree, W.C., Bardin, C.W. and Lipsett, M.B. (1969). *J. clin. Invest.* **48**, 1809.
Hilgar, A.G. and Hummel, D.J. (1964). *In* "U.S. Dept. of Health, Education and Welfare Publication, Issue 1".
Horton, R. and Tait, J.F. (1966). *J. clin. Invest.* **45**, 301.
Horton, R., Romanoff, E. and Walker, R. (1966). *J. clin. Endocr.* **26**, 1267.
Horton, R. and Tait, J.F. (1967). *J. clin. Endocr.* **27**, 79.
Huq, M.S., Pfaff, M., Jespersen, D., Zucker, I.R. and Kirschner, M.A. (1976). *J. clin. Endocr. (in press)*.
Ito, R. and Horton, R. (1971). *J. clin. Invest.* **50**, 1621.
Jacobs, J.B. (1969). *Radiology* **92**, 885.

Kirschner, M.A., Bardin, C.W. and Hembree, W.C. (1970). *J. clin. Endocr.* **30**, 727.
Kirschner, M.A. and Bardin, C.W. (1972). *Metabolism* **21**, 667.
Kirschner, M.A. and Jacobs, J.B. (1971). *J. clin. Endocr.* **33**, 199.
Kirschner, M.A., Sinhamahapatra, S., Zucker, I.R., Loriaux, L. and Nieschlag, E. (1973). *J. clin. Endocr.* **37**, 183.
Kirschner, M.A. (1975). *In* "Nuclear Medicine in Vitro" (B. Rothfeld, ed) pp.164-178. J.B. Lippincott Co., Philadelphia.
Kirschner, M.A., Zucker, I.R. and Jespersen, D. (1976) *(submitted for publication)*.
Lloyd, C.W. (1968). *In* "Textbook of Endocrinology" (R.H. Williams, ed). Saunders, N.Y.C.
Lloyd, C.W. *et al.* (1971). *J. clin. Endocr.* **32**, 155.
Longcope, C., Layne, D.S. and Tait, J.F. (1968). *J. clin. Invest.* **47**, 93.
Lopez, J.M., Migeon, C.J. and Jones, G.E.S. (1967). *Am. J. Obstet. Gynec.* **98**, 749.
Mahoudeau, I.A., Bardin, C.W. and Lipsett, M.B. (1971). *J. clin. Invest.* **50**, 1338.
Mahesh, V.B., Greenblatt, R.B., Aydar, C.K. *et al.* (1964). *J. clin. Endocr.* **24**, 1283.
Newmark, S.R., Dluhy, R.G., Williams, G.H., Pochi, P. and Rose, L.I. (1975). *Clin. Res.* **23**, 240A.
Nieschlag, E. and Loriaux, D.L. (1972). *Z. Klin. Chem. Klin. Biochem.* **10**, 164.
Oake, R.J., Davies, S.J., McLachlan, M.S.F. and Thomas, J.P. (1974). *Quart. J. Med.* **43**, 603.
Reforzo-Membrives, J., Rocca, D.Z., Euriori, C.L. *et al.* (1967). *Acta endocr. (Copenh.)* **55**, 685.
Rivarola, M.A., Saez, J.M., Jones, H.W., Jones, G.S. and Migeon, C.J. (1976). *Johns Hopkins Med. J.* **121**, 82.
Rosenfeld, R.S., Hellman, L., Roffwarg, H. *et al.* (1971). *J. clin. Endocr.* **33**, 87.
Ruder, H.J., Guy, R.L. and Lipsett, M.B. (1972). *J. clin. Endocr.* **35**, 219.
Serio, M., Dell'Acqua, S., Calabresi, E., Fiorelli, G., Forti, G., Cattaneo, S., Lucisano, A., Lombardi,, G., Pazzagli, M. and Borrelli, D. (1975). This volume, pp.471-479.
Stahl, N.L., Teeslink, C.R. and Greenblatt, R.B. (1973). *Am. J. Obstet. Gynec.* **117**, 194.
Strott, C.A., Yoshimi, T., Ross, G.T. and Lipsett, M.B. (1969). *J. clin. Endocr.* **29**, 1157.
Tait, J.F. and Horton, R. (1966). *In* "Steroid Dynamics" (G. Pincus, T. Nakao and J.F. Tait, eds) p.393. Academic Press, New York.
Vaitukaitus, J.S., Dole, S.L. and Melby, J.C. (1969). *J. clin. Endocr.* **29**, 1443.
Wieland, R.G., de Courcy, L., Levy, R.P., Zala, A.B. and Hirschmann, H. (1965). *J. clin. Invest.* **44**, 159.

PLASMA ANDROGENS IN PATIENTS WITH HIRSUTISM

V.H.T. James, A.E. Rippon and H.S. Jacobs

*Department of Chemical Pathology, St. Mary's Hospital Medical School
London W2 1PG*

The androgens present in peripheral blood in the adult human female originate from both the ovary and the adrenal cortex. These compounds are C_{19} steroids, some of which are biologically active themselves (e.g., testosterone) and others, such as androstenedione, which owe their importance principally to their role as pre-hormones or precursors. In this chapter, we shall discuss some of the data which document plasma androgen levels in normal women and will also consider the differences which occur in patients with hirsutism.

The important C_{19} steroids present in plasma are listed in Fig.1. These are testosterone, dihydrotestosterone, androstenedione, dehydroepiandrosterone (DHA) and its sulphate, androstenediol and androstanediol. With the exception of dihydrotestosterone and androstanediol, there is evidence for direct secretion of all these steroids by both the ovary and the adrenal cortex (Wieland et al., 1965; Gandy and Peterson, 1968; Baird, 1971). In addition though, some of these steroids attain their blood concentrations partly or wholly by conversion from other steroids. Androstenedione is a particularly important pre-hormone in this respect and up to 60% of the blood production rate of testosterone in women is contributed by this steroid (Horton and Tait, 1966), and most of the dihydrotestosterone originates from androstenedione (Ito and Horton, 1970).

Peripheral levels and production rates of individual steroids differ considerably. It is not feasible to measure directly the rate of secretion of individual steroids, except in very special situations at surgery or by cannulation, and so the available data are indirect and are based upon measurements derived from peripheral blood levels and calculated rates of clearance. These data allow the calculation of a blood production rate. This is usually done by making one or more measurements of the peripheral concentration of the steroid in question and measuring the metabolic clearance rate. For this calculation of blood production rate to be meaningful, it has to be assumed that neither the metabolic clearance

	Mean Level (µg/100 ml)	MCR (l/24 h)	Blood Production Rate (mg/24 h)
Testosterone	0.035 ± .022	600	0.21
Dihydrotestosterone	0.020 ± .008	200	0.04
Androstenedione	0.105 ± .036	2,100	2.2
Androstenediol	0.041 ± .014	–	–
5α-Androstanediol	0.002 ± .001	–	–
Dehydroepiandrosterone	0.48 ± .32	1,640	8.0
Dehydroepiandrosterone sulphate	60 ± 40	22	13.0

Fig.1 C_{19} steroids in female plasma.

rate, nor the plasma concentration of the steroid has varied too widely throughout the 24 hours. These assumptions may not be justified.

The relative contributions to the androgen pool by the ovary and the adrenal are not easily determined, and the evidence is largely indirect and has been obtained by studies of peripheral androgen levels in women who have had either their adrenal glands or their ovaries removed, or who have been given an amount of glucocorticoid sufficient to cause suppression of ACTH secretion and thus, presumably inhibition of adrenocortical androgen secretion. From studies of this type, it has been concluded that in the premenopausal female almost all the DHA and DHA sulphate originates from the adrenal gland, only small amounts being secreted by the ovary. Androstenedione is secreted about equally by both glands and about half of the peripheral testosterone level is contributed by the ovary, the rest by the adrenal (Abraham, 1974).

So far, there have been relatively few studies which have set out to investigate the extent to which plasma androgen levels vary in any one subject. Vermeulen (this volume) has shown that in post-menopausal women, mean androgen levels were lower than in premenopausal subjects. Figure 2 shows plasma androstenedione levels in relation to age; the levels may fall slightly, as age increases, but no marked change occurs and testosterone shows a similar pattern (Fig.3). Changes in plasma androgen levels occur through the menstrual cycle, although not all investigators have agreed on details. Abraham (1974) concluded that the ovarian contribution to peripheral levels of testosterone, androstenedione and DHA sulphate reached a maximum at mid-cycle, but recently Valette and colleagues (1975) found no significant change in peripheral testosterone levels at mid-cycle.

Investigation of short-term changes shows that androgen levels fluctuate significantly throughout the 24 hours. We have shown elsewhere (James and Andre, 1974) that the well-established changes in plasma cortisol, due to episodic ACTH secretion, are exactly paralleled by androstenedione, and this suggests that these variations are due to pulsatile adrenocortical secretion. The changes in testosterone levels are less marked, and the lack of any direct relationship between the levels of testosterone and androstenedione, in spite of the fact that androstenedione is a major precursor of testosterone, is presumably due to the time taken for conversion of androstenedione to testosterone to occur. DHA levels also fluc-

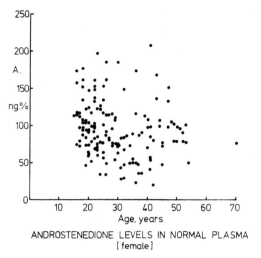

Fig.2 Plasma androstenedione levels in normal women in relation to age.

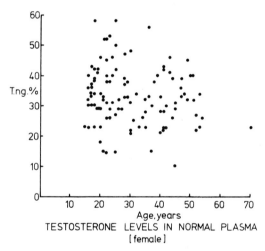

Fig.3 Plasma testosterone levels in normal women in relation to age.

tuate in parallel with cortisol levels, again suggesting ACTH dependent secretion, but DHA sulphate levels do not, because of the long plasma half-life of this steroid.

Thus, one of the major controlling factors of adrenal androgen levels is ACTH and after administration of ACTH, plasma androstenedione levels increase rapidly (Fig.4). However, although the picture with androstenedione looks straightforward, the effect of ACTH on testosterone levels is complex. In some women, as for example is shown by the study in Fig.4, there is an increase of plasma androstenedione and plasma cortisol levels which is followed exactly by a rise in plasma testosterone, and this may be due to direct secretion of testosterone, although it seems more likely that it represents conversion from androstenedione.

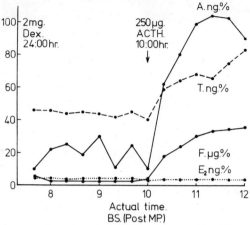

Fig. 4 Effect of ACTH on plasma steroid levels in a post-menopausal woman. 2 mg of dexamethasone was given at midnight, and 250 μg of ACTH was injected at 10.00 h. Androstenedione (A), testosterone (T), cortisol (F) and oestradiol (E_2) plasma levels are shown.

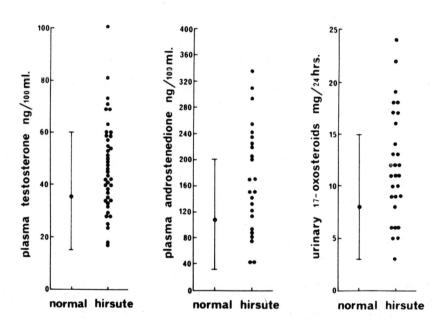

Fig. 5 Plasma androstenedione and testosterone levels, and urinary 17-oxosteroids in normal and hirsute women. Means and ranges for normal women are shown. (Reproduced with permission from James and Andre, 1974.)

In this study, a single injection of ACTH was given. In other women, whilst both cortisol and androstenedione levels increase, there are no significant changes in testosterone levels. Even very large amounts of ACTH may not be effective. Thus, whilst on average, injection or infusion of ACTH produces a slight increase in plasma testosterone levels (Andre and James, 1974), the pattern of response is variable from subject to subject and it is unclear why this should be so.

In normal women, plasma testosterone levels fluctuate very little through the 24 hours, whereas androstenedione levels vary markedly and show a characteristically ACTH-dependent, reproducible pulsatile pattern.

Turning now to patients with hirsutism, there have been numerous studies which have attempted to define the alterations in androgen metabolism which occur in these patients. Our own data are shown in Fig.5. These data are from an unselected group of patients complaining of hirsutism, none of whom was receiving treatment. In agreement with many other investigators, plasma testosterone levels are often within normal limits, although occasional subjects show levels outside the normal range. However, it has also been shown that the concentration of binding protein in the plasma of patients with hirsutism is reduced compared to normal women, and if the unbound testosterone levels are estimated, these are much more consistently abnormal (Vermeulen et al., 1971). It has also been shown that testosterone production rates are increased (Bardin and Lipsett, 1967). Plasma androstenedione levels follow a similar pattern to testosterone, although many more subjects show levels which exceed the normal range. However, since androstenedione is not bound to a specific protein, and since the metabolic clearance rate of this steroid is reported to be normal in hirsute patients, the plasma levels presumably reflect androstenedione production more accurately than testosterone levels reflect testosterone production. If so, these data certainly suggest that many hirsute patients may have elevated androstenedione production rates. It has been reported that androstenedione production rates are elevated in hirsute subjects (Bardin and Lipsett, 1967), but these data are based upon measuring clearance rate and instantaneous levels of androstenedione. Since we have seen that androstenedione levels fluctuate markedly through the 24 hours this method of calculation is clearly invalid.

Dihydrotestosterone levels, and androstenediol levels are not markedly abnormal (James and Andre, 1974); again, this may reflect altered binding to plasma protein. Plasma androstanediol levels are, in our experience, very low in normal women. We have no data for hirsute women, but data from other investigators suggest that they may be high in some women with virilising conditions.

Almost all the information so far available on plasma androgen levels in patients with hirsutism relate to blood levels at a single point in time, and there is little work on day-to-day changes, or changes through the 24 hours. Recently, Abraham (1975) reported a study of steroid hormone levels measured daily through the menstrual cycle in which very considerable fluctuations of testosterone levels occurred, but serial studies with frequent blood sampling do not appear to have been made. In the work described here, a small group of hirsute women were studied by frequent blood sampling, with measurement of plasma LH, FSH, testosterone, cortisol and androstenedione levels.

The purpose of the study was to discover how plasma androgen levels varied over short periods of time, and the effect on these levels of minimal doses of glucocorticoid, sufficient to achieve suppression of plasma cortisol levels. Large doses were deliberately avoided since it was intended that the dose of steroid employed should be continued therapeutically, and because the data of Kirschner

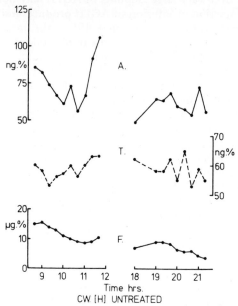

Fig. 6 Plasma androstenedione (A), testosterone (T) and cortisol levels (F) in patient CW with hirsutism, pre-treatment. Blood samples were taken at the times shown during the morning and evening of one day.

and Jacobs (1971) suggest that large amounts of dexamethasone may inhibit ovarian androgen secretion.

Five patients were studied, and all had moderate to severe hirsutism and amenorrhoea or oligomenorrhoea. Each was examined by laparoscopy. Two had ovaries of normal appearance, whilst the other three had cystic ovaries. The patients were all admitted to hospital for the study and serial blood samples were obtained from 0600 to 1000 h and from 1800 to 2200 h. They then commenced daily treatment with 0.25 mg of dexamethasone at night and two days later, the blood sampling was repeated. The results are shown in the figures.

Figures 6 and 7 show the results of the studies of the patients with ovaries of normal appearance.

Patient CW. (Fig.6) This patient has normal steroid levels although occasionally testosterone levels exceed the upper normal limit. However, it is noticeable that although there are no fluctuations in cortisol levels during the period of study, very marked fluctuations occur in plasma androstenedione levels and these clearly cannot be due to ACTH stimulation of the adrenal cortex. There are also quite marked fluctuations in testosterone levels.

Patient DC. (Fig.7) Similar comments apply to this patient, but the changes are even more marked. Quite large swings occur in both androstenedione and testosterone levels, again with no relationship to changes in cortisol levels.

Patient EG. (Fig.8) Turning to the first of the patients with cystic ovaries, whilst cortisol and androstenedione levels are normal, testosterone is well above normal.

Patient RF. (Fig.9) This lady also has an elevated and fluctuating level of testosterone and there are marked changes in androstenedione independent of any change in cortisol.

Plasma Androgens in Patients with Hirsutism 463

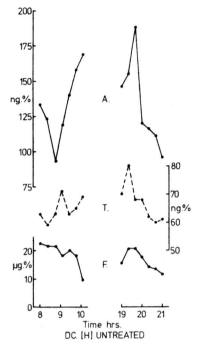

Fig. 7 Plasma steroid levels in patient DC, with hirsutism, untreated. Key as for Fig. 6.

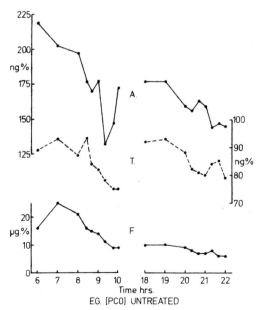

Fig. 8 Plasma steroid levels in patient EG, polycystic ovarian disease, untreated. Key as for Fig. 6.

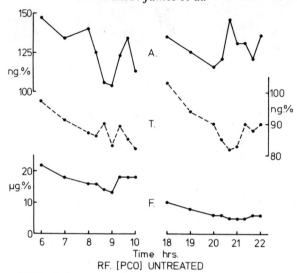

Fig. 9 Plasma steroid levels in patient RF, polycystic ovarian disease, untreated. Key as for Fig. 6.

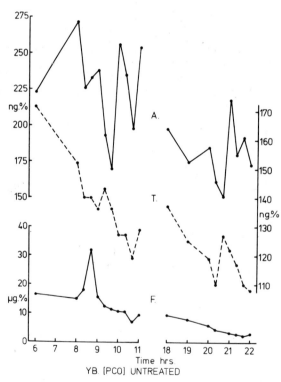

Fig. 10 Plasma steroid levels in patient YB, polycystic ovarian disease, untreated. Key as for Fig. 6.

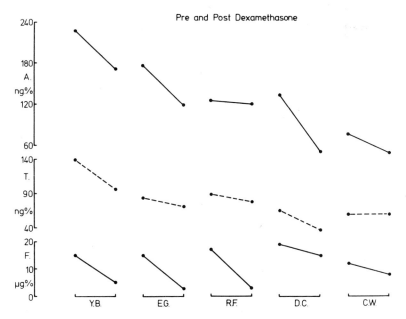

Fig.11 Mean morning plasma steroid levels before and after treatment with dexamethasone in the five hirsute patients studied. Key as for Fig.6.

Patient YB. (Fig.10) This patient shows very striking abnormalities. Testosterone levels are very high, and fluctuate between 170 and 110 ng/100 ml. Androstenedione levels are abnormally high and show many large secretory episodes. It is particularly noticeable that these are independent of cortisol.

The effect of treatment with dexamethasone is shown in Fig.11. It shows the mean morning plasma androgen levels before and after dexamethasone. The responses vary somewhat. The two patients with presumed normal ovaries did not show complete suppression of plasma cortisol, whereas all the patients with cystic ovaries did. In one case (YB), cortisol, androstenedione and testosterone all decreased quite markedly, which would be compatible with removing a significant adrenocortical component. On the other hand, patient RF showed good suppression of cortisol but virtually no change in androgen levels. However, it is very difficult to conclude very much from these mean figures because of the major fluctuations which occur, and which can be seen in the individual figures.

Taking the patients in the same order, Patient CW (Fig.12) now demonstrates very clearly the continuing changes in androgen levels which occur even during glucocorticoid suppression of ACTH. Both testosterone and androstenedione levels fluctuate rapidly. Curiously though, the minimal changes which occur in cortisol levels appear to correlate with testosterone. The results from Patient DC (Fig.13) are equally striking. Here androstenedione levels increase, whilst cortisol levels fall.

The patients with cystic ovaries show much more marked suppression of cortisol. In patient EG (Fig.14) testosterone levels appear to follow androstenedione levels well, but this is not a consistent finding as the next patient (RF; Fig. 15) shows. The last patient's results are similar (Fig.16) - again there are still

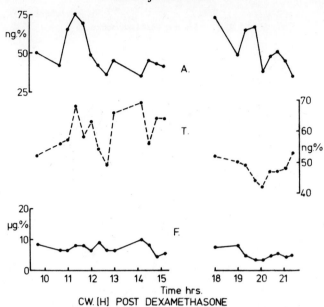

Fig.12 Plasma steroid levels in patient CW, after 2 days treatment with 0.25 mg daily dexamethasone given at night. Key as for Fig.6.

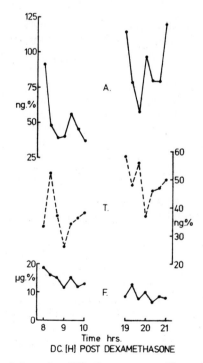

Fig.13 Plasma steroid levels in patient DC, after treatment. Key as for Fig.6.

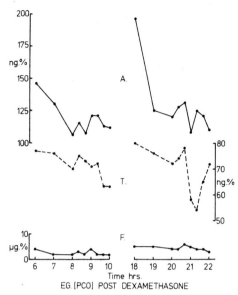

Fig.14 Plasma steroid levels in patient EG, after treatment. Key as for Fig.6.

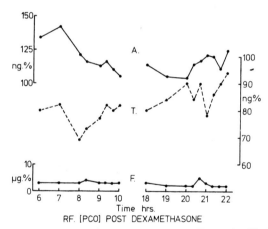

Fig.15 Plasma steroid levels in patient RF, after treatment. Key as for Fig.6.

marked changes in androgen levels which are independent of cortisol.

Thus, in this group of patients, there are a number of differences from normal women. The most striking features are the rapid and well-defined changes in testosterone and androstenedione levels. Unlike the situation in normal women, these major peaks of androstenedione are completely unrelated to plasma cortisol levels, and are not diminished in amplitude by administration of glucocorticoid. It seems clear that these episodes are likely to be totally independent of ACTH. We were also unable to discover any relationship with plasma gonadotrophin levels.

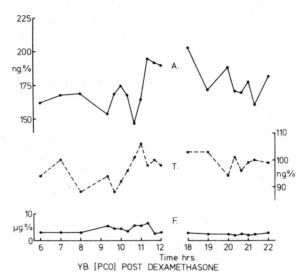

Fig. 16 Plasma steroid levels in patient YB, after treatment. Key as for Fig.6.

The data though, do not allow us to distinguish between the various possible explanations. It seems unlikely that the changes in plasma androgen are due to rapid alterations in metabolic clearance rate, and pulsatile secretion by the adrenal cortex or the ovary is a more attractive explanation. On the basis that the changes occur independently of ACTH, it seems more likely that pulsatile secretion by the ovary is the cause of the changes, but adrenocortical secretion, independently of ACTH is not excluded. Neither is it suggested that this phenomenon is peculiar to, or necessarily characteristic of patients with hirsutism. In a study in which a normal woman was given a suppressive dose of dexamethasone, we have found marked spikes in plasma testosterone levels, again suggesting pulsatile ovarian secretion.

However, it seems clear that the hirsute patients have much more marked swings of androgen levels than occur in normal women and it is also clear that a single blood sample cannot be at all representative of the events through the 24 hours. Accurate calculation of blood production rates is thus not possible on this basis.

The question of the relationship of plasma androgen levels to gonadotrophin secretion is unclear. Studies of the patients reported here did not reveal any relationship with LH and FSH plasma levels, but abnormalities of these hormones have been known to occur in some patients with polycystic ovarian disease for many years. Whether these changes are causal or a consequence of the ovarian disorder is a matter for discussion. In an attempt to investigate this problem, we have treated one of these patients (YB) with Danazol, a semi-synthetic derivative of ethisterone, which is known to suppress gonadotrophin secretion, but to have no intrinsic gonadal steroid-like peripheral actions. Figure 17 shows the effect of this compound, in a dose of 800 mg/day given during dexamethasone treatment for the preceding week and during the study day, on steroid and gonadotrophin levels. Androstenedione concentrations fell to a level lower than that seen at any other time on this patient, but paradoxically, LH and FSH concentrations were

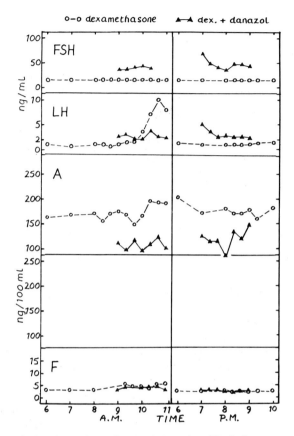

Fig. 17 Plasma steroid and gonadotrophin levels in patient YB during treatment with dexamethasone (0.25 mg daily) alone (o — o) and then with addition of Danazol (800 mg daily) (▲—▲) FSH, LH, androstenedione (A) and cortisol (F) levels are shown. Cortisol levels are in µg/100 ml. (Reproduced with permission from Hull *et al.*, 1976).

actually higher when she was taking the Danazol, compared to the study period on dexamethasone alone. Problems of cross-reaction have prevented the measurement of testosterone in this experiment, but the data would be compatible with a direct action of Danazol on an ovary whose excessive secretion of androgen was independent of prevailing gonadotrophin secretion.

The remaining patients, received only dexamethasone treatment. Even after several months, none reported any improvement in hirsutism, but one patient started to menstruate regularly after a prolonged period of amenorrhoea.

In the investigation and treatment of patients with hyperandrogenic states, there is still a need for methods which will distinguish reliably between ovarian and adrenal sources of androgen, and for drugs which can selectively diminish androgen production. More detailed studies of plasma androgen levels and the development of drugs resembling Danazol may help achieve this goal.

ACKNOWLEDGEMENT

Figure 5 is reproduced from James, V.H.T. and Andre, C.M. (1974) by permission of C.R.C. Press Inc.

REFERENCES

Abraham, G.E. (1974). *J. clin. Endocr.* **39**, 340.
Abraham, G.E. (1975). *Obstet. Gynec.* **46**, 29.
Andre, C.M. and James, V.H.T. (1974). *J. Endocr.* **61**, xxxi.
Baird, D.T. (1971). *In* "Control of Gonadal Steroid Secretion", Pfizer Medical Monographs 6, pp.175-191.
Bardin, C.W. and Lipsett, M.B. (1967). *J. clin. Invest.* **46**, 891.
Gandy, H.M. and Peterson, R.E. (1968). *J. clin. Endocr.* **28**, 949.
Horton, R. and Tait, J.F. (1966). *J. clin. Endocr.* **27**, 79.
Hull, M.G.R., Jacobs, H.S. and James, V.H.T. (1976). *Proc. Roy. Soc. Med. (in press).*
Ito, T. and Horton, R. (1970). *J. clin. Endocr.* **31**, 362.
James, V.H.T. and Andre, C.M. (1974). *In* "Androgen Metabolism in the Human Female" (A.S. Curry and J.V. Hewitt, eds) pp.23-40. CRC Press, Ohio, USA.
Kirschner, M.A. and Jacobs, J.B. (1971). *J. clin. Endocr.* **33**, 199.
Valette, A., Seradour, B. and Boyer, J. (1975). *J. clin. Endocr.* **40**, 160.
Vermeulen, A., Stoica, T. and Verdonck, L. (1971). *J. clin. Endocr.* **33**, 759.
Wieland, R.G., De Courcy, C., Levy, R.P., Zala, P. and Hirschmann, H. (1965). *J. clin. Invest.* **44**, 159.

ANDROGEN SECRETION BY THE HUMAN OVARY: MEASUREMENT OF ANDROGENS IN OVARIAN VENOUS BLOOD

M. Serio, S. Dell'Acqua, E. Calabresi, G. Fiorelli, G. Forti, S. Cattaneo
A. Lucisano, G. Lombardi, M. Pazzagli and D. Borrelli

Endocrinology Unit and Department of Surgery, University of Florence
and
Department of Obstetrics and Gynecology, Università Cattolica S. Cuore
Rome, Italy

Physiological studies have demonstrated that the human ovary secretes estrogens (estradiol and estrone), progesterone and its congeners (20α-dihydroprogesterone, 17OH-progesterone) and androgens. These hormones may also be produced at extraglandular sites from precursors secreted by both the ovaries and adrenal. The characteristics of ovarian secretion are different in relation to age. In fact, an adult pattern of hormonal secretion (cyclical secretion of estrogens and progesterone) begins only at puberty and, unlike testicular function, ovarian activity ceases after three to four decades. However after the ovary has lost its cyclic function, some women produce significant amounts of estrogens of extraglandular origin (Siiteri and MacDonald, 1973).

The normal ovary produces three androgens: androstenedione, testosterone and dehydroepiandrosterone (Van Der Molen, 1969; Lloyd et al., 1971), dehydroepiandrosterone sulphate is produced almost entirely by the adrenal cortex (Kalliala et al., 1970; Nieschlag et al., 1975). In the present chapter three problems related to androgen secretion by the human ovary will be discussed:

A) *Secretion of Androgens by the "Active" and "Inactive" Human Ovary During the Follicular and Luteal Phase*
It is difficult to correlate the data from "in vitro" studies with those from "in vivo" work. Savard et al. (1965) incubating ovarian stroma *in vitro* found greater androgen production in comparison to the follicles and corpus luteum, Smith and Zucherman (1973), incubating the "active" and contralateral human

Table I.

CASE	YEARS	DAY OF THE CYCLE	DIAGNOSIS	FSH (ng/ml)	LH (ng/ml)
1065	40	2	Uterine Fibroid	2.1	2.0
11	42	4	Uterine Fibroid	1.0	0.9
2	41	5	Uterine Fibroid	2.7	4.0
6	37	10	Uterine Fibroids	1.8	0.9
4	41	11	Ovarian Cysts	2.7	4.0
5	44	13	Uterine Fibroid	3.6	1.6
1064	32	14	Uterine Fibroid	6.8	12.6
227	35	17	Uterine Fibroid	3.6	4.1
1058	37	17	Uterine Fibroid	—	—
1	36	18	Uterine Fibroid	1.2	0.7
213	31	20	Uterine Fibroids	2.3	2.5
1059	36	20	Uterine Fibroid	0.9	0.6
14	41	21	Uterine Fibroid	1.1	2.4
1060	44	22	Uterine Fibroid	—	—
8	37	22	Uterine Fibroid	—	—
17	41	22	Uterine Fibroids	—	—
1061	34	23	Uterine Fibroids	4.0	1.9
3	31	23	Uterine Fibroid	—	—
219	41	26	Uterine Fibroid	0.8	1.3
1062	45	26	Uterine Fibroid	1.0	0.7
13	42	28	Uterine Fibroid	1.0	0.7
16	41	28	Uterine Fibroid	1.6	0.7

ovary during follicular and luteal phase. By contrast De Jong et al. (1974) have recently found that the concentrations of androstenedione, 17β-estradiol, estrone and progesterone in the ovarian venous plasma from the ovary containing the Graafian follicle are consistently higher than those in plasma from the contralateral ovary. Similar results have been obtained by the same authors during the luteal phase.

The second problem discussed in this paper is:

B   *The Secretion of 5α-dihydrotestosterone (DHT) by the Human Ovary*

Interest in DHT has been stimulated by the recent suggestion that this potent androgen is an intracellular effector of testosterone action (Wilson, 1972). The concentration in the peripheral venous plasma of normal women is about 15 ng/ 100 ml (Ito and Horton, 1970; Tremblay et al., 1972). The production rate of DHT in normal women is about 60 μg/24 h, whilst the metabolic clearance rate is about 170 L/M^2/24 h (Ito and Horton, 1971; Tremblay et al., 1972; Saez et al., 1972). The sources of plasma DHT in normal women have been studied by Ito and Horton, Tremblay et al. and Saez et al..

Androstenedione appears to be the major prehormone for blood DHT in the adult female, while testosterone is the major source of plasma DHT in the male. The relative contribution of androstenedione and testosterone to the blood production rate of DHT is consistent with an absence of a direct secretion of DHT in

Table II. Radioimmunoassays.

STEROID	ANTISERUM	PAPER CHROMATOGRAPHY	BLANK VALUES (pg per tube)	COEFFICIENTS OF VARIATION within-assay	COEFFICIENTS OF VARIATION between-assay
DHEA	DHEA-7-BSA	Bush A system	1.2 ± 2.0	12.1%	16.6%
DHT	DHT-3-BSA	Bush B_3 system	1.2 ± 1.8	4.0%	9.4%
Δ	Δ_4-6-BSA	Bush A system	0.5 ± 1.0	5.6%	7.7%
E_1	E_2-17-BSA	Bush B_3 system	0.3 ± 0.6	5.6%	8.5%
E_2	E_2-17-BSA	Bush B_3 system	0.5 ± 0.8	10.5%	16.0%
P	11-OHP-11-BSA	Bush B_3 system	4.0 ± 2.6	3.4%	9.9%
T	T-3-RSA	Bush A system	1.9 ± 2.2	8.9%	9.6%

normal women. However, 5α-reductase activity has been found in normal human ovaries during the luteal phase (Smith *et al.*, 1974).

The last problem discussed in this chapter is:

C) *Androgen Secretion in the Polycystic Ovarian Syndrome*

This syndrome is characterized by obesity, hirsutism, menstrual irregularity, infertility and enlarged polycystic ovaries. Increased androgen secretion by the ovaries in the subjects affected by polycystic ovarian disease has been clearly demonstrated (Kirschner and Jacobs, 1971; Stahrl *et al.*, 1973). However the DHT concentration in ovarian venous blood coming from patients affected by this syndrome has not been measured.

MATERIALS AND METHOD

Androstenedione (Δ), testosterone (T), dihydrotestosterone (DHT), progesterone (P), 17β-estradiol (E_2), estrone (E_1) and dehydroepiandrosterone (DHEA) have been measured by radioimmunoassay in ovarian blood samples drawn from 22 women (Table I) during minor surgery of the uterus and from 3 women affected by polycystic ovarian disease. The time from the first day of the last menstrual period was used to estimate the time of the cycle. After induction of anaesthesia the ovaries were examined by naked eye for the presence of follicles and corpora lutea, then plasma samples were taken almost simultaneously from the cubital vein and ovarian vein according to the technique described by Baird and Fraser (1974).

In 9 cases ovarian blood samples were also taken from the contralateral ovary. Heparinized blood samples were cooled on ice after collection, centrifuged at 3000 r.p.m. for 10 minutes at 4°C and the plasma stored at −20°C until analysis.

The radioimmunoassay was performed after paper chromatography. The blank value, the within-assay and the between-assay precision and the chromatographic systems used are reported in Table II. The accuracy of the methods has been evaluated by recovery experiments and dilution experiments. Good linearity was always observed. The specificity has been evaluated by measuring after paper chromatography the specific steroid in plasma samples loaded with interfering analogues. No interference was present even at high concentration of competing steroids.

Table III. Concentrations (ng/100 ml) of steroids in 4 women during follicular phase. P: peripheral venous plasma; O: ovarian venous plasma from ovary containing the ripe follicle; C: ovarian venous plasma from contralateral ovary. In brackets the day of the cycle.

CASE	Δ			DHEA			T			DHT			E_1			E_2			P		
	O	C	P	O	C	P	O	C	P	O	C	P	O	C	P	O	C	P	O	C	P
4 (11)	2500	686	210	3260	–	703	56.2	34.7	22.6	21.0	15.0	17.0	124	19.4	11.2	1028	283	35.8	1346	296	74
5 (13)	2442	345	164	569	406	191	142.0	17.0	18.0	16.0	9.0	8.0	166	13.5	14.5	1188	55	27.0	3191	335	15
6 (10)	2176	222	146	2180	1520	1040	92.0	29.0	25.0	23.0	16.0	16.0	4	0.5	0.5	127	26	4.7	577	78	–
11 (4)	771	257	125	672	544	541	33.0	28.0	14.7	9.0	7.0	7.0	20	9.5	7.2	255	105	14.6	429	55	–
Mean	1972	377	161	1670	823	618	80.8	27.1	20.0	17.2	11.8	11.9	78	10.6	8.2	650	117	20.5	1385	191	44
± S.D.	813	212	36	1290	–	352	47.4	7.3	4.6	6.	4.3	5.3	79	8.1	6.3	536	115	13.6	1268	145	–

Table IV. Concentrations (ng/100 ml) of steroids in 5 women during luteal phase. P: peripheral venous plasma; O: ovarian venous plasma from ovary containing the corpus luteum; C: ovarian venous plasma from contralateral ovary. In brackets the day of the cycle.

CASE	Δ			DHEA			T			DHT			E_1			E_2			P		
	O	C	P	O	C	P	O	C	P	O	C	P	O	C	P	O	C	P	O	C	P
1062 (26)	889	419	185	–	1000	–	40	51	29	7	5	8	50	1.3	–	285	8.0	–	58500	1950	511
1 (18)	787	90	209	1180	1180	–	44	28	34	8	5	6	16	0.7	–	208	1.6	–	4100	394	364
8 (22)	750	165	480	1660	1290	282	141	12	6	29	10	9	156	2.0	9.5	2040	9.0	12.0	149900	760	1580
227 (17)	2167	910	151	1780	–	–	65	59	79	15	14	28	21	1.6	15.2	196	52.	32.4	1120	165	144
14 (21)	1600	996	–	1257	1230	953	90	103	32	26	24	19	5	6.7	3.2	101	15.2	5.5	11220	410	630
Mean	1238	516	256	1471	1175	617	76	50	36	17	11	14	50	2.4	9.3	568	7.8	16.6	45000	735	645
± S.D.	623	419	151	292	125	–	41	34	26	4	7	9	62	2.4	–	825	5.0	–	63000	711	552

RESULTS AND DISCUSSION

The steroid concentrations found in venous plasma samples coming from the active and the contralateral ovary and in peripheral venous plasma during the follicular and luteal phase are reported in Tables III and IV. It is clear that ovaries not containing a developing follicle or a corpus luteum secrete only small amounts of androstenedione, 17β-estradiol, estrone, progesterone and dehydroepiandrosterone. Testosterone and 5α-dihydrotestosterone are not secreted by the "inactive" ovary. By contrast the "active" contralateral ovary secretes significant amounts of androstenedione, testosterone and dehydroepiandrosterone.

Our results are in agreement with those of De Jong *et al.* (1974). In conclusion, the "in vivo" studies are to a large extent in disagreement with "in vitro" investigations; thus the meaning of the "in vitro" experiments in the study of human ovarian steroidogenesis remains uncertain.

The concentrations of steroids in 22 ovarian venous plasma samples draining the "active" ovary during the follicular and luteal phase are reported in Tables V and VI. The gradients of DHT between ovarian and peripheral plasma are insignificant; these results are consistent with an absence of DHT secretion in women, in agreement with previous observations obtained with different techniques (Tremblay *et al.*, 1972; Saez *et al.*, 1972) and with the data of Ito and Horton (1971) who measured DHT in three ovarian venous plasma samples.

However, during the luteal phase in some cases (numbers 1060, 8, 1061, 16) with high levels of testosterone in the ovarian plasma, a small gradient is present. Therefore we can suppose that 5α-reductase activity in the human ovary depends on the amounts of testosterone present. In fact, the 5α-dihydrotestosterone in ovarian venous plasma is significantly correlated with testosterone in the luteal phase ($r = 0.60$; $P < 0.05$), while a similar correlation with androstenedione is absent ($r = 0.31$). Therefore it may be that in the ovarian compartment, testosterone is the most important precursor for 5α-dihydrotestosterone, while in the peripheral blood most of this steroid comes from peripheral transformation of androstenedione.

In relation to other androgens secreted by the human ovary our results confirm that androstenedione is the major one. Testosterone and dehydroepiandrosterone secretion is much less important. In an attempt to determine the cellular origin of various androgens (Δ, DHEA, T and DHT) correlation coefficients with 17β-estradiol and progesterone were calculated for their concentrations in ovarian venous plasma. In the follicular phase, significant correlations were found between concentration of androstenedione and 17β-estradiol ($r = 0.75$), androstenedione and progesterone, testosterone and progesterone ($r = 0.71$); in the luteal phase significant correlations were found between concentration of androstenedione and 17β-estradiol ($r = 0.69$), androstenedione and progesterone ($r = 0.67$). Therefore androstenedione seems secreted cyclically by the ovary as is shown by the correlation of this steroid and 17β-estradiol and progesterone.

Recent studies on variations of androgen levels during the menstrual cycle have demonstrated a relative midcycle increase of androstenedione and testosterone in peripheral venous plasma (Judd and Yen, 1973; Abraham, 1974; Kim *et al.*, 1974); this midcycle increase remained evident even after suppression of the adrenal cortex with dexamethasone (Abraham, 1974; Kim *et al.*, 1974). These results, in the opinion of the authors, are consistent with an ovarian origin of the

Table V. Concentrations (ng/100 ml) of steroids in ovarian and peripheral venous plasma during the follicular phase. O: ovary containing the ripe follicle; P: peripheral. *In brackets the day of the cycle.*

CASE	Δ		DHEA		T		DHT		E_1		E_2		P	
	O	P	O	P	O	P	O	P	O	P	O	P	O	P
1065 (2)	204	157	3950	1090	46	31	17.0	24.0	3	2.3	12	8.0	103	162
11 (4)	771	125	672	541	33	15	9.0	6.7	20	7.2	225	14.6	429	–
2 (5)	812	103	–	–	23	9	17.0	13.0	19	4.0	357	5.4	95	101
6 (10)	2176	146	2180	1040	92	25	23.0	16.0	38	0.5	127	4.7	577	–
4 (11)	2500	210	3260	703	56	23	21.0	17.0	124	11.2	1028	36.0	1346	74
5 (13)	2442	164	569	191	142	18	16.0	8.0	166	14.5	1188	27.0	3191	15
1064 (14)	588	346	–	–	88	73	9.0	13.5	–	–	–	–	532	–
Mean	1356	178	2126	713	68.5	28	16	14	55.7	6.6	494	15.9	896	87.8
±S.D.	976	80	1512	371	41.0	21	5	6	70.7	5.4	491	12.8	1094	61.0

Table VI. Concentrations (ng/100 ml) of steroids in ovarian and peripheral venous plasma during the luteal phase. O: ovary containing the corpus luteum; P: peripheral. *In brackets the day of the cycle.*

CASE	Δ		DHEA		T		DHT		E_1		E_2		P	
	O	P	O	P	O	P	O	P	O	P	O	P	O	P
227 (17)	2167	480	1780	–	65	79	15.0	28.0	21.0	15.0	196	32.0	1200	143
1058 (17)	1300	211	1770	–	71	34	21.0	18.0	0.5	–	329	7.3	7380	150
1 (18)	787	–	1180	–	44	34	8.0	6.0	16.0	–	208	–	4100	364
213 (20)	794	556	–	–	78	33	25.0	26.0	5.2	–	243	–	15540	493
1059 (20)	1100	121	–	–	110	32	7.0	6.7	34.0	–	336	20.5	10654	553
14 (21)	1600	151	1267	953	90	32	26.0	19.0	5.0	3.2	101	5.5	11220	629
1060 (22)	8000	106	–	–	116	18	7.8	3.7	48.6	–	3310	–	662500	2300
8 (22)	750	209	1660	282	141	7	29.0	9.0	156.0	9.5	2400	12.0	149900	1580
17 (22)	360	152	–	–	83	22	9.0	5.0	2.3	2.0	166	2.0	5000	110
1061 (23)	8700	197	–	–	229	29	42.0	17.0	189.0	–	1279	–	86900	294
3 (23)	3200	197	2110	740	140	27	31.0	14.0	147.0	3.0	1466	32.0	53980	2513
219 (26)	854	158	–	–	259	14	30.0	17.0	2.2	1.7	33	10.5	1840	514
1062 (26)	889	185	–	–	40	29	7.4	8.0	50.0	–	295	–	58500	511
13 (28)	493	279	870	256	76	25	7.0	2.0	3.6	–	25	–	16010	333
16 (28)	1900	174	1430	605	415	20	31.0	9.0	8.3	4.0	50	2.7	8220	364
Mean	2192	226	1508	567	134	29	19.7	12.6	45.9	5.5	672	13.9	72894	723
±S.D.	2168	131	397	299	100	15	11.6	8.0	63.9	5.0	949	11.7	168282	767

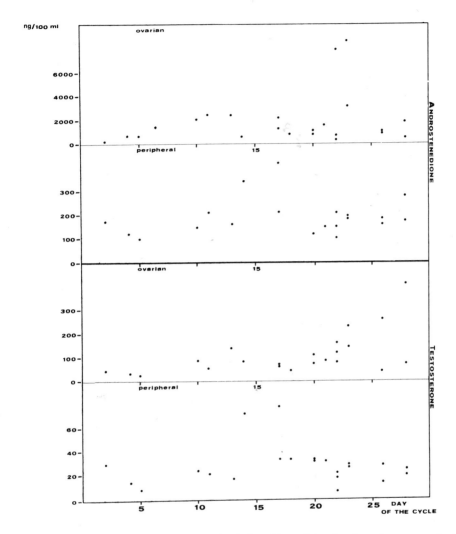

Fig.1 Androstenedione and testosterone levels found in peripheral and ovarian venous plasma.

midcycle increase. In Fig.1 the peripheral and ovarian concentrations of androstenedione and testosterone found in our cases are reported. In peripheral plasma the highest levels are at midcycle, while in ovarian plasma the highest levels are at the end of the luteal phase. This phenomenon is a little surprising and the disagreement between our results and the supposed ovarian origin of the relative midcycle increase of androstenedione and testosterone is evident. However, a large number of cases is needed to elucidate completely this phenomenon. Finally, we would like to discuss briefly the data obtained on ovarian and peripheral venous plasma of 3 subjects affected by the polycystic ovarian syndrome. In Fig.2, these data are compared with those obtained in 3 women during the early follicular phase

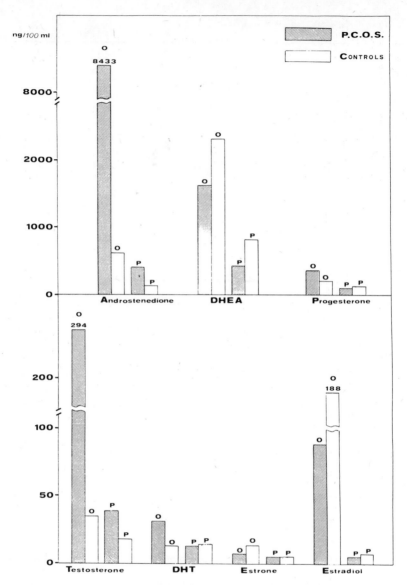

Fig. 2 The values represent the mean steroid concentrations found in ovarian (o) and peripheral (p) venous plasma of 3 women affected by polycystic ovary syndrome (P.C.O.S.). Three women in early follicular phase (see text) were used as controls.

(cases 1065, 2, 11). In agreement with previous results a larger secretion of androstenedione and testosterone is present in the polycystic ovarian syndrome. Moreover, ovarian secretion of 5α-dihydrotestosterone might be partly due to the high concentration of testosterone in ovarian venous plasma.

ACKNOWLEDGEMENT

This work was supported by a grant from the Administrative Board of the University of Florence (Item XI/B).

REFERENCES

Abraham, G.E. (1974). *J. clin. Endocr. Metab.* **39**, 340-346.
Baird, D.T. and Fraser, I.S. (1974). *J. clin. Endocr. Metab.* **38**, 1009-1017.
De Jong, F.H., Baird, D.T. and Van Der Molen, H.J. (1974). *Acta endocr. (Copenh.)* **77**, 575-587.
Ito, T. and Horton, R. (1970). *J. clin. Endocr. Metab.* **31**, 362-368.
Ito, T. and Horton, R. (1971). *J. clin. Invest.* **50**, 1621-1626.
Judd, H.L. and Yen, S.S.C. (1973). *J. clin. Endocr. Metab.* **36**, 475-481.
Kalliala, K., Laatikainen, T., Lukkainen, T. and Vihko, R. (1970). *J. clin. Endocr. Metab.* **30**, 533-535.
Kim, N.H., Hassenian, A.H. and Dupon, C. (1974). *J. clin. Endocr. Metab.* **39**, 706-711.
Kirshner, M.A. and Jacobs, J.B. (1971). *J. clin. Endocr. Metab.* **33**, 199-209.
Lloyd, C.W., Lobatsky, J., Baird, D.T., McCracken, J.A. and Weisz, J. (1971). *J.clin. Endocr. Metab.* **32**, 155-166.
Mikhail, G. (1970). *Gynec. Invest.* **1**, 5-20.
Nieschlag, E., Lariaux, D.L., Ruder, H.J., Zucker, I.R., Kirschner, M.A. and Lipsett, M.B. (1973). *J. Endocrinol.* **57**, 123-134.
Saez, J.M., Forest, M.G., Morera, A.M. and Bertrand, J. (1972). *J. clin. Invest.* **51**, 1226-1234.
Savard, K., Marsh, J.M. and Rice, B.F. (1965). *Rec. Progr. Horm. Res.* **21**, 285-294.
Siiteri, P.K. and MacDonald, P.C. (1973). *In* "Handbook of Physiology Secretion" 7-Vol.II, part 1 Female Reproductive System (R.O. Greep, ed) pp.615-629. American Physiological Society.
Smith, O.W. and Zuckerman, N.G. (1973). *Steroids* **22**, 379-399.
Smith, O.W., Ofner, P. and Vena, R.L. (1974). *Steroids* **24**, 311-315.
Stahrl, N.L., Teeslink, C.R. and Greenblatt, R.B. (1973). *Am. J. Obstet. Gynec.* **117**, 194-200.
Tremblay, R.R., Kowarsk, H., Park, I.J. and Migeon, C.J. (1972). *J. clin. Endocr. Metab.* **35**, 101-107.
Van Der Molen, H.J. (1969). *In* "Progress in Endocrinology" (C. Gual, ed) pp.894-901. Excerpta Medica Foundation, Amsterdam.
Wilson, J.D. (1972). *New Engl. J. Med.* **287**, 1284-1290.

TESTOSTERONE 5α-REDUCTION IN HUMAN SKIN AS AN INDEX OF ANDROGENICITY

P. Mauvais-Jarvis, F. Kuttenn and F. Gauthier-Wright

Biochemical Department, Faculty of Medicine, Pitie-Salpetriere, 75634 Paris Cedex 13, France

INTRODUCTION

The stimulation of the development of secondary sexual characteristics in skin by androgens has long been recognized. The work of Strauss and Pochi (1963) on the stimulation of sebaceous activity by androgens has further established that skin is a target organ for androgens. In addition, the androgen dependence of human skin is of endocrinological importance since the absence of sexual hair growth at puberty is usual in male hypogonadism whereas its excess in women is generally observed in cases of overproduction of androgens by the adrenal gland or the ovaries. In addition, there are many dermatological diseases which seem to be related to hyperresponsiveness of the skin to androgens such as acne vulgaris, male alopecia, etc.. Recent studies of steroid metabolism in human skin have demonstrated the transformation of several steroids in this tissue. In particular, the 5α-reduction of testosterone to 17β-hydroxy-5α-androstan-3-one (dihydrotestosterone DHT) which seems to be an essential prerequisite for the action of testosterone in most androgen target cells (Bruchovsky and Wilson, 1968; Anderson and Liao, 1968) have been observed in the skin (Northcutt et al., 1969; Wilson and Walker, 1969; Voigt et al., 1970). As for the prostate no 5β-reduction of testosterone was observed (Ofner, 1969). In addition, specific receptors for DHT have been isolated in human skin fibroblasts (Keenan et al., 1974) and in sebaceous glands of hamsters (Adachi, 1974; Adachi and Kano, 1972). The intranuclear binding of DHT has been recently observed by Takayasu and Adachi (1975) in sebaceous gland of the hamster. Human skin may therefore be considered as a target organ for androgen like the male accessory glands, since all the events involved in testosterone action have been observed in this tissue. It has also been reported that human skin was capable of converting dehydroepiandrosterone (Gallegos and Berliner, 1967) and androstenedione (Gomez and Hsia, 1968) into testosterone, i.e. transforming a biologically inert precursor into a potent androgen. Skin thus appears to participate not only in the catabolism of steroids, but also in the formation of active hormones from inert steroid precursors supplied through the blood. The transformation of testosterone by DHT by human skin has been further localized in sebaceous glands and hair follicles (Adachi et al., 1970). More recently Voigt et al. (1970) have reported that the 5α-reductase enzyme was essentially present in microsome preparations and required the presence of NADPH for its activity.

In our laboratory (Mauvis-Jarvis et al., 1970a) the reduction of testosterone to DHT and 5α-androstane-3α, 17β-diol (androstanediol) by human skin has been already suggested since in normal men, testosterone was a better precursor of androstanediol when the labelled testosterone was administered through the skin than when the radioactive precursor was injected in the peripheral circulation. In addition, from experiments performed in normal subjects and patients with abnormal sex differentiation (Kuttenn and Mauvais-Jarvis, 1975), it was postulated that 5α-reductase activity present in the skin of human subjects varies according to human male sex differentiation. Regional variations in cutaneous conversion of testosterone into DHT were examined in detail by Wilson and Walker (1969); these authors observed that 5α-reductase activity was higher in the skin of perineal sites than in miscellaneous sites. Same results were found by Mulay et al. (1973). However the results of Flamigny et al. (1971) did not show any statistically significant difference in the production of DHT from radioactive testosterone by the corresponding areas of skin from healthy men and women. Similarly Jenkins and Ash (1973) did not find any difference in the yield of DHT formed from radioactive testosterone incubates with slices of suprapubic skin from normal men and women. If one compares the results obtained by different investigators from incubation of human skin with radioactive testosterone, the results appear to be very conflicting. We think that this is mainly due to the different experimental procedures used. In particular, some results were obtained from slices incubated without cofactor (Thomas and Oake, 1974). Flamigny et al. (1971) used minces and added not only NADPH but also NAD, NADH and NADP. In addition, the length of incubation differs from one investigator to another and they did not always measure the recovery of 3α- and 3β-androstanediol. These steroids are, indeed, formed essentially from DHT and represent after 6 h of skin incubation the main 5α-reduced metabolites of testosterone (Kuttenn and Mauvias-Jarvis, 1975).

Also, the studies reported in this chapter were performed in a standardized manner to ensure that the results obtained only reflect the 5α-reductase capacity in human skin and to provide a comparison between subjects studied.

TESTOSTERONE 5α-REDUCTION IN THE SKIN OF NORMAL ADULT SUBJECTS

Suprapubic skin was obtained either at laparotomy for urological or gynaecological disease or at biopsy under anesthesia from 40 subjects.

All skin specimens were immediately stored at $-20°C$ for approximately 2 weeks. Approximately 200 mg of each skin sample was used for incubation with

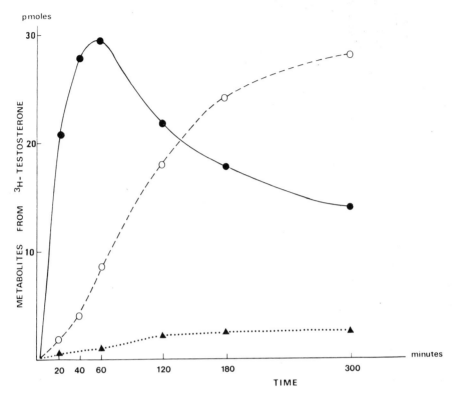

Fig.1 Time course of the formation of various metabolites of ³H-testosterone after incubation with 200 mg of human pubic skin from a normal man (●—●: DHT; ○---○: androstanediols; ▲ ... ▲: androstenedione).

[1,2-³H]-testosterone (SA: 42.7 Ci/mmol - New England Nuclear Corp.). The procedures of homogenization and isolation of radioactive steroid have been previously described (Kuttenn and Mauvais-Jarvis, 1975; Mauvais-Jarvis et al., 1974). Except in one case (see later), all incubations were performed for 1 h at 37°C. As above mentioned, the amount of radioactive DHT, 3α- and 3β-androstanediol recovered after incubation of a suprapubic skin sample with [³H]-testosterone varies according to the length of incubation (Fig.1). The formation of DHT is optimal after 1 h, thereafter it decreased progressively whereas that of 3α and 3β-androstanediol increased in parallel. As for androstenedione arising from testosterone, no significant variation was noted at the different times of incubation. From these results, it was decided that the subsequent incubations of normal and pathological skin would be performed for 1 h and that the calculations of metabolites recovered from radioactive testosterone would include the sum of DHT + 3α- and 3β-androstanediol. The 40 healthy adults studied were endocrinologically normal and their ages ranged from 22 to 50 years (Fig.2). The mean recovery of 5α-reduced metabolites from 20 normal men averaged 14.9 ± 3.4% (SD) of incubated testosterone whereas in 20 women it was only 3.6 ± 1.4%. These 2 values are significantly different (p < 0.001).

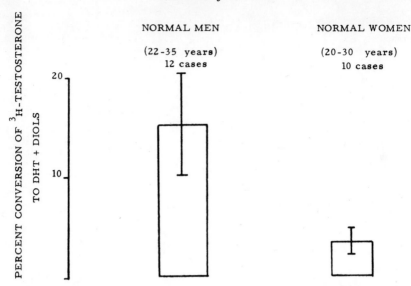

Fig. 2 Conversion rate to DHT + androstanediols (Diols) of radioactive testosterone incubated with 200 mg of pubic skin from normal adults.

Table I. *In vitro* and *in vivo* data on androgens in idiopathic hirsutism. *In vitro:* conversion rates by pubic skin of radioactive testosterone (^{3}H-T) to DHT + androstanediols (Diols). *In vivo:* concentration in plasma of testosterone (T), androstenedione (Δ_4) and DHT - Excretion rate per 24 h of urinary androstanediol (Adiol).

	IN VITRO	IN VIVO			
	% Conversion ^{3}H-T to DHT + Diols by skin	Plasma : ng/100 ml			Urine: µg/24 h
		T	4	DHT	Adiol
Normal Men (20)	14.9 ± 3.4	600 ± 142	109 ± 35	30 ± 16	120 ± 30
Normal Women (20)	3.6 ± 1.4	33 ± 12	133 ± 30	23 ± 11	20 ± 5
Idiopathic Hirsutism (30)	13.5 ± 5.5	54 ± 20	220 ± 50	36 ± 15	75 ± 20

HIRSUTE FEMALES (Table I)

The same investigation was carried out in 30 women with idiopathic hirsutism. These women had abnormal hair growth of various degree of density and extent. All patients had normal ovulatory cycles; ovaries were normal at pelvic or culdoscopic examination. Plasma testosterone and DHT levels were only slightly higher

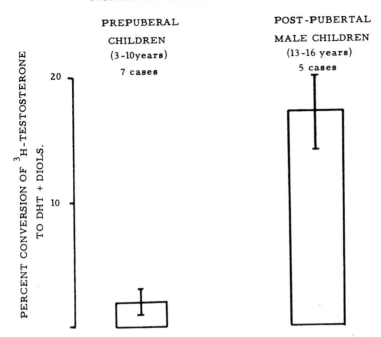

Fig.3 Conversion rate to DHT + androstanediols of radioactive testosterone incubated with 200 mg of pubic skin from children.

than those observed in normal subjects. The only abnormalities observed were an elevated level of plasma androstenedione: (normal value in women: 133 ± 30 ng %) and of urinary androstanediol (N: 20 ± 50 µg/24 h). In these hirsute women the recovery of 5α-reduced metabolites of testosterone in skin homogenates was 13.5 ± 5.5%. This value is significantly higher than that of normal women (p < 0.001) and close to that of normal men. In other words, ovaries or adrenals of women with idiopathic hirsutism do not seem to secrete testosterone at a level higher than that of normal women. The slight increase of plasma testosterone may result from the *in vivo* conversion of androstenedione to testosterone. In these patients, there exists an overproduction of androstenedione by either gonads or adrenals (Bardin *et al.*, 1968). In fact, the main abnormality resulting from our *in vivo* and *in vitro* data seems to consist in an exaggerated "utilization" of circulating androstenedione by androgen target cells of male accessory organs and particularly by the skin. In that tissue, indeed, androstenedione may be converted to testosterone (Gomez and Hsia, 1968) and the exaggerated 5α-reductase activity observed in the skin of hirsute patients may be responsible for an elevated conversion of testosterone to DHT *in vitro* and for an abnormally high urinary excretion of androstanediol *in vivo* (Mauvais-Jarvis *et al.*, 1973).

CHILDREN AND HYPOGONADAL MEN (Figs 3 and 4)

The 5α-reductase activity of the skin of children before puberty is very low since in these subjects the *in vitro* formation by skin of DHT + androstanediol

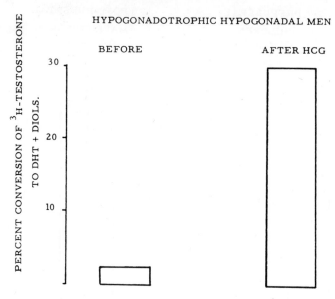

Fig. 4 Conversion rate to DHT + androstanediols of radioactive testosterone incubated with 200 mg of pubic skin from hypogonadotropic hypogonadal men.

ranges from 0.6 to 3.4% of incubated testosterone. These values are significantly lower than that of normal women ($p < 0.001$). However it is very interesting to note that male puberty is accompanied by an abrupt increase of 5α-reductase activity of skin from the sexual area. The same observation was made in the case of hypogonadotropic hypogonadal males treated for 3 weeks with HCG (1500 IU daily). In these subjects before the appearance of clinical symptoms of male puberty (i.e. pubic hair growth and sebaceous gland activity), the skin content in 5α-reductase activity increases dramatically (Kuttenn and Mauvais-Jarvis, 1975). This result indicates that 5α-reductase activity at least in sexual areas of skin is androgen dependent and could be an important step in the control of androgen action. This was first postulated by Moore and Wilson (1973) who showed that testosterone treatment of castrated rats results in a 7-fold increase in the total activity of the 5α-reductase enzyme above the normal level in the prostate.

Conversely, estrogen depresses the skin capacity to convert testosterone to DHT + androstanediols in spite of the presence of NADPH. This confirms our *in vivo* results showing that the conversion of radioactive testosterone to 3α-androstanediol as well as the excretion rate of this metabolite is considerably depressed in men treated with estrogen (Mauvais-Jarvis et al., 1969b). Such an observation was also made by Leav et al. (1971). These investigators had shown that treatment of dogs with estradiol-17β-cyclopentane pripionate was responsible, after 30 days, for morphological alterations in the prostate cytoplasm and nucleus and also for modifications in testosterone metabolism which was mainly due to the oxidation of incubated testosterone to androstenedione whereas the 5α-reduction to DHT was abolished. Even supplementation of the tissue minces with NADPH did not

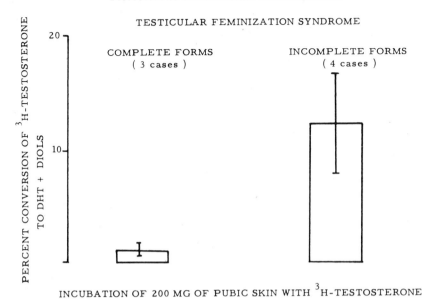

Fig.5 Conversion rate to DHT + androstanediols of radioactive testosterone incubated with 200 mg of pubic skin from Testicular Feminization Syndrome.

affect this shift. It is hardly likely that estrogen might be a competitor for 5α-reductase (Voigt et al., 1970). The fact that NADPH does not restore the capacity of testosterone to be reduced to DHT eliminates a direct effect of estrogen upon the cofactor for 5α-reductase, as has been proposed for testosterone, which possibly controls via the pentose cycle, the amount of NADPH inside the androgen target cell (Adachi et al., 1970). Thus, the way in which estrogens impede the 5α-reduction of testosterone to DHT remains unclear and further experiments will be necessary to know the exact site at which estrogen interferes with androgen metabolism in target cells.

TESTOSTERONE 5α-REDUCTION IN THE SKIN OF PATIENTS WITH TESTICULAR FEMINIZATION SYNDROME (Fig.5)

Male pseudohermaphroditism is an interesting human model with which to study the mode of action of androgen at the level of target cells, since this inherited syndrome seems to be related to a primary defect in one or several of the events necessary for testosterone action in androgen dependent tissues (Mauvias-Jarvis, 1974). Although, in the complete form of testicular feminization syndrome, an absence of cytoplasmic receptor necessary for conveying DHT to the nucleus of androgen target cells has been observed in human beings (Keenan et al., 1974) as in the rat (Bullock et al., 1971) or other animal species with inherited testicular feminization syndrome (Lyon and Hawkes, 1970; Ohno and Lyon, 1970; Goldstein and Wilson, 1972), some discrepancies have also been observed by different investigators in the capacity by the skin of patients with feminizing testis to 5α-reduced testosterone (Northcutt et al., 1969; Wilson and Walker, 1969; Jenkins and Ash, 1973). However, an inherited deficiency in the ability to form

DHT in androgen target tissues seems to be capable *per se* of giving phenotypic aspect of male pseudohermaphroditism. Wilson (1975) has recently reported that patients with familial incomplete male pseudohermaphroditism, type 2, an autosomal recessive disorder of phenotypic sexual differentiation, showed a marked deficiency in the capacity to form DHT by cultured fibroblasts from foreskin. Such a defect in testosterone 5α-reduction may be responsible during embryogenesis for a complete failure of the DHT-mediate events, i.e. the virilization of the urogenital sinus and the anlage of the external genitalia. By contrast, DHT formation by scrotum fibroblasts from patients with familial incomplete male pseudohermaphroditism, type 1 (incomplete form of testicular feminization), was normal (Wilson, 1975). The same observation was made in our laboratory from homogenates of pubic skin of similar patients. In these last subjects, indeed, the defect in testosterone activity seems to be further localized to the 5α-reduction of testosterone, possibly at the level of the cytoplasmic receptor for DHT. If the physiopathology of male pseudohermaphroditism, type 1 and 2, seems to be clearer now, the presence or the absence of testosterone 5α-reduction in the skin of patients with the complete form of testicular feminization is not presently solved. Wilson (1975) recently reported in a limited study with fibroblasts grown from the labia majora of 4 XY individuals with testicular feminization that the DHT formation observed during these experiments was intermediate between that of type 1 and type 2 of male pseudohermaphroditism. In our series of 3 cases, homogenates of suprapubic skin were unable to convert incubated radioactive testosterone into measurable amounts of DHT, 3α- and 3β-androstanediol. Such a result confirmed our previous *in vivo* results showing that radioactive testosterone when percutaneously applied was unable to be 5α-reduced during its passage through the skin (Mauvais-Jarvis *et al.*, 1970a). The absence of testosterone 5α-reduction by the skin of patients with the complete form of testicular feminization was also observed by Northcutt *et al.* (1969) and Hinrichs *et al.* (1969). By contrast, Jenkins and Ash (1971, 1973) have observed normal conversion of radioactive testosterone in 4 patients with a complete form of testicular feminization after incubation of pubic skin for 6 h. However, the results of these investigators did not indicate the recovery of androstanediols in their skin homogenates. This would be interesting to know, since in normal men after 6 h of incubation, androstanediols represent the main metabolites of testosterone. Furthermore, the discrepancies observed by different investigators in the capacity to 5α-reduce testosterone by the skin from the sexual area originating from patients with feminizing testes may be due to the different methods used, as we mentioned above.

The different data obtained from the study of 5α-reductase capacity contained in the skin from the sexual areas of patients with male pseudohermaphoroditism permit the following conclusion: if the defective DHT formation in the androgen target cells is not responsible for all the clinical features of unresponsiveness of male accessory organs to testosterone, this defect by itself may be the source of important abnormalities in male sexual differentiation. For these reasons, the 5α-reduction of testosterone to DHT may be considered as an essential step in androgen action in male target tissues.

INHIBITION OF TESTOSTERONE 5α-REDUCTION TO DHT IN TARGET CELLS

The substances which antagonize the effects of testosterone upon target organs act by preventing the binding of the active structures of this androgen to speci-

Table II. Inhibitory effect of various steroids on 5α-reduction of testosterone (from Voigt, Fernandez and Hsia (1970). *J. biol. Chem.* 245, 5594 - reference 6).

Additions	Inhibition
———	0
Androstenedione	76.4
5α-DHT	4.5
Corticosterone	23.7
Deoxycorticosterone	84.7
Epitestosterone	23.5
Estradiol	2.8
17α-Methyl-B-Nortestosterone	0.8
19-Nortestosterone	55.4
Progesterone	93.3

fic receptor sites of the target structures. In the field of androgen antagonists, this approach has led to the discovery of several effective compounds. Most of the antiandrogens used at the present time are known to interfere with the cytoplasmic protein binding specifically DHT in male accessory cells (Mainwaring, 1969; Liao and Fang, 1969), thus preventing the transfer of the DHT-cytoplasmic protein complex to the nuclear chromatin (Mainwaring and Peterken, 1971). However, at least in male accessory organs, the cytoplasmic protein has a particularly high binding affinity for DHT and not for testosterone (Mainwaring, 1969). As a result, the 5α-reduction of testosterone to DHT being an essential prerequisite in androgen action, the inhibition of this metabolic event may be considered from different points of view as antiandrogenic. Voigt *et al.* (1970) had found in 1970 that a number of natural steroids inhibited the 5α-reduction of testosterone by microsomal preparations of human skin. Among these, the most potent inhibitor was progesterone (Table II) which competes effectively with testosterone for the active site of the enzyme. In addition, although to a lesser degree than cyproterone, progesterone is also known to interfere with DHT for the binding sites of the specific cytoplasmic receptor in androgen dependent tissues (Mainwaring, 1969; Fang *et al.*, 1969).

The antiandrogenic activity of progesterone is well documented. Indeed, Dorfman (1971) has previously shown that the inunction of the chick comb with progesterone inhibited its stimulation by testosterone whereas progesterone diminished the growth of male accessory organs of mice when administered subcutaneously with testosterone.

From results obtained by Voigt *et al.* (1970) during incubation of skin microsomes with radioactive testosterone, in the presence and absence of progesterone, it appeared likely that progesterone may act as a competitive inhibitor of the 5α-reduction of testosterone, i.e. the same enzyme may utilize both steroids as substrates and may reduce respectively testosterone to DHT and progesterone to 5α-pregnane-3,20-dione (dihydroprogesterone). However, the 5α-reductase present in the skin utilizes progesterone more efficiently than testosterone (Voigt *et al.*, 1970). Similar results have been obtained in our laboratory from homogenates or suprapubic skin incubated with radioactive testosterone in the

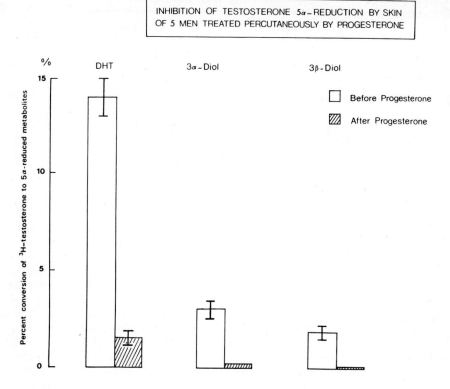

Fig. 6 Inhibition of testosterone 5α-reduction to DHT + androstanediols in human homogenates from normal men after topical administration of 100 mg of progesterone.

absence and presence of unlabeled progesterone (Mauvais-Jarvis et al., 1974). Similarly, Massa and Martini (1972) have demonstrated that progesterone may interfere with the conversion of testosterone into DHT both at the anterior pituitary and prostatic level in rats. In addition Jenkins and McCaffery (1974) have observed that progesterone was a much more potent inhibitor of human prostatic 5α-reductase *in vitro* than estradiol and that the degree of inhibition was such that progesterone could have a direct effect on the metabolism of testosterone by the prostate when it was administered *in vivo*.

In our laboratory, many studies have been performed *in vivo* and *in vitro* with testosterone and progesterone. *In vivo*, the 5α-reduction of both steroids by skin was postulated since in normal men testosterone and progesterone were better precursors of their respective 5α-metabolites: 3α-androstanediol and 5α-pregnanediol when labeled testosterone or progesterone were administered through the skin (in other words by the percutaneous route) than when the radioactive precursors were injected in the peripheral circulation (Mauvais-Jarvis et al., 1969a; Mauvais-Jarvis et al., 1969c). Thus, theoretically, the inhibition of testosterone 5α-reduction by progesterone might be considered as an antiandrogenic effect of endocrinological and especially of dermatological significance since progesterone administered percutaneously is largely metabolized *in situ* (Mauvais-Jarvis *et al.*, 1969a) before reaching the peripheral circulation. Indeed, the possibility for progesterone to pass through the skin and to be retained in this tissue is of particular interest from a dermatological point of view and leads us to propose the topical application of progesterone in the treatment of skin diseases such as acne vulgaris and seborrhea which are related to hyperresponsiveness of the skin to androgens. In patients with acne vulgaris, Sansone and Reisner (1971) have reported that skin samples from acne areas were 5 to 20 times more effective in reducing testosterone to DHT than skin samples from the corresponding areas of normal people.

In our laboratory experiments were carried out in order to obtain information on the antiandrogenic activity of progesterone administered percutaneously. Five normal men, 20-30 years old, were studied. A suprapubic skin sample was obtained before and 24 h after topical application to the same public area of 100 mg of unlabeled progesterone dissolved in absolute ethanol. The total application of progesterone was fractionated into 4 percutaneous applications of progesterone for 24 h between 2 biopsies of pubic skin. Approximately 100 mg of skin was used for homogenization and incubation with [^{3}H]-testosterone (Mauvais-Jarvis *et al.*, 1974). Plasma progesterone, testosterone and DHT were evaluated before and 24, 26 and 48 h after the beginning of the treatment.

No significant change in plasma testosterone and DHT was observed during the topical application of progesterone. Plasma progesterone slightly increases from a mean value of 0.52 ng/ml to 2.14 ng % ml 24 h after the beginning of the experiment. In contrast, in skin samples obtained 24 h after the topical administration of progesterone a very important fall in the conversion of radioactive testosterone to DHT and androstanediols was observed (Fig.6). This fall averaged for the 5 experiments 75.2 ± 11.1% when compared with the conversion rate observed before the experiment. More recently, Voigt and Hsia (1973) have reported a very interesting study on the antiandrogenic activity of 4-androsten-3-one-17β-carboxylic acid. This compound dissolved in acetone, and topically applied to the flank organs of hamsters inhibited locally the enlargement of the sebaceous glands provoked by testosterone, whereas the 5α-reduction of testosterone to DHT and androstanediol was blocked *in vitro*. This study confirms the possibility of the inhibition of testosterone 5α-reduction by compounds typically applied. The steroid used by this group (Voigt and Hsai, 1973) was interesting because it had no known hormonal activity and was not metabolized *in situ* by the 5α-reductase. However, Voigt and Hsia did not mention if such a compound may be effective in humans when percutaneously administered. Indeed, no data are given upon the absorption rate of 4-androsten-3-one-17β-carboxylic acid through the skin. By contrast, the total absorption of progesterone through human skin has been calculated both *in vitro* (Feldman and Maibach, 1969) and *in vivo* (Mauvais-Jarvis *et al.*, 1969a) and averages 10% of the administered dose. The clinical use of progesterone as an antiandrogen was criticized by Voigt and Hsia (1973) because this steroid might be a precursor for active androgens in the testes or adrenals after its passage through the blood stream. Our results showing that plasma testosterone does not increase after topical administration of progesterone do not agree with this assumption. In addition, the increase in plasma progesterone after the topical administration of progesterone is so poor that it is not sufficient to cause, even in woman, systemic effects in particular at endometrial level. On the flank organ of the hamster, progesterone, when topically applied, caused the same

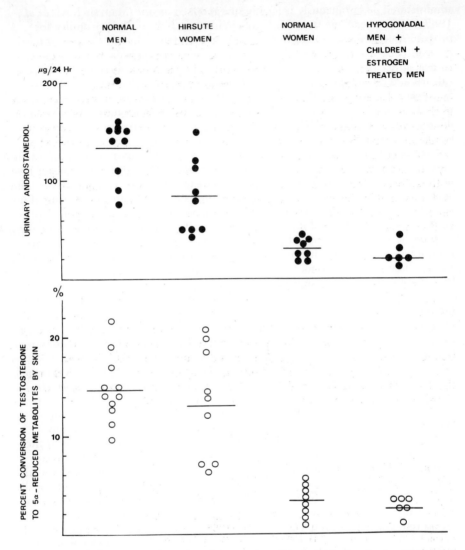

Fig. 7 Relationship between *in vivo* data (urinary excretion per 24 h of androstanediol) and *in vitro* data (percent conversion by pubic skin of ^{3}H-testosterone to DHT + androstanediols) in normal subjects and patients with abnormal sex development (individual values and mean).

histological modifications of the sebaceous glands as those reported by Voigt and Hsia (1973) with 4-androstan-3-one-17β-carboxylic acid (unpublished observation). In addition, there are now available clinical data which suggest that the percutaneous application of progesterone is effective in skin disease related to an excess of seborrhea (Cherif-Cheik and Lignières, 1974; Fayolle, 1975).

CONCLUSIONS

In human beings, the 5α-reduction of testosterone to DHT is an essential prerequisite for the action of androgens in most target cells, particularly the skin. In that tissue, this enzymatic activity seems to be androgen dependent: its genetic absence is responsible for a defect in the virilization of the external genitalia during embryogenesis and in sexual hair growth at puberty; an increased concentration in this enzyme is observed in cases where an extensive development of pilosebaceous glands in sexual areas is noted. In other words, there is a satisfactory correlation between hair growth from sexual areas and the extent of testosterone 5α-reduction in this tissue.

In addition, the formation of DHT + androstanediol by pubic skin incubated with radioactive testosterone varies in the same way as the rate of 3α-androstanediol recovered in the urine of the same subjects (Fig.7). This might be expected, since in target cells DHT which is not utilized for binding to nuclear protein is further metabolized into 3α-androstanediol which is excreted in the urine. Circulating DHT is also reduced by the liver to androstanediol. As well as in target cells, 3β-androstanediol formed from DHT seems to be largely transformed to 3α-androstanediol after its passage in the peripheral circulation; in our laboratory, it was observed that 80% of injected labeled 3β-androstanediol was converted to urinary metabolites having a 3α-configuration, particularly to 3α-androstanediol (Mauvais-Jarvis et al., 1970b). Thus, the urinary excretion of 3α-androstanediol may be considered as mainly reflecting the production of both DHT and 3β-androstanediol (Mauvais-Jarvis et al., 1973). Moreover, these *in vitro* and *in vivo* data corroborate our previous hypothesis (Kuttenn and Mauvais-Jarvis, 1975; Mauvais-Jarvis et al., 1973; Mauvais-Jarvis et al., 1969; Mauvais-Jarvis et al., 1968) namely that the determination in biological fluids of the end products of testosterone metabolism in biological fluid provides useful information in clinical practice.

REFERENCES

Adachi, K. (1974). *J. Invest. Derm.* 62, 217.
Adachi, K. and Kano, M. (1972). *Steroids* 19, 567.
Adachi, K., Takayasu, S., Takashima, I., Kano, M. and Kondo, S. (1970). *J. Soc. Cosm. Chem.* 21, 901.
Anderson, K.M. and Liao, S. (1968). *Nature (London)* 219, 272.
Bardin, C.W., Hembree, W.C. and Lipsett, M.B. (1968). *J. clin. Endocr.* 28, 1300.
Bruchovsky, N. and Wilson, J.D. (1968). *J. biol. Chem.* 243, 2012.
Bullock, L.P., Sherins, R.J. and Bardin, C.W. (1971). *J. clin. Invest.* 50, 15a (Abs.).
Cherif-Cheik, J.L. and de Lignières, B. (1974). *Sem. Hôp. Paris* 50, 489.
Dorfman, R.I. (1971). *Proc. of Third International Congress on Hormonal Steroids* (V.H.T. James and L. Martini, eds) p.995. Excerpta Medica, Amsterdam.
Fang, S., Anderson, K.M. and Liao, S. (1969). *J. biol. Chem.* 244, 6584.
Fayolle, J. (1975). *Lyon Med.* 233, 1303.
Feldman, R.J. and Maibach, H.I. (1969). *J. Invest. Derm.* 52, 89.
Flamigni, C., Collins, W.P., Koullapnis, E.N., Craft, I., Dewhurst, C.J. and Sommerville, I.F. (1971). *J. clin. Endocr.* 32, 737.
Gallegos, A.J. and Berliner, D.L. (1967). *J. clin. Endocr.* 27, 1214.
Goldstein, J.L. and Wilson, J.D. (1972). *J. clin. Invest.* 51, 1647.
Gomez, E.C. and Hsia, S.L. (1968). *Biochemistry* 7, 24.
Hinrichs, W.L., Karznia, R., Wyss, R. and Hermann, W.L. (1969). *Clin. Res.* 17, 143.
Jenkins, J.S. and Ash, S. (1971). *J. Endocrinol.* 49, 515.

Jenkins, J.S. and Ash, S. (1973). *J. Endocrinol.* 59, 345.
Jenkins, J.S. and McCaffery, V.M. (1974). *J. Endocrinol.* 63, 517.
Keenan, B.S., Meyer, W.J., Hadjian, A.J., Jones, H.W. and Migeon, C.J. (1974). *J. clin. Endocr.* 38, 1143.
Kuttenn, F. and Mauvais-Jarvis, J. (1975). *Acta endocr. (Copenh.)* 79, 164.
Leav, I., Morfin, R.F., Ofner, P., Cavazos, L.F. and Leads, L.E. (1971). *Endocrinology* 89, 465.
Liao, S. and Fang, S. (1969). *Vitam. Horm.* 27, 18.
Lyon, M.F. and Hawkes, S.G. (1970). *Nature (London)* 227, 1217.
Mainwaring, W.I.P. (1969). *J. Endocrinol.* 45, 531.
Mainwaring, W.I.P. and Peterken, B.M. (1971). *J. biol. Chem.* 125, 285.
Massa, R. and Martini, L. (1972). *Gynec. Invest.* 2, 253.
Mauvais-Jarvis, P. (1974). *In* "The Endocrine Function of the Human Testis" (V.H.T. James, M. Serio and L. Martini, eds) p.125. Academic Press, New York and London.
Mauvais-Jarvis, P., Floch, H.H. and Bercovici, J.P. (1968). *J. clin. Endocr.* 28, 381.
Mauvais-Jarvis, P., Baudot, N. and Bercovici, J.P. (1969a). *J. clin. Endocr.* 29, 1580.
Mauvais-Jarvis, P., Bercovici, J.P. and Floch, H.H. (1969b). *Rev. Franc. Études Clin. Biol.* 14, 169.
Mauvais-Jarvis, P., Bercovici, J.P. and Gauthier, F. (1969c). *J. clin. Endocr.* 29, 417.
Mauvais-Jarvis, P., Bercovici, J.P., Crepy, O. and Gauthier, F. (1970a). *J. clin. Invest.* 49, 31.
Mauvais-Jarvis, P., Guillemant, S., Corvol, P., Floch, H.H. and Bardou, L. (1970b). *Steroids* 16, 173.
Mauvais-Jarvis, P., Charransol, G. and Bobas-Masson, F. (1973). *J. clin. Endocr.* 36, 452.
Mauvais-Jarvis, P., Kuttenn, F. and Baudot, N. (1974). *J. clin. Endocr.* 38, 142.
Moore, R.J. and Wilson, J.D. (1973). *Endocrinology* 93, 581.
Mulay, S., Filkenstein, R., Pinsky, L. and Solomon, S. (1973). *J. clin. Endocr.* 34, 133.
Northcutt, R.C., Island, D.P. and Liddle, E.W. (1969). *J. clin. Endocr.* 29, 422.
Ofner, P. (1969). *Vitam. and Horm.* 28, 237.
Ohno, S. and Lyon, M.F. (1970). *Clin. Genet.* 1, 121.
Sansone, G. and Reisner, R.H. (1971). *J. Invest. Derm.* 56, 366.
Strauss, J.S. and Pochi, P.E. (1963). *Rec. Progr. Horm. Res.* 19, 385.
Takayasu, S. and Adachi, K. (1975). *Endocrinology* 96, 525.
Thomas, J.P. and Oake, R.J. (1974). *J. clin. Endocr.* 38, 19.
Voigt, W. and Hsia, S.L. (1973). *Endocrinology* 92, 1216.
Voigt, W., Fernandez, E.P. and Hsia, S.L. (1970). *J. biol. Chem.* 245, 5594.
Wilson, J.D. (1975). *J. biol. Chem.* 250, 3498.
Wilson, J.D. and Walker, J.D. (1969). *J. clin. Invest.* 48, 371.

ANDROGEN METABOLISM IN THE SKIN OF HIRSUTE WOMEN

R.J. Oake and J.P. Thomas

*Department of Medicine, Welsh National School of Medicine
University Hospital of Wales, Cardiff, UK*

Male type distribution of body hair in the female is not only a cosmetic defect but may also be a manifestation of endocrine disease. The hirsutism is a sign of excess androgen production or of increased sensitivity of the hair follicles to normal circulating androgen levels. Two groups of hirsute women are recognised (Fig.1). The first display hirsutism in the presence of other signs of virilism including menstrual irregularities, increased muscle size, clitomegaly and deepening of the voice. Hirsutism may appear at any age and often progresses rapidly. The second group usually exhibits no other signs of virilism and hirsutism is often first observed around puberty. In the former group plasma testosterone levels and testosterone production rates are invariably found to be elevated, whilst in the latter category, plasma testosterone levels are found to be normal in about one third of patients. Testosterone production rates however are still found to be elevated in the majority (Fig.2). The fact that hirsutism can be seen in the presence of normal circulating androgen levels has led to investigations of the role of skin as a target tissue. The physiological responses of skin to androgens have long been known. The response is not confined to hair growth. The total collagen content of skin is greater in men than in women, and is increased by exogenous androgens (Black et al., 1970). It has also been shown that skin from hirsute women has an increased total collagen content (Shuster and Bottoms, 1963). Burton et al. (1972) demonstrated that skin from hirsute females showed increased sebum secretion (Fig.3), and increased rate of pharmacologically induced sweat excretion (Fig.4). The rates of secretion in hirsute women approached those of normal males of the same age. These findings suggest that the whole skin is virilised in hirsutism, presumably due to androgenic stimulation. It has been postulated that some of the androgen may be of cutaneous origin and this may help to account for the absence of systemic virilisation in idiopathic hirsutism.

Tumour, Adrenal Hyperplasia	Polycystic Ovaries/Idiopathic Hirsutism
Excess body hair	Excess body hair
Evidence of virilism	No evidence of virilism
1. Increased muscle size	Some menstrual irregularities
2. Temporal baldness	Onset of disease often at puberty
3. Clitoromegaly	Slow progression of disease
4. Deepening of voice	
5. Menstrual irregularities	
Onset of disease at any time	
Rapid progression of disease	

Fig. 1 Clinical differentiation of hirsute subjects.

	NORMAL	HIRSUTE
Plasma testosterone	20–68 ng/100 ml	Normal (1/3) to elevated (2/3)
Plasma free testosterone	0.2–0.75 ng/100 ml	0.41–5.61 ng/100 ml
	mean 0.35 ng/100 ml	mean 1.79 ng/100 ml
Testosterone Production	130–330 µg/day	550–2640 µg/day
Metabolic Clearance Rate	530–760 L/day	590–1750 L/day

Fig. 2 Steroid levels in idiopathic hirsutism.

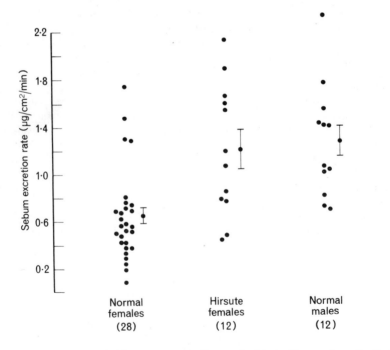

Fig. 3 Sebum excretion rates in hirsute women and control subjects. The number of patients in each group is shown in parenthesis. The vertical lines indicate the mean ± S.E.M. (From Burton, Johnson, Libman and Shuster, 1971)'

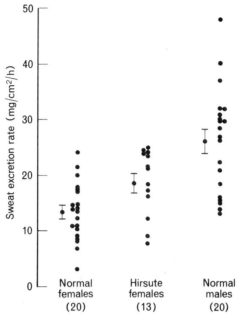

Fig. 4 Sweat rates in hirsute women and control subjects. The number of patients in each group is shown in parenthesis. The vertical lines indicate the mean ± S.E.M. (From Burton, Johnson, Libman and Shuster, 1971.)

Table I. Histochemically demonstrated enzymes in skin. (From Baillie, Calman and Milne, 1965.)

Enzyme	Degree of Activity
3β Hydroxysteroid dehydrogenase	Intense
16β Hydroxysteroid dehydrogenase	Intense
3α Hydroxysteroid dehydrogenase	Moderage
11β Hydroxysteroid dehydrogenase	Moderate
17β Hydroxysteroid dehydrogenase	Moderate
6β Hydroxysteroid dehydrogenase	Trace
20β Hydroxysteroid dehydrogenase	Trace

That metabolism of steroids occurs in skin has been known for many years. Baillie *et al.* (1965) demonstrated histochemically the presence of hydroxysteroid dehydrogenases in skin (Table I). Of the many different dehydrogenases, the 3β and 16β hydroxysteroid dehydrogenases gave the strongest histochemical reaction, with slightly less activity from 3α, 11β and 17β hydroxysteroid dehydrogenases. These enzymes are important in steroid biosynthesis. 3β hydroxysteroid dehydrogenase converts pregnenolone to progesterone and dehydroepiandrosterone to androstenedione. 17β hydroxysteroid dehydrogenase converts androstenedione to testosterone (Fig.5). Several groups of workers have identified androgenic steroids on the surface of skin in man. Carrie and Ruhrmann (1955) and Dubovyi (1960)

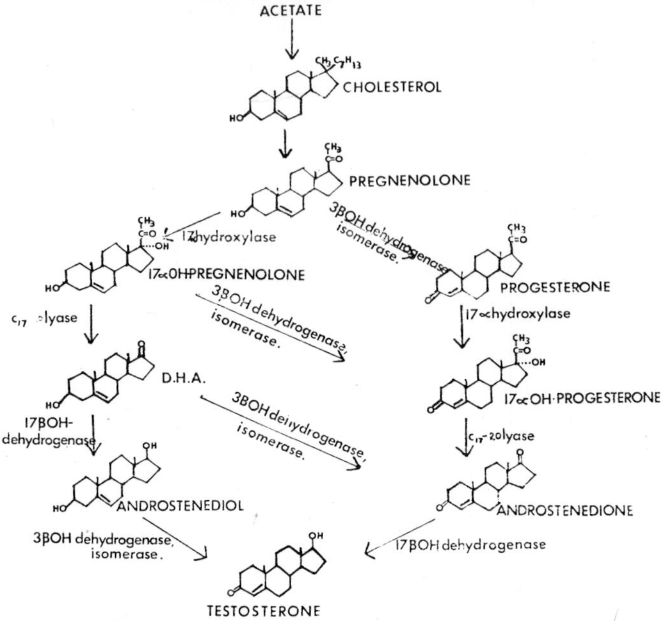

Fig. 5 Biosynthetic pathways of steroids.

have reported finding Zimmerman positive chromogens in skin surface lipids and Julesz *et al.* (1966) have noted the same in skin and hair.

In vivo studies such as these do not prove that the skin actively participates in the metabolism of steroids, since it may simply be passively excreting preformed steroids. There have been several studies indicating that skin can metabolise androgens *in vitro*. The first such study by Wotiz *et al.* (1956) demonstrated that skin metabolised more testosterone than any other normal tissue studied. Rongone (1966) identified four specific end-products of testosterone metabolism by mammary skin, 5α androstan-3βol-17-one, 5α androstan-3,17,dione, 4-androstene 3,17 dione and 5β-androstan-3αol-17 one. His study confirmed the presence of hydroxysteroid dehydrogenases shown histochemically by Baillie. The most important transformation in skin is the reduction of testosterone to dihydrotestosterone. Gomez and Hsai (1968) identified dihydrotestosterone as one of the metabolic products of incubation of skin with testosterone or androstenedione. Using foreskin or skin from various other sites they found 15-80% of added testosterone was metabolised, the principal metabolic product was dihydrotestosterone. These studies were further extended by Wilson and Walker (1969) who studied *in vitro* metabolism in skin from various body sites. The highest rate of formation of dihydrotestosterone was shown in perineal skin. With the capacity to form dihydrotestosterone now firmly established, the concept of local formation of active androgens from weakly androgenic precursors was investigated. We have studied the metabolism of labelled dehydroepiandrosterone and testosterone in normal and hirsute skin (Thomas and Oake, 1974). The utilisation and metabolism of both these steroids was assessed in a small group of hirsute women and compared

Androgen Metabolism in the Skin of Hirsute Women 499

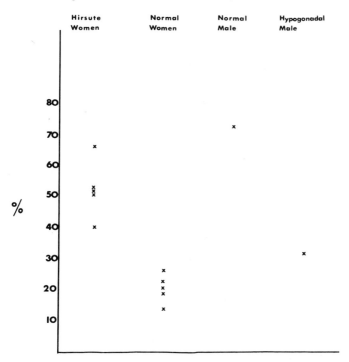

Fig.6 Utilisation of D.H.A.

with normal women, one normal man and one hypogonadal man. Four of the five women had exhibited hirsutism from puberty and none showed any other evidence of virilism. Skin was obtained from the linea alba in all patients under general anaesthesia and placed in chilled saline. After removal of cutaneous fat, slices of skin 0.5 mm thick were prepared by hand. Separate skin specimens were then incubated with radioactive dehydroepiandrosterone and testosterone in 2 ml Krebs-Ringer phosphate buffer under an atmosphere of 95% O_2 :5% CO_2. Incubation was for four hours at 37°C after which time reactions were stopped by addition of acetone. To the incubates were added 50 μg amounts of ten predicated androgen metabolites to allow for losses during extraction and subsequent chromatography. After extraction the residues were applied to thin layer chromatography plates and chromatographed in a preliminary system which separated the component steroids into groups. Each of these groups was subsequently chromatographed on several other thin layer or paper systems to isolate the component steroids, using a radiochromatogram scanner to locate radioactive areas. Identification of each of the androgens measured was indicated by its behaviour on at least three systems, its change in mobility on acetylation or oxidation and finally, by its retention time on gas liquid chromatography. The gas liquid chromatogram also served to determine the recovery of each androgen through the purification procedure. Radioactivity associated with each of the purified androgens was determined, and this coupled with the known recovery allowed calculation of the percentage conversion of precursor to each metabolite.

The utilisation of dehydroepiandrosterone was significantly greater in hirsute

Table II. Percentage conversions (mean ± S.E.) of D.H.A. metabolism in skin.

	Hirsute Women (5)	Normal Women (5)
Utilisation	52 ± 4	19 ± 2.1
Testosterone	4.2 ± 0.82	0.52 ± 3.6
Dihydrotestosterone	2.6 ± 0.49	2.4 ± 0.60
Androstanediol	1.0 ± 0.52	0.84 ± 0.53
Total 17-hydroxy	7.8 ± 0.61	3.7 ± 1.1
Androstenedione	5.0 ± 3.0	2.1 ± 0.67
Androstanedione	6.0 ± 1.6	0.55 ± 0.35
Androsterone	3.5 ± 1.1	0.94 ± 0.42
Total 17-keto	14.6 ± 1.9	3.6 ± 0.88

Table III. Percentage conversion (mean ± S.E.) of testosterone metabolism in skin.

	Hirsute Women	Normal Women
Utilisation	36 ± 2.7	31 ± 4.2
Dihydrotestosterone	8.0 ± 1.8	5.4 ± 1.8
Androstanediol	1.7 ± 0.62	2.6 ± 0.58
Total 17-hydroxy	9.7 ± 1.5	8.0 ± 2.0
Androstenedione	6.2 ± 1.2	6.4 ± 1.7
Androstanedione	4.2 ± 1.1	2.9 ± 1.8
Androsterone	5.0 ± 1.1	1.7 ± 0.91
Total 17-keto	15.4 ± 2.2	11.0 ± 3.2

than in normal women and was comparable to that of the normal man (Fig.6). Utilisation by the hypogonadal man was similar to that in the normal females. Corresponding with the increased utilisation in the hirsute skin more androgen metabolites were formed and although the 17-ketosteroids predominated the sum of active 17 hydroxysteroids formed was also significantly greater than in normal skin (Table II). Testosterone was the major 17-hydroxysteroid in hirsute skin, there being little difference in the rate of formation of dihydrotestosterone and androstanediol. There was little difference between hirsute and normal skin on incubation with testosterone (Fig.7, Table III). Other workers notably Jenkins and Ash (1974) (Table IV) have shown more clearly enhanced metabolism of testosterone in hirsute skin, in particular the reduction to dihydrotestosterone. This difference may well be related to addition by those other workers of NADPH as cofactor, since Mauvais-Jarvis et al. (1974) demonstrated that optimum reduction of testosterone took place in the presence of NADPH (Table V). Reduction of testosterone to dihydrotestosterone is important since the latter is now generally thought to be the active form of testosterone in target tissues. First evidence of its importance came as a result of studies of the intracellular localisation of ^{3}H

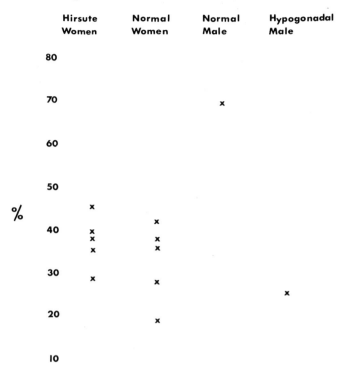

Fig. 7 Utilisation of testosterone.

testosterone in target tissues. Wilson and Loeb (1965) demonstrated that radioactivity deriving from ^{3}H testosterone was bound to a protein component of the duck preen gland. Further analysis in rat tissue demonstrated the uptake of radioactivity by target tissues but not by non-target tissues like liver and small intestine. Dihydrotestosterone was the principal steroid present in nuclei. Bruchovsky and Wilson (1968) demonstrated that prostatic nuclei contain an enzyme which converts testosterone to dihydrotestosterone. This finding induced considerable interest in the cutaneous metabolism of testosterone in diseases where end organ defects are suspected. In the testicular feminisation syndrome, there appears to be no lack of 5α reductase enzyme in the skin and yet, despite normal circulating androgen levels, there is a lack of body hair. There is some evidence for deficiency of the cytosol binding protein necessary for transfer of dihydrotestosterone to the nucleus. The possibility thus arises that excessive hair growth may be a function of abnormal interaction of androgen with a cellular receptor.

The mechanism of action of testosterone has been investigated primarily in the prostate (Fig.8). Unbound circulating testosterone is thought to enter the cell where it is probably bound to an intracellular binding protein, as yet uncharacterised. Regardless of whether it is bound to the cytosol protein or not, testosterone is then reduced in the endoplasmic reticulum or on the nuclear membrane, and bound to a specific high affinity protein, which gains access to the nucleus. There it is bound to chromatin and initiates the transcription of messen-

Table IV. Metabolism of ^{14}C testosterone in suprapubic skin. % radioactivity recovered. (From Jenkins and Ash, 1973.)

	Dihydrotestosterone	Androstenedione
Normal Females (12)		
Mean	6.7	4.6
Range	2.8 − 10.6	1.6 − 9.4
Normal Males (7)		
Mean	7.2	5.7
Range	2.3 − 11.8	3.8 − 8.7
Hirsute Females		
1.	9.7	4.6
2.	12.3	9.8
3.	19.2	15.3
4	16.1	20.7
5.	18.6	15.1
6.	8.7	10.4

Table V. Effect of cofactors on metabolism of testosterone by human skin. % radioactivity recovered. (From Mauvais-Jarvis, Kuttenn and Baudot, 1974.)

Cofactors Added	% Dihydrotestosterone	% Androstenedione
None	0.8	0.1
NADP	0.56	0.8
NADP + Glucose-6-phosphate + glucose-6-phosphate dehydrogenase	8.4	22.9
NADPH		
1	12.7	3.5
2	13.6	1.8
3	15.3	2.4
4	10.6	6.2

ger R.N.A., and ultimately the complex metabolic changes of androgen action.

There are a variety of ways in which the amount of active androgen available to the target tissue can be influenced. One can reduce the overall secretion of testosterone by administration of suppressive agents like prednisone and dexamethasone. The results obtained with such agents are variable and any benefit obtained only on continued long-term use. Another possibility is the use of oestro-

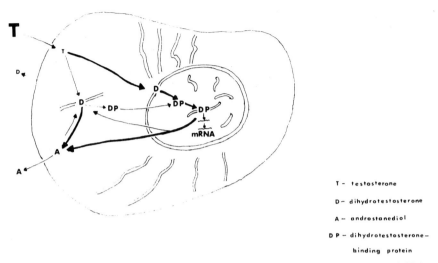

Fig. 8 Schematic representation of the mechanism by which testosterone accelerates R.N.A. synthesis within target cells.

T — testosterone
D — dihydrotestosterone
A — androstanediol
DP — dihydrotestosterone-binding protein

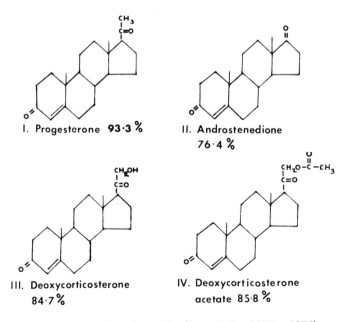

I. Progesterone 93·3 %

II. Androstenedione 76·4 %

III. Deoxycorticosterone 84·7 %

IV. Deoxycorticosterone acetate 85·8 %

Fig. 9 Steroids tested for anti-androgenic activity (from Hsai and Voigt, 1974).

gens. These are known to increase the concentration of circulating testosterone binding globulin and consequently reduce the amount of free steroid available to the target tissues. Perhaps the most promising treatment is the use of cutaneous anti-androgens. Inhibition of testosterone action is achieved in two ways. By pre-

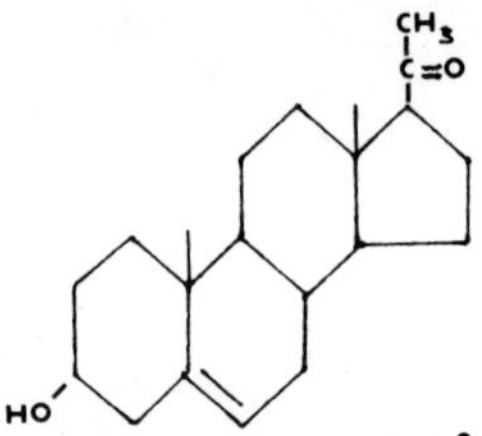

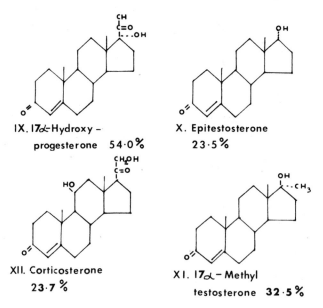

Fig.10 Steroids tested for anti-androgenic activity (from Hsai and Voigt, 1974).

V. Pregnenolone 3·3%
VI. Dehydroepiandrosterone 1·9%
VII. 4-Androsten-3β,17β-diol 4·3%
VIII. 5α-Dihydrotestosterone 4·5%

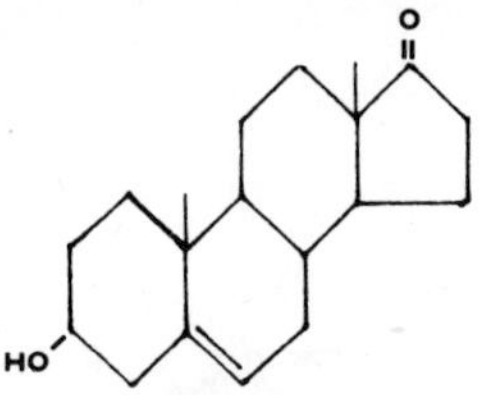

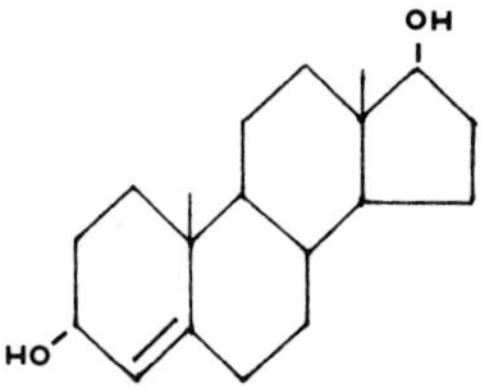

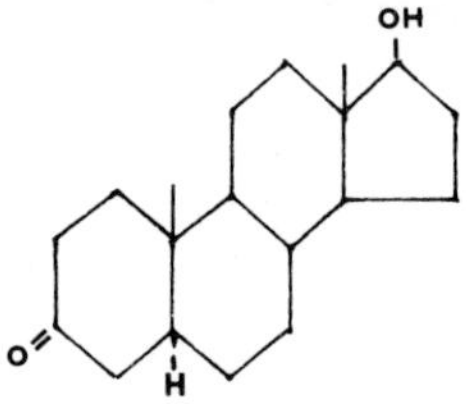

IX. 17α-Hydroxyprogesterone 54·0%
X. Epitestosterone 23·5%
XII. Corticosterone 23·7%
XI. 17α-Methyltestosterone 32·5%

Fig.11 Steroids tested for anti-androgenic activity (from Hsai and Voigt, 1974).

venting testosterone binding to the cytoplasmic receptor protein or by inhibition of 5α reductase activity. Compounds which inhibit testosterone reduction do so presumably by competition and have certain structural limitations placed upon them. Hsai and Voigt (1974) have studied the structural requirements in some

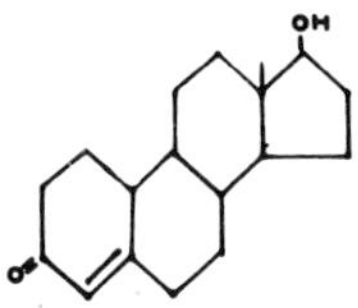

Androgen Metabolism in the Skin of Hirsute Women 505

XIII. 19-Nortestosterone
55·4 %

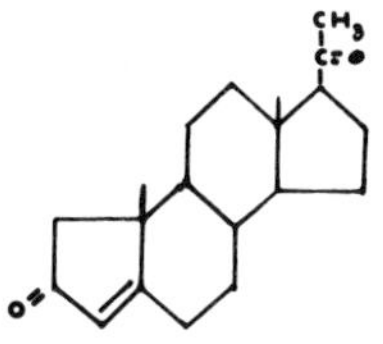

XIV. A-norprogesterone
0 %

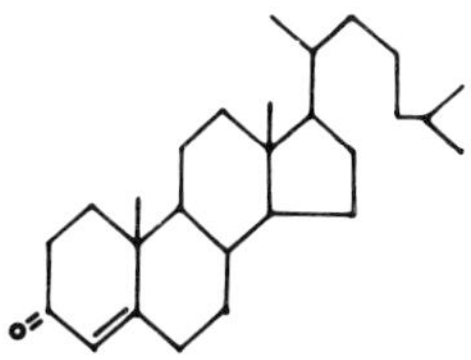

XV. Cholestenone
10·0 %

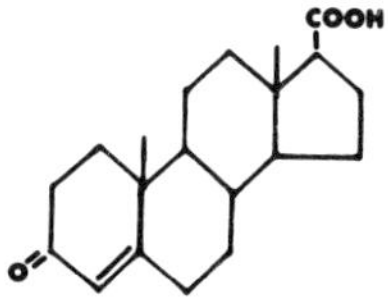

XVI. 4-Androsten-3-one-
17β-carboxylic acid 87·7 %

Fig.12 Steroids tested for anti-androgenic activity (from Hsai and Voigt, 1974).

detail. They reported on the inhibitory activity of sixteen steroids. Progesterone, androstenedione, deoxycorticosterone and deoxycorticosterone acetate (Fig.9) are all effective inhibitors. They have in common a 3-keto Δ^4 structure in ring A, and this structural similarity to testosterone apparently enables them to compete with the latter for binding with the 5α reductase. Four compounds differing slightly in structure from the first four (Fig.10), show little inhibitory activity. Changing from 3-keto Δ^4 to 3β hydroxy Δ^5 as in pregnenolone and dehydroepiandrosterone, simply replacing the 3-keto by a 3β hydroxy group as in 4-androstene-3β-17β-diol, or saturating the A ring to dihydrotestosterone eliminates inhibitory activity. The 3-keto Δ^4 structure is a rigid requirement for effective inhibition. The next four compounds (Fig.11) although having the required 3-keto Δ^4 structure are less effective inhibitors and they have in common D ring substituents in the α position or an 11β group as in corticosterone. The reduced inhibitory activity of these compounds may be due to the spatial distribution of these substituents sterically hindering the binding with the enzyme. The final four compounds (Fig.12) illustrate the adverse effect of alteration of the steroid nucleus and also the presence of a hydrocarbon chain at carbon 17. It would also appear that the 17β side chain needs to be hydrophilic for effective inhibition, although within limits the size of this side chain may vary. Ideally, a therapeutic inhibitor should have no undesirable hormonal effects nor should it be a potential precursor of steroid hormones. These factors preclude the use of several potential anti-androgens. Thus, deoxycorticosterone is precluded by its mineralocorticoid activity. Since androstenedione can be converted to testosterone it is also precluded. Ethers and esters of testosterone, because of possible release of free testosterone on hydrolysis, are unsatisfactory.

Hsai and Voigt investigated testosterone 17β carboxylic acid (Fig.12) in detail since this compound fulfils all the structural requirements, has no known hormonal activity and is not found in any natural steroidogenic pathway. This steroid inhibited the androgenic effect of cutaneous testosterone propionate in the hamster flank organ. When testosterone propionate was replaced by dihydrotestosterone, inhibition of the androgen action could not be achieved, indicating that testosterone 17β carboxylic acid acts in the flank organ by inhibiting dihydrotestosterone formation. *In vitro* experiments with flank organ homogenates supported this explanation.

Hsai and Voigt excluded progesterone as a cutaneous anti-androgen on the grounds that it might be a precursor of androgens. Our own studies have shown that progesterone is converted predominantly to non-androgenic steroids in skin. Mauvais-Jarvis,*et al.* (1974) studied the inhibitory activity of progesterone in male skin. They were able to show both *in vivo* and *in vitro* that progesterone was an effective inhibitor of 5α reduction of testosterone. These authors reported that more than 70% of absorbed progesterone was metabolised in the skin, the principal products being 5α progesterone and allopregnanediol. These compounds are not known to have an androgenic effect, nor was any trace found of them in blood after percutaneous administration of progesterone. Many other steroid anti-androgens have been described. Compounds like cyproterone acetate and 17α methyl-B-nortestosterone were found to be quite successful in reducing hair growth, but doubt regarding their side effects has limited their usefulness. Work on the structural requirements of effective inhibitors of testosterone 5α reduction enables more accurate prediction of anti-androgenic activity of compounds, and may spend the discovery of an effective cutaneous anti-androgen with minimal side effects.

IN SUMMARY

1. Skin is a target tissue for circulating male androgen, and in this tissue conversion of testosterone to dihydrotestosterone appears to be a pre-requisite for androgen action.
2. The skin is capable of synthesis of testosterone and dihydrotestosterone from weak androgen precursors; in hirsutism synthesis is enhanced.
3. Dihydrotestosterone is the active androgen which initiates protein synthesis and subsequent hair growth.
4. Cutaneous anti-androgens can inhibit dihydrotestosterone formation by competing for the 5α reductase enzyme.
5. The structure of effective anti-androgens has been studied and certain rigid requirements noted.

REFERENCES

Baillie, A.H., Calman, K.C. and Milne, J.A. (1965). *Br. J. Dermatol.* 77, 610-616.
Black, M.M., Shuster, S. and Bottoms, E. (1970). *Br. Med. J.* iv, 773-774.
Bruchovsky, N. and Wilson, J.D. (1968). *J. biol. Chem.* 243, 2012-2021.
Burton, J.L., Johnson, C., Libman, L. and Shuster, S. (1972). *J. Endocrinol.* 53, 349-354.
Carrie, C. and Ruhrmann, H. (1955). *Hautarzt* 6, 404-407.
Dubovyi, M.I. (1960). *Vestn. Derm. Vener.* 34, 10-19.
Gomez, E.C. and Hsai, S.L. (1968). *Biochemistry* 7, 24-32.

Hsai, S.L. and Voigt, W. (1974). *J. Invest. Dermatol.* **62**, 224-227.
Jenkins, J.S. and Ash, S. (1973). *J. Endocrinol.* **59**, 345-351.
Julesz, M., Faredin, I. and Toth, I. (1966). *Acta Med. Acad. Sci. Hung.* **22**, 25-34.
Mauvais-Jarvis, P., Kuttenn, F. and Baudot, N. (1974). *J. clin. Endocr. Metab.* **38**, 142-147.
Rongone, E.L. (1966). *Steroids* **7**, 489-504.
Shuster, S. and Bottoms, E. (1963). *Clin. Sci.* **25**, 487-491.
Thomas, J.P. and Oake, R.J. (1974). *J. clin. Endocr. Metab.* **38**, 19-22.
Wilson, J.D. and Loeb, P.M. (1965). *In* "Nineteenth Annual Symposium on Funadmental Cancer Research", pp.375-391. Williams and Wilkins Company, Baltimore.
Wilson, J.D. and Walker, J.D. (1969). *J. clin. Invest.* **48**, 371-379.
Wotiz, H.H., Mescon, H., Doppel, H. and Lemon, H.M. (1956). *J. Invest. Dermatol.* **26**, 113-120.

SUBJECT INDEX

A

Acetate-1-^{14}C
  incorporation into steroids, 39, 41
  incubation studies with the corpus luteum, 4
    with the ovary, 26, 28, 41
    with the stroma, 38
Acromegaly and sex hormone binding globulin, 153
ACTH effects on adrenal secretion, 453
  on ovarian secretion, 453
  response to in hirsutism, 461
  stimulation in post menopausal women, 237-243
Addison's disease
  androstenedione in, 130
  estradiol in, 130
  FSH in, 130
  LH in, 130
  progesterone in, 130
Adenine incorporation into cyclic AMP, 27
Adenylate cyclase
  activity in the corpus luteum, 43
  assay of, 183
  effect of estrogen on, 189
  HCG binding and, 188
Adenylate cyclase, steroidogenesis and, 33
Adrenal
  catheterization procedure, 446
  secretion
    effect of ACTH on, 453
    effect of HCG on, 454
    of various steroids, 128
  venous effluent, steroid levels in, 448-451
Alkaline phosphatase
  in ovarian tissues, 3
Amenorrhea
  anorexia nervosa and, 291-293
  body weight and, 271, 291
  bromocryptine treatment of, 307-308, 327, 330
  chromosome anomalies in, 262, 269
  evaluation of, 267-273
  functional disorders with, 271-273
  FSH levels in, 121
  galactorrhea and, 250-251, 264, 271-272
  hypothyroidism and, 251
  investigation of, 261-265
  LH levels in, 121
  LH-RH response in, 121, 276-289
    tests in, 263-264
  organic causes of, 267, 271
  pituitary radiology and, 249, 250
  pneumopelvographic patterns in, 302
  primary, 261-265
    developmental defects and, 262
    hypothalamic-hypophyseal involvement, 270-271
    LRH-TRH response in, 224, 324
  secondary
    diagnosis of, 245-259
    estradiol levels in, 251-252
    estrogen production in, 253-254
    functional classification of, 246
    gonadotropin levels in, 246-249
    prolactin levels in, 251
    protocol for the investigation of, 257
"Amplifier Hypothesis"
  sex hormone binding globulin and, 148-149
Androblastoma, 401
Androgenic effects
  in the mouse, 159-173
  in the rat, 159-173
Androgens, see also specific steroids
  binding by non-receptor proteins, 159-160
  binding protein in the germinal epithelium, 159

509

Subject Index

hirsutism and, 437-442, 457-470
hypothalamo-hypophyseal axis and, 169
metabolism in the skin, 495-506
ovarian venous blood levels, 471-479
polycystic ovarian tissue in, 437-438
production in the ovary, 10
secretion
    by the "active ovary", 471
    by the "inactive ovary", 471
    theca cells, 40
    in polycystic ovary syndrome, 473, 477
synthesis in the follicle, 39
target cell metabolism, 160-165
target organs, 481
Androstenedione
  in Addison's disease, 130
  bone age and, 209
  in children, 204
  conversion to estrone, 136-138
  in idiopathic hirsutism, 441
  ovarian feedback control and, 350
  ovarian secretion, 128-131
  in ovariectomised women, 238
  in polycystic ovary syndrome, 376, 440
  in post menopausal women, 237-243
  production in pathological conditions, 139
  radioimmunoassay of, 199
  sexual maturation and, 220
  synthesis
    in the embryonic testicle, 6
    in the follicle, 26, 39
Anorexia nervosa
  clomiphene response in, 295
  estradiol levels in, 253-254, 293
  estrone levels in, 293
  FSH levels in, 294
  hormonal studies in, 293-294
  hypothalamic-pituitary-ovarian disorders in, 291-304
  LH levels in, 294
  LH-RH response in, 121, 254, 295
  pneumopelvographic patterns in, 299-302
  prolactin levels in, 302
  testosterone levels in, 302-303
  urinary gonadotropins and, 294
Anovulation, chronic, 373-381
  hyperprolactinaemia and, 327-330
Anti-androgen activity, studies on, 503
Antibodies to LH-RH, 393
Arginine, response to infusion in polycystic ovary syndrome, 381
Arrhenoblastomas, 2, 401
Aspirin and prostaglandin synthesis, 54
Assay
  of adenylate cyclase activity, 183
  of ovarian receptor binding activity, 183
  radioreceptor for HCG, 191
ATP activity in oocytes, 67

Atresia, follicular, 15-16, 105
Autotransplantation of the ovary, 126

B

Balbiani's vitelline body, 11
Binding of HCG to ovarian receptors, 181-194
Biosynthetic pathway of steroids, 498
Blood flow
  measurement in ovarian artery, 72
  role in ovarian function, 71-79
Blood production rates of steroids, 128
Body temperature and progesterone levels, 256
  weight and amenorrhea, 271, 291
Bone age and hormone levels, 200-217
Bonnet monkey *(M. Radiata)* and SHBG, 152-153
Bound:unbound ratios for DHT and testosterone, 162
Bovine serum albumin effect on Graafian follicles, 31
Brain, as an endocrine organ, 334
Breast development and steroid hormones, 218-219
Bromocryptine
  galactorrhea and, 307, 327, 330
  menstrual cycle and, 325
  post partum period and, 309
  treatment of
    amenorrhea, 307-308, 327, 330
    oligomenorrhea, 307-308

C

CNS-pituitary system and PCO syndrome, 383
Cancer, ovarian, 397-415
  effect on LH, 82
  effect on FSH, 82
Catecholamines, role in neuroendocrine regulation, 334
Catheterization
  of adrenal vein, 446
  of ovarian vein, 126, 446
Cells, hilus, 7
Children
  craniopharyngioma in, 225-227
  FSH levels in, 203
  hypothalamo-pituitary-gonadal axis in, 195-222
  LH levels in, 203
  in vitro and *in vivo* data of androgens in, 485
Cholesterol, *in vivo* ovarian perfusion studies and, 428

Chromosomes
  anomalies and amenorrhea, 262, 269
  in sex differentiation, 3-4
Cirrhosis and SHBG, 153
Clomiphene
  ovulation and, 254
  progesterone levels and, 254-255
  tests in amenorrhea, 262, 263
  treatment of short luteal phase, 390
  response and the hypothalamic-pituitary-ovarian axis, 351
    in anorexia nervosa, 295
    in polycystic ovary syndrome, 376-377
Cofactors and testosterone metabolism, 8
Common blastema, differentiation, 3
Computerised scanning and tomography, 234
Constant infusion technique for MCR, 127
Control of ovarian secretion, 131-132
Corpus luteum, 16-20, 40
  adenylate cyclase activity in, 43
  cyclic AMP and, 42
  hyperemia, 16
  incubation studies with, 41
  involution of, 20
  LH and, 106
  maturity of, 16
  of pregnancy, 18
  proliferation, 16
  radioactive microspheres, uptake of, 75
  steroid biosynthesis in, 40
    secretion, 131
  synthesis of progesterone, 40
  transformation, 16
  vascularization, 16
Corticoid treatment of the short luteal phase, 387-392
Craniopharyngioma, 225-227
Cyclic AMP
  concentration in the corpus luteum, 42
  isolation and purification, 27
  ovulation and, 25-35
  preovulatory changes in, 26-27, 33
  prostaglandins and, 25-35
  steroids and, 25-35
Cyproterone acetate and SHBG in men, 148

D

Deaths from ovarian cancer, 398
Dehydroepiandrosterone
  follicular synthesis, 39
  idiopathic hirsutism and, 441
  ovarian secretion, 128-131
  in ovariectomised women, 238
  polycystic ovary syndrome and, 374, 376, 440

  in post menopausal women, 237-243
  sulphate and PCOS, 374
Δ^4-pathway
  in the follicles, 39-40
  in the corpus luteum, 42
Δ^5-pathway in the follicle, 39-40
Developmental defects and amenorrhea, 262
Dexamethasone
  response to hirsutism, 462, 467
  suppression in post menopausal women, 237-243
20α dihydroprogesterone in ovarian venous blood, 430
Dihydrotestosterone
  binding in tissues, 160-161
  ovarian secretion, 472
  in ovariectomised women, 238
  in post menopausal women, 237-243
  radioimmunoassay of, 473
  receptor-protein synthesis, 155
  skin and, 481
  target cell formation, 164
Dissociation constant, 187, 319
DNA repair in oocytes, 67
Doppler ultrasonic blood flow measurement, 72, 77
Dynamic tests in anorexia nervosa, 295-299
  of hypothalamic-pituitary-ovarian axis, 351
Dysfunctional uterine bleeding, 352-353

E

Electron microscopy of ovarian tissues, 411
Embryonic testicle
  androstenedione synthesis in, 6
  testosterone synthesis in, 6
Embryo, 3β and 17β hydroxysteroid dehydrogenase in, 5
Endocrine activity
  in developing follicles, 14-15
  in the foetal ovary, 6
Endometrial progesterone receptors, 179
Environmental factors and ovarian cancer, 409
Enzymes in the skin, 497
Equilibrium studies of gonadotropin-receptor interactions, 185-188
Estradiol
  blood levels
    in amenorrhea, 251-252
    in anorexia nervosa, 253-254, 293
    in menstrual cycle, 361
  body weight and, 253-254
  levels in ovarian venous blood, 430
    polycystic ovary syndrome, 374, 376 440
  ovarian feedback control and, 350

Estrogens, see also specific substances
 in anorexia nervosa, 293
 effect on
  adenylate cyclase, 189
  HCG binding, 189
  LH secretion, 81
  LH in post menopausal women, 116
  FSH secretion, 81
  FSH in post menopausal women, 116
  ovarian progesterone, 189
  progesterone receptors, 177
 extraglandular, in health and disease, 135-140
 gonadotropin secretion and, 81-88
 menopausal women and, 117
 production in secondary amenorrhea, 253-254
  polycystic ovary syndrome, 138-139
 secretion from theca cells, 40
  in post menopausal women, 136-137
  in pre menopausal women, 155
 synthesis in the follicle, 39
Estrone
 in anorexia nervosa, 293
 breast development and, 219
 conversion from androstenedione, 136-138
 follicular synthesis of, 26
 incorporation into acetate-1-^{14}C, 39, 41
 levels and bone age, 212, 215-217
  in children, 207, 214-216
  in menstrual cycle, 361
  in polycystic ovary syndrome, 374, 376, 440
 ovarian secretion, 128-131
 in ovariectomised women, 238
 in post menopausal women, 237-243
 radioimmunoassay of, 199
 sexual maturation and, 220
 testes volume and, 219
Ethinylestradiol
 effect on
  FSH in post menopausal women, 116
  LH in post menopausal women, 116
  SHBG in men, 148
  testosterone in men, 148
 response to, the hypothalamic-pituitary-ovarian axis, 351

F

Feedback control
 hypothalamic-pituitary-gonadal axis and, 103, 350-353
 systems and gonadotropin release, 345
Female hypogonadism, 245-259
Fibroblasts of the stroma, 8

Fibrocytes of the stroma, 8
Follicle stimulating hormone (FSH)
 in Addison's disease, 130
 effect on Graafian follicles, 31
  of estradiol benzoate on, 82, 100
  of progesterone on, 119
 levels after castration, 82
  in amenorrhea, 121, 248
  in anorexia nervosa, 294
  and bone age, 201, 205-206
  in children, 203
  in infants, 199
  in the menstrual cycle, 361
  in polycystic ovary syndrome, 376
 metabolic clearance rate, 94
 ovarian feedback control and, 350
 in post menopausal women, 98
 prolactin and, 102
 response to LH-RH, 110, 281
 secretion, 81, 83-86, 89-108, 125
Follicular
 atresia, 15-16, 105
 phase and gonadotropin levels, 114
 steroid secretion, 131

G

Galactorrhea
 amenorrhea and, 250-251, 264, 271-272
 bromocryptine and, 307, 327, 330
 HCG response in, 307
 HMG (human menopausal gonadotropin) response in, 307
Germinal cells, maturation of, 1
 epithelium, 6, 159
Glucose-6-phosphate dehydrogenase activity in oocytes, 66
Gonadal blastema, somatic cells of, 5
 function and prolactin, 264
Gonadotropic activity in the puerperium, 325-327
 hormones and ovarian blood flow, 74
Gonadotropin, human chorionic (HCG)
 basal levels in normal women, 112
 binding and adenylate cyclase, 188
  effect on estrogen, 189
  and ovarian progesterone, 188
  to ovarian receptors, 181-194
 dissociation constants, 187
 effect on adrenal secretion, 454
  cyclic AMP in the corpus luteum, 43
  steroid synthesis in the corpus luteum, 41
 galactorrhea and, 307
 Hill plot of binding data, 188
 levels in children, 203
  early follicular phase, 114
  menopausal women, 117

Subject Index

secondary amenorrhea, 246-249
mechanism of action in the corpus
    luteum, 42-43
ovarian venous blood studies and, 430
pinealoma and, 232
prostaglandin and, 31
receptor interactions, 185-188
receptors, 192
release after LH-RH stimulation, 275-290
    feedback control systems, 345
releasing hormone tests in amenorrhea,
    263, 264
response to LH-RH in pathological
    conditions, 119-122
secretion and estrogens, 81-88
    neuroendocrine regulation of, 333-338
    by the pituitary, 335
    polycystic ovary syndrome and, 374-377
sexual maturation and, 218-219
specific binding in the corpus luteum, 44
stimulation tests in post menopausal
    women, 237-243
in urine by radioreceptor assay, 191
Graafian follicle, 12
    ability to synthesise steroids, 26
    effect of
        FSH on, 31
        prolactin on, 31
    prostaglandins in, 13, 28, 32
    rat, 30
use as experimental model, 25
Granulosa cells, 18
    binding studies with, 317, 320
    secretion of progesterone from, 40
    steroid synthesis in the follicle and, 40
        theca cell tumours, 39
Growth hormone
    binding studies with, 317
    craniopharyngioma and, 227
Guinea pig uterus, progesterone receptors in,
    175-180

H

Hamartoma and puberty, 233
Hill plot of binding data, 188
Hilus cell
    tissue, incubation studies, 401
    tumour, 402
    of the ovary, 7
Hirsutism
    androgens in, 437-442, 457-470
    clinical differentiation, 496
    idiopathic, 443
        in vitro and *in vivo* data of androgens
            in, 484

response to ACTH in, 461
    dexamethasone in, 462-467
    and SHBG, 153-154
steroid levels in, 496
Histiocytes of the stroma, 8
Histology of the ovary, 1-24
Hormonal
    changes in lactating women, 311
        non-lactating women, 311-315
    control in the menstrual cycle, 359-372
        of progesterone receptors, 176
    levels in abnormal menstrual cycles, 370-371
    response to arginine releasing factor, 381
        thyroid releasing factor, 381
    state and the menopause, 366
    status of the short luteal phase, 388
    studies in luteoma of pregnancy, 417-425
Hydroxydehydrogenases in the skin, 497
    steroid dehydrogenase in human embryos, 5
        effect of testosterone propionate on, 165
Hydroxyprogesterone
    follicular synthesis, 26, 39
    incorporation of acetate-1-^{14}C into, 39, 41
    levels in ovarian venous blood, 430
    ovarian secretion, 128-131
    in ovariectomised women, 238
    in post menopausal women, 237-243
Hyperprolactinaemia, 306
    amenorrhea and, 251
    anovulation and, 251
    the menstrual cycle and, 324
    pituitary tumours and, 251
Hypogonadal men, *in vitro* and *in vivo* data of
    androgens in, 485
Hypogonadism, female, 245-259
Hypogonadotropic hypogonadism, FSH
    response to LH-RH in, 281
Hypophysectomised rats, transcortin levels in,
    170
Hypothalamic
    control of gonadotropin release, 333-338
        pituitary function, 89-93
    hypophyseal involvement in amenorrhea,
        270-271
    pituitary gonadal axis, feedback control, 95
    pituitary ovarian disorders in anorexia
        nervosa, 291-304
    pituitary system and PCO syndrome, 373-392
    synaptosomes, superfusion of, 58
    tumours, effect on puberty, 223-236
Hypothalamo-hypophyseal axis and androgens,
    169
Hypothalamo-pituitary-gonadal axis in
    children, 195-222
Hypothalamus
    hypophyseal-ovarian feedback mechanisms,
        109-124

pituitary-ovarian axis,
    disorders of, 351
    dynamic tests of, 351
    feedback mechanisms, 350-351, 353
    response to
        clomiphene, 351
        ethinyl estradiol, 351
    prolactin and, 305
    release of prostaglandin from, 56-59
Hypothyroidism and amenorrhea, 251

I

Idiopathic hirsutism, 443
    androgen levels in, 441
    in vitro and *in vivo* data of androgens in, 484
Incubation studies with suprapubic skin, 483
Indomethacin
    ovulation and, 52
    prostaglandins and, 29, 54
Infants
    estradiol levels in, 203
    FSH levels in, 199
    LH levels in, 199
    testosterone levels in, 199
Interstitial gland
    development of, 9
    cells, differentiation from theca cells, 9
Involution of corpus luteum, 20
Iodination of LH and HCG, 182

K

Kinetic studies of gonadotropin-receptor interactions, 185-188
Krukenberg tumours, 408

L

Lactate dehydrogenase
    synthesis in mouse oocytes, 66
Lactation,
    hormonal changes in women and, 311
    prolactin in women and, 310-311
Lactotropic activity in the puerperium, 325-327
Lipoid cell tumours, 406
Luteal phase, short
    clomiphene treatment of, 390
    corticoid treatment of, 387-392
    hormonal status of, 388
Luteinizing hormone (LH), 93-95
    in Addison's disease, 130
    binding to ovarian receptors, 181-194
    after castration, 82
    corpus luteum and, 106
    craniopharyngioma and, 227
    dissociation constant for, 187
    effect
        of estradiol benzoate on, 82, 100
        on cyclic AMP, 28
        of progesterone on, 119
        on steroid synthesis in the corpus luteum, 41
    in the Graafian follicle, 25, 28
    Hill plot of binding data on, 188
    in vivo ovarian perfusuion studies and, 428
    iodination of, 182
    levels
        in amenorrhea, 121
        anorexia nervosa, 294
        and bone age, 200, 204, 206
        in children, 203
        in infants, 199
        in the menstrual cycle, 361-368
        in ovarian failure, 247
        in polycystic ovary syndrome, 374-376
        in secondary amenorrhea, 247
    mechanism of action in the corpus luteum, 42-43
    metabolic clearance rate of, 94
    ovarian feedback control and, 350
    pituitary chromophobe adenoma and, 224
    in post menopausal women, 98
    primary amenorrhea and, 224
    prolactin and, 102
    prostaglandins and, 31
    receptors, 192
        in the human ovary, 193
    response to LH-RH, 110, 224, 335, 341
    secretion, 93-95, 125
        control in women of, 89-108
        effects of
            estrogens on, 81
            progesterone on, 83-86, 100
    Scatchard plot of binding data, 186-187
Luteinizing hormone-releasing hormone (LH-RH), 109
    antibodies to, 393
    craniopharyngioma and, 226
    effect
        on FSH secretion, 110
        on LH secretion, 110
        on menopausal women, 118
        in normal women, 112
        on physiological conditions, 110
    endogenous, 344-345
    FSH response to, 281
    inhibition
        in vitro, 395
        in vivo, 394
    levels during reproductive cycle, 395
    pinealoma and, 232

pituitary chromophobe adenoma and, 224
primary amenorrhea and, 224
radioimmunoassay of, 395
response
  in amenorrhea, 121, 324
  in anorexia nervosa, 121, 254, 295
  of LH to, 335, 340-341
role of ovulation, 393-396
self-priming effect, 343-344
stimulation and
  amenorrhea, 276-289
  gonadotropin release, 275-290
  polycystic ovary syndrome, 376-377
Luteinizing releasing factor (LRF), see
  luteinizing hormone-releasing hormone
Luteinizing releasing hormone (LRH), see
  luteinizing hormone-releasing hormone
Luteoma
  histopathology of, 49
  of pregnancy
    androgen levels in, 422
    hormonal studies in, 417-425
    progesterone levels in, 422

M

Macromolecule synthesis in oocytes, 68
Macrophages of the stroma, 8
Mastocytes of the stroma, 8
Medroxyprogesterone, response in amenorrhea, 252
Medulla of the ovary, 7
Medullary cords, 2
Melanocyte stimulating hormone, see MSH
Menarche, age of in PCO syndrome, 382
Menopausal women
  effect of
    estrogens in, 117
    progestogens in, 119
  gonadotropin response to LH-RH in, 118
Menopause, hormonal state in the, 366
Menstrual corpus luteum, 16-18
Menstrual cycles
  abnormal, 370-371
    hormonal levels in, 370-371
  age and, 363
  bromocryptine and, 325
  hormonal control in, 359-372
  hyperprolactinaemia and, 324
  prolactin levels in, 323-325
  in young women, 361
Menstruation, pituitary-ovarian relationships in, 341-357
Metabolic clearance rate, 126
  constant infusion technique for, 127
  of FSH, 94
  in hirsutism, 438
  of LH, 94
  SHBG and, 146-147, 150

Metabolism
  of mammalian oocyte, 63-70
  of ovarian steroids in women, 125-133
Mihalcovicz's organ, formation of, 2
Mono-amine oxidase, activity in human plasma, 337
Mouse
  androgenic effects in the, 159-173
  kidney binding of DHT and testosterone in, 161
Mullerian ducts
  differentiation in the female, 6
  regression in the male, 6
Myometrial progesterone receptors, 179
α-MSH and testosterone activity, 168

N

Neonatal imprinting of testosterone effects, 170
Neuroendocrine regulation of gonadotropin secretion, 333-347
  role of catecholamines in, 334
New born, sex hormone binding globulin and the, 151-153
Non-lactating women, hormonal changes in, 311-315
Non-receptor proteins, binding by androgens, 159-160

O

Oestradiol, see estradiol
Oestrogens, see estrogens
Oestrone, see estrone
Oligomenorrhea, 276, 283
  bromocryptine treatment and, 307-308
Oocyte
  activation of, 68
  ATP activity in, 67
  control of maturation, 68
  development, 64
    and RNA, 64
  DNA repair in, 67
  glucose-6-phosphate dehydrogenase activity in, 66
  lactate dehydrogenase synthesis in, 66
  maturing, 67
  metabolism, 63-70
  in parthenogenesis, 68
  succinic dehydrogenase activity in, 67
  synthesis of macromolecules in, 68
Ovary
  androgen production by, 10
  arterial blood flow velocity measurement, 72
  arterial cross-sectional area measurement, 72
  autotransplantation, 126
  blood flow
    following intrauterine injection of prostaglandin, 76
    radioactive microsphere method of measurement, 73
    serum gonadotropic hormone levels and 74

serum progesterone levels and, 74
catheterization procedure, 446
cancer
   biochemistry of, 397-415
   deaths from, 398
   environmental factors and, 409
   hormonal steroid levels and, 399
development, 2
failure
   causes of, 249
   LH levels in, 247
feedback control, 350
foetal, 6
function
   blood flow and, 71-79
   neural regulation, 21
   post menopausal, 237-244
   prolactin and, 305-321
histology, 1-24
incubation with acetate-1-^{14}C, 26, 28, 41
in vivo studies, 37-45
medulla, 7
   formation of, 2
organelles, 11-20
   changes in, 1
perfusion studies, 428
progesterone
   effect of estrogen on, 189
   HCG binding and, 188
receptors
   assay of binding activity of, 183
   binding to
      HCG, 181-194
      LH, 181-194
   preparation of, 182
secretion of various steroids, 128-131, 472
steroid metabolism in women, 125-133
   secretion, control of, 131-132
      in women, 125-133
steroidogenesis, 37-45, 428-436
steroids, feedback effects, 350, 356
stroma, 7-11
tissues and talc, 411
tumours, classification of, 398
uptake of radioactive microspheres, 75
vein catheterisation, 126
venous blood, androgen levels in, 471-479
      effluent, steroid levels in, 448-451
      shunting, 73, 76, 78
Ovariectomised women, steroid levels in, 237-243
Ovine growth hormone, binding studies with, 317
Ovine prolactin, binding studies with, 317-320
Ovulation, 25, 105
   clomiphene treatments and, 254
   cyclic AMP and, 25-35
   hypothetical model, 34
   role of LH-RH in, 393-396

prolactin and, 323-331
prostaglandins in, 25-35
steroids and, 25-35
17-oxosteroid levels, urinary, in women, 460

P

Parthenogenesis of the oocyte, 69
Pathological conditions, androstenedione and, 139
Pathophysiology of chronic anovulation, 380
Perfusion studies, ovarian, 428
Pineal tumours, precocious puberty and, 229
Pinealoma, gonadotropins and, 232
Pituitary
   adenoma, 223-225
   function, hypothalamic control of, 89-93
   hormone levels in pseudocyesis, 336
   hormones and testosterone activity in
      target cells, 165
   ovarian relationships
      dysfunctional uterine bleeding and, 352, 353
      menstruation and, 349-357
      polycystic ovary syndrome and, 354
   radiology and amenorrhea, 249-250
   release of prostaglandins from, 56-59
   reserve
      estradiol and, 341
      progesterone and, 341
   response to LH-RH at puberty, 111
   secretion of gonadotropins, 335
   sensitivity, 341-343
   tissue, superfusion of, 54
   tumours
      effect of puberty, 223-236
      prolactin levels and, 251, 306
Pneumopelvography in
   amenorrhea, 302
   anorexia nervosa, 399-302
Polycystic ovarian disease, *see* polycystic ovary syndrome
   tissue, androgens in, 437-438
Polycystic ovary syndrome (PCOS), 262, 269
   age of menarche in, 382
   androgen secretion in, 473, 477
   CNS-pituitary system and, 383
   estrogen production and, 138-139
   FSH levels in, 374
   genesis of, 381-384
   hypothalamic pituitary system and, 373-392
   LH levels in, 374
   pituitary-ovarian relationships and, 354
   puberty and, 382-383
   response to
      clomiphene in, 376-377
      LH-RH in, 376-377

Subject Index

Post menopausal women
  FSH in, 98, 116
  LH in, 98, 116
  steroids in, 237-243
Post partum period
  bromocryptine and, 309
  hormonal changes in, 309
  prolactin and, 309
Porcine ovaries, *in vivo* studies of, 316-320
  granulosa cells, binding studies with, 317-320
Pregnancy
  sex hormone binding globulin and, 151-153
  progesterone receptors and, 176
Pregnenolone
  incorporation into acetate-1-^{14}C, 39, 41
  ovarian secretion, 128-131
  synthesis in the follicle, 39
Pre-menopausal, estrogen secretion, 135
Primary amenorrhea, *see* amenorrhea
Primary follicle, 11
Primordial follicle, 11
  germ cells, ovarian development and, 2
Progesterone
  in Addison's disease, 130
  basal body temperature and, 256
  effect on
    FSH secretion, 83-86, 100
    LH secretion, 83-86, 100
    progesterone receptors, 178
  follicular synthesis, 26, 39
  incorporation into acetate-1-^{14}C, 41
  levels
    and clomiphene treatment, 255
    in luteoma of pregnancy, 422
    in ovarian venous blood, 430
    in polycystic ovary syndrome, 376
    and uterine receptor sites, 176
  metabolism in the skin, 12
  ovarian feedback control and, 350
    secretion, 128-131
  in ovariectomised women, 238
  pituitary reserve and, 341
  in post menopausal women, 237, 243
  radioimmunoassay of, 473
  receptors
    effect of estrogen on, 177
    effect on progesterone, 178
    endometrial, 179
    in guinea pig uterus, 175-180
    hormonal control and, 176
    myometrial, 179
    oestrous cycle and, 175
    in pregnancy, 176
    protein synthesis and, 177
    RNA and, 177
    stability of, 179
  secretion from corpus luteum at term, 41
    granulosa cells, 40
  serum levels following intrauterine injection of prostaglandin, 76
  and ovarian blood flow, 74
  synthesis in the corpus luteum, 40, 42
Progestogens, effect on FSH and LH, 119
Prolactin
  amenorrhea and, 261-265
  effect on Graafian follicles, 31
  FSH and, 102
  galactorrhea and, 307
  gonadal function and, 264
  hypothalamus and, 305
  in vitro studies and, 316-320
  levels
    in anorexia nervosa, 302
    in lactating women, 310-311
    during menstrual cycle, 323-325
  LH and, 102
  menstrual disturbances and, 307-308
  ovarian function and, 305-321
  ovine, binding studies with, 317-320
  ovulation and, 323-331
  pituitary tumours and, 250, 306
  in the post partum period, 309
  response to TRH
    in primary amenorrhea, 224
    in pituitary chromophobe adenoma, 224
  secondary amenorrhea and, 251
Prostaglandins
  cyclic-AMP and, 25-35
  cyclic changes in, 47
  gonadotropin secretion and, 48-50, 59
  in Graafian follicles, 13, 28, 30
  HCG and, 31
  indomethacin and, 29, 54
  LH and, 31
  ovarian blood flow and, 76
  ovulation and, 25-35
  pre ovulatory changes, 28, 31, 33
  post ovulatory changes, 31
  release from
    the hypothalamus, 56-59
    the pituitary, 56-59
  role in gonadotropin secretion, 47-61
  separation into E and F types, 29
  steroids and, 25-35, 76
  synthesis
    and aspirin, 54
    in vitro inhibition of, 54
    in vivo inhibition of, 51
Prostatic tissue
  binding of DHT in, 161
  binding of testosterone in, 161
Protein
  synthetic activity in oocytes, 65

synthesis and progesterone receptors, 177
Pseudocyesis, pituitary hormone levels in, 336
Puberty
   effect of
      hypothalamic tumours on, 223-236
      pituitary tumours on, 223-236
   FSH and, 111
   hamartoma and, 233
   LH and, 111
   LH-RH and, 111
   precocious, 229-234
      pineal tumours in, 229
   polycystic ovary syndrome and, 382-383
Puerperium
   gonadotropic activity in, 325-327
   lactotropic activity in, 325-327
      androstenedione, 461, 467
Pulsatile secretion of LH, 336-338

R

Radioactive microsphere
   measurement of arterial blood flow, 73, 77-78
   uptake by the corpus luteum, 75
      sheep's ovaries, 75
Radioimmunoassay (RIA)
   of LH-RH, 395
   of steroids, 199, 473
Radioreceptor assay of HCG and LH, 190
Rat
   androgenic effects in, 159-173
   prostaglandins, 30
   uterus binding of
      DHT, 161
      testosterone, 161
Relaxin in luteal cells, 19
RNA
   oocyte development, 64
   progesterone receptors and, 177
   synthetic activity in oocytes, 64
Reproductive cycle, LH-RH levels in, 395
   system, 103-107
Rhesus monkey (M. Mulatta), SHBG and, 152-153

S

Scatchard plot, of binding data, 186-187, 318
Sebum, excretion rates, 496
Secondary amenorrhea, see amenorrhea
Sex differentiation, 2, 4
   chromosomes in, 3-4
Sex hormone binding globulin, SHBG, 141-158
Skin
   androgen metabolism in, 495-506
   incubation studies with, 483, 498-500
   progesterone metabolism in, 506

   suprapubic, incubation studies with, 483
   testosterone 5α-reduction in, 481-494
Steroids
   binding subunits, evidence for, 160
   biosynthetic pathways, 399, 498
   follicular secretion of, 131
   interrelationship with
      cyclic AMP, 25-35
      prostaglandins, 25-35
   ovarian secretion in women, 125-133
   secretion
      by corpus luteum, 131
      the stroma and, 131
   urinary metabolites of, 126
Steroidogenesis
   adenylate cyclase and cyclic-AMP in, 33
   ovarian, 21, 37-45, 428-436
   preovulatory changes in, 26, 33
Steroids, see specific names
Stroma
   morphological characteristics, 8
   transformation into
      steroidogenic cells, 8-9
      theca interna cells, 13
   steroid secretion, 131
   synthesis of steroids in, 38, 39
Succinic dehydrogenase (SDH), activity in oocytes, 67
Superfusion
   of hypothalamic synaptosomes, 58
   of pituitary tissue, 54, 56
Suprapubic skin
   incubation studies with, 483
   testosterone metabolism in, 502
Sweat, excretion rates, 497
Synaptosomes, superfusion of, 58

T

Talc, in ovarian tissue, 411
Target cell
   formation of DHT, 164
   metabolism of androgen, 160-165
   pituitary hormone activity in, 165
   testosterone activity in, 165
   testosterone 5α-reduction in, 488
Testes volume
   estradiol and, 218
   estrone and, 218
   testosterone and, 219
Testicular feminization syndrome
   testosterone 5α-reduction in, 487
Testosterone
   binding, 160
      in mouse kidney, 161
      in prostatic tissue, 161
      in rat uterus, 161
      with SHBG, 143

breast development and, 218
effect of ethinyl estradiol on, 148
effects, neonatal printing and, 170
follicular synthesis of, 26
incorporation into acetate-1-^{14}C, 39
levels
    in anorexia nervosa, 302-303
    and bone age, 208
    in children, 203
        idiopathic hirsutism, 441
        infants, 200
        ovarian venous blood, 430
        polycystic ovary syndrome, 376, 440
α-MSH and, 168
metabolism, effect of cofactors on, 502
origin in women, 445, 452
ovarian secretion, 128-131
in ovariectomised women, 238
production rates in women, 444
propionate, 165
in post menopausal women, 237-243
radioimmunoassay of, 199, 473
5α-reduction in
    the skin, 481-484
    target cells, 488
    testicular feminization syndrome, 487
sexual maturation and, 220
synthesis in the embryonic testicle, 6
testes volume and, 218
Theca cells
    steroid synthesis and, 40
Theca internal cell, transformation from stroma cell, 13
    lutein cells, 17
Thecoma, 400
Thyroid hormones
    SHBG and, 151
    stimulating hormone (TSH), craniopharyngioma and, 227
    releasing hormone (TRH), craniopharyngioma and, 227
        response in
            amenorrhea, 224
            pituitary chromophobe adenoma, 224
            polycystic ovary syndrome, 381
Thyrotoxicosis, SHBG and, 153
Transcortin
    levels in hypophysectomised rats, 170
Tumours
    ovarian, 398
    pineal, 229
    of the 3rd ventricle, 227
    thecoma, 400

U

Urinary
    estrogens in anorexia nervosa, 293
    gonadotropins in anorexia nervosa, 294
    steroid excretion and LH-RH, 282-286
Urogenital connections, formation of, 2
Uterine receptor sites, progesterone levels and, 176

V

3rd Ventricle, tumours of, 227-229
Virilization in women, 443
Virilizing luteoma of pregnancy, 417-425

W

Wolffian ducts
    regression in the female, 6
    transformation in the male, 6